Lecture Notes in Physics

Springer
Berlin
Heidelberg
New York
Barcelona
Hong Kong
London
Milan
Paris
Tokyo

Physics and Astronomy ONLINE LIBRARY

http://www.springer.de/phys/

Editorial Policy

The series *Lecture Notes in Physics* (LNP), founded in 1969, reports new developments in physics research and teaching -- quickly, informally but with a high quality. Manuscripts to be considered for publication are topical volumes consisting of a limited number of contributions, carefully edited and closely related to each other. Each contribution should contain at least partly original and previously unpublished material, be written in a clear, pedagogical style and aimed at a broader readership, especially graduate students and nonspecialist researchers wishing to familiarize themselves with the topic concerned. For this reason, traditional proceedings cannot be considered for this series though volumes to appear in this series are often based on material presented at conferences, workshops and schools (in exceptional cases the original papers and/or those not included in the printed book may be added on an accompanying CD ROM, together with the abstracts of posters and other material suitable for publication, e.g. large tables, colour pictures, program codes, etc.).

Acceptance

A project can only be accepted tentatively for publication, by both the editorial board and the publisher, following thorough examination of the material submitted. The book proposal sent to the publisher should consist at least of a preliminary table of contents outlining the structure of the book together with abstracts of all contributions to be included.
Final acceptance is issued by the series editor in charge, in consultation with the publisher, only after receiving the complete manuscript. Final acceptance, possibly requiring minor corrections, usually follows the tentative acceptance unless the final manuscript differs significantly from expectations (project outline). In particular, the series editors are entitled to reject individual contributions if they do not meet the high quality standards of this series. The final manuscript must be camera-ready, and should include both an informative introduction and a sufficiently detailed subject index.

Contractual Aspects

Publication in LNP is free of charge. There is no formal contract, no royalties are paid, and no bulk orders are required, although special discounts are offered in this case. The volume editors receive jointly 30 free copies for their personal use and are entitled, as are the contributing authors, to purchase Springer books at a reduced rate. The publisher secures the copyright for each volume. As a rule, no reprints of individual contributions can be supplied.

Manuscript Submission

The manuscript in its final and approved version must be submitted in camera-ready form. The corresponding electronic source files are also required for the production process, in particular the online version. Technical assistance in compiling the final manuscript can be provided by the publisher's production editor(s), especially with regard to the publisher's own Latex macro package which has been specially designed for this series.

Online Version/ LNP Homepage

LNP homepage (list of available titles, aims and scope, editorial contacts etc.):
http://www.springer.de/phys/books/lnpp/

LNP online (abstracts, full-texts, subscriptions etc.):
http://link.springer.de/series/lnpp/

J. Frauendiener H. Friedrich (Eds.)

The Conformal Structure of Space-Time

Geometry, Analysis, Numerics

Springer

Editors

Jörg Frauendiener
Universität Tübingen
Theoretische Astrophysik
Auf der Morgenstelle 10
72076 Tübingen, Germany

Helmut Friedrich
Albert-Einstein-Institut
MPI für Gravitationsphysik
Am Mühlenberg 1
14476 Golm, Germany

Cover Picture: (The figure on the cover is a rendering of the finite representation of space-like infinity as a cylinder. For more details see the articles by Friedrich and by Kroon. The figure was obtained by ray-tracing.)

Cataloging-in-Publication Data applied for

A catalog record for this book is available from the Library of Congress.

Bibliographic information published by Die Deutsche Bibliothek

Die Deutsche Bibliothek lists this publication in the Deutsche Nationalbibliografie; detailed bibliographic data is available in the Internet at http://dnb.ddb.de

CR Subject Classification (1998): D.4.5, E.3, C.2.0, H.2.0, K.6.5, K.4.4
ISSN 0075-8450
ISBN 3-540-44280-4 Springer-Verlag Berlin Heidelberg New York

Springer-Verlag Berlin Heidelberg New York
a member of BertelsmannSpringer Science+Business Media GmbH

http://www.springer.de

Typesetting: Camera-ready by the authors/editor
Camera-data conversion by Steingraeber Satztechnik GmbH Heidelberg
Cover design: *design & production*, Heidelberg

Printed on acid-free paper
SPIN: 10893706 54/3141/du - 5 4 3 2 1 0

Preface

Controlling the behaviour of the solutions to Einstein's field equations on large scales is still the most important technical task in classical general relativity. The long standing problem of 'cosmic censorship' will not be resolved without a sufficiently general and deep understanding of the solutions. The unexpected emergence of the Bartnik-McKinnon solutions and Choptuik's disclosure of critical collapse phenomena show us that Einstein's theory still has surprises in store.

The last two discoveries also demonstrate clearly the important role of numerical techniques in the analysis of specific solutions and in the study of the manifold of solutions. The interplay of analytical with numerical methods is bound to become the most important strategy in exploring the content of the theory.

Moreover, since various projected gravitational wave antennae will soon become operational, templates with gravitational waves forms will urgently be needed for analysing the recorded data. This makes the interaction between analytical and numerical techniques a most exciting and important project, because it may help open new vistas of our universe. The fact that the calculation of the form of gravitational waves generated by astrophysical processes turned out so much more difficult than expected hints at a lack of insight into the analytical basis of the theory or at an insufficient exploitation of the present analytical knowledge.

Any general analysis of asymptotically flat space-times in the large needs to take into account the causal and the underlying null cone or conformal structure of the field. These structures should thus also be critical in the semi-global or global numerical calculation of space-times. Not surprisingly, the two numerical techniques presently used to perform such calculations, the characteristic method and the method based on the conformal field equations, employ basic features of the conformal structure in the definition of their procedures.

The analytical and the numerical methods both offer possibilities not accessible to the other one, and each of them asks questions and poses problems likely to initiate interesting research with the other method. For researchers in the fields to get an insight into the potential and the problems of the other field, we therefore organized a workshop on analytical, geometrical, and numerical studies which make explicit use of conformal or related structures. This book contains extended versions of the contributions to this workshop.

Following the suggestion of the publisher we tried to avoid the traditional style of proceedings and aimed at a book which will help a newcomer find access to the field, which offers new results to the experienced researcher, and which provides a comprehensive source of references. In particular, one of us (H.F.) wrote an extensive introductory chapter in which he tries to introduce the newcomer to the field and to provide a general perspective by pointing our the relations to the other studies represented here.

Because this perspective may be clouded by the author's ignorance and personal taste, however, we do urge the reader to understand these references as an encouragement to carefully study those articles themselves. The intricate net of relations between the different parts of the work discussed in this book will then become evident. A complete picture of the present situation can only be obtained by trying to understand the full scope of those articles and the specific views of their author's. Were different opinions occur the reader is invited to search for solutions of the corresponding open problems.

While we have tried to maintain to some extent the division of this area of research into the three subfields indicated in the title of this book, it is clear that the assignment of a single article to any of these subfields is not sharply defined. This fact should be seen as a virtue since it was the expressed purpose of the workshop to have researchers in different areas interact with each other and see how they can profit by viewing their subject from different perspectives.

There remains the pleasant task to thank the speakers and the participants for helping create an inspiring atmosphere at the workshop and the contributors of these proceedings for helping create a picture of the present situation of the field which illustrates its richness and its potential.

We also thank the Deutsche Forschungsgemeinschaft and the 'Unibund' of the University of Tübingen for their financial support. It is a pleasure to thank Prof. Hanns Ruder for his support and Heike and Bettina Fricke for their help in administrative matters during and after the workshop. Finally, we acknowledge help from T. Müller, A. King and M. King during the preparation of this book.

Tübingen and Golm,
June 2002

Jörg Frauendiener
Helmut Friedrich

List of Contributors

Lars Andersson
Department of Mathematics
University of Miami
Coral Gables, FL 33124
USA
larsa@math.miami.edu

Robert Bartnik
School of Mathematics and Statistics
University of Canberra,
ACT 2601, Australia
bartnik@ise.canberra.edu.au

Adrian Butscher
MPI für Gravitationsphysik
Albert-Einstein-Institut
14476 Golm
Germany
butscher@aei-potsdam.mpg.de

Piotr Chruściel
Département de Mathématiques
Faculté des Sciences
F 37200 Tours
France
piotr@gargan.math.univ-tours.fr

Sergio Dain
MPI für Gravitationsphysik
Albert-Einstein-Institut
14476 Golm
Germany
dain@aei-potsdam.mpg.de

José A. Font
Dep. de Astronomía y Astrofísica
Universidad de Valencia
46100 Burjassot (Valencia)
Spain
j.antonio.font@uv.es

Jörg Frauendiener
Institut für Theoretische Astrophysik
Universität Tübingen
72076 Tübingen
Germany
joergf@tat.physik.uni-tuebingen.de

Helmut Friedrich
MPI für Gravitationsphysik
Albert-Einstein-Institut
14476 Golm
Germany
hef@aei-potsdam.mpg.de

Simonetta Frittelli
Physics Department
Duquesne University
Pittsburgh, PA 15282
USA
simo@mayu.physics.duq.edu

Greg Galloway
University of Miami
Coral Gables
FL 33124
USA
galloway@math.miami.edu

David Garfinkle
Department of Physics
Oakland University
Rochester, Michigan 48309
USA
garfinkl@vela.acs.oakland.edu

Sascha Husa
MPI für Gravitationsphysik
Albert-Einstein-Institut
14476 Golm
Germany
shusa@aei-potsdam.mpg.de

Niky Kamran
Mathematics Department
McGill University
Montreal
Canada

Heinz-Otto Kreiss
Department of Mathematics
University of California
Los Angeles, CA 90095
USA
kreiss@math.ucla.edu

Juan Valiente Kroon
MPI für Gravitationsphysik
Albert-Einstein-Institut
14476 Golm
Germany
jav@aei-potsdam.mpg.de

Luis Lehner
Department of Physics and Astronomy
The University of British Columbia
Vancouver, BC V6T 1Z1
Canada
luisl@sgi1.physics.ubc.ca

Ezra T. Newman
Department of Physics and Astronomy
University of Pittsburgh
Pittsburgh PA 15260
USA
newman@phyast.pitt.edu

Andrew Norton
School of Mathematics and Statistics
University of Canberra,
ACT 2601, Australia

Omar E. Ortiz
Universidad Nacional de Córdoba
Facultad de Matemática,
Astronomía y Física,
(5000) Córdoba
Argentina
ortiz@fis.uncor.edu

Roger Penrose
Mathematical Institute
University of Oxford
24–29 St. Giles'
Oxford OX1 3LB
England

Bernd G. Schmidt
MPI für Gravitationsphysik
Albert-Einstein-Institut
14476 Golm
Germany
bernd@aei-potsdam.mpg.de

Walter Simon
Institut für theoretische Physik
der Universität Wien
Boltzmanngasse 5
A-1090 Wien
Austria
simon@ap.univie.ac.at

K. P. Tod
Mathematical Institute
University of Oxford
24–29 St. Giles'
Oxford OX1 3LB
England
tod@maths.ox.ac.uk

Contents

1 Conformal Einstein Evolution

Helmut Friedrich

Albert Einstein Institut, Max Planck Institut für Gravitationsphysik,
Am Mühlenberg 1, 14476 Golm, Germany

Abstract. We discuss various properties of the conformal field equations and their consequences for the asymptotic structure of space-times.

1.1 Introduction

The causal structure, which decides whether a signal can be sent between two space-time points, is of fundamental importance for all of field theory. In the case of Einstein's theory of gravitation, general relativity, it acquires a particular significance, because the null cone structure underlying the causal relations is itself a basic unknown of the field equations and as such subject to the causal relations in the small.

From the beginning of general relativity the effect of 'deflection of light in gravitational fields', demonstrated nowadays so clearly in the gravitational lensing phenomena, has been considered theoretically and observationally as one of the clearest manifestations of the deviation of the light cone field in general relativity from that of special relativity. As seen in the article by G. Galloway in this volume, which shows, among other things, the ubiquity of caustics on wave fronts in solutions to Einstein's field equations, it is not only a local effect but also related to the global structure of space-times.

This relation to the global structure of space-time becomes quite pronounced in the case of black holes, which seem to be on the verge now of also becoming 'observed' phenomena. The observational situation as well as various theoretical aspects of the idea are addressed directly in P. Chruściel's article in this volume. However, tacitly or explicitly, the related questions of gravitational collapse, merger of black holes, generation of gravitational radiation, singularities of gravitational fields, etc. motivate to a large extent all studies in this volume.

The dual and critical role of the light cone structure in Einstein's theory and in physics in general led some researchers to speculate that the null cone structure might be a classical space-time reflection of a fundamental feature of a quantum theory of gravitation. This motivated efforts to recast the classical theory into forms which put the null cone structure, or associated structures like wave fronts, into the foreground. The article by S. Frittelli, N. Kamran, and E. Newman as well as the article by R. Penrose in this volume are aiming into this direction. Trying instead to derive qualitative and quantitative results about the solutions, the remaining articles explore and exploit, in one way or other, the light cone structure along more conventional lines.

That in studies of the global structure of gravitational fields the light cone structure acquires a particular technical interest, has been demonstrated most impressively by the proofs of the singularity theorems ([51], [67], cf. also [52]). It provides the most important tool for analyzing space-times in the large. While these theorems show that under reasonable assumptions space-times are singular in the sense that they contain non-complete causal geodesics, they do not provide any information on the specific nature of the singularities. It is expected that in general curvature singularities will arise, however, the theorems give not the slightest hint whether this will be the case, not to mention the causal nature of the singularities or their strength.

In spite of various attempts to obtain more detailed results, the related problem of cosmic censorship (cf. [72], [71], [79]) still poses some of the most important outstanding questions of classical general relativity. While the limited use of the field equations made in the singularity theorems contributes to their generality, the same fact is responsible for the scarcity of details supplied by them. In view of the enormous richness of pathological behaviour observed on the level of pure space-time theory (cf. [22] and the literature given there) it is not surprising that attacks at the problem of cosmic censorship which rely mainly on the topological methods used in the proof of the singularity theorems met little success.

To disclose the more subtle secrets of Einstein's theory we need a deeper understanding of the causal structure in the context of Einstein's field equations. In any serious investigation of the problem the consequences of these equations, which represent the principal restrictive principle of the theory, must be taken into account in all completeness. Getting control on the evolution process defined by Einstein's equations still remains the most important technical task in classical general relativity.

There arises the question of how to combine the analysis of the field equations with that of the causal structure. In Einstein's theory these structures are particularly closely related. By the local causality requirement the causal structure is determined by the null cones of the metric. Conversely, provided certain pathologies are excluded, the causal structure can be used to reconstruct the null cone structure. Moreover, the physical characteristics of the field equations, which largely govern the evolution process, are given by the null hypersurfaces defined by the solutions. These tight relations between causal structure, null cone structure, and characteristics structure clearly offer possibilities for the desired analysis, but it may be noted that they are also responsible for the *quasi-linearity* of the equations which, together with their *gauge hyperbolicity*, is the main source of our difficulties.

The null cone structure is equivalent to the conformal structure, given by the information which is retained if the metric is subject to arbitrary conformal rescalings $g \to \Omega^2 g$ with conformal factors $\Omega > 0$. This may allow us to relate the null cone structure in a new way to the structure of the field equations and to the large scale behaviour of their solutions.

However, to what extent can we draw from the behaviour of the field equations under conformal rescalings of the metric conclusions on the large scale behaviour of the solution? The present article is trying to give some answer to this question. The latter might be rather aimless were it not for a concrete proposal pointing precisely into this direction. Motivated by earlier studies of the important physical question of how to characterize gravitational radiation without relying on approximations, R. Penrose suggested to analyse the asymptotic behaviour of gravitational fields in terms of extensions of the conformal structure to null infinity ([65], [66]).

The subsequent attempt ([23], [24]) to resolve the controversy concerning the assumptions made in [65] and to understand the smooth interplay between the conformal structure and the field equations suggested in [66] disclosed the very peculiar behaviour of Einstein's field equations under conformal rescalings. They turned out to be *conformally regular* and assume with this property a very special position between *conformally singular* equations like the massive Klein-Gordon equations on a given background and *conformally invariant* equations like the Maxwell or Yang-Mills equations in four dimensions. It will be the purpose of this article to explain the origin, the properties, and the consequences of the resulting *conformal field equations*.

Almost all the results described in the following have been obtained by considering as a guiding problem the question to what extent the assumptions underlying [65] can be justified. We are beginning to see now that this question proved fruitful because these assumptions refer to the key geometrical structure and draw such a sharp line between what is true and what isn't that they required, and still require, a most careful and detailed analysis. The results obtained so far indicate that in the most interesting case, that of a vanishing cosmological constant, the situation is much too subtle to simply allow us to say that the assumptions of [65] are right or wrong.

In the following we shall make a particular effort to sort out the delicate issues which arise here, hoping to put a controversy, whose 'arguments' too often relied more on personal taste than on insight, onto a rational basis. Such an effort is appropriate because due to the results which are available now the mathematical analysis is coming close to a point where the more subtle mathematical and physical questions concerning the foundations of the theory of isolated self gravitating systems become amenable to detailed discussions.

The investigations are now way beyond a situation which would allow us to arrive at reliable statements by investigations of the linearized equations, 'physical arguments' etc. In fact, it is difficult to develop any intuition on physical grounds for the extreme questions about solutions of Einstein's equations which are being asked here. Consequently, most results are stated in the form of existence theorems (referred to in the following only in a non-technical way). The physicist mainly interested in wave forms and observational facts, who conceives of this as merely interesting mathematics, may be reminded that abstract statements about the existence of general classes of solutions with certain global properties usually only epitomize collections of results about details of the solutions, particular features of the equations, and new techniques with many more applications.

More importantly, it may be pointed out to him that the results alluded to above become of a practical use now, which was not anticipated in [65]. As discussed in the articles by J. Frauendiener and S. Husa in this volume they form a basis for numerical calculations covering asymptotic parts of space-time. Most remarkably, though still to be explored in numerical work, they open the unique possibility to calculate numerically on finite grids maximal, globally hyperbolic, asymptotically flat solutions to Einstein's field equations, including their asymptotics in null directions and the radiation field at null infinity. These solutions will be determined solely and uniquely from standard Cauchy data (subject to certain, quite weak, conditions near space-like infinity) and require besides the idealization of asymptotic flatness no further approximation.

Working out the details of such truly global solutions, either analytical or with numerical methods, will still require a considerable amount of work and inventiveness. However, the main point to be made here is that both possibilities are open now and both approaches have a lot to gain from each other. Each of them will pose questions which will shed new light on problems of the other approach and the answers obtained in one of them will suggest new solutions to the other one.

A main open problem in this context is concerned with the question to what extent it will be possible to approach 'time-like infinity' within a numerical calculation if the solution develops singularities and event horizons. This problem has hardly been touched yet. Even for basic solutions like the Schwarzschild or the Kerr solution a careful and detailed analysis of time-like infinity in the conformal picture is not available yet. Some of these questions and also the related problem of gauge conditions will be addressed at the end of this article.

It is clear that there is still a long way to go before we will be able to obtain wave forms of sufficient generality or arrive at insights which will bring us closer to a resolution of the problem of cosmic censorship. Nevertheless, it appears that the interaction between numerical and analytical methods referred to above offers the best prospects for further progress.

Any work on general classes of solutions to the Einstein equations is necessarily of a very technical nature. In the following we shall give a survey on certain results and the current analytical work based on the conformal fields equations which emphasizes the general ideas. With the exception of the somewhat detailed discussion of the conformal field equations, technicalities will be kept at a minimum.

1.2 The Conformal Field Equations

The core of Einstein's theory of gravitation is represented by his field equations

$$\tilde{R}_{\mu\nu} - \lambda\,\tilde{g}_{\mu\nu} = \kappa\,(T_{\mu\nu} - \frac{1}{2}\,T\,\tilde{g}_{\mu\nu}), \tag{1.1}$$

for the metric $\tilde{g}_{\mu\nu}$, where λ denotes the cosmological constant and $T_{\mu\nu}$ the energy momentum tensor. Most properties of the gravitational field should follow from these equations. Those which don't and have to be 'put in by hand' characterize general features as well as specific properties of the physical situation to be modeled. For instance, in the case of cosmological space-times one may assume compact time slices while in the description of isolated gravitating systems one will require the existence of Cauchy hypersurfaces with asymptotically flat ends, but in both cases one has the freedom to prescribe initial data in agreement with these assumptions. In the following we will mainly be interested in the second of these scenarios.

Equations (1.1) are known to be *gauge hyperbolic* in the sense that they are hyperbolic if suitable *gauge conditions* are imposed. Expressed in terms of coordinates x^μ satisfying the harmonicity condition $\tilde{\nabla}_\nu \tilde{\nabla}^\nu x^\mu = 0$, equations (1.1) take the form

$$\tilde{g}^{\alpha\beta} \partial_\alpha \partial_\beta \, \tilde{g}_{\mu\nu} = Q_{\mu\nu}(\tilde{g}, \partial\tilde{g}, T) \tag{1.2}$$

of a system of wave equations for the metric coefficients $\tilde{g}_{\mu\nu}$, where the right hand side depends on the derivatives of these coefficients up to first order and on the energy momentum tensor.

The form (1.2) reveals the double role played by the light cone structure or, equivalently, the conformal structure. On the one hand the conformal structure is part of the unknown since it determines the metric up to a positive point dependent factor, on the other hand it defines the characteristic cone $\{\xi \in T^*M |\, \tilde{g}^{\alpha\beta} \, \xi_\alpha \, \xi_\beta = 0\}$ of the form (1.2) of Einstein's equations (for this and various other general notions and results concerning the field equations we refer the reader to [45]). The characteristics strongly govern the local propagation properties of Einstein's equations. In fact, a more detailed argument involving (1.2) implies that equations (1.1) respect the local causality requirement of relativity, provided the equations for the matter fields entering the energy momentum tensor are chosen appropriately. This is related to the fact that the characteristic cone determines the domains of uniqueness for equation (1.1).

The tight relations between the structure of characteristics, the causal structure, and the light cone structure or, equivalently, the conformal structure, all of which depend on the solution at hand, is a very distinctive feature of Einstein's theory. However, the theory proposed by Einstein gives no reason to expect that the field equations would be particularly well behaved under conformal rescalings $\tilde{g}_{\mu\nu} \to g_{\mu\nu} = \Omega^2 \tilde{g}_{\mu\nu}$. The equations are chosen to determine isometry classes of solutions and are thus not conformally covariant. Formally, this is seen from the transformation law of the Ricci tensor, given in n dimensions by

$$R_{\nu\rho}[g] = \tilde{R}_{\nu\rho}[\tilde{g}] - \frac{n-2}{\Omega} \nabla_\nu \nabla_\rho \Omega - g_{\nu\rho} \, g^{\lambda\delta} \left(\frac{1}{\Omega} \nabla_\lambda \nabla_\delta \Omega - \frac{n-1}{\Omega^2} \nabla_\lambda \Omega \nabla_\delta \Omega \right), \tag{1.3}$$

where on the right hand side ∇ denotes the connection associated with the metric g.

If Einstein's equation (1.1) is assumed in this relation and the conformal factor Ω is considered as a given function on the solution manifold, we obtain an equation for the conformal metric $g_{\nu\rho}$. It should be possible to analyse this equation by applying the techniques known for Einstein's equation, since the principal parts of the differential operators occurring in both equations, i.e. the parts containing the derivatives of highest order, are identical. For our intended application this equation for $g_{\nu\rho}$ has two deficiencies. The most conspicuous one is the occurrence of factors $1/\Omega$, which makes the equation singular at the set $\mathscr{I}' = \{\Omega = 0\}$. Since this set will indicate some kind of infinity of the solution space-time, it will be of particular interest to us. The second problem arises because the function Ω, in fact also the manifold underlying the solution and the set $\mathscr{I}'$, are not given a priori. They are related to the global geometry of the solution and must be determined jointly with the metric.

Remarkably, Einstein's equations allow us to resolve these problems. They turn out to be *conformally regular* in the sense that there exist conformal representations of the equation which do not contain factors Ω^{-1} (resp. factors Ω in the principal part) and for which the conformal factor is determined by the equations themselves. To discuss this further we recall a few facts of conformal geometry.

1.2.1 Conformal Geometry

A smooth conformal structure $\mathcal{C}$ on a manifold M is given by a family of smooth locally defined metrics of a given signature whose domains of definition exhaust M and which are related on the intersections of these domains by conformal rescalings with positive conformal factors. A smooth pseudo-Riemannian space (M, g) uniquely determines a smooth conformal structure on M which we denote by $\mathcal{C}_g$. Using a suitable partition of unity, it can be shown that any conformal structure can be obtained this way and we shall often represent a conformal structure in terms of a globally defined metric in the conformal class.

A Lorentzian conformal structure $\mathcal{C}$ determines a null cone at each point of M. Conversely, the null cones fix $\mathcal{C}$ uniquely (cf. [52]). As a consequence 'space- and time-like' are well defined for $\mathcal{C}$ and a few further important space-time structures are implied. A null hypersurface for $\mathcal{C}$ is a smooth (say) hypersurface which at each of its points is tangent to the null cone field. The unique directions of tangency can be integrated to obtain null curves which rule the null hypersurface. Considered as point sets, these curves coincide with null geodesics of the metrics in the conformal class and any null geodesic can locally be obtained this way. Extending the null geodesics generating a null hypersurface in any of their affine parameterizations we obtain an unique extension. The resulting set, called a *wave front*, will in general not any longer be a smooth hypersurface, because families of null geodesics tend to develop caustics, i.e. form envelopes or intersections. The wave fronts and their caustics are also invariants of the conformal structure.

On the level of local differential geometry the most important invariant of a conformal structure $\mathcal{C}_g$ is its conformal Weyl tensor $C^{\mu}{}_{\nu\lambda\rho}$ (observe the index positions), which is the trace-free part in the decomposition

$$R^{\mu}{}_{\nu\lambda\rho} = 2\,\{g^{\mu}{}_{[\lambda}\,L_{\rho]\nu} - g_{\nu[\lambda}\,L_{\rho]}{}^{\mu}\} + C^{\mu}{}_{\nu\lambda\rho}, \tag{1.4}$$

of the curvature tensor of g, where the trace parts are represented by the tensor

$$L_{\mu\nu} = \frac{1}{n-2}\left(R_{\mu\nu} - \frac{1}{2\,(n-1)}\,R\,g_{\mu\nu}\right). \tag{1.5}$$

In terms of the Levi-Civita connection ∇ of g and some coordinate basis ∂_μ our conventions are such that $R^\mu{}_{\nu\lambda\rho}\,X^\nu = (\nabla_\lambda\,\nabla_\rho - \nabla_\rho\,\nabla_\lambda)\,X^\mu$. We denote by $R_{\mu\nu} = R^\rho{}_{\mu\rho\nu}$ and R the Ricci tensor and the Ricci scalar of g respectively.

The class of torsion free covariant derivative operators naturally associated with a conformal class $\mathcal{C}$ comprises but is larger than the set of Levi-Civita operators for the metrics representing $\mathcal{C}$. A *Weyl connection* for $\mathcal{C}_g$ is a torsion free connection $\hat{\nabla}$ on M which respects the conformal class $\mathcal{C}_g$ in the following sense. For any C^1 curve $x(\lambda)$ defined on a neighbourhood of the origin of $\mathbb{R}$ and any frame field e_k, $k = 1, \ldots, n$, which is parallelly transported along $x(\lambda)$ with respect to $\hat{\nabla}$, there exists a function $\Omega_\lambda > 0$ along $x(\lambda)$ such that $g(e_i, e_k)|_{x(\lambda)} = \Omega^2_\lambda\,g(e_i, e_k)|_{x(0)}$. It follows from this that

$$\hat{\nabla}_\rho\,g_{\mu\nu} = -2\,f_\rho\,g_{\mu\nu}, \tag{1.6}$$

with some arbitrary 1-form f on M and that the connection coefficients defined by $\nabla_{\partial_\mu}\,\partial_\nu = \Gamma_\mu{}^\rho{}_\nu\,\partial_\rho$ and $\hat{\nabla}_{\partial_\mu}\,\partial_\nu = \hat{\Gamma}_\mu{}^\rho{}_\nu\,\partial_\rho$ are related by

$$\hat{\Gamma}_\mu{}^\rho{}_\nu = \Gamma_\mu{}^\rho{}_\nu + S(f)_\mu{}^\rho{}_\nu, \quad \text{with} \quad S(f)_\mu{}^\rho{}_\nu \equiv \delta^\rho{}_\mu\,f_\nu + \delta^\rho{}_\nu\,f_\mu - g_{\mu\nu}\,g^{\rho\lambda}\,f_\lambda. \tag{1.7}$$

In the special case where the 1-form f is exact such that it can be written on suitable open sets U in the form

$$f = -\Omega^{-1}\,d\,\Omega, \tag{1.8}$$

with some smooth function $\Omega > 0$, the connection $\hat{\nabla}$ coincides on U with the Levi-Civita connection $\tilde{\nabla}$ of the metric $\tilde{g}$ satisfying $g_{\mu\nu} = \Omega^2\,\tilde{g}_{\mu\nu}$.

The curvature tensors of the connections $\hat{\nabla}$ and ∇ satisfying (1.7) are related by

$$\hat{R}^\mu{}_{\nu\lambda\rho} - R^\mu{}_{\nu\lambda\rho} = 2\left\{\nabla_{[\lambda}\,S_{\rho]}{}^\mu{}_\nu + S_\delta{}^\mu{}_{[\lambda}\,S_{\rho]}{}^\delta{}_\nu\right\} = 2\left\{g^\mu{}_{[\rho}\,\nabla_{\lambda]}\,f_\nu \right.$$
$$\left. + \nabla_{[\rho}\,f^\mu\,g_{\lambda]\nu} - g^\mu{}_\nu\,\nabla_{[\rho}\,f_{\lambda]} - g^\mu{}_{[\rho}\,f_{\lambda]}\,f_\nu + g_{\nu[\rho}\,f_{\lambda]}\,f^\mu + g^\mu{}_{[\rho}\,g_{\lambda]\nu}\,f_\delta\,f^\delta\right\},$$

were indices are raised or lowered with respect to g. This implies for the Ricci tensor $\hat{R}_{\nu\rho} = \hat{R}^\mu{}_{\nu\mu\rho}$ and the Ricci scalar $\hat{R} = g^{\nu\rho}\,\hat{R}_{\nu\rho}$

$$\hat{R}_{\nu\rho} = R_{\nu\rho} - (n-1)\,\nabla_\rho\,f_\nu + \nabla_\nu\,f_\rho + (n-2)\,f_\nu\,f_\rho - g_{\nu\rho}\,(\nabla_\lambda\,f^\lambda + (n-2)\,f_\lambda\,f^\lambda),$$

$$\hat{R} = R - 2\,(n-1)\,\nabla_\lambda\,f^\lambda - (n-1)\,(n-2)\,f_\lambda\,f^\lambda.$$

Setting here $f = -\Omega^{-1}\,d\,\Omega$, we obtain for the metrics g, $\tilde{g}$ which are related by the rescaling $g = \Omega^2\,\tilde{g}$ the relation (1.3) of their Ricci tensors, which implies for their Ricci scalars the relation

$$4\,\frac{n-1}{n-2}\,\nabla_\mu\,\nabla^\mu\,\theta - R[g]\,\theta = -\tilde{R}[\tilde{g}]\,\theta^{\frac{n+2}{n-2}}. \tag{1.9}$$

For later convenience we set here $\theta = \Omega^{-\frac{n-2}{2}}$.

The analogue of the decomposition (1.4) takes the form

$$\hat{R}^{\mu}{}_{\nu\lambda\rho} = 2\,\{g^{\mu}{}_{[\lambda}\,\hat{L}_{\rho]\nu} - g^{\mu}{}_{\nu}\,\hat{L}_{[\lambda\rho]} - g_{\nu[\lambda}\,\hat{L}_{\rho]}{}^{\mu}\} + C^{\mu}{}_{\nu\lambda\rho}, \tag{1.10}$$

where the trace parts of the curvature tensor are represented by the tensor

$$\hat{L}_{\mu\nu} = \frac{1}{n-2}\left(\hat{R}_{(\mu\nu)} - \frac{n-2}{n}\,\hat{R}_{[\mu\nu]} - \frac{1}{2\,(n-1)}\,g_{\mu\nu}\,\hat{R}\right). \tag{1.11}$$

It is related to (1.5) by

$$\nabla_{\mu}\,f_{\nu} - f_{\mu}\,f_{\nu} + \frac{1}{2}\,g_{\mu\nu}\,f_{\lambda}\,f^{\lambda} = L_{\mu\nu} - \hat{L}_{\mu\nu}. \tag{1.12}$$

By a straightforward calculation it can be verified that (1.5), (1.11) satisfy the identity

$$\hat{\nabla}_{\rho}\,\hat{L}_{\mu\nu} - \hat{\nabla}_{\mu}\hat{L}_{\rho\nu} = \nabla_{\rho}\,L_{\mu\nu} - \nabla_{\mu}\,L_{\rho\nu} + f_{\lambda}\,C^{\lambda}{}_{\nu\rho\mu}. \tag{1.13}$$

Conformal geodesics. We have seen that null geodesics, considered as point sets, are invariants of a Lorentzian conformal structure. With any conformal structure $\mathcal{C}_{\tilde{g}}$ is associated in fact a much larger class of distinguished curves called *conformal geodesics*. We shall only give an outline of their properties here, referring the reader to [38], [41], and [46] for details and further results. A conformal geodesic curve $x(\tau)$ is obtained, together with a 1-form $b = b(\tau)$ along the curve, as a solution to the system of equations

$$(\tilde{\nabla}_{\dot{x}}\dot{x})^{\mu} + S(b)_{\lambda}{}^{\mu}{}_{\rho}\,\dot{x}^{\lambda}\,\dot{x}^{\rho} = 0, \tag{1.14}$$

$$(\tilde{\nabla}_{\dot{x}}b)_{\nu} - \frac{1}{2}\,b_{\mu}\,S(b)_{\lambda}{}^{\mu}{}_{\nu}\,\dot{x}^{\lambda} = \tilde{L}_{\lambda\nu}\,\dot{x}^{\lambda}, \tag{1.15}$$

where $\tilde{L}$ and S are as in (1.5), (1.7). There are more conformal geodesics than geodesics for any metric in the conformal class because for given initial data $x_* \in M$, $\dot{x}_* \in T_{x_*}M$, $b_* \in T^*_{x_*}M$ there exists a unique conformal geodesic $x(\tau)$, $b(\tau)$ near x_* satisfying for given $\tau_* \in \mathbb{R}$

$$x(\tau_*) = x_*, \quad \dot{x}(\tau_*) = \dot{x}_*, \quad b(\tau_*) = b_*. \tag{1.16}$$

The sign of $\tilde{g}(\dot{x},\dot{x})$ is preserved along a conformal geodesic but not its modulus. Conformal geodesics, considered as point sets, are in general different from the metric geodesics of a metric in the conformal class. Furthermore, they admit general fractional linear maps as parameter transformations. Initial data with $\tilde{g}(\dot{x}_*,\dot{x}_*) = 0$, $\dot{x}_* \neq 0$, determine null curves which coincide, as point sets, with the null geodesics of the metrics in the conformal class. The fact that we shall mainly make use of time-like conformal geodesics in the following, should not be taken as an indications that null or space-like conformal geodesics can not be useful.

Conformal geodesics are *conformal invariants* in the following sense. Denote by f a smooth 1-form field. Then, if $x(\tau)$, $b(\tau)$ solve the conformal geodesics equations (1.14), (1.15), the pair $x(\tau)$, $b(\tau)-f|_{x(\tau)}$ solves the same equations with $\tilde{\nabla}$ replaced by the connection $\hat{\nabla} = \tilde{\nabla} + S(f)$ and L by $\hat{L}$, i.e. $x(\tau)$, in particular its parameter τ, is independent of the Weyl connection in the conformal class which is used to write the equations.

Assume that there is given a smooth congruence of conformal geodesics solving (1.14), (1.15) which covers an open set U of our manifold such that the associated 1-forms b define a smooth field on U. Denote by $\hat{\nabla}$ the torsion free connection on U which defines with the connection $\tilde{\nabla}$ entering (1.14), (1.15) the difference tensor $\hat{\nabla} - \tilde{\nabla} = S(b)$ and denote by $\hat{L}$ the tensor (1.11) derived from $\hat{\nabla}$. Comparing with (1.12), we find that equations (1.14), (1.15) can be written

$$\hat{\nabla}_{\dot{x}}\, \dot{x} = 0, \tag{1.17}$$

$$\hat{L}_{\lambda\nu}\, \dot{x}^{\lambda} = 0. \tag{1.18}$$

Let e_k be a frame field satisfying

$$\hat{\nabla}_{\dot{x}}\, e_k = 0, \tag{1.19}$$

along the congruence, and suppose that S is a hypersurface transverse to the congruence which meets each of the curves exactly once and on which $\tilde{g}(e_i, e_k) = \Theta_*^2 \eta_{ik}$ with some function $\Theta_* > 0$. Following the argument which leads to (1.6) we see that $\tilde{g}(e_i, e_k) = \Theta^2\, \eta_{ik}$ on U with a function Θ satisfying

$$\hat{\nabla}_{\dot{x}}\, \Theta = \Theta < \dot{x}, b >, \quad \Theta|_S = \Theta_*. \tag{1.20}$$

1.2.2 Derivation of the Conformal Field Equations

The curvature tensor always satisfies the Bianchi identity $\nabla_{[\delta}\, R^{\mu}{}_{|\nu|\lambda\rho]} = 0$. Observing (1.4), we can write it in $n \geq 3$ dimensions in the form

$$\nabla_{[\delta}\, C^{\mu}{}_{|\nu|\lambda\rho]} = 2\, \{g_{\nu[\lambda}\, \nabla_{\delta}\, L_{\rho]}{}^{\mu} - g^{\mu}{}_{[\lambda}\, \nabla_{\delta}\, L_{\rho]\nu}\}, \tag{1.21}$$

which gives after a contraction

$$\nabla_{\mu}\, C^{\mu}{}_{\nu\lambda\rho} = (n-3)\, \{\nabla_{\lambda}\, L_{\rho\nu} - \nabla_{\rho}\, L_{\lambda\nu}\}. \tag{1.22}$$

We assume now that the *physical metric* $\tilde{g}$ satisfies Einstein's field equation (1.1) and try to express these equations in terms of the *conformal metric*

$$g = \Omega^2\, \tilde{g}, \tag{1.23}$$

and the conformal factor $\Omega > 0$ such that the difficulties pointed out above are avoided. Although our discussion can be generalized to include certain matter fields (cf. for instance [32], [53] and also, for an analysis with a different intention and different results, P. Tod, this volume), we assume for simplicity that matter fields are absent, i.e. $T_{\mu\nu} = 0$. Equation (1.22) reduces then to

$$\tilde{\nabla}_\mu \, C^\mu{}_{\nu\lambda\rho} = 0. \tag{1.24}$$

This equation is of particular interest for us because due to the identity

$$\nabla_\mu \, (\Omega^{3-n} \, C^\mu{}_{\nu\lambda\rho}) = \Omega^{3-n} \, \tilde{\nabla}_\mu \, C^\mu{}_{\nu\lambda\rho},$$

which holds for any conformal factor, it has a certain conformal covariance. It implies for

$$d^\mu{}_{\nu\lambda\rho} = \Omega^{3-n} \, C^\mu{}_{\nu\lambda\rho}, \tag{1.25}$$

the *Bianchi equation*

$$\nabla_\mu \, d^\mu{}_{\nu\lambda\rho} = 0, \tag{1.26}$$

which, in a sense, represents the core of all the systems to be considered in the following. We consider now equation (1.4) for the conformal metric g and write it in the form

$$R^\mu{}_{\nu\lambda\rho} = 2\,\{g^\mu{}_{[\lambda} \, L_{\rho]\nu} - g_{\nu[\lambda} \, L_{\rho]}{}^\mu\} + \Omega^{n-3} \, d^\mu{}_{\nu\lambda\rho}. \tag{1.27}$$

Using (1.22) also for g and observing (1.25), (1.26), we obtain for $L_{\mu\nu}$, which represents the Ricci tensor of the conformal metric, the equation

$$\nabla_\lambda \, L_{\rho\nu} - \nabla_\rho \, L_{\lambda\nu} = \Omega^{n-4} \, \nabla_\mu \Omega \; d^\mu{}_{\nu\lambda\rho}. \tag{1.28}$$

For later discussion we note how (1.27), (1.28) generalize if we admit besides rescalings also transitions to Weyl connections which are related to the Levi-Civita connection ∇ of g by difference tensors

$$\hat{\nabla} - \nabla = S(f), \tag{1.29}$$

with smooth 1-forms f. Since $\nabla - \tilde{\nabla} = S(\Omega^{-1} \nabla \Omega)$, it follows that $\hat{\nabla} - \tilde{\nabla} = S(\Omega^{-1} \, d)$ with the smooth 1-form $d_\mu = \Omega \, f_\mu + \nabla_\mu \Omega$. By (1.10) equation (1.27) is then replaced by

$$\hat{R}^\mu{}_{\nu\lambda\rho} = 2\,\{g^\mu{}_{[\lambda} \, \hat{L}_{\rho]\nu} - g^\mu{}_\nu \, \hat{L}_{[\lambda\rho]} - g_{\nu[\lambda} \, \hat{L}_{\rho]}{}^\mu\} + \Omega^{n-3} \, d^\mu{}_{\nu\lambda\rho}, \tag{1.30}$$

while (1.28) and (1.13) imply the equation

$$\hat{\nabla}_\rho \, \hat{L}_{\mu\nu} - \hat{\nabla}_\mu \hat{L}_{\rho\nu} = \Omega^{n-4} \, d_\mu \, d^\mu{}_{\nu\lambda\rho}. \tag{1.31}$$

By admitting rescalings and transitions to Weyl connections we have artificially introduced a conformal gauge freedom into the equations. If these equations are to be used to derive information about the physical metric $\tilde{g}$ by analyzing boundary value problems, this freedom has to be removed again by imposing gauge conditions such that (i) the resulting equations ensure that $\tilde{g}$, derived by (1.23) from Ω and g, satisfies the field equations, (ii) interesting information on $\tilde{g}$ can be gained by suitably fixing Ω and f. Two quite different methods have been discussed so far ([23], [24], [35]).

The metric conformal field equations. We consider only conformal rescalings of the metric and try to get equations which allow us to specify the conformal factor in a useful way. If we observe (1.1) in the form $\tilde{R}_{\mu\nu} = \frac{1}{n}\,\tilde{R}\,\tilde{g}_{\mu\nu}$ in (1.3), the resulting equation, which we read now as equation for Ω, can be written in the form

$$\nabla_\mu\,\nabla_\nu\,\Omega = -\Omega\,L_{\mu\nu} + s\,g_{\mu\nu}, \tag{1.32}$$

with

$$s = \frac{1}{n}\,\nabla_\lambda\,\nabla^\lambda\,\Omega + \frac{1}{2\,n\,(n-1)}\,R\,\Omega.$$

Equation (1.32) is obviously overdetermined. If we apply ∇_ρ to both sides, commute ∇_ρ with ∇_μ, contract over ρ and ν, and use the twice contracted Bianchi identity, we obtain the integrability condition

$$\nabla_\mu\,s = -\nabla^\nu\,\Omega\,L_{\nu\mu}, \tag{1.33}$$

which we read as an equation for s.

To see the role of the cosmological constant, we observe that the transformation law of the Ricci scalar gives

$$\lambda = (n-1)\,(2\,\Omega\,s - \nabla_\rho\,\Omega\,\nabla^\rho\,\Omega). \tag{1.34}$$

Equations (1.32), (1.33) then imply for the quantity on the right hand side

$$\frac{1}{2}\,\nabla_\lambda\,(2\,\Omega\,s - \nabla_\rho\,\Omega\,\nabla^\rho\,\Omega) = \nabla_\lambda\,\Omega\,s + \Omega\,\nabla_\lambda\,s - \nabla^\rho\,\Omega\,\nabla_\rho\,\nabla_\lambda\,\Omega$$

$$= \nabla_\lambda\,\Omega\,s + \Omega\,(-\nabla^\nu\,\Omega\,L_{\nu\lambda}) - \nabla^\rho\,\Omega\,(-\Omega\,L_{\rho\lambda} + g_{\rho\lambda}\,s) = 0,$$

which reflects the well known fact that the twice contracted Bianchi identity implies that λ must be constant. It follows that (1.34) will be a consequence of the other equations if it is arranged to hold at one point by a suitable choice of the initial data.

We consider (1.26), (1.27), (1.28), (1.32), (1.33), (1.34) as equations for the unknown tensor fields

$$g_{\mu\nu},\ \Omega,\ s,\ L_{\mu\nu},\ d^\mu{}_{\nu\lambda\rho},$$

and refer to them as the *metric conformal field equations*. These equations hold for any vacuum solution $\tilde{g}$ and any conformal factor Ω.

A way to impose a gauge condition on the conformal factor is suggested by the following observation. If g is any metric and R^* a given function, the metric $g^* = \theta^{\frac{4}{n-2}}\,g$ will have by (1.9) the Ricci scalar $R[g^*] = R^*$, if θ solves the equation

$$4\,\frac{n-1}{n-2}\,\nabla_\mu\,\nabla^\mu\,\theta - R[g]\,\theta = -R^*\,\theta^{\frac{n+2}{n-2}}.$$

Since locally this equation can always be solved for a positive function θ, it follows that locally the Ricci scalar R of the conformal metric can be prescribed in the metric conformal field equations as a *gauge source function* for the conformal scaling. Its choice will depend on the given situation and the choices of the other gauge dependent quantities. We can set $R = 0$ but, depending on the circumstances, there may be better choices. In any case this leaves the freedom to specify suitable boundary data for the conformal factor.

From the way we obtained the metric conformal field equations it is clear that they are equivalent to the vacuum field equations (ignoring questions of smoothness) where Ω is positive but that they make sense also where Ω vanishes. While the system is regular in this sense, it is also complicated and highly overdetermined. We shall see below that the formal regularity of the equations does not necessarily imply that the data must be regular nor does it ensure by itself that the asymptotic behaviour of the solutions can be controlled conveniently. To get an insight into this one needs to work out suitable initial value problems in great detail and control the *constraints* as well as the *evolution*. We now take a look at the various parts of the system.

The Bianchi equation. For a solution $\tilde{g}$ to the vacuum field equations the Bianchi identity (1.21) takes the form

$$\tilde{\nabla}_\delta \, C^\mu{}_{\nu\lambda\rho} + \tilde{\nabla}_\rho \, C^\mu{}_{\nu\delta\lambda} + \tilde{\nabla}_\lambda \, C^\mu{}_{\nu\rho\delta} = 0. \tag{1.35}$$

Contracting with $\tilde{\nabla}_\delta$, commuting derivatives, and observing (1.24), we obtain

$$\tilde{\nabla}_\delta \, \tilde{\nabla}^\delta \, C^\mu{}_{\nu\tau\rho} = 2 \, \{ C^\mu{}_{\pi\delta\rho} \, C^\pi{}_{\nu\tau}{}^\delta + C^\mu{}_{\pi\tau\delta} \, C^\pi{}_{\nu\rho}{}^\delta + C^\mu{}_{\nu\pi\delta} \, C^\pi{}_{\tau\rho}{}^\delta \} + 2 \, \lambda \, C^\mu{}_{\nu\tau\rho}.$$

Thus, in the Lorentzian case, which will be considered in the following, we find the well known fact that the Bianchi identity implies a system of hyperbolic equations for the conformal Weyl tensor in any dimension $n \geq 4$.

In $n = 4$ dimensions we can define the left and right dual of the conformal Weyl tensor, given by ${}^*C_{\mu\nu\tau\rho} = \frac{1}{2} \, \tilde{\epsilon}_{\mu\nu}{}^{\delta\pi} \, C_{\delta\pi\tau\rho}$ and $C^*_{\mu\nu\tau\rho} = \frac{1}{2} \, \tilde{\epsilon}_{\tau\rho}{}^{\delta\pi} \, C_{\mu\nu\delta\pi}$ respectively, and it is well known that they coincide. It follows that

$$\tilde{\epsilon}_\pi{}^{\delta\tau\rho} \tilde{\nabla}_\delta \, C^\mu{}_{\nu\tau\rho} = 2 \, \tilde{\nabla}_\delta \, C^{*\mu}{}_{\nu\pi}{}^\delta = 2 \, \tilde{\nabla}_\delta \, {}^*C^\mu{}_{\nu\pi}{}^\delta = - \, \tilde{\epsilon}^\mu{}_\nu{}^{\tau\rho} \, \tilde{\nabla}_\delta \, C^\delta{}_{\pi\tau\rho},$$

which implies that the contracted Bianchi identity (1.24) is equivalent to the Bianchi identity (1.35). It follows that (1.24) and, by the same argument equation (1.26), imply hyperbolic equations.

In dimensions $n \geq 5$ the situation is quite different as can be seen by a simple function counting. The conformal Weyl tensor has in n dimensions

$$\frac{1}{12} \, n^2 \, (n^2 - 1) - \frac{1}{2} \, n \, (n + 1),$$

independent components while the contracted Bianchi identity (1.24) provides

$$\frac{1}{3} \, n \, (n - 2) \, (n + 2),$$

independent equations. Moreover, if we express (1.24) in Gauss coordinates with x^0 the affine parameter on the geodesics orthogonal to the space-like hypersurfaces $\{x^0 = \text{const.}\}$, it follows that the equations $\tilde{\nabla}_\delta \, C^\delta{}_{0\lambda\rho} = 0$ involve only derivatives in the directions $\partial_{x^1}, \ldots, \partial_{x^{n-1}}$ and thus provide

$$\frac{1}{2}\,n\,(n-1)$$

constraints. Consequently, there are not enough equations in (1.24) if $n \geq 6$ while for $n = 5$ there are as many equations as unknowns but not sufficiently many evolution equations. *It follows that a system of conformal equations which includes as equation for the rescaled conformal Weyl tensor only the Bianchi equation provides a good evolution system only in four dimensions.*

If we use in $n \geq 5$ dimensions the full Bianchi identities to derive a system of wave equations for $d^{\mu}{}_{\nu\lambda\rho}$, it appears that not all terms of the form $\Omega^{-1}\,\nabla_{\mu}\,\Omega$ can be removed from the equations. It would be interesting to know whether there exist other conformally covariant equations which could replace the Bianchi equation in higher dimensions. If it turns out that a conformally regular representation for the Einstein equation can only be obtained in four dimensions, it would also be interesting to know whether there is a principal difference in the asymptotic behaviour of the fields in four and higher dimensional space-times.

We shall restrict our discussion in the following to the case of four dimensions.

Instead of expressing the intrinsic hyperbolicity of the Bianchi equation in terms of a system of wave equations we work directly with the first order system. The benefits of this will be seen later. There are known two quite different possibilities to make the hyperbolicity of the equations manifest, both of them have been used in numerical calculations (cf. J. Frauendiener and S. Husa, this volume). It has been shown in [37] how to obtain hyperbolic equations from the Bianchi equation in the ADM representation of the metric. The other method, which is based on a frame formalism, has many advantages and can be adapted to various geometrical situations. We can use real orthonormal frames (cf. [44]) but most convenient are pseudo-orthonormal frames or spin frames. In the spin frame formalism the equations split almost automatically into constraints and (symmetric hyperbolic) evolution equations (cf. [23], [27], [32]). We shall use this fact here to discuss certain features of the Bianchi equation.

In the spin frame formalism we have a frame field $e_{aa'}$, $a, a' = 0, 1$, with $e_{00'}$, $e_{11'}$ real null vector fields and $e_{01'}$, $e_{10'}$ complex vector fields which are conjugate to each other. Their only non-vanishing scalar products are given by $g(e_{00'}, e_{11'}) = 1$, $g(e_{01'}, e_{10'}) = -1$. In the following discussion it is useful to think of these vector fields as being obtained in the form

$$e_{00'} = \frac{1}{\sqrt{2}}\,(e_0 + e_3), \quad e_{11'} = \frac{1}{\sqrt{2}}\,(e_0 - e_3),$$
$$e_{01'} = \frac{1}{\sqrt{2}}\,(e_1 + i\,e_2), \quad e_{10'} = \frac{1}{\sqrt{2}}\,(e_1 - i\,e_2),$$

from a frame field e_k, $k = 0, \ldots, 3$, satisfying $g(e_i, e_k) = \eta_{ik}$.

The rescaled conformal Weyl tensor is represented by a completely symmetric spinor field ϕ_{abcd}, and the Bianchi equation by the equation

$$0 = \Lambda_{abca'} \equiv \nabla^{f}{}_{a'}\, \phi_{abcf}, \tag{1.36}$$

where $\nabla_{aa'}$ denotes the covariant derivative on spinor fields in the direction of $e_{aa'}$ and the index is raised with the antisymmetric spinor field ϵ^{ab} which satisfies $\epsilon^{01} = 1$.

If we set $\Lambda_{abcd} = \Lambda_{abca'}\, \tau_d{}^{a'}$ with $\tau_a{}^{a'} = \delta_a{}^1\, \delta_{0'}{}^{a'} - \delta_a{}^0\, \delta_{1'}{}^{a'}$ we get the decomposition $\Lambda_{abcd} = \Lambda_{(abcd)} - \frac{3}{4}\, \epsilon_{d(c}\, \Lambda_{ab)f}{}^{f}$ into its irreducible parts

$$-2\, \Lambda_{(abcd)} = P\, \phi_{abcd} - 2\, \mathcal{D}_{(d}{}^{f}\, \phi_{abc)f}, \qquad \Lambda_{abf}{}^{f} = \mathcal{D}^{ef}\, \phi_{abef}.$$

Here $P = \tau^{aa'}\, \nabla_{aa'} = \sqrt{2}\, \nabla_{e_0}$ and $\mathcal{D}_{ab} = \tau_{(a}{}^{a'}\, \nabla_{b)a'}$ denote covariant directional derivatives such that $\mathcal{D}_{00} = -\nabla_{01'}$, $\mathcal{D}_{01} = \mathcal{D}_{10} = \frac{1}{\sqrt{2}}\, \nabla_{e_3}$, $\mathcal{D}_{11} = \nabla_{10'}$ (cf. [32], [35] for more details of the underlying space-spinor formalism).

In a Cauchy problem one will in general assume e_0 to be the future directed normal to the initial hypersurface S. The operators $\mathcal{D}_{ab}$ then involve only differentiation in directions tangent to S and the equations $\Lambda_{abf}{}^{f} = 0$ are interior equations on S. They represent the six real constraint equations implied on S by the Bianchi equation.

The remaining equations for ϕ_{abcd} should be evolution equations. Multiplying the equations by suitable binomial coefficients (and considering the frame field and the connection coefficients as given), we find that the system

$$-\binom{4}{a+b+c+d}\, \Lambda_{(abcd)} = 0, \tag{1.37}$$

is in fact *symmetric hyperbolic*, i.e. it has the form

$$A^{\mu}\, \partial_{\mu}\, u = H(x, u),$$

with a $\mathbb{C}^N$-valued unknown u, a $\mathbb{C}^N$-valued function $H(x,u)$, and $N \times N$-matrix-valued, possibly u-dependent functions A^{μ} which are hermitian, i.e. ${}^{T}\bar{A}^{\mu} = A^{\mu}$, and for which there exists at each point x a covector ξ_{μ} such that $A^{\mu}\xi_{\mu}$ is positive definite. There exists a well developed theory for such systems (cf. [48], [58], [77], and also H.-O. Kreiss and O. Ortiz, this volume).

While the constraints implied on a given space-like hypersurface are determined uniquely, there is a large freedom to select useful evolution systems. In the present case it turns out that any system of the form

$$
\begin{aligned}
0 &= 2\,a\,\Lambda_{(0001')},\\
0 &= (c-d)\,\Lambda_{(0011')} - 2\,a\,\Lambda_{(0000')},\\
0 &= (c+d)\,\Lambda_{(0111')} - (c-d)\,\Lambda_{(0010')},\\
0 &= 2\,e\,\Lambda_{(1111')} - (c+d)\,\Lambda_{(0110')},\\
0 &= -2\,e\,\Lambda_{(1110')},
\end{aligned}
$$

with $a, c, e > 0$ and $-(2\,e+c) < d < 2\,a+c$, is symmetric hyperbolic (the system (1.37) occurs here as a special case). This flexibility allows the Bianchi equation to be adapted to various geometrical situations such as characteristic initial value problems ([23], [24], [25], [28], [57]), standard Cauchy problems ([30], [29], [32]), and initial boundary value problems ([35], [44]). Note that for these systems the set of *characteristics*, given by the hypersurfaces of the form $\{f = 0, df \neq 0\}$ with $f \in C^1(M)$ on which $\det(A^\mu\, f_{,\mu}) = 0$ is satisfied at each of their points, does not only comprise the null hypersurfaces and can in fact be quite complicated.

The constraints. Related to the gauge freedom is the fact that, written in suitably adapted coordinates (and frame field, if used), Einstein's equations as well as the metric conformal field equations imply inner equations, i.e. constraints, on any time-like, null, or space-like hypersurface. We will consider in the following the latter two cases, because they are particularly important in initial value problems. (Time-like hypersurfaces are of interest in initial-boundary value problems, however, even in general analytical discussions of this problem a detailed discussion of the constraints on these hypersurfaces is not necessarily required [44].)

Null hypersurfaces representing the *physical characteristics* of the Einstein equations, the system of inner equations is quite different in nature from those arising in the non-null case. There are on the one hand the intrinsic equations arising because of the gauge freedom and on the other hand the characteristic inner equations arising due to the intrinsic wave equation character of the field equations and these equations combine in a very peculiar way. Let us consider a situation where data are to be prescribed on a null hypersurface N and another hypersurface which has a space-like intersection Σ with N. Then some of the unknowns can be prescribed *completely freely* on N. All the remaining unknowns can then be calculated on N in suitably adapted coordinates from certain data on Σ by solving a certain hierarchy of inner equations on N, each member of which reduces to a system of ODEs along the null generators at their subsequent integration. Those equations which are not used in this procedure will then be satisfied on N if they hold on Σ (cf. [75] for the vacuum case).

The enormous simplification of the equations resulting from these facts in a suitably adapted gauge and the possibility to integrate the equations along characteristics 'out to infinity' led a number of groups to develop very efficient *characteristic numerical computer codes* (cf. R. Bartnik and A. Norton, J. Font, L. Lehner, this volume).

Though being more complicated, the inner equations induced on null hypersurfaces by the metric conformal field equations have a similar structure and, besides the data required to fix the conformal gauge, the free data are, of course, up to a rescaling the same ([24]).

In the following we will mainly be concerned with Cauchy problems and therefore consider now space-like initial hypersurfaces. The standard constraints induced on space-like hypersurfaces by the vacuum field equations form a system of underdetermined elliptic quasi-linear equations for the data induced by the 4-metric on the initial hypersurface, i.e. the inner metric $\tilde{h}_{\alpha\beta}$ and the extrinsic curvature $\tilde{\chi}_{\alpha\beta}$. There exists a detailed theory for these equations which, under the assumption of constant mean extrinsic curvature, reduces the problem of solving the constraints to a problem of solving a linear elliptic system and a semi-linear elliptic scalar equation (cf. [12], [13]).

The metric conformal field equations imply a system of *conformal constraints* ([26], [35]) on a space-like hypersurface S which comprises the usual vacuum constraints but which is more complicated. This is not only due to the artificial scaling freedom which has been introduced into the vacuum equations but also because the equations include integrability conditions and the Gauss-Codazzi equations. There appear to be two ways of dealing with these equations. The first method solves essentially the standard vacuum constraints and calculates from the vacuum data the conformal data after introducing a suitable conformal rescaling. This method, which has been analyzed with a certain completeness and underlies the present numerical calculations, introduces certain numerical difficulties (cf. J. Frauendiener and S. Husa, this volume). A second possibility would be to solve the conformal constraints directly. First steps in a systematic investigation of this approach are reported in the article by A. Butscher in this volume (cf. also the discussion in [33]).

Any of these approaches has to deal with a problem which is not considered in the usual discussions of the standard vacuum constraints. In the situations which are most interesting for our applications the initial hypersurface contains a subset Σ, not belonging to the physical space-time, where the conformal factor vanishes. As seen in detail in the next sections, this set can be a 2-dimensional boundary which represents points at null infinity (cf. L. Andersson, this volume) or consist of a number of discrete compactification points, $\Sigma = \{i_1, i_2, \ldots i_N\}$, representing space-like infinities (cf. S. Dain, this volume). In principle, Σ could consist of a union of such sets or more complicated sets. In any of these situations there arises the question about the precise smoothness of the data near Σ in suitably chosen coordinates.

It turns out that in general, even under the strongest smoothness assumptions on the 'free' data, there may occur terms of the form $r^k \log^l r$ near Σ, where r denotes the (non-physical) distance from Σ. However, it also turns out that under mild additional assumptions logarithmic terms will be absent near Σ. We will discuss some of the subtle questions arising here later.

Gauge conditions, hyperbolicity, and local evolution. In principle everything known about Einstein's equations also applies to our system. If we introduce a fictitious energy-momentum tensor by writing $\kappa T_{\mu\nu} = 2(L_{\mu\nu} - L g_{\mu\nu})$ with a constant $\kappa \neq 0$ (introduced here to emphasize the similarity, we could set $\kappa = 1$), our equations imply after a contraction of (1.27) the system

$$R_{\mu\nu} - \frac{1}{2} R g_{\mu\nu} = \kappa T_{\mu\nu},$$

$$\nabla_{[\lambda} T_{\rho]\nu} - \frac{1}{3} \nabla_{[\lambda} T g_{\rho]\nu} = \frac{1}{\kappa} \nabla_\mu \Omega \, d^\mu{}_{\nu\lambda\rho},$$

$$\nabla_\mu d^\mu{}_{\nu\lambda\rho} = 0,$$

$$\nabla_\mu \nabla_\nu \Omega = -\frac{\kappa}{2} (T_{\mu\nu} - \frac{1}{3} T g_{\mu\nu}) \Omega + s g_{\mu\nu},$$

$$\nabla_\mu s = -\frac{\kappa}{2} (T_{\mu\nu} - \frac{1}{3} T g_{\mu\nu}) \nabla^\nu \Omega,$$

which can be read as Einstein equations with certain matter fields, in which $R = -\kappa T$ represents a gauge source function. We note that in boundary value problems for these equations it has to be made sure, by the way the problems are posed and the data are given, that equation (1.27), which gives geometrical meaning to the fields $L_{\mu\nu}$ and $d^\mu{}_{\nu\lambda\rho}$, will be satisfied.

Following the discussion of the Bianchi equation above (or that of [37], if one prefers to consider the metric instead of a frame field as a basic variable), it is not difficult to extract symmetric hyperbolic evolution equations, the *reduced equations*, from the last four subsystems. Of course, any choice here will leave us with constraint equations. It remains to discuss the first equation. In principle one can bring to bear here all the methods developed for the ADM equations if one feels most comfortable with these equations. However, it should be kept in mind that this is only one of the many possibilities of dealing with equation (1.27). In the references given in this article there can be found many others which have been employed successfully under various geometrical assumptions.

If one is dealing with the Einstein equation above or directly with equation (1.27), in either case one will have to make the decision whether to use the metric coefficients $g_{\mu\nu}$ or frame coefficients $e^\mu{}_k$ as basic variables. Furthermore, one needs to choose gauge conditions, i.e. a way to specify the coordinates and, in the second case, the frame field. Usually this is done in terms of a *gauge system* of differential equations and suitable initial conditions. A large set of possibilities is known by now.

Basic objectives in choosing the gauge are the following (cf. also [45]): (i) The gauge should be adapted to the type of problem (Cauchy problem, initial-boundary value problem, characteristic initial value problem), (ii) The evolution equations should be simple, manifestly hyperbolic, or of some other form (e.g. 'elliptic-hyperbolic') which allows us to handle them in existence proofs or in numerical calculations (cf. the discussion by H.-O. Kreiss and O. Ortiz, this volume). There are available gauge condition which are governed by implicit wave equations like the harmonic gauge or the conditions for frame or spin frame fields employed in [27], [32], and there are known explicit conditions, in particular certain geometric gauge conditions, which lead to a reduction of the number of the unknowns (cf. [35], [37] and the discussion below) and which allow us to control to some extent the development of a degeneracy of the gauge. (iii) The gauge conditions as well as the constraints should be preserved during the evolution. (iv) The gauge should allow us to control its behaviour during the evolution in numerical or in analytical studies. (v) It should have a sufficiently long 'life time'.

While (ii) and (iii) are in the case of hyperbolic equations local properties which are accessible to direct calculations, (iv) and (v) depend very much on the given situation and cannot be handled in a general way. These points belong to the most difficult ones in the large scale evolution problem in numerical as well as in analytical studies.

There are known now various ways to perform *hyperbolic reductions* (cf. [37]) for initial value problems for the metric conformal field equations: given a suitable solution to the constraints, we find a *reduced system* of hyperbolic evolution equations for which the local existence and uniqueness of solutions follow from standard theorems, and, by invoking the available integrability conditions, we can derive a *subsidiary system* which allows us to show that all constraints and gauge conditions are preserved during the evolution if they are satisfied initially. *The significance of the conformal field equations is determined to a large extent by the fact that they admit hyperbolic reductions irrespective of the sign of the conformal factor.*

After we use the large freedom to choose the gauge system and the reduced system, the subsidiary system is essentially fixed. For the analytical discussions of solutions, the latter is only required to possess the uniqueness property, which ensures that trivial data imply the trivial solution. However, since in the numerical treatment of the Einstein equations the constraints are only satisfied up to a numerical error, the detailed propagation properties of the subsidiary system may become important in that context. A systematic analysis of how changes in the gauge system and the reduced system affect the propagation properties of the subsidiary system and whether this dependence can be used to choose reduced systems with improved evolution of the constraints has not been given yet.

The general conformal field equations. We will now consider a system of conformal field equations which employs the additional freedom gained by admitting general Weyl connections. Combined with suitable gauge conditions it will lead to a system of evolution equations which is quite different from and in fact much simpler than the evolution systems obtained from the metric conformal field equations.

Assume that $\tilde{g}$ solves Einstein's vacuum field equations with cosmological constant λ and let $g = \Theta^2 \, \tilde{g}$ be given a conformal metric where Θ is some conformal factor. Furthermore, let b be a smooth 1-form. We denote by $\tilde{\nabla}$ and ∇

the Levi-Civita connection of $\tilde{g}$ and g respectively and by $\hat{\nabla}$ the Weyl connection for $\tilde{g}$ which satisfies $\hat{\nabla} - \tilde{\nabla} = S(b)$. It follows that $\hat{\nabla} = \nabla + S(f)$ with $f = b - \Theta^{-1}\,\nabla\Theta$. We define a further 1-form by setting $d_\mu = \Theta\, f_\mu + \nabla_\mu \Theta$.

Let e_k be a frame field satisfying $g(e_i, e_k) = \eta_{ik}$. We denote by ∇_k and $\hat{\nabla}_k$ the covariant derivative in the direction of e_k with respect to ∇ and $\hat{\nabla}$ and define the connection coefficients $\hat{\Gamma}_i{}^j{}_k$ of $\hat{\nabla}$ in this frame field by $\hat{\nabla}_i e_k = \hat{\Gamma}_i{}^j{}_k\, e_k$. If we express now all tensor fields (except the e_k) as well as equations (1.30), (1.31), (1.26) and the condition that $\hat{\nabla}$ be torsion free in terms of the frame field and the connection coefficients, we get for the unknown

$$u = (e^\mu{}_k, \quad \hat{\Gamma}_i{}^j{}_k, \quad \hat{L}_{jk}, \quad d^i{}_{jkl}), \tag{1.38}$$

the system of equations

$$[e_p, e_q] = (\hat{\Gamma}_p{}^l{}_q - \hat{\Gamma}_q{}^l{}_p)\, e_l, \tag{1.39}$$

$$\begin{aligned} e_p(\hat{\Gamma}_q{}^i{}_j) - e_q(\hat{\Gamma}_p{}^i{}_j) - \hat{\Gamma}_k{}^i{}_j(\hat{\Gamma}_p{}^k{}_q - \hat{\Gamma}_q{}^k{}_p) + \hat{\Gamma}_p{}^i{}_k\hat{\Gamma}_q{}^k{}_j - \hat{\Gamma}_q{}^i{}_k\hat{\Gamma}_p{}^k{}_j \quad (1.40)\\ = 2\,\{g^i{}_{[p}\,\hat{L}_{q]j} - g^i{}_j\,\hat{L}_{[pq]} - g_{j[p}\,\hat{L}_{q]}{}^i\} + \Theta\, d^i{}_{jpq}, \end{aligned}$$

$$\hat{\nabla}_p\,\hat{L}_{qj} - \hat{\nabla}_q\,\hat{L}_{pj} = d_i\, d^i{}_{jpq}, \tag{1.41}$$

$$\nabla_i d^i{}_{jkl} = 0. \tag{1.42}$$

In the last equation the connection ∇ is used. This poses no problem because the connection coefficients of ∇ in the frame e_k are related to those of $\hat{\nabla}$ by the purely algebraic relation $\hat{\Gamma}_i{}^j{}_k = \Gamma_i{}^j{}_k + \delta^j{}_i\, f_k + \delta^j{}_k\, f_i - \eta_{ik}\,\eta^{jl}\, f_l$, whence $f_i = 1/n\,\hat{\Gamma}_i{}^k{}_k$.

No differential equations are given above for the fields Θ and $d_k = \Theta\, f_k + \nabla_k\Theta$, they reflect the artificially introduced conformal gauge freedom. Note that they cannot be prescribed quite arbitrarily. For a solution which extends smoothly into a domain containing a set where $\Theta = 0$, we should have $d = \nabla\Theta$ on this set.

To obtain the metric conformal field equations, (1.26), (1.27), (1.28) have been complemented by equations for the conformal factor which contain basic information on Einstein's equations. Finding a suitable gauge to obtain useful evolution equations was then considered a different issue. In the case of the equations above we will combine these two steps into one. The extended gauge freedom will be used here in a very special way. The question whether there are other useful possibilities has not been investigated.

We begin by describing the construction of a gauge. Let $\tilde{S}$ be a space-like hypersurface in the given vacuum solution $(\tilde{M}, \tilde{g})$. We choose on $\tilde{S}$ a positive 'conformal factor' Θ_*, a frame field e_{k*}, and a 1-form b_* such that $\tilde{g}(e_{i*}, e_{k*}) = \Theta_*^2\,\eta_{ik}$ and e_{0*} is orthogonal to $\tilde{S}$. Then there exists through each point $x_* \in \tilde{S}$ a unique conformal geodesic $x(\tau)$, $b(\tau)$ with $\tau = 0$ on $\tilde{S}$ which satisfies there the initial conditions $\dot{x} = e_{0*}$, $b = b_*$. These curves define a smooth caustic free congruence in a neighbourhood U of $\tilde{S}$ if all data are smooth. Furthermore, b defines a smooth 1-form on U which supplies a Weyl connection $\hat{\nabla}$ as described above. A smooth frame field e_k and conformal factor Θ are then obtained on U by solving (1.19), (1.20) for given initial conditions $e_k = e_{k*}$, $b = b_*$ on $\tilde{S}$. The frame field is orthonormal for the metric $g = \Theta^2 \tilde{g}$, i.e. $g_{ik} \equiv g(e_i, e_k) = \mathrm{diag}(1, -1, -1, -1)$. Dragging along local coordinates x^α, $\alpha = 1, 2, 3$ on $\tilde{S}$ with the congruence and setting $x^0 = \tau$ we obtain a coordinate system. In this gauge we have on U

$$\dot{x} = e_0 = \partial_\tau, \quad \hat{\Gamma}_0{}^j{}_k = 0, \quad \hat{L}_{0k} = 0. \tag{1.43}$$

Coordinates, a frame field, and a conformal gauge as above will be referred to as a *conformal Gauss system.*

We could now derive from (1.20) and the conformal geodesics equations an equation for $d_k = \Theta\, b_k = \Theta\, f_k + \nabla_k \Theta$ and use it together with (1.20) to complement the system (1.39) to (1.42). Surprisingly, however, the fields Θ and d_k can be determined explicitly ([35]): *If $\tilde{g}$ is a solution to Einstein's vacuum equations with cosmological constant λ, the fields Θ and d_k are given in our gauge by the explicit expressions*

$$\Theta = \Theta_* \left(1 + \tau < b_*, \dot{x}_* > + \frac{\tau^2}{2} \left(\Theta_*^{-2} \frac{\lambda}{6} + \frac{1}{2}\, g^\sharp(b_*, b_*)\right)\right), \tag{1.44}$$

$$d_0 = \dot{\Theta}, \quad d_a = < b_*, \Theta_*\, e_{a*} >, \quad a = 1, 2, 3, \tag{1.45}$$

where the quantities with a subscript star are considered as constant along the conformal geodesics and $g^\sharp$ denotes the contravariant version of g.

With these expressions (1.39), (1.40), (1.41), (1.42) provide a complete system for u. In spite of the fact that we use a special gauge, we refer to these equations as the *general conformal field equations* to indicate that they employ the full gauge freedom preserving a conformal structure.

One might wonder why in the present case it should be sufficient to give the explicit expression for Θ and b while in the case of the metric conformal field equation we had to use Einstein's equations again to obtain equations for the conformal factor. However, the occurrence of λ shows that information about Einstein's equations is encoded in (1.44), (1.45). We note the following important consequence. If we have a sufficiently smooth solution to the metric conformal field equations in a domain comprising a non-empty set $\mathscr{I} = \{\Omega = 0, d\,\Omega \neq 0\}$, it follows from (1.34) that $\nabla_\rho\, \Omega\, \nabla^\rho\, \Omega = -\lambda/3$ on $\mathscr{I}$, i.e. the causal nature of $\mathscr{I}$ is determined by the field equations ([65]). The analogous result

$$\nabla_k\, \Theta\, \nabla^k\, \Theta = -\frac{1}{3}\, \lambda \quad \text{where} \quad \Theta = 0, \tag{1.46}$$

follows in the present case from (1.44), (1.45) and the definition of b if the congruence does not degenerate where Θ vanishes ([35]).

Our gauge is not only distinguished by the fact that it is provided by the conformal structure itself and the explicit information on Θ and b, but also by a remarkable simplicity of the resulting evolution equations. Setting $p = 0$ and observing the gauge conditions (1.43) we obtain from (1.39)–(1.42)

$$\partial_\tau e^\mu{}_q = -\hat{\Gamma}_q{}^l{}_0 \, e^\mu{}_l,$$
$$\partial_\tau \hat{\Gamma}_q{}^i{}_j = -\hat{\Gamma}_k{}^i{}_j \, \hat{\Gamma}_q{}^k{}_0 + g^i{}_0 \, \hat{L}_{qj} + g^i{}_j \, \hat{L}_{q0} - g_{j0} \, \hat{L}_q{}^i + \Theta \, d^i{}_{j0q},$$
$$\partial_\tau \hat{L}_{qj} + \hat{\Gamma}_q{}^k{}_0 \, \hat{L}_{kj} = d_i \, d^i{}_{j0q},$$
$$\nabla_i d^i{}_{jkl} = 0.$$

Extracting from the Bianchi equation by the methods discussed above a symmetric hyperbolic system, we get symmetric hyperbolic reduced equations for those components of u which are not determined explicitly by the gauge conditions. It has been shown in [35] for such a choice of reduced equations (assuming (1.44), (1.45)) that any solution which satisfies the constraints on a suitable space-like hypersurface does indeed satisfy the complete set of field equations in the part of the domain of dependence of the initial data set where Θ is positive.

A system of similar simplicity is obtained if we use metric Gauss coordinates and the vacuum Bianchi identity to solve the vacuum field equations directly ([24]). While it is well known, however, that in general Gauss coordinates tend to develop caustics quickly, it will be seen below that conformal Gauss coordinates can be arranged to cover large space-time domains or even entire maximal globally hyperbolic space-times in non-trivial cases.

Our main purpose in reformulating the field equations was to obtain results about their solutions. In fact, all the properties above have been used to establish properties of solutions by formulating and analyzing boundary value problems which take into account as far as possible the geometric nature of the situations under study. In the following we shall indicate the kind of questions which have been asked and the type of results obtained so far.

1.3 The Penrose Proposal

Experience with other theories governed by hyperbolic field equations suggests that Einstein's theory should admit wave phenomena. For the same reason for which some work is required to bring to light the hyperbolic nature of Einstein's equation, we need to make an effort to exhibit the wave phenomena. Since the theory only supplies the metric and the associated curvature field, we need to explain which part of the gravitational field shows wave-like behaviour with respect to which reference structure and to what extent wave phenomena observed in mathematical expressions are of any physical significance. The standard textbook discussion of gravitational radiation bypasses this problem by postulating some ad hoc background solution and studying solutions to the linearized equations on it. This method allows one to apply the machinery developed in other areas of field theory, but specific features of Einstein's theory arising from its non-linearity are suppressed. Nowadays there exist much more sophisticated approximation methods, which also take into account insights from the approach to be described below (cf. [8], [78]).

Work initiated in the late 1950's by Pirani, Trautman, Bondi and others led to a concept of radiation which involves an 'approximation to nature', namely the idealization of an *isolated system*, but which does not require for its definition any mathematical approximations in the given framework. Clearly, the gravitational field can be expected to develop a wave-like character only sufficiently far away from the generating system. The idea then is to consider a gravitating system, consisting of stars, black holes, or bounded systems of stars like galaxies, which is so far away from other such systems that the latter can essentially be ignored and taken into account only in terms of the gravitational radiation emitted by them. Thus one is led to study the *far field* of this system.

Following a light ray which escapes from the system to infinity we may expect the field to become weaker and weaker and in fact *asymptotically flat* in the sense that the metric field approaches that of a limit Minkowski space. If some substructure, e.g. some component of the Weyl curvature, implies a non-trivial field on this limit space it might attain a natural interpretation as the radiation escaping from the system.

To convert this idea into a workable definition two things are needed: a precise description of the limit procedure and detailed information on the decay of the metric and the curvature field. Bondi et al. [9], Newman and Penrose [61], and Sachs [74] specify the limits in terms of so-called *Bondi coordinates*, which are generated by distinguished families of outgoing null hypersurfaces which extend in the future smoothly to infinity. On the hypersurfaces a convenient parameter is specified on the generating null geodesics which themselves are parameterized by spherical coordinates. This choice of coordinates is quite natural for analyzing radiation phenomena, since for hyperbolic field equations perturbations of the field essentially travel along characteristics.

The most delicate part of the whole project is to gain information on the decay. Obviously, we cannot prepare the systems of interest and let them go to observe the decay of the field. In fact, at present the main interest in our whole subject is to obtain information on wave forms mathematically to possibly help identify gravitational radiation in the signals supplied by the measuring devices. To prepare the systems and study their evolution for an infinite time by purely mathematical means was out of the question at a time when not even the Cauchy problem local in time was completely understood. Therefore the authors above had to rely on their experience with a few model examples, linearized theory, assumptions of simplicity, physical intuition etc. to make a reasonable guess.

In [9], [74] the asymptotic behaviour was imposed by requiring that certain metric coefficients behave in Bondi coordinates similarly as in the case of Minkowski space while in [61] fall-off conditions for certain components of the Weyl curvature were specified. In both cases the assumptions proved consistent with the first few steps of a formal expansion involving the field equations. The authors could associate certain asymptotic data convincingly with the gravitational radiation field, and conclude from the behaviour of the *Bondi-energy-momentum* that outgoing gravitational waves carry only positive energy ([9], [74]).

Soon afterwards a new idea brought into the open the basic geometric structure underlying this analysis. In [65], [66] Penrose proposed to *characterize the fall-off behaviour of gravitational fields in terms of the extensibility of the conformal structure through null infinity.*

To illustrate the idea we begin by describing certain conformal embeddings of the simply connected, conformally flat standard solutions to the vacuum field equations $\mathrm{Ric}[\tilde{g}] = \lambda\,\tilde{g}$ into the *Einstein cosmos* (cf. also [52]). The latter is given by the manifold $\bar{M} = \mathbb{R} \times S^3$, endowed with the product line element $g = d\,s^2 - d\,\omega^2$ where $d\,\omega^2 = d\,\chi^2 + \sin^2\chi\,d\,\sigma^2$ denotes the standard line element on the 3-sphere with $d\,\sigma^2 = d\,\theta^2 + \sin^2\theta\,d\,\phi^2$ the standard line element on the 2-sphere. We shall in the following be mainly interested in the conformal extension of Minkowski space and shall use the other two examples only to illustrate certain arguments.

The case $\lambda = 0$: We express the metric of *Minkowski space* $(\tilde{M} \simeq \mathbb{R}^4, \tilde{g} = \eta_{\mu\nu}\,d\,x^\mu\,d\,x^\nu)$ in spherical coordinates such that $\tilde{g} = dt^2 - (dr^2 + r^2\,d\,\sigma^2)$ and define a diffeomorphism Φ of $\tilde{M}$ onto the open subset $\Phi(\tilde{M}) = \tilde{M}' \equiv \{|s + \chi| < \pi,\ \ |s - \chi| < \pi,\ \ \chi \geq 0\}$ of $\bar{M}$ such that Φ^{-1} is given by

$$t = \frac{\sin s}{\cos s + \cos\chi}, \qquad r = \frac{\sin\chi}{\cos s + \cos\chi}.$$

It follows that $\Omega^2\,\Phi^{-1*}\tilde{g} = g$ on $\tilde{M}'$ with the conformal factor $\Omega = \cos\,s + \cos\,\chi$ which is positive on $\tilde{M}'$. The embedded Minkowski space acquires then a *conformal boundary* $\mathscr{I}$ in $\bar{M}$, which is defined as the subset $\mathscr{I} \equiv \partial\,\tilde{M}' = \mathscr{I}^- \cup \mathscr{I}^+ \cup \{i^-, i^0, i^+\}$, of M, where the hypersurfaces $\mathscr{I}^\pm = \{\tau \pm \chi = \pm\pi,\, 0 < \chi < \pi\}$, which can be thought of as being generated by endpoints of null geodesics, represent *future* resp. *past null infinity*, the points $i^\pm = \{\chi = 0, \tau = \pm\pi\}$ represent *future* resp. *past time-like infinity* and the point $i^0 = \{\chi = \pi, \tau = 0\}$ represents *space-like infinity*. The key point here is that the embedded Minkowski space together with its conformal structure and the conformal factor Ω can be extended smoothly to the manifold with boundary $M = \tilde{M}' \cup \mathscr{I}^+ \cup \mathscr{I}^-$ such that $\Omega = 0, d\,\Omega \neq 0$ on the null hypersurfaces $\mathscr{I}^\pm \simeq \mathbb{R} \times S^2$.

It may be of interest to note that H. Weyl already in 1923 considers the conformal embedding of Minkowski space into a quadric in the five dimensional real projective space in a relativity inspired discussion of the Möbius group and refers to the set which corresponds to $\mathscr{I}$ as to the 'sphere at infinity' ([80]). The fact that Weyl did not see, in spite of its beautifully geometric nature, the generalization proposed in [65], indicates to what extent the kind of questions being asked and, in fact, our whole view of the subject have changed since then.

The case $\lambda < 0$: The *de Sitter space* is given by $(\tilde{M} = \mathbb{R} \times S^3, \tilde{g} = dt^2 - \cosh^2 t\,d\,\omega^2)$. The map $\tilde{M} \ni (t,\ \vartheta) \xrightarrow{\Phi} (s = \arctan e^t - \frac{\pi}{4},\ \vartheta) \in \bar{M}$, where ϑ denotes the points in S^3, maps de Sitter space diffeomorphically onto the open subset $\tilde{M}' = [-\pi/4, \pi/4] \times S^3$ of the Einstein cosmos. It satisfies $\Omega^2\,\Phi^{-1*}\tilde{g} = g$ on $\tilde{M}'$ with the positive conformal factor $\Omega = \cos(2\,s)$. Again the embedded space together with its conformal structure and the conformal factor Ω can be extended smoothly to a manifold with boundary $M = \tilde{M}' \cup \mathscr{I}^+ \cup \mathscr{I}^-$ such that $\Omega = 0, d\,\Omega \neq 0$ on the space-like hypersurfaces $\mathscr{I}^\pm = \{\pm\pi/4\} \times S^3$ of M.

The case $\lambda > 0$. In spherical coordinates we obtain *anti-de Sitter covering space* in the form ($\tilde{M} = \mathbb{R}^4, \tilde{g} = \cosh^2 r\, dt^2 - dr^2 - \sinh^2 r\, d\sigma^2$). The map $\tilde{M} \ni (t, r, \theta, \phi) \stackrel{\Phi}{\to} (s = t, \chi = 2\arctan(e^r) - \frac{\pi}{2}, \theta, \phi) \in \bar{M}$ defines a diffeomorphism of $\tilde{M}$ onto the subset $\tilde{M}' = \{\chi < \pi/2\}$ of the Einstein cosmos which satisfies $\Omega^2\, \Phi^{-1*}\tilde{g} = g$ on $\tilde{M}'$ with the positive conformal factor $\Omega = \cos\chi$. The embedded space together with its conformal structure and the conformal factor Ω can be extended smoothly to a manifold with boundary $M = \tilde{M}' \cup \mathscr{I}$ such that $\Omega = 0, d\,\Omega \neq 0$ on the time-like hypersurface $\mathscr{I} = \{\chi = \pi/2\} \simeq \mathbb{R} \times S^2$ of M.

The conformal extensions above are of interest already because they provide detailed information about the global causal relations in the three space-times and allow us to discuss the asymptotic behaviour of solutions to conformally invariant equations such as the Maxwell equations easily and with any desired precision. For our present discussion they are of particular interest because the three conformally embedded spaces and the given conformal factors provide *solutions of the metric conformal field equations which extend, as solutions, through the conformal boundaries.* The question whether this fact can be considered as a model for the behaviour of more general gravitational fields leads us (in hindsight) to the generalization proposed in [65].

The embedding formulae given above are distinguished by their simplicity but may look otherwise somewhat arbitrary. It turns out that we do not have much freedom. For given $\vartheta \in S^3$ the curve $\mathbb{R} \ni \tau \to (s = 2\arctan\frac{\tau}{2},\ \vartheta) \in \bar{M}$ is a conformal geodesic on the Einstein cosmos. If $\chi(\vartheta) \neq 0, \pi$, this curve intersects the conformal boundary $\mathscr{I}^{\pm}$ of the embedded Minkowski space for values $\tau_{\pm}(\vartheta) \neq 0$ with $\tau_- = -\tau_+$. By specifying suitable initial data on the hypersurface $\{t = 0\}$, this curve can be constructed completely in terms of Minkowski space and we find that the conformal factor Θ obtained by solving (1.20) along this conformal geodesic remains positive on Minkowski space and vanishes precisely as $\tau \to \tau_{\pm}$, where $\dot{\Theta}(\tau_{\pm}) \neq 0$. It follows that, apart from a coordinate transformation and a rescaling with a conformal factor ω which is positive on $\bar{M}$, the conformal extension can be constructed completely by integrating a suitable congruence of conformal geodesics and an associated conformal factor on Minkowski space and extending the resulting formulae by continuity (cf. [41] for details).

We could have surmised the possibility for this already when we encountered formula (1.44). It shows that for suitable initial data Θ has precisely two zeros. This also raises a question. Does the fact that the formula is valid on a general vacuum solutions suggest that a similar procedure works under much more general assumptions? We are thus led again to the generalization proposed in [65].

Stating in general terms what has been found above, one arrives at the following definition.

Definition 1.1. A smooth space-time $(\tilde{M}, \tilde{g})$ is called asymptotically simple if there exists a smooth, oriented, time-oriented, causal space-time (M, g) and on M a smooth function Ω such that:
(i) M is a manifold with boundary $\mathscr{I}$,
(ii) $\Omega > 0$ on $M \setminus \mathscr{I}$ and $\Omega = 0$, $d\,\Omega \neq 0$ on $\mathscr{I}$,

(iii) there exists an embedding Φ of $\tilde{M}$ onto $\Phi(\tilde{M})=M\backslash\mathscr{I}$ such that $\Omega^2\,\Phi^{-1*}\tilde{g}=g$.
(iv) each null geodesic of $(\tilde{M},\tilde{g})$ acquires two distinct endpoints on $\mathscr{I}$.

This definition, which only involves the conformal structure of (M,g), reflects the differential geometric aspects of the situations considered above. It implies restrictions on the global structure of the space-time. Condition (iv), for instance, implies that all null geodesics are complete. The set $\mathscr{I}$ thus represents a boundary at null infinity which can be thought of as being generated by ideal endpoints of null geodesics. Conditions (i), (ii) are somewhat redundant, since (ii) implies already that $\mathscr{I}$ is a smooth hypersurface. However, this is not the only purpose of (ii). Together with (iii) it specifies how precisely $\Phi^{-1*}\tilde{g}$ is to be rescaled to obtain a smooth, non-degenerate metric. The definition above was motivated by the following new idea ([65]).

Penrose proposal: *Far fields of isolated systems behave like asymptotically simple space-times in the sense that they can be smoothly extended to null infinity, as indicated above, after suitable conformal rescalings.*

It should be noted that we have only tried to indicate the idea here. It should be applied with understanding and the definition above may require modifications in certain situations. For instance, in applying the definition to physically realistic situations it may be advisable to replace (iv) by some other completeness conditions (cf. [50] for an example) to allow for the possibility that some of the null geodesics enter black holes. The standard example here is the Schwarzschild-Kruskal solution, which has two asymptotic ends which admit smooth conformal extensions (cf. B. Schmidt, this volume). We consider any such changes as minor adaptations. As we shall see below, there may be required much more subtle ones.

While in the earlier description of fall-off conditions coordinates play a crucial role and while even in case of asymptotically simple space-times it is often convenient to use Bondi coordinates, the present characterization is completely independent of any distinguished coordinate systems and therefore much more flexible than the earlier ones. But the beauty of the proposal is that it brings out clearly the key geometrical structure.

Central and critical is the requirement (iii) and the smoothness assumption. The smoothness properties of the rescaled field g are an expression for the fall-off behaviour of the physical field $\tilde{g}$. In view of the Penrose proposal definition (1.1) acquires suddenly a much stronger interest. It is now in competition with another requirement and involves an extremely sharp characterization of the decay of fields which are governed by the quasi-linear, gauge hyperbolic field equations.

The question whether the assumption that the field equations hold near $\mathscr{I}$ is consistent with the smoothness and conditions (i) – (iii), or whether the smoothness assumptions made in definition (1.1) need to be relaxed has been and, as seen below, still is a matter of debate. The question is not whether C^∞ should be replaced by C^k for some large k. The question is whether solutions to the field equations admit conformal extensions of class C^k where k can be chosen large enough to make the proposal meaningful and, related to this, whether the conformal Weyl tensor tends to zero on the conformal boundary.

Penrose addresses this problem in [66]. For later comparison we recall the assumptions and give (a version of) the main argument, using the spin frame formalism. Consider a smooth solution to the metric conformal field equations with $\lambda = 0$ on the physical manifold $\tilde{M}$ and *assume that it admits a conformal extension to a manifold* $M = \tilde{M} \cup \mathscr{I}^+$ *of class* C^4 *such that* g *and* Ω *extend to fields in* $C^3(M)$. The conformal Weyl spinor Ψ_{abcd} is then in $C^1(M)$.

By definition, $\Omega = 0$, $d\Omega \neq 0$ and, as observed earlier, $\nabla_{aa'}\Omega\,\nabla^{aa'}\Omega = 0$ on $\mathscr{I}^+$. We assume that there is a space-like surface Σ in $\mathscr{I}^+$ which intersects each null generator of $\mathscr{I}^+$ exactly once. In a neighbourhood of Σ there are then C^3 spinor fields o^a, ι^a with $\iota^a\, o_a = 1$ such that $\nabla^{aa'}\Omega = -o^a\bar{o}^{a'}$ on $\mathscr{I}^+$ and the complex vector field $m^{aa'} = o^a\bar{\iota}^{a'}$ is tangent to Σ. Note that the set of such spin frames defines a reduction $U(\Sigma)$ of the bundle of normalized spin bases over Σ to the subgroup $U(1)$ of $SL(2,\mathbb{C})$, where $e^{i\frac{\theta}{2}} \in U(1)$ acts on the spin frames by $(o,\iota) \rightarrow (e^{i\frac{\theta}{2}}\,o, e^{-i\frac{\theta}{2}}\iota)$. A $2:1$ bundle morphism maps $U(\Sigma)$ onto the bundle $O^+(\Sigma)$ of oriented frames which are orthonormal for the induced metric h on Σ such that we can write $m = 1/\sqrt{2}\,(r_1 + i\,r_2)$ with $r \equiv \{r_A\}_{A=1,2} \in O^+(\Sigma)$. Denote by u the generator of the structure group $SO(2)$ of $O^+(\Sigma)$ with components $u^A{}_B = \delta^A{}_1\,\delta^2{}_B - \delta^A{}_2\,\delta^1{}_B$ and by Z the vertical vector field on $O^+(\Sigma)$ generated by u.

Equation (1.32) reduces on $\mathscr{I}^+$ to $\nabla_{aa'}\nabla_{bb'}\Omega = s\,\epsilon_{ab}\,\epsilon_{a'b'}$. Since $o^a\nabla_{aa'}$ is an interior operator on $\mathscr{I}^+$ it follows that $o^a\nabla_{aa'}(o_b\bar{o}_{b'}) = -s\,o_b\,\epsilon_{a'b'}$ on $\mathscr{I}^+$ from which we obtain by suitable contractions $\iota^a\nabla_m\,o_a + \bar{\iota}^{a'}\nabla_m\,\bar{o}_{a'}$ whence $\gamma \equiv g(\nabla_m m, \bar{m}) = -2\,\iota^a\nabla_m\,o_a$. Given γ, the connection $\mathcal{D}$ of h is known, because $\mathcal{D}_m\,m = -\gamma\,m$, $\mathcal{D}_m\,\bar{m} = \gamma\,\bar{m}$.

With $\phi_{abcd} = \Omega^{-1}\Psi_{abcd}$ the Bianchi equation (1.36) takes on $\tilde{M}$ the form

$$\nabla^a{}_{a'}\,\Omega\,\Psi_{abcd} - \Omega\,\nabla^a{}_{a'}\,\Psi_{abcd} = 0. \tag{1.47}$$

(We note that in the case where $0 \neq \lambda = -3\,\nabla_{ff'}\Omega\,\nabla^{ff'}\Omega$ on $\mathscr{I}$ this implies at each point of $\mathscr{I}$ immediately that $0 = -3\,\nabla_{ea'}\Omega\,\nabla^{aa'}\Omega\,\Psi_{abcd} = \lambda\,\Psi_{ebcd}$.) It follows that $\Psi_{abcd} = \alpha\,o_a\,o_b\,o_c\,o_d$ on $\mathscr{I}^+$ with $\alpha = \Psi_{abcd}\,\iota^a\,\iota^b\,\iota^c\,\iota^d$. *If in addition it is assumed that*

$$\Omega\,\nabla_{ee'}\,\nabla^a{}_{a'}\,\Psi_{abcd} \rightarrow 0 \quad \text{at} \quad \mathscr{I}^+, \tag{1.48}$$

then equation (1.47) implies, after taking a covariant derivative, symmetrizing in a' and e', and performing some spinor algebra, the equation $o^e\,\nabla_{ee'}\,\Psi_{abcd} = 0$ on $\mathcal{J}^+$. With the special form of the conformal Weyl spinor on $\mathscr{I}^+$ it follows after contraction with $\bar{\iota}^{b'}$

$$\nabla_m\,\alpha - 2\,\gamma\,\alpha = 0 \quad \text{on} \quad \mathscr{I}^+. \tag{1.49}$$

By its definition the C^1 function α can be considered as a 0-form on $O^+(\Sigma)$ of spin weight -2, i.e.

$$Z\,\alpha = -2\,i\,\alpha \quad \text{on} \quad O^+(\Sigma). \tag{1.50}$$

Equation (1.49) says that the covariant differential of α on $O^+(\Sigma)$ (cf. [21]) vanishes on vector fields which project onto m. It can thus be expressed in the form

$$(H_1 + i\,H_2)\,\alpha = 0 \quad \text{on} \quad O^+(\Sigma), \tag{1.51}$$

where for $A = 1, 2$ we denote by H_A the horizontal vector field on $O^+(\Sigma)$ which projects at the point $r \in O^+(\Sigma)$ onto the tangent vector r_A of Σ.

Suppose now that Σ is diffeomorphic to the two-sphere S^2. Because S^2 carries only one conformal structure the conformal scaling of g can be chosen near Σ such that (Σ, h) can be identified isometrically with $(S^2, -d\,\sigma^2)$. This allows us to identify $O^+(\Sigma)$ with $SO(3)$, the isometry group of S^2, such that there are left invariant vector fields Y_i, $i = 1, 2, 3$, on $SO(3)$ which satisfy under this identification $Y_1 = H_2$, $Y_2 = H_1$, $Y_3 = Z$ and $[Y_i, Y_j] = \epsilon_{ijk}\,Y_k$, where ϵ_{ijk} is totally anti-symmetric with $\epsilon_{123} = 1$. Equations (1.50), (1.51) then imply the distributional equation

$$-\frac{1}{2}(Y_1^2 + Y_2^2 + Y_3^2)\,\alpha = \alpha \quad \text{on} \quad O^+(\Sigma).$$

Since the operator on the left hand side is the Casimir operator with respect to the Killing form on $SO(3)$, the solution space of this equation is well known (cf. [76]). It is spanned by the matrix elements of an irreducible three-dimensional unitary representation of the group $SU(2)$ (in the notation of [38] the functions $T_2{}^i{}_j$), which descends to $SO(3)$. Because the only function in this space which satisfies (1.50) is $\alpha = 0$, it follows that the Weyl spinor field vanishes on Σ. Thus, ϕ_{abcd} extends in a continuous fashion to $\mathscr{I}^+$ and, as a consequence, satisfies Sachs' *peeling property* ([66], [73], [74]).

Remarkable as it is that such a conclusion can be drawn from the special conformal covariance of the Bianchi equations, the fact that we are dealing with a spin-2 field, and the topological assumption on Σ, there remains the question whether it follows from the nature of the long time evolution by the field equations that the smoothness assumptions made above and (1.48) can be satisfied and whether this is in fact the case for a sufficiently large class of solutions.

There are two different issues here. We start with a discussion of the situation where the vacuum field equations with a cosmological constant λ are satisfied *locally near a point p of $\mathscr{I}$* and where Ω and g are smooth. As seen earlier, the conformal field equations then imply $\nabla_\rho\,\Omega\,\nabla^\rho\,\Omega = -\lambda/3$. In particular, $\mathscr{I}$ is a null hypersurface if $\lambda = 0$. In that case we can expect $\mathscr{I}$ to have two components $\mathscr{I}^\pm$ and $p \in \mathscr{I}^+$. To get some idea of the class of solutions which are compatible with conditions (i) – (iii), we consider the *asymptotic characteristic initial value problem* for the metric conformal field equations where data are prescribed on two intersecting hypersurfaces which are to assume, by the arrangement of the initial data on them, the meaning of an outgoing null hypersurface $\mathcal{N}$ which meets $\mathscr{I}^+$ in a space-like 2-surface Σ and the part $\mathscr{I}^+_*$ of $\mathscr{I}^+$ in the past of Σ. This is essentially the problem underlying the formal expansions studied in [9], [61], [74].

All solutions for which the rescaled conformal Weyl tensor $d^i{}_{jkl}$ has a smooth limit on $\mathscr{I}^+_*$ can be characterized in the analytical as well as in the C^∞ case

([25], [57]) in the sense that *for suitably given data the existence of a unique local solution near Σ can be shown.* The freedom to prescribe data on $\mathcal{N}$ and $\mathscr{I}_*^+$, which are in each case two components of $d^i{}_{jkl}$ referred to as the *null data*, turns out to be the same as in the standard characteristic initial value problem for Einstein's vacuum field equations (corresponding to the special geometric nature of the hypersurface $\mathscr{I}_*^+$, there is a slight difference with the latter problem concerning the data which may be prescribed in addition to the null data on the intersection of the two null hypersurfaces). If, as above, the tensor $d^i{}_{jkl}$ is represented by the symmetric spinor ϕ_{abcd}, then $\phi_{abcd}\,\iota^a\,\iota^b\,\iota^c\,\iota^d$ and $\phi_{abcd}\,o^a\,o^b\,o^c\,o^d$ denote the null data on $\mathcal{N}$ and $\mathscr{I}^+$ respectively. It is quite a remarkable property of the conformal field equations that they define the conformal boundary $\{\Omega = 0\}$ in terms of one of their unknowns without restricting the freedom to specify smooth data there.

The null data $\phi_{abcd}\,o^a\,o^b\,o^c\,o^d$ on $\mathscr{I}^+$ have a natural interpretation as the *outgoing radiation field.* Further important physical concepts can be associated with the hypersurface $\mathscr{I}^+$ or subsets of it and the questions of interpretation have been extensively analyzed (cf. also [5], [49]). All indications suggest that we are on the right track.

The situation is similar in the de Sitter-type case $\lambda < 0$. Here null infinity is space-like. We ask whether we can characterize the solutions which admit a smooth conformal extension at past null infinity $\mathscr{I}^-$ under the (natural) assumption that this hypersurface is orientable and compact as in the case of de Sitter space and the further assumption that $d^i{}_{jkl}$ extends smoothly to $\mathscr{I}^-$. Again, *for suitably given data on $\mathscr{I}^-$ there can be shown the existence of a unique solution to the conformal field equations.* The freedom to specify data on $\mathscr{I}^-$ is essentially the same as in a standard Cauchy problem with the exception that the mean intrinsic curvature of the initial hypersurface cannot be prescribed ([29]). This appears natural, because $\mathscr{I}^-$ is a very particular hypersurface with respect to the solution. Another peculiar feature consists in the fact that the Hamiltonian constraint becomes trivial on $\mathscr{I}^-$.

The situation where $\lambda > 0$, in which case null infinity is time-like, differs in various aspects from the other cases. In this case the assumption of the smoothness of this limit has a different status than in the other cases. It appears that in general there would not even exist complete null geodesics if the field $d^i{}_{jkl}$ tended to blow up sufficiently fast 'far out'. However, *there is available a detailed discussion of the existence of solutions for which $d^i{}_{jkl}$ has a smooth limit to $\mathscr{I}$* ([35]).

The solutions studied in these three situations are semi-global in the sense that in terms of the physical metric their domains are of infinite extent and even comprise asymptotic regions. However, this is not sufficient to answer the question posed above. Here comes in the second issue concerning the compatibility of asymptotic simplicity with the field equations. *What we need to control is the behaviour of the solutions as they evolve towards null infinity.* Thus, we cannot assume anything about the field $d^i{}_{jkl}$ (except in the anti-de Sitter-type solutions where boundary data must be given at space-like and null infinity to characterize the solutions uniquely) and it is not clear a priori whether it remains bounded near null infinity. We need to find out whether it behaves, together with the other unknowns in the conformal field equations, such that a conformal extension (M, g) of sufficient smoothness *can be constructed.*

There exists a model case which illustrates the type of result we would like to have. In the case of the de Sitter space considered above ($\lambda = -3$) we consider the hypersurface $S = \{t = 0\}$. The Cauchy data induced by de Sitter space on S are given by the standard metric on S^3 and a vanishing extrinsic curvature. It turns out that *data for the equation* $\mathrm{Ric}[\tilde{g}] = \lambda \tilde{g}$ *which are sufficiently close to the de Sitter data evolve into solutions which are asymptotically simple in the sense of definition* (1.1), or, in other words, *asymptotic simplicity is non-linearly stable* ([30], [32]). Here $d^i{}_{jkl}$ *attains a smooth limit at the conformal boundary as a consequence of the evolution process and the assumptions on the initial data.* The fact that the Schwarzschild-de Sitter solution has time slices diffeomorphic to $S^2 \times S^1$ but develops horizons, singularities, and only patches of a smooth boundary shows that neither the compactness of the initial hypersurface nor the sign of the cosmological constant is sufficient for this behaviour. Remarkably, it has been shown recently that a solution with a compact Cauchy hypersurface can admit smooth conformal boundaries in the past as well as in the future only if the Cauchy hypersurface satisfies the topological restriction that its fundamental group be finite ([4], cf. G. Galloway, this volume, for further results concerning de Sitter-type solutions).

An analogous stability result in the Minkowski-type case $\lambda = 0$ is still lacking. We shall see that a result of this simplicity cannot be expected.

1.4 Asymptotic Behaviour of Vacuum Fields with Vanishing Cosmological Constant

For physical applications one would like to know how the sources interact with each other and with the gravitational field and to what extent properties of the far field reflect this interaction. For a first mathematical analysis of the general behaviour of the far fields details of the sources are irrelevant and it is natural to ignore them in a first study. Thus we are led to ask whether there exist other solutions to the Einstein equations $\mathrm{Ric}[\tilde{g}] = 0$ than Minkowski space which satisfy the requirements of definition (1.1).

Two different types of initial value problems have been studied in this context. The most important one is the *standard Cauchy problem* where Cauchy data are prescribed on a space-like hypersurface S with one or several asymptotically Euclidean ends. If there are several such ends we can expect black holes to develop and condition (*iv*) to be violated. Therefore, and because it is the evolution near an end we want to focus on, we will concentrate on the case $S \simeq \mathbb{R}^3$ with one end. Standard examples are the hypersurface $\{t = 0\}$ in Minkowski space with the induced data or perturbations thereof.

The other type of problem is the *hyperboloidal initial value problem* where data are prescribed on a hypersurface H with boundary Σ, which is thought of as a hypersurface in an asymptotically simple space-time which extends to future null infinity $\mathscr{I}^+$, intersects $\mathscr{I}^+$ in Σ, and is space-like everywhere. The boundary Σ can have several components, examples exist in the Schwarzschild-Kruskal space-time (cf. B. Schmidt, this volume). Again we will concentrate on the case where only one asymptotic region exists such that $H \simeq \{x \in \mathbb{R}^3 |\, |x| \leq 1\}$ and $\Sigma \simeq S^2$. The standard example of such a hypersurface is given by the extension of the space-like hyperboloid $\{t^2 - |x|^2 = 1, t > 0\}$ in Minkowski space to $\mathscr{I}^+$. Its interior inherits from Minkowski space a metric of constant negative curvature which can be rescaled to become Euclidean.

While the hyperboloidal initial value problem has been introduced initially as an intermediate step towards a treatment of the standard Cauchy problem ([26]), it has gained in recent years an interest in its own as a basis for numerical calculations (cf. J. Frauendiener and S. Husa, this volume). It is intrinsically non-time-symmetric. The future evolution will extend under suitable assumptions to null infinity while the past time evolution develops a Cauchy horizon which corresponds to a part N of an outgoing wave front. The domain covered by its evolution coincides with that which can be determined by the characteristic initial value problem where data are prescribed on N (if this set is sufficiently smooth).

1.4.1 The Hyperboloidal Initial Value Problem

Since, for reasons which will become clear later, the hyperboloidal initial value problem is much easier to analyze in the context of asymptotic simplicity than the standard Cauchy problem, it has been analyzed much earlier and much more is known about it. If one wants to construct its solutions by solving the conformal field equations, the first task is to provide conformal initial data. In the analyses available so far, in which a non-vanishing constant mean extrinsic curvature is assumed, certain data can be prescribed on H arbitrarily, certain data are obtained by solving elliptic equations which degenerate on Σ, and the remaining data are obtained by differentiation and algebraic operations which involve divisions by the conformal factor, which vanishes on Σ. We give an outline of the results needed in the following discussion (for further information cf. L. Andersson, this volume).

The subtleties arise at Σ. In [1], [3] free data are prescribed which extend smoothly to Σ. It turns out that (a) in general the remaining data have expansions in terms of powers of r and $\log r$, where r denotes the conformal distance from Σ (i.e. they admit *polyhomogeneous expansions* at Σ), while (b) all data will be smooth if the free data satisfy certain *regularity conditions* which require that a few functions derived from the free data vanish on Σ. A much more general class of data is considered in [2]. In particular, (c) free data are prescribed which admit polyhomogeneous expansions at Σ and it is shown that the remaining data also admit such expansions.

The conformal field equations have been used to study the evolution of some of these data. *In the case* (b) *the data evolve into solutions which satisfy the first three conditions of definition* (1.1) *and which admit in fact a conformally* *regular point i^+ at time-like infinity, analogous to the point i^+ in the case of Minkowski space, if the data are sufficiently close to Minkowskian hyperboloidal data* ([26]), [30], [32]). Recently also *the existence of solutions to the conformal field equations for data of the type* (a) *has been shown on a manifold which contains a boundary hypersurface $\mathscr{I}'$ on which the unknown Ω of the conformal field equations vanishes.* The solution is 'rough' at $\mathscr{I}'$ and it is expected that it admits a polyhomogeneous expansion there ([60], cf. also [14]). The evolution of data of type (c) has not been studied in detail.

While the basic idea underlying the Penrose proposal, namely to characterize the asymptotic behaviour of solutions in terms of conformal rescalings, is still being used in this study of asymptotically rough data by employing the conformal field equations, this situation certainly forbids any straightforward application of definition (1.1). The result seems to indicate that the smoothness assumption in definition (1.1) may be too strong. However, before we can arrive at a conclusion here we need to answer the following question: of which physical relevance are data of type (a) and (c), are they needed to model the systems of interest and which are the 'systems of interest'?

A clear answer could be given, if it could be shown that standard Cauchy data by necessity develop into solutions containing hyperboloidal data of type (a) or (c). The results for data of type (b) tell us that the decision about the smoothness of the conformal structure near null infinity is essentially made near space-like infinity. Thus we need to analyze the behaviour of the solution to the standard Cauchy problem there.

1.4.2 On the Existence of Asymptotically Simple Vacuum Solutions

We shall see below that the general analysis of the conformal structure of solutions to the Einstein equations near space-like infinity is quite complicated. On the other hand, static or, more generally, stationary asymptotically flat vacuum solutions are known to admit smooth conformal extensions through null infinity (cf. [18]). Thus, to avoid the difficulties discussed below one could try to construct asymptotically flat standard Cauchy data (on $\mathbb{R}^3$ say) which agree with stationary data near space-like infinity. The time evolution of these data would then be stationary near space-like infinity and clearly contain smooth hyperboloidal hypersurfaces. One could then hope to combine this with the results of [30] to show the existence of solutions to the Einstein equations which satisfy all conditions of definition (1.1).

The first attempt at this was made by Cutler and Wald who succeeded in constructing smooth initial data for the Einstein-Maxwell equations with arbitrarily small ADM mass which agree with Schwarzschild data near space-like infinity ([17]). Using the results of [32], they were able to show for the first time the existence of smooth solutions to the Einstein-Maxwell equations which satisfy the conditions of asymptotic simplicity. The vacuum case remained difficult.

More recently Corvino developed a gluing construction by which an arbitrary smooth, time-symmetric, asymptotically flat solution $(\mathbb{R}^3, h)$ to the constraints can be deformed on a certain annulus $\{x \in \mathbb{R}^3 | 0 < R < |x| < 2\,R\}$, such that it extends with a given finite differentiability to Schwarzschild data outside $2R$ while remaining unchanged inside the radius R and still satisfying the constraints everywhere ([16]). It turned out difficult, however, to keep the radius R fixed or to control its growth sufficiently if the construction is applied to families of initial data for which the ADM mass m is going to zero. This makes it impossible to obtain the desired result by an immediate application of [30].

Quite recently Chruściel and Delay obtained a similar result based on a modification of Corvino's technique *in which the radius* R *as well as the differentiability class* C^k *may be kept fixed while* $m \to 0$ *if they base their construction on initial data* $(\mathbb{R}^3, h)$ *such that in the standard coordinates on* $\mathbb{R}^3$ *the parity condition* $h(x) = h(-x)$ *is satisfied by the 3-metric* ([15]). In this case it is easy to construct in a systematic way hyperboloidal hypersurfaces in the time evolution of the data which contain the part $\{x \in \mathbb{R}^3 | \, |x| < 2R\}$ of the initial Cauchy hypersurface as a subset and which extend into to future of the initial hypersurface in the Schwarzschild part of the evolution. It can be arranged that the induced data approach Minkowskian hyperboloidal data if h approaches the Euclidean metric on $\mathbb{R}^3$ (cf. [41] for a construction based on conformal Gauss systems and [38] for the discussion of a smooth limit of Schwarzschild solutions to the Minkowski solution near space-like infinity). Since this can be done with sufficient smoothness, the results of [30] and the fact that the data are time-symmetric then imply *the existence of non-trivial solutions to the vacuum field equations which satisfy the conditions of definition* (1.1) *with a finite differentiability of the rescaled metric which can be chosen to be large. For these solutions the rescaled conformal Weyl tensor is differentiable on the conformally extended space and there exist conformal extensions containing regular points* $i^{\pm}$ *which represent future and past time-like infinity.* This resolves a question which remained open and controversial for forty years.

While it is reassuring to see that we are not discussing a class of solutions which consists only of Minkowski space, it is clear that the result does not answer all our questions. The solutions obtained here are very special. Being Schwarzschildean near space-like infinity, their NP constants vanish ([42], [62]). As a consequence the rescaled conformal Weyl tensor $d^i{}_{jkl}$ vanishes at $i^{\pm}$ ([46], [63]). However, from the numerical calculation of solutions with a regular i^+ we know that in general $d^i{}_{jkl}$ does not vanish at i^+ ([54]). This suggests that modifying initial data sets to become Schwarzschild near infinity considerably reduces their radiation content.

Hopefully, there will be developed generalizations of the method which will allow us to glue other stationary data to given data such that we prescribe multipoles moments and non-vanishing NP constants. Then the argument above will not be available any longer and the data will gain in interest in the modeling of physical systems.

But this will still not help us answer the decisive question: what degree of restriction is implied by 'asymptotic simplicity' in the class of all ('reasonably' defined) asymptotically flat solutions.

1.4.3 The Regular Finite Cauchy Problem

Under the conformal embedding of Minkowski space into the Einstein cosmos described above, the Cauchy hypersurface $\tilde{S} = \{t = 0\} \simeq \mathbb{R}^3$ with its interior Euclidean metric $\tilde{e}$ is mapped conformally into the unit 3-sphere, $\tilde{S} \to S \equiv \{s = 0\} = \tilde{S} \cup \{i\}$, with its standard metric $d\omega^2$ and a point i with coordinates $s = 0$, $\chi = \pi$ which represents space-like infinity for $(\tilde{S}, \tilde{e})$. If more general data $\tilde{h}$, $\tilde{\chi}$ satisfying the vacuum constraints are prescribed on $\tilde{S}$ which are asymptotically Euclidean in the sense that $\tilde{h}$ suitably approaches the metric $\tilde{e}$ near space-like infinity, it appears reasonable to conformally compactify $(\tilde{S}, \tilde{h})$ by suitably rescaling $\tilde{h} \to h = \Omega^2 \tilde{h}$ and adding a point to $\tilde{S}$ as above. Depending on $\tilde{h}$, there may not exist coordinates x^α on S near i and a conformal factor Ω such that the metric coefficients $h_{\alpha\beta}$ will be smooth at i. But even if the data $\tilde{h}$, $\tilde{\chi}$ are chosen such that h extends smoothly to i, in fact, even if $\tilde{\chi} = 0$, it turns out that in coordinates adapted to h the rescaled conformal Weyl tensor behaves as

$$d^\mu{}_{\nu\lambda\rho} = O(r^{-3}) \quad \text{as} \quad r \to 0, \tag{1.52}$$

where r denotes the h-distance from i, unless the ADM mass of the data vanishes, i.e. $\tilde{h}$ is flat.

To let things not look utterly hopeless, we note that $d^\mu{}_{\nu\lambda\rho}$ is not an arbitrary tensor with Weyl symmetry satisfying (1.52), that its structure near i is restricted by the assumption that $\tilde{h}$ and $\tilde{\chi}$ be asymptotically Euclidean, satisfy the constraints, and that $d^\mu{}_{\nu\lambda\rho}$ is given by a specific formula as a functional of the initial data. We refer to [38] for explicit general expressions.

In the case of Minkowski space the hypersurfaces $\mathscr{I}^\pm$, which are swept out by the future resp. past directed null geodesics emanating from the point i^0 (naturally identified with $i \in S$), are characteristics for the conformal field equations in the conformally extended space-time. If we assume for a moment that a similar picture can be established for the space-time which develops from non-trivial data $\tilde{h}$, $\tilde{\chi}$, then (1.52) makes us wonder why this strong singularity at i does not spread along the characteristics $\mathscr{I}^\pm$ and destroy the smoothness of the conformal structure there in the first place ! If such a contradiction does not arise, it can only be due to (a) specific features of the field equations and/or (b) specific requirements on the data which are analogous to the regularity conditions observed in the hyperboloidal initial value problem.

If features as referred to in (a) exist, they must lie beyond conformal regularity. If requirements such as those referred to in (b) exist we may wonder what kind of conditions could, in a situation where the conformal extension is only given by a point, replace the vanishing of certain functions on a 2-surface as required in the hyperboloidal problem. Moreover, if one tries to analyze an initial value problem for the conformal field equations for data as singular as (1.52) the choice of a gauge becomes a delicate problem and there is little lee-way for rough estimates.

Somewhat unexpectedly it turns out that for suitably chosen asymptotically flat initial data *the standard Cauchy problem can be reformulated as a regular finite initial value problem for the general conformal field equations with data given on a compact manifold with boundary.* Moreover, the underlying setting also allows us to analyze to some extent the questions raised above.

The underdetermined quasi-linear system of constraints imposes very weak conditions on the data. In particular, it requires very little with respect to the smoothness of the data at space-like infinity. Thus it is clear that we have to make assumptions concerning the precise behaviour of the data near i. In [38], were the setting has been developed and worked out in detail as described below, it was assumed that the data are time-symmetric, that h extends smoothly to i, and that h is in fact analytic in normal coordinates at i. It follows in particular that $\tilde{\chi} = 0$ and that the conformal factor Ω which relates h to the physical metric $\tilde{h} = \Omega^{-2}\,h$ satisfies $\Omega \in C^2(\bar{S}) \cup C^\infty(\bar{S} \setminus \{i\})$ with $\Omega = 0$, $d\,\Omega = 0$, Hess $\Omega = c\,h$ with $c \neq 0$ at i, but Ω is not smooth at i unless the data have vanishing ADM mass. As a consequence we still have (1.52). It should be emphasized that our assumptions have been made for convenience because the structure of the data near space-like infinity needs to be known in all details (cf. the remarks below). In particular, the assumption of analyticity can be dropped and replaced by a C^∞ condition.

The cylinder at space-like infinity. In the following we shall describe the main considerations leading to the formulation of the regular finite initial value problem near space-like infinity. The general conformal field equations will be used here for several reasons. We want to obtain a formulation which is leading to statements which depend as far as possible on the conformal structure itself and as little as possible on conditions 'put in by hand'. The nature of the singularity at i forces us to carefully distinguish between the dependence of the fields on radial and angular directions at i. The frame field in the formalism is readily adapted to this situation. Much more importantly, we want to avoid any gauge which depends on implicit wave equations, because these might transport non-intrinsic singularities along null infinity. An example for this occurs if the conformal gauge in the metric conformal field equations is removed by prescribing the Ricci scalar. Clearly, space-like infinity is not met by time-like conformal geodesics starting at points of $\tilde{S} = S \setminus \{i\}$ and conformal Gauss systems should thus be unaffected by the singularity at i. Finally, the location of the set where the conformal factor vanishes will be known a priori.

In [38] the general conformal field equations have been used in the spin frame formalism, because this leads to various algebraic simplifications. Here we will discuss the standard frame formalism, suppressing many details for which we refer the reader to [38].

Assume a fixed choice of the conformal scaling for the initial data. Choose a fixed oriented h-orthonormal frame e_a, $a = 1, 2, 3$ at i. Any other such frames at i is then obtained in the form $e_a(s) = s^c{}_a\, e_c$ with $s = (s^c{}_a) \in SO(3)$. *For given s, we distinguish $e_3(s)$ as the radial vector at i* and assume $e_a(s)$ to be parallelly transported with the Levi-Civita connection of h along the geodesics with tangent vector $e_3(s)$ at i. We denote by ρ the affine parameter on these geodesics which vanishes at i and denote the frame obtained from $e_a(s)$ at the value ρ by $e_a(\rho, s)$. It will be assumed that $|\rho| < a$, where a is chosen such that the metric ball B centered at i with radius a is a convex normal neighbourhood for h. Then $]-a, a[\times SO(3) \ni (\rho, s) \to e_a(\rho, s) \in SO(S)$ defines a smooth embedding of a 4-dimensional manifold into the bundle $SO(S)$ of oriented orthonormal frames over S.

We denote by $\hat{B}$ the image of the set $[0, a[\times SO(3)$, by I^0 its boundary $\{\rho = 0\} \simeq SO(3)$, and by π the restriction to $\hat{B}$ of the projection of $SO(S)$ onto S. For given $s \in SO(3)$ the vectors $e_3(s\,s')$ are identical for $s' \in SO(2)$, the subgroup of $s = (s^c{}_a) \in SO(3)$ for which $s^c{}_3\, e_c = e_3$, and all these vectors are parallelly transported along the same geodesic. It follows that the set $\hat{B}$ inherits from $SO(S)$ an action of the subgroup $SO(2)$ which implies a factorization $\hat{B} \overset{\pi'}{\to} B' = \hat{B}/SO(2) \overset{\pi''}{\to} B$ of the map π such that π'' maps the set $\pi'(I^0) \simeq S^2$ onto i while it implies a diffeomorphism of $B' \setminus \pi'(I^0)$ onto the punctured ball $B \setminus \{i\}$ which we could use to identify these sets.

It turns out more convenient, however, to pull the initial data on B via π back to $\hat{B}$, use this set as the initial manifold and ρ and s as 'coordinates' on it. Questions of smoothness are then easily discussed and certain expressions simplify considerably. Since all fields have a well defined transformation behaviour (spin weight) under the action of $SO(2)$ and this action commutes with the evolution defined by the propagation equations, no problem will arise from this. In [38] it is shown how to prescribe on the part of $\hat{B}$ where $\rho > 0$ a frame field X, $c_a(\rho, s)$, $a = 1, 2, 3$, where X is generated by the action of $SO(2)$, the vector fields $c_a(\rho, s)$ satisfy $T(\pi)c_a(\rho, s) = e_a(\rho, s)$, and a frame formalism can be applied. In the coordinates ρ and s on $\hat{B}$, which are essentially spherical coordinates, some frame vector fields and certain data derived from them are singular at $\rho = 0$.

With a suitable choice of the initial data for the conformal geodesics the conformal factor and the 1-form d_k take the form

$$\Theta = \frac{\Omega}{\kappa_*}\left(1 - \tau^2\,\frac{\kappa_*^2}{\omega_*^2}\right), \qquad d_0 = -2\,\tau\,\frac{\kappa_*\,\Omega_*}{\omega_*^2}, \qquad d_a = \frac{1}{\kappa_*}\,(e_a(\Omega))_* \tag{1.53}$$

where $\omega^2 = 2\,\Omega\,|h^{\alpha\beta}\,\Omega_{,\alpha}\,\Omega_{,\beta}|^{-1/2}$ and κ is a function on $\hat{B}$ which we are free to choose. Any function f_* in (1.53) is considered to be constant along the conformal geodesics and to agree with its given initial value f. In the conformal extension of Minkowski space in which space-like infinity is represented by the point i^0 we would have $\kappa = O(1)$ as $\rho \to 0$. For our purpose it will be much more useful to choose κ in the form $\kappa = \rho\,\kappa'$ with some smooth scalar (i.e. $SO(2)$-invariant) function κ' on $\hat{B}$ with $\kappa' = 1$ on I^0 such that $\Theta|_{\tau=0} = \frac{\Omega}{\kappa_*} = O(\rho)$ as $\rho \to 0$. *In this particular conformal gauge all data extend smoothly to $I^0 \subset \hat{B}$ and the coefficient functions Θ, d_k in the general conformal field equations are smooth as well.*

We are free to choose κ'. For definiteness we set $\kappa = \omega$ such that $\Theta = 0$ for $\tau = \pm 1$. If the resulting coordinates remain regular long enough, future and past null infinity will be represented near space-like infinity by the sets $\mathscr{I}^{\pm} = \{\tau = \pm 1\}$, while space-like infinity will now be represented by the *cylinder at space-like infinity* which is given by the set $I = \{\rho = 0,\ |\tau| < 1\}$. It 'touches' null infinity at the sets $I^{\pm} = \{\rho = 0,\ \tau = \pm 1\}$. What in the beginning was a Cauchy problem now looks like an initial boundary value problem. It turns out, however, that the boundary I is of a very special nature.

If we write $w = (e^{\mu}{}_{k}, \hat{\Gamma}_{i}{}^{j}{}_{k}, \hat{L}_{jk})$ and the unknown (1.38) in the form $u = (w, z)$ with $z = (d^{i}{}_{jkl})$, the symmetric hyperbolic reduced equations take the form

$$A^{\mu}(w)\,\partial_{\mu}\,z = H(w)\,z, \qquad \partial_{\tau}\,w = F(w, z), \tag{1.54}$$

(where we suppress the dependence of F on the conformal factor Θ and the 1-form d_k). Extending the data on $\hat{B}$ as well as the evolution equations (preserving symmetric hyperbolicity) smoothly into the domain $\rho < 0$, standard results imply the existence of a smooth local solution in a neighbourhood of the initial hypersurface which contains a piece of I. The fact that $\Theta = 0$ on I then allows us to conclude from the details of the equations above that

$$A^{\rho} = 0 \quad \text{on} \quad I. \tag{1.55}$$

This implies that I is a *total characteristic* in the sense that the complete system (1.54) reduces on I to a symmetric hyperbolic system of inner equations on I. It follows that u is determined uniquely by its values on I^0, that the solution obtained for $\rho \geq 0$ does not depend on the chosen extension of the data through I^0, and that u extends smoothly to I (which can not be taken for granted in general initial boundary value problems, cf. [35], [44]).

The construction of I can be considered as attaching a boundary at space-like infinity to the physical space-time. However, we wish to emphasize that not the definition of a boundary but the analysis of the field equations is our main interest. We do not impose any conditions on the time evolution of the solutions except the field equations. The construction is defined in terms of general properties of conformal geometry, the conformal field equations, and the choice of the initial data for the gauge.

While everything extends smoothly to I near I^0, it should be observed that I is not a geometrical entity but that it is attained as a certain limit of geometric objects, namely the conformal geodesics defining our gauge. It is not a time-like hypersurface and some of the metric coefficients do in fact diverge at I in the given coordinates. This is consistent with the smoothness of the frame coefficients and the vanishing of the matrix A^{ρ} on I. Since the wave equation for the rescaled conformal Weyl tensor would degenerate at I, it turns out important here that we extract the hyperbolic system from the first order Bianchi equation. It is remarkable that the blow-up procedure $i^0 \to I$, the coordinate ρ which is adapted to the conformal scaling defined on S by Ω, and the conformal gauge defined by Θ conspire to produce a regular finite problem. By a simple rescaling (different choice of κ) and a corresponding redefinition of the coordinates the representation of space-like infinity in terms of the cylinder I can be converted into the conventional representation of space-like infinity by a point i^0 ([38]).

The transport equations on the cylinder at space-like infinity. The procedure described above, by which the point i^0 is replaced by the cylinder I, does not only give us regular data and equations but also leads to an unfolding of the evolution process near space-like infinity which allows us to analyze the process there at arbitrary order and in all details. Taking formal derivatives of the evolution equations with respect to ρ, restricting to the cylinder, and observing (1.55), one obtains transport equations on I for all the functions $u^p = \partial_\rho^p u|_I$, $p = 0, 1, \ldots$, on the cylinder. Expanding $u^p = u^p(\tau, s)$ in terms of a certain functions system $T_{2m}{}^i{}_k$ (where $m = 0, 1, 2, \ldots$ and $i, k = 0, \ldots, 2m$), on $SO(3)$ (resp. on $SU(2)$ in the case of the spin frame formalism used in [38]), these equations reduce to a hierarchy of ODEs along the curves $]-1, 1[\ni \tau \rightarrow (\tau, \rho = 0, s) \in I$, where the differential operator depends on p but not on the data while the right hand side of the equation for u^{p+1} depends quadratically on $u^0, \ldots, u^p$ and on the coordinates.

The calculation of u on I allows us to determine $A^\mu(u)$ on I, which gives in particular

$$A^\tau = \mathrm{diag}(1 + \tau, 1, 1, 1, 1 - \tau) \quad \text{on} \quad I.$$

While the second of equations (1.54) is an ODE which is regular at $I^\pm$, the matrix A^τ occurring in the first of equations (1.54) degenerates precisely on $I^\pm$. *All the remaining open questions concerning the regular finite initial value problem near space-like infinity are related to this fact.*

This degeneracy (which is removed at the next order in ρ) has important consequences. *It turns out that in general the quantities u^p, in particular the expansion coefficients of the rescaled conformal Weyl tensor, contain terms which behave as $(1-\tau)^k \log^j(1-\tau)$ at I^+.* In the expressions which so far have been calculated explicitly only terms with the powers $j = 0, 1$ occur, but due to the non-linearity of the equations other powers should arise as well. Because of the hyperbolicity of the reduced equations these singularities can be expected to spread along $\mathcal{J}^\pm$. The powers k and j observed in the explicit calculations then suggest that in general not all components of $d^i{}_{jlm}$ will be bounded at null infinity. In fact, in the spinor notation used in (1.47), (1.48) it appears that in general the rescaled Weyl spinor field will have an unbounded component $\phi_{abcd}\, \iota^a\, \iota^b\, \iota^c\, \iota^d$ which behaves as $\log(1-\tau)$ at $\mathscr{I}^+$, where $\tau \rightarrow 1$.

This suggests that in general the conformal Weyl tensor will vanish on $\mathscr{I}^+$ but will neither be differentiable there nor satisfy the assumption (1.48).

Some care is required here, because the hyperbolicity does not extend to $I^\pm$. However, if the whole setting developed in [38] is linearized at Minkowski space, a complete analysis of the situation can be given which confirms the conclusion (cf. J. Valiente-Kroon, this volume).

Due to the geometric nature and the conformal invariance of our gauge conditions we are discussing here intrinsic features of the solutions. If the null infinities of the solutions would admit smooth extensions of the conformal structures, then the conformal geodesics, the conformal factor, and the frame would extend smoothly through null infinity. Also the components of the rescaled conformal Weyl tensor in that frame would then extend as smooth functions of τ

through null infinity. Thus the observations above suggest that *even under assumptions on the conformal data which cannot be improved from the point of view of* smoothness *there will in general occur logarithmic singularities at null infinity due to the evolution process.*

A conjecture. The class of logarithmic singularities determined in [38] have been found by solving the transport equations on I explicitly for certain components u'^p of the functions u^p, $p = 0, 1, 2, \ldots$, in the decomposition with respect to the functions $T_{2m}{}^i{}_k$. Inspecting the initial data u'^p_* for these quantities on I^0 and relating them to the free datum h on $\tilde{S}$, we find that the logarithmic terms determined in [38] for general data in our class do not occur in u'^p for $p \leq q_* + 2$ if and only if h satisfies the *regularity condition*

$$\mathcal{S}(D_{i_1}, \ldots, D_{i_q} B_{jk}(i)) = 0 \quad \text{for} \quad q = 0, 1, \ldots, q_*, \tag{1.56}$$

where D denotes the connection and $B_{jk} = \frac{1}{2} \epsilon_j{}^{il} D_i L_{lk}$ the Cotton tensor of h while $\mathcal{S}$ denotes the operation of taking the symmetric trace-free part of a tensor.

Consistent with the fact that the free initial datum is given by the conformal structure of h, condition (1.56) is conformally invariant for given $q_* \geq 0$. The condition has been found earlier in the quite different, though related, context of vanishing ADM mass ([31]). Although the derivation is much easier in that case and related to a geometric picture, we are still lacking a geometrical interpretation of condition (1.56).

Since the condition is conformally invariant and it is easy to construct data which are conformally flat near space-like infinity, there exist many non-trivial data satisfying the condition. With the techniques of [15] and [16] such data can even be constructed so that they agree with prescribed data on a compact set. Moreover, there exists a large class of data which satisfy the condition in a non-trivial way ([38]). In fact, all data which are conformal to static data near space-like infinity satisfy (1.56) for all q_* ([31]). Since static data are analytic near space-like infinity ([6], [59]), it follows that condition (1.56) can be satisfied with Cotton tensors which do not vanish near space-like infinity ([38]). This is consistent with the fact that the static solutions represent the largest class of time-symmetric solutions which are known to admit near space-like infinity a smooth structure at null infinity. It would be interesting to know whether the results of [15] and [16] can be generalized to glue static ends to given data.

For various reasons we expect condition (1.56) to be of a much wider significance.

Conjecture: *There exists an integer $k_* \geq 0$ such that for given $k \geq k_*$ the time evolution of an asymptotically Euclidean, time-symmetric, conformally smooth initial data set admits a conformal extension to null infinity of class C^k near space-like infinity, if the Cotton tensor satisfies condition* (1.56) *for a certain integer $q_* = q_*(k)$.*

If this conjecture can be shown to be correct, solutions for data which satisfy (1.56) for all q_* will contain smooth hyperboloidal hypersurfaces. We expect that the corresponding hyperboloidal initial data approach Minkowskian hyperboloidal data in a continuous way if the Cauchy data approach Minkowski data. This then can be used in conjunction with the results of [30] about the hyperboloidal Cauchy problem to show the existence of solutions satisfying all the requirements of definition (1.1). Because of the degeneracy of the equations at $I^{\pm}$ the proof of the conjecture is not a standard problem and requires new ideas. However, we can split off an important and interesting subproblem.

Subconjecture: *If condition* (1.56) *holds for a given* $q_* \geq 0$, *the functions* u^p, $p \leq q_* + 2$, *extend smoothly to* $I^{\pm}$.

In the formalism employed in [38] the validity of the statement above can in principle be checked by a recursive calculation of the functions u^p. In fact, such a calculation shows that the u^p are smooth at $I^{\pm}$ for $p \leq 3$ if the corresponding regularity conditions hold ([38], [42]). The general proof essentially reduces to controlling the algebraic structure of the terms on the right hand side of the ODEs governing the transport of the u^p, which, with increasing order, becomes quite complicated (cf. the discussion in [42], [43]). It is most desirable to determine terms of higher order by an explicit calculation, possibly with an algebraic computer program, to obtain an insight into the general structure of the functions u^p. This will help with a general analytic proof of the subconjecture and also provide indispensable information for solving the regular finite initial value problem numerically.

Problems and prospects. Condition (1.56) suggests that data for solutions satisfying the requirements of definition (1.1) have to satisfy at i conditions at all orders. Therefore, a generalization of the regularity condition to the case of data with non-trivial extrinsic curvature requires a general and very detailed investigation of the solutions to the constraints near space-like infinity. Such a study has recently been carried out (cf. S. Dain, this volume) and the generalization of the analysis of [38] to a much larger class of data, which also requires new considerations concerning the gauge, is possible.

The proof of our conjecture, or its generalization to the case of non-trivial extrinsic curvature, requires to show the existence of solutions to the finite regular initial value problem near I whose underlying manifolds contain a piece of the form $\{|\tau| < 1, 0 \leq \rho < \rho_0\}$. Moreover, we need to obtain sufficient control on the smoothness of the solutions near $\mathscr{I}^{\pm}_{\rho_0} = \{|\tau| = 1, 0 \leq \rho < \rho_0\}$. The specification of the precise values of k_* and $q_*(k)$, possibly referring to Sobolev spaces, should be part of this proof. In principle it is not only of interest to see whether the conjecture can be justified but we would also like to know how the logarithmic terms propagate along $\mathscr{I}^{\pm}_{\rho_0}$, if they do so, and what the role and meaning of these terms is.

It would also be desirable to sort out for more general classes of free data the type of non-regularities which arise from solving the constraints and from solving the evolution equations. Since free data can be quite 'rough' at space-like infinity while still being asymptotically Euclidean, there has to be made a reasonable choice. There is no point in striving for a generality which is not required and of no use in modeling physical systems. However, already for quite natural looking free data the constraints can imply terms of the form $\rho^k \log^j \rho$ on $\tilde{S}$ (cf. S. Dain, this volume). Clearly it is of interest to see to what extent our analysis can be generalized to handle such data even if our conjecture and its possible generalizations to more general data bear out. It will be the only way to assess the generality and the role of the solutions which admit a smooth structure at null infinity in the class of all solutions arising from asymptotically Euclidean data.

If the smoothness of the conformal structure obtained at null infinity turns out to be much weaker than expected in systems which appear of 'physical importance'[1], we will have to reconsider definition (1.1). While it may not be applicable any longer in the desired generality, the underlying idea is viable because the conformal field equations may be used to control also solutions with a 'rough' asymptotic behaviour ([14], [60], cf. also P. T. Chruściel, this volume). If the solutions for suitable data admit polyhomogeneous expansions in τ, we can expect to relate the coefficients of this expansion, which can be considered as functions on $\mathscr{I}^+$, again to the initial data. This should allow us to give some interpretation for these terms.

In all such studies the insight into the equations gained by trying to establish the extremely sharp characterization of the fall-off behaviour of gravitational fields in terms of asymptotic simplicity will prove important. For instance, the general analytic proof of the subconjecture stated above will be of particular interest because it will provide a further understanding of the peculiar features of the field equations. The precise structure of the lower order terms in the equations, which is essentially determined by the fact that we are dealing with a soldered gauge theory (cf. the discussion of the general conformal field equations in the context of normal conformal Cartan connections in [35]), becomes critical here. Understanding its role in the context of the data dependent conformal singularity at space-like infinity may shed light also on the mechanism of the field equations in other singular situations.

The concept of an isolated system is not part of the general theory but a matter of expediency (cf. the discussion in [36]). The final criterion for its definition is the possibility to model physical systems in a way which allows us to extract interesting information about physical processes. If the conjecture above bears out, the resulting space-times will allow us to perform very detailed discussions of all their physical aspects and the proof of the conjecture will provide information which reduces many discussions to direct calculations.

An example for this is given by the calculation of the NP-constants in terms of the initial data in [42]. As neatly discussed in [20], these provide interesting restrictions on the global structure of solutions. Another example is the reducibility of the BMS group to the Poincaré group found in [42]. So far both results still rely on the correctness of our conjecture. In fact, many earlier studies of

[1] Besides getting control on the mathematical possibilities, finding out which data are of 'physical importance' may turn out to be the most interesting part of the whole exercise. The interpretation of initial data is still a wide open field.

physical concepts associated with the asymptotic structure will be put on a firm analytical footing.

If logarithmic singularities at null infinity (or the coefficients of the logarithmic terms) turn out to be of any physical significance this will not only create problems for the concept of asymptotic simplicity but also for the conventional method of calculating gravitational radiation by solving numerically initial boundary value problems for Einstein's field equations. In that approach the singularities and what might be lost by ignoring them cannot even be seen, unless the time-like boundary is moved in subsequent runs further and further out to infinity. While the initial data for the initial boundary value problem can be arranged not to be affected by the modification of the data near space-like infinity, with our present knowledge there appears to be no way, whatever boundary treatment in the numerical calculation is used, to control whether one is calculating a solution where the influence of such a modification enters through the boundary into the numerical solution (cf. [44] for the amount of freedom and control available in the analytic treatment of the initial boundary value problem for Einstein's field equations).

On the other hand if the latter method makes any sense at all, it may for practical purposes be quite sufficient and of advantage to calculate numerically asymptotically simple solutions for data with a Schwarzschild end *if the procedures in* [15], [16] *are amenable to a numerical treatment.* It is quite simple to identify in the evolution of such data hyperboloidal hypersurfaces which intersect $\mathscr{I}^+$ in the Schwarzschild part of the evolution and which extend with the implied data smoothly as solutions to the conformal constraints across $\mathscr{I}^+$ ([41]). In fact, the corresponding hyperboloidal data for the conformal field equations can be calculated by solving a system of ODEs ([38]). This would resolve a number of difficult problems occurring in the present numerical codes (cf. J. Frauendiener and S. Husa, this volume).

Again, while this in the end may turn out sufficient for practical purposes, we have no possibility to decide on that before the existence of asymptotically simple solutions has been discussed under sufficiently general assumptions. In that case quite general complete maximal hyperbolic solutions together with their asymptotics may become calculable from standard Cauchy data and the interpretational framework associated with the concept of asymptotic simplicity may allow us to draw the desired conclusion.

It appears that presently the conformal field equations offer the only way to perform such calculations. In the case of the characteristic method, which also allows one to calculate numerically semi-global solutions, there may arise problems because null hypersurfaces have the tendency to develop caustics (an intrinsic feature which cannot be overcome by smoothing procedures, cf. [47] for a discussion of caustics in the context of the characteristic initial value problem and G. Galloway, this volume, for an illustration of the ubiquity of conjugate points on null geodesics). There are also problems in prescribing appropriate initial data. Assuming past time-like infinity to be represented by a regular point, one could think of prescribing null data on the set $\mathscr{I}^- \cup \{i^-\}$ which only has a very special and well understood 'caustic' at i^-. However, at present it is by no means clear how to prescribe these data such that they develop into a solution which admits an asymptotically flat Cauchy hypersurface. Similar difficulties occur if one tries to invent other ways of calculating complete space-times from characteristic data.

1.4.4 Time-Like Infinity

It is a remarkable feature of the conformal Einstein equations, which should not be taken for granted, that they force the null generators of $\mathscr{I}^+$ to meet in precisely one regular point i^+ of the conformal extension if the solution exists for long enough in a sufficiently regular fashion and the initial data are prescribed appropriately ([30]). In such a situation 'time-like infinity' has a clear-cut meaning, it is locally similar to that of Minkowski space, and it can 'readily' be attained in numerical calculations ([54], cf. also S. Husa, this volume). The calculation of i^+ provides an opportunity to test numerical codes in long range calculations of space-times and to perform studies of the non-linear interaction of gravitational waves under well controlled conditions, however, it is of course not a purpose in itself.

To calculate the gravitational radiation emitted by the merger of two black holes we will have to consider solution which develop singularities and horizons. While it may appear wise in a numerical calculation to avoid, if possible, entering the horizon and hitting the singularity, such an attitude does not help in an analytical investigation and we do, in fact, consider it necessary to attack the situation head on. In fluid mechanics there exists a long tradition of studying singularities (shocks) and there have been developed methods to handle them in numerical calculations. The singularities observed in general relativity, however, are in general much more complicated and severe from a conceptional point of view as well as from the point of view of actual analytical expressions. In spite of a considerable effort to come to grips with them analytically and, more recently, numerically (cf. D. Garfinkle, this volume, and also [7] and the literature given there, which discusses some remarkable achievements), we are still far from a situation where we could attain the desired information in sufficiently general situations. In fact, the 'well-known' Schwarzschild-Kruskal or Kerr solutions still pose difficulties.

There exist several ways to discuss 'singular points of space-times' in an abstract way (cf. [71] and the references given there), however, there are no concepts available which would allow us to perform under general assumptions an analysis of singularities in the context of the field equations. While it is a central problem in singular situations to obtain detailed statements which are independent of any specific gauge, it is at least as difficult to develop gauge conditions which allow us to arrive at any interesting statement at all. The latter problem certainly does not have an universal solution. The best we can hope for is to find manageable conditions which by their general nature and by experience with specific solutions may be expected to work for certain classes of solutions.

We have seen above that conformal Gauss systems have many nice local properties and the discussion of the regular finite Cauchy problem suggests that they also behave nicely near space-like infinity. It can also be shown that they can be used in calculations of i^+ from hyperboloidal Cauchy data which are sufficiently close to Minkowskian hyperboloidal data. However, in all these situations the gravitational field must be considered as weak and one would like to know whether conformal Gauss systems can also exist globally and with good asymptotic properties on asymptotically flat space-times which posses regions of strong curvature or even singularities and horizons.

Caustics of congruences of metric geodesics arise because the latter obey equations of second order. Congruences of conformal geodesics are governed by equations of third order (cf. [64]) and are thus even more prone to caustic formation, in fact, different conformal geodesics may become tangent to each other. However, the same structural property which may be responsible for such severe caustics allows us to show that *there exist conformal Gauss systems on the Schwarzschild-Kruskal solution which smoothly cover the entire space-time and extend smoothly and without degeneracy through $\mathscr{I}^{\pm}$ into any smooth extension of the space-time through null infinity* ([41]).

These systems are characterized by initial data for the conformal geodesics on the hypersurface $\tilde{S}$ of the Schwarzschild-Kruskal solution which extends analytically the hypersurface $\{t = 0\}$ of the Schwarzschild space. The data considered in [41] are invariant under the space reflection symmetry in $\tilde{S}$ and the rotational symmetry and the congruence is orthogonal to $\tilde{S}$. On $\tilde{S}$ are given the standard spherical coordinates ϕ, θ and the isotropic radial coordinate r, $0 < r < \infty$, which is related to the radial coordinate $\bar{r}$ of the standard Schwarzschild line element $g_S = (1 - \frac{2\,m}{\bar{r}})\,dt^2 - (1 - \frac{2\,m}{\bar{r}})^{-1}\,d\bar{r}^2 - \bar{r}^2\,d\,\sigma^2$ on the subset $\bar{r} < 2\,m$ by $r = 1/2(\bar{r} - m + \sqrt{\bar{r}\,(\bar{r} - 2\,m)})$. The two asymptotically flat ends of $\tilde{S}$ are at $r = 0$ and $r = \infty$. The coordinates ϕ, θ, r are dragged along with the congruence and define together with the parameter τ of the conformal geodesics a smooth coordinate system (ignoring here the singularity inherent in the spherical coordinates which in the present situation can be easily avoided) on the Schwarzschild-Kruskal space with $\tilde{S} = \{\tau = 0\}$. The initial data are prescribed such that the conformal factor takes the form $\Theta = (r - \frac{m}{2})^2(r + \frac{m}{2})^{-2}(f(r)^2 - \tau^2)$ where $f(r) \equiv r/(r^2 - \frac{m^2}{4})$.

The coordinates stay regular, because of the specific choice of the 1-form b on $\tilde{S}$. The acceleration of the conformal geodesics, which is determined by b, tends by our choice to spread out the curves such that no caustics can develop. This latter fact is established by analyzing an equation satisfied by the connecting vectors of the congruence of conformal geodesics which generalizes the Jacobi equation known for metric geodesics. We consider the fact, that such a geometric analysis is possible as particularly useful.

It is well known that the Schwarzschild-Kruskal solution admits global double null coordinates (cf. B. Schmidt, this volume), however, these coordinates, or coordinates derived from them by simple algebraic relations, do not extend smoothly through $\mathscr{I}^{\pm}$ and they are very specific for the Kerr family. The discussion above supports the hope that conformal Gauss systems can be constructed in much more general situations. Of course, while there are a few general considerations which may help here, a good choice of the initial data for the conformal geodesics will depend on the specific solution and one has to gather experience with this, as with any choice of gauge. If conformal Gauss systems are used in the context of the general conformal field equations to perform numerical calculations, the tensor fields entering the generalized Jacobi equations are essentially calculated as one goes along and the possible tendency of the underlying congruence of conformal geodesics to form caustics can be monitored during the calculation.

It should be noted that Gauss systems for which the initial tangent vector field on $\tilde{S}$ and the pull back of the 1-form field b to $\tilde{S}$ are identical, define the same congruence of curves, if these are considered as point sets. The parameterizations of these curves, however, may be different and the systems will thus define different foliations. This freedom has hardly been explored yet.

Since the space reflection in $\tilde{S}$ is given by $r \to m^2/4\,r$, it is sufficient to discuss the conformal geodesics starting at the subset of $\tilde{S}$ where $r \geq m/2$, which cover a full $\bar{r} \geq 2\,m$ part and two 'halves' (one in the past and one in the future of $\tilde{S}$) of the $\bar{r} < 2\,m$ part of the Schwarzschild solution. The conformal geodesics with $\phi =$ const., $\theta =$ const., $r = r_+ \equiv (3+\sqrt{5})\,m/4$, stay on a hypersurface $\{\bar{r} = 5\,m/2\}$ of the Schwarzschild-Kruskal space and approach time-like infinity, denoted by i^+, as $\tau \to \tau_i = 2/\sqrt{5}\,m$. The conformal geodesics starting with $r > r_+$ run out to null infinity and arrive there as $\tau \to \tau_{\mathscr{I}} = f(r)$, which is consistent with the expression above of the conformal factor.

The curves starting with $\frac{m}{2} \leq r < r_+$ hit the singularity at $\bar{r} = 0$ as $\tau \to \tau_{\text{sing}} \equiv f(r)\,H(K(k),k)$, where $k(r) \equiv \sqrt{2}\left(1 + \sqrt{m\,r/2}\,(r - \frac{m}{2})\right)^{-1/2}$, $K(k) \equiv \int_0^{\pi/2}(1-k^2\sin^2 x)^{-1/2}\,d\,x$, and

$$H(x,k) \equiv \tanh\left(\sqrt{\frac{2-k^2}{2}}\int_0^x \frac{dn^2(v,k) - (1-k^2)}{dn^2(v,k) + (1-k^2)}\,d\,v\right),$$

with Jacobi's elliptic function dn. They pierce the horizon $\mathcal{H}^+ = \{\bar{r} = 2\,m\}$ at $\tau = \tau_{\text{hor}} \equiv f(r)\,H(x_h,k)$, with $x_h \equiv \int_0^{\arcsin(k^2/(4-3k^2))}(1-k^2\sin^2 x)^{-1/2}\,d\,x$. As to be expected, we find that $\tau_{scri} \to \tau_i$ as $r \to r_+$ with $r > r_+$ and also $\tau_{\text{sing}}, \tau_{\text{hor}} \to \tau_i$ as $r \to r_+$ with $r < r_+$.

Signals periodic in proper time of an observer approaching the horizon are received on $\mathscr{I}^+$ at Bondi-times which grow only logarithmically[2]. Consequently, a high resolution is required in a numerical calculation which aspires to probe the field near i^+ by following outgoing null geodesics which start on a Cauchy hypersurface and arrive at null infinity at very late Bondi-times. Therefore we

[2] I owe much to discussions with P. Hübner about this point. However, we both have to acknowledge the priority of W. Allen's formula: *eternity is very long, especially near the end.*

consider it particularly important to acquire a detailed understanding of the characteristic features of the field near i^+.

It is customary, and natural from the point of view of the causal structure (cf. [71]), to denote time-like infinity in the schematic (conformally compactified) causal diagrams by one or several ideal points i^+, but the specific structure of the field in the region close to them is little understood. In our representation of the Schwarzschild-Kruskal space the behaviour of the term $(\Theta\,\bar{r})^2\,d\,\sigma^2$ of the metric $\Theta^2 g_S$ in the limit as $\tau \to \tau_i$ on the hypersurface $\{\bar{r} = \hat{r}\}$ or $\{\bar{r} = 2\,m\}$ suggests that we should consider i^+ as a point with coordinates $\tau = \tau_i$, $r = r_+$ (the coordinates ϕ and θ loosing their meaning there as at the origin of a polar coordinate system). Without further analysis, however, this can only be considered as a way of speaking. Contrary to what is observed in Minkowski space, i^+ is not only approached by time-like geodesics and by the null generators of $\mathscr{I}^+$, but also by null geodesics originating from the interior of the space-time, namely those generating the event horizon and those which stay on the set $\{\bar{r} = 3\,m\}$ or approach it in the future. Such phenomena do not occur for regular conformal structures and thus i^+ cannot be considered as a point in a suitably defined regular conformal extension.

It is reasonable to expect that conformal geodesics of a regular conformal structure which approach a point p in the underlying manifold M can be extended, perhaps after an admissible reparametrisation, through that point as a smooth solution to the conformal geodesic equation, and that the curve will be time-like at p if it was time-like before (while such a statement is known to hold for metric geodesics it still has to be shown for conformal geodesics). Assuming this to be true, we can observe an effect of the conformal singularity near i^+ in our global conformal Gauss system. Since $\partial\,\tau_{\mathscr{I}}/\partial\,r$ converges to a finite negative number as $r \to r_+$ with $r > r_+$, we see that $\mathscr{I}^+$ approaches, as to be expected, the hypersurface $\{r = r_+\}$ near i^+ under a certain angle. In contrast, we find that $\partial\,\tau_{\text{sing}}/\partial\,r \approx (r_+ - r)^{-\frac{\sqrt{2}-1}{\sqrt{2}}}$ as $r \to r_+$ with $r < r_+$, such that the graph of τ_{sing}, and with it also the graph of τ_{hor}, becomes tangent to $\{r = r_+\}$ at i^+. As a consequence, the conformal geodesics generating $\{r = r_+\}$ become tangent to the null geodesics generating $\mathcal{H}^+$ when they approach i^+.

Though this may look like a subtlety, one may wonder about its consequences if one tries to observe the Courant-Friedrichs-Lewy condition in a numerical calculation near i^+. Moreover, the speed with which the singularity and the horizon approach the hypersurface $\{r = r_+\}$ may require special measures to sufficiently resolve these structures numerically.

There are of course much more drastic effects. While the rescaled conformal Weyl tensor (expressed in suitably adapted coordinates or frame fields) extends smoothly to $\mathscr{I}^+$, it diverges along the null generators of the horizon like Θ^{-1} near i^+, and blows up like $1/\bar{r}^3$ near the singularity (if the conformal factor approaches a positive value there, as it does in our case), resulting in a non-uniformity with strong gradients near i^+. Moreover, slight changes in the data may lead to a Kerr solution with drastically different behaviour near i^+, including a Cauchy horizon, a completely different singularity structure, and causality violations (cf. [11]). In recent years considerable work has been done to understand what happens to these structures under perturbations of the data (cf. [56] and the references given there) but these studies supply surprisingly little precise information on the behaviour of the field in the close vicinity of i^+.

One may object to our discussion that we create unnecessary problems by insisting on conformal rescalings in a region where the field is usually considered as weak. But the phenomena we discuss are present anyway, they are just brought out most succinctly by suitable rescalings. It is not so clear, though, what 'suitable' may mean here. In the case of space-like infinity we have seen that there are representations which allow us to see properties which are hard to analyse with other choices of the gauge.

This raises the question whether there exists a kind of 'resolution of the singularity at time-like infinity' which makes the situation more amenable to a detailed analysis and to a numerical treatment. The situation near i^+ is clearly quite different from that near i^0. There is nothing near i^0 which compares with the singularity '$\{\bar{r} = 0\}$' and the horizon $\mathcal{H}^+$ and the way they meet with $\mathscr{I}^+$ at i^+. Because of these structures the singularity obtained in a conformal representation which makes time-like infinity a point is much less isotropic than the singularity obtained if space-like infinity is represented by a point. Moreover, i^+ is causally differently related to the physical space-time than i^0. A 'resolution of the singularity at i^+' must be compatible with a time evolution which essentially takes into account the complete information given by the initial data. It should be emphasized that we are not interested here in a formal definition of a space-time boundary which relies on more or less implicit assumptions of the time evolution of the field. We seek a construction which will help us analyse the latter. This construction can only be fruitful if it follows from the structure of the field equations in a similar sense as the cylinder at space-like infinity does.

Looking again at the complicated behaviour of the fields near time-like infinity suggested in [56], this task may appear hopeless. However, it seems to be the accepted view that in the case of a gravitational collapse to a black hole some kind of universal behaviour can be expected near i^+ in the sense that the field will in general approach near time-like infinity somehow that of a Kerr solution ([68]). The recent proof of the Penrose inequality ([10], [55]) lends some support to this, because the original argument leading to this inequality relied on the 'establishment viewpoint' indicated above ([69]).

References

1. L. Andersson, P.T. Chruściel. *On hyperboloidal Cauchy data for the vacuum Einstein equations and obstructions to the smoothness of scri.* Comm. Math. Phys. **161** (1994) 533–568.
2. L. Andersson, P.T. Chruściel. *Solutions of the constraint equations in general relativity satisfying hyperboloidal boundary conditions.* Dissertationes Mathematicae Polska Akademia Nauk, Inst. Matem., Warszawa, 1996.

3. L. Andersson, P.T. Chruściel, H. Friedrich. *On the regularity of solutions to the Yamabe equation and the existence of smooth hyperboloidal initial data for Einstein's field equations.* Comm. Math. Phys. **149** (1992) 587–612.
4. L. Andersson, G. J. Galloway. *DS/CFT and spacetime topology.* `http://xxx.lanl.gov/archive/hep-th/0202161`
5. A. Ashtekar. *Asymptotic properties of isolated systems: Recent developments.* In: B. Bertotti et. al (eds.) *General Relativity and Gravitation* Dordrecht, Reidel, 1984.
6. R. Beig, W. Simon. *Proof of a multipole conjecture due to Geroch.* Comm. Math. Phys. **79** (1981) 581–589.
7. B. K. Berger. *Numerical Approaches to Spacetime Singularities.* `http://xxx.lanl.gov/archive/gr-qc/0201056`
8. L. Blanchet. *Post-Newtonian Gravitational Radiation.* In: B. Schmidt (ed.), *Einstein's Field Equations and Their Physical Implications.* Berlin, Springer, 2000.
9. H. Bondi, M. G. J. van der Burg, A. W. K. Metzner. *Gravitational waves in general relativity VII. Waves from axi-symmetric isolated systems.* Proc. Roy. Soc A **269** (1962) 21–52.
10. H. L. Bray. *Proof of the Riemannian Penrose conjecture using the positive mass theorem.* J. Diff. Geom. to appear. `http://xxx.lanl.gov/math.DG/9911173`
11. B. Carter. *Global structure of the Kerr family of gravitational fields.* Phys. Rev. **174** (1968) 1559–1571.
12. Y. Choquet-Bruhat, J. W. York. *The Cauchy problem.* In: A. Held (ed.) *General Relativity and Gravitation*, Vol. 1. New York, Plenum, 1980.
13. Y. Choquet-Bruhat, J. Isenberg, J. W. York. *Einstein constraints on asymptotically Euclidean manifolds.* Phys. Rev. D **61** (2000) 084034.
14. P.T. Chruściel, O. Lengard. *Solutions of wave equations in the radiation regime.* Preprint (2002). `http://xxx.lanl.gov/archive/math.AP/0202015`
15. P.T. Chruściel, E. Delay. *Existence of non-trivial, vacuum, asymptotically simple space-times.* Class. Quantum Grav. **19** (2002) L 71 - L 79. *Erratum* Class. Quantum Grav. **19** (2002) 3389.
16. J. Corvino. *Scalar curvature deformation and a gluing construction for the Einstein constraint equations.* Comm. Math. Phys. **214** (2000) 137–189.
17. C. Cutler, R. M. Wald. *Existence of radiating Einstein-Maxwell solutions which are C^∞ on all of $\mathscr{I}^-$ and $\mathscr{I}^+$.* Class. Quantum Grav. **6** (1989) 453–466.
18. S. Dain. *Initial data for stationary space-times near space-like infinity.* Class. Quantum Grav. **18** (2001) 4329–4338.
19. S. Dain, H. Friedrich. *Asymptotically Flat Initial Data with Prescribed Regularity.* Comm. Math. Phys. **222** (2001) 569–609. `http://xxx.lanl.gov/archive/gr-qc/0102047`
20. S. Dain, J. A. Valiente-Kroon. *Conserved quantities in a black hole collision.* Class. Quantum Grav. **19** (2002) 811–815.
21. J. Dieudonné. *Treatise on Analysis*, Vol IV. Academic Press, New York, 1974.
22. G. F. R. Ellis, B. G. Schmidt. *Classification of singular space-times.* GRG **10** (1979) 989–997.
23. H. Friedrich. *On the regular and the asymptotic characteristic initial value problem for Einstein's vacuum field equations.* Proceedings of the 3rd Gregynog Relativity Workshop on Gravitational Radiation Theory MPI PEA/Astro 204 (1979) 137–160 and Proc. Roy. Soc., A **375** (1981) 169–184.
24. H. Friedrich. *The asymptotic characteristic initial value problem for Einstein's vacuum field equations as an initial value problem for a first-order quasilinear symmetric hyperbolic system.* Proc. Roy. Soc., A **378** (1981) 401–421.

25. H. Friedrich. *On the Existence of Analytic Null Asymptotically Flat Solutions of Einstein's Vacuum Field Equations.* Proc. Roy. Soc. Lond. A **381** (1982) 361–371.
26. H. Friedrich. *Cauchy Problems for the Conformal Vacuum Field Equations in General Relativity.* Comm. Math. Phys. **91** (1983) 445–472.
27. H. Friedrich. *On the hyperbolicity of Einstein's and other gauge field equations.* Comm. Math. Phys. **100** (1985) 525–543.
28. H. Friedrich. *On purely radiative space-times.* Comm. Math. Phys. **103** (1986) 35–65.
29. H. Friedrich. *Existence and structure of past asymptotically simple solutions of Einstein's field equations with positive cosmological constant.* J. Geom. Phys. **3** (1986) 101–117.
30. H. Friedrich. *On the existence of n-geodesically complete or future complete solutions of Einstein's equations with smooth asymptotic structure.* Comm. Math. Phys. **107** (1986) 587–609.
31. H. Friedrich. *On static and radiative space-times.* Comm. Math. Phys., **119** (1988) 51–73.
32. H. Friedrich. *On the global existence and the asymptotic behaviour of solutions to the Einstein-Maxwell-Yang-Mills equations.* J. Diff. Geom., **34** (1991) 275–345.
33. H. Friedrich. *On the existence of hyperboloidal initial data.* In: Z. Perjés (Ed.): *Relativity Today.* New York: Nova Science, 1992.
34. H. Friedrich. *Asymptotic structure of space-time.* In: A.Janis, J.Porter (Eds.): *Recent Advances in General Relativity.* Boston, Basel, Berlin: Birkhäuser, 1992.
35. H. Friedrich. *Einstein equations and conformal structure: existence of anti-de Sitter-type space-times.* J. Geom. Phys., **17** (1995) 125–184.
36. H. Friedrich. *Einstein equations and conformal structure.* In: *Geometric Issues in the Foundations of Science* P. Tod et al. (eds.), Oxford University Press, Oxford, 1996.
37. H. Friedrich. *Hyperbolic Reductions for Einstein's Equations.* Class. Quantum Gravity **13** (1996) 1451–1469.
38. H. Friedrich. *Gravitational fields near space-like and null infinity.* J. Geom. Phys. **24** (1998) 83–163.
39. H. Friedrich. *Evolution equations for gravitating ideal fluid bodies in general relativity.* Phys. Rev. D **57** (1998) 2317–2322.
40. H. Friedrich. *Einstein's equations and geometric asymptotics.* In: N. Dadhich, J. Narlikar (Eds.): *Gravitation and Relativity at the Turn of the Millennium.* Inter-University Centre for Astronomy and Astrophysics, Pune, India, 1998. `http://xxx.lanl.gov/archive/gr-qc/9804009`
41. H. Friedrich. *Conformal geodesics on vacuum space-times.* Preprint 2002. `http://xxx.lanl.gov/archive/gr-qc/0201006`
42. H. Friedrich, J. Kánnár *Bondi systems near space-like infinity and the calculation of the NP-constants.* J. Math. Phys. **41**, (2000), 2195–2232.
43. H. Friedrich, J. Kánnár. *Calculating asymptotic quantities near space-like and null infinity from Cauchy data.* Ann. Phys. (Leipzig) **9**, (2000), 321–330. `http://xxx.lanl.gov/archive/gr-qc/9911103`
44. H. Friedrich, G. Nagy. *The initial boundary value problem for Einstein's vacuum field equations.* Comm. Math. Phys. **201** (1999) 619–655.
45. H. Friedrich, A. Rendall. *The Cauchy Problem for the Einstein Equations.* In: B. Schmidt (ed.), *Einstein's Field Equations and Their Physical Implications.* Berlin, Springer, 2000.
46. H. Friedrich, B. G. Schmidt. *Conformal geodesics in general relativity.* Proc. Roy. Soc. Lond. A **414** (1987) 171–195.

47. H. Friedrich, J. Stewart. *Characteristic Initial Data and Wave Front Singularities in General Relativity.* Proc. Roy. Soc. Lond. A **385** (1983) 345–371.
48. K. O. Friedrichs. *Symmetric hyperbolic linear differential equations.* Comm. Pure Appl. Math. **7** (1954) 345–392.
49. R. Geroch. *Asymptotic structure of space-time.* In: F. P. Esposito, L. Witten (eds.) *Asymptotic Structure of Space-Time.* New York, Plenum, 1977.
50. R. P. Geroch, G. Horowitz. *Asymptotically simple does not imply asymptotically Minkowskian.* Phys. Rev. Lett. **40** (1978) 203–206.
51. S. W. Hawking, R. Penrose. *The singularities of gravitational collapse and cosmology.* Proc. Roy. Soc. A **314** (1970) 529–548.
52. S. W. Hawking, G. F. R. Ellis. *The Large Scale Structure of Space-Time.* Cambridge, Cambridge University Press, 1973.
53. P. Hübner. *General relativistic scalar field models and asymptotic flatness.* Class. Quantum Grav. **12** (1995) 791–808.
54. P. Hübner. *From now to timelike infinity on a finite grid.* Class. Quantum Grav. **18** (2001) 1871–1884.
55. G. Huisken, T. Ilmanen. *The Riemannian Penrose inequality.* Int. Math. Res. Not. **20** (1997) 1045–1058.
56. W. Israel. *Internal structure of black holes.* In: F. W. Hehl. et al. (eds.) *Black Holes: Theory and Observation.* Berlin, Springer, 1998
57. J. Kannar *On the existence of C^∞ solutions to the asymptotic characteristic initial value problem in general relativity.* Proc. Roy. Soc. A **452** (1996) 945–952.
58. T. Kato. *The Cauchy problem for quasi-linear symmetric hyperbolic systems.* Arch. Ration. Mech. Anal., **58** (1975) 181–205.
59. D. Kennefick, N. O'Murchadha. *Weakly decaying asymptotically flat static and stationary solutions to the Einstein equations.* Class. Quantum. Grav. **12** (1995) 149–158.
60. O. Lengard. *Solution of the Einstein equation, wave maps, and semilinear waves in the radiation regime.* Ph. D. thesis, Université de Tours, 2001. `http://www/phys.uni-tours.fr/~piotr/papers/batz`
61. E. T. Newman, R. Penrose. *An approach to gravitational radiation by a methods of spin coefficients.* J. Math. Phys. **3** (1962) 566–578.
62. E. T. Newman, R. Penrose. *10 exact gravitationally conserved quantities.* Phys. Rev. Lett. **15** (1965) 231–233.
63. E. T. Newman, R. Penrose. *New conservation laws for zero rest-mass fields in asymptotically flat space-times.* Proc. Roy. Soc. A **305** (1968) 175–204.
64. K. Ogiue. *Theory of conformal connections.* Kodai Math. Sem. Rep. **19** (1967) 193–224.
65. R. Penrose. *Asymptotic properties of fields and space-time.* Phys. Rev. Lett., **10** (1963) 66–68.
66. R. Penrose. *Zero rest-mass fields including gravitation: asymptotic behaviour.* Proc. Roy. Soc. Lond., A **284** (1965) 159–203.
67. R. Penrose. *Gravitational collapse and space-time singularities.* Phys. Rev. Lett., **14** (1965) 57–59.
68. R. Penrose. *Gravitational collapse: the role of general relativity.* Riv. Nuovo Cimento, **1** (1965) 252–276.
69. R. Penrose. *Naked singularities.* Ann. N. Y. Acad. Sci., **224** (1973) 125–134.
70. R. Penrose. *Relativistic symmetry groups.* In: A. O. Barut (ed.), *Group Theory in Non-Linear Problems.* New York, Reidel, 1974.
71. R. Penrose. *Singularities of space-time.* In: N. R. Leibowitz et al. (eds.), *Theoretical Principles in Astrophysics and Relativity.* University of Chicago Press, 1978.

72. R. Penrose. *Singularities and time-asymmetry.* In: S. W. Hawking, W. Israel (eds.), *General Relativity.* Cambridge University Press, 1979.
73. R. K. Sachs. *Gravitational waves in general relativity VI. The outgoing radiation condition.* Proc. Roy. Soc A **264** (1961) 309–338.
74. R. K. Sachs. *Gravitational waves in general relativity VIII. Waves in asymptotically flat space-time.* Proc. Roy. Soc A **270** (1962) 103–126.
75. R. K. Sachs. *On the characteristic initial value problem in gravitational theory.* J. Math. Phys. **3** (1962) 908–914.
76. M. Sugiura. *Unitary Representations and Harmonic Analysis.* North Holland, Amsterdam, 1990.
77. M. E. Taylor. *Partial Differential Equations*, Vol. 3. Berlin, Springer, 1996
78. K. S. Thorne. *The theory of gravitational radiation: An introductory review.* In: N. Deruelle, T. Piran (eds.) *Gravitational Radiation.* Amsterdam, North-Holland, 1983.
79. R. M. Wald. *Gravitational collapse and cosmic censorship.* In: B. R. Iyer, B. Bhawal, *Black Holes, Gravitational Radiation and the Universe.* Dordrecht, Kluwer Academic Publishers, 1999.
80. H. Weyl. *Mathematische Analyse des Raumproblems.* Springer, Berlin, 1923. Reprinted in: *Das Kontinuum.* Chelsea Publ. Comp., New York, 1960.

2 Some Global Results for Asymptotically Simple Space-Times

Gregory J. Galloway

University of Miami, Coral Gables, FL 33124, USA

Abstract. A uniqueness theorem for Minkowski space and de Sitter space associated with the occurrence of null lines (inextendible globally achronal null geodesics) is presented. This result is obtained as a consequence of the null splitting theorem, which is also discussed.

2.1 Introduction

The aim of this paper is to present some rigidity results for asymptotically simple space-times. By a *space-time* we mean a smooth connected time oriented Lorentzian manifold (M, g) of dimension ≥ 3, having signature $(- + \cdots +)$. We use standard notation for causal theoretic notions, e.g., for $A \subset M$, $I^+(A, M)$, the time-like future of A in M (resp., $I^-(A, M)$, the time-like past of A in M), is the set of all points that can be reached from A by a future directed (resp. past directed) time-like curve in M. For other basic definitions and results in space-time geometry and causal theory we refer the reader to the standard references [14, 19].

Penrose's treatment of infinity [18, 20] in asymptotically flat space-times (and space-times with other asymptotic structures, as well) is based on his notion of *asymptotic simplicity*. A 4-dimensional, chronological space-time (M, g) is asymptotically simple provided there exists a smooth space-time-with-boundary $(\tilde{M}, \tilde{g})$ such that,

(a) M is the interior of $\tilde{M}$, and hence, $\tilde{M} = M \cup \mathscr{I}$, where $\mathscr{I} = \partial\tilde{M}$,
(b) $\tilde{g} = \Omega^2 g$, where Ω is a smooth function on $\tilde{M}$ such that (i) $\Omega > 0$ on M and (ii) $\Omega = 0$ and $d\Omega \neq 0$ along $\mathscr{I}$, and
(c) every inextendible null geodesic in M has a past and future end-point on $\mathscr{I}$.

Condition (c) is a strong global assumption, which implies that M is null geodesically complete, and ensures that $\mathscr{I}$ includes all of null infinity. But it also rules out space-times with singularities, and black holes, etc. To treat such cases, condition (c) must be suitably weakened, cf. [3, 14, 22].

In this paper, we will be primarily interested in asymptotically simple space-times which obey the vacuum Einstein equation with cosmological term,

$$\mathrm{Ric} = \lambda g\,. \tag{2.1}$$

If $\lambda = 0$ (the asymptotically flat case) then $\mathscr{I}$ is necessarily a smooth null hypersurface, which decomposes into two parts, $\mathscr{I}^+$, future null infinity, and

$\mathscr{I}^-$, past null infinity. If $\lambda > 0$ (the asymptotically de Sitter case) then we have a similar decomposition, except that $\mathscr{I}$ is space-like.

We present here a uniqueness theorem for Minkowski space and for de Sitter space associated with the occurrence of *null lines*. A null line in a space-time (M, g) is an inextendible null geodesic which is globally achronal (meaning that no two points can be joined by a time-like curve). Arguments involving null lines have arisen in many situations, such as the Hawking-Penrose singularity theorems [14], results on topological censorship [4, 11], the Penrose-Sorkin-Woolgar approach to the positive mass theorem [21, 23] (and related results on gravitational time delay [13]), and most recently results concerning the AdS/CFT Correspondence [12].

Every null geodesic in Minkowski space and de Sitter space is a null line. At the same time, each of these space-times, and indeed any space-time satisfying equation (2.1) (regardless of the sign of λ) obeys the *null energy condition*, $\mathrm{Ric}\,(X, X) = R_{ij}X^iX^j \geq 0$, for all null vectors X. In general, it is difficult for complete null lines to exist in space-times which obey the null energy condition. The null energy condition tends to focus congruences of null geodesics, which can lead to the occurrence of null conjugate points. A null geodesic containing a pair of conjugate points cannot be achronal. Thus we expect a space-time which satisfies the null energy condition and which contains a complete null line to be special in some way. The following theorem supports this point of view.

Theorem 2.1. *Suppose M is an asymptotically simple space-time satisfying the vacuum Einstein equation (2.1) with $\lambda \geq 0$. If M contains a null line then M is isometric to Minkowski space (if $\lambda = 0$) or de Sitter space (if $\lambda > 0$).*

Let us give an interpretation of Theorem 2.1 in terms of the initial value problem for the vacuum Einstein equation, in the case $\lambda > 0$. According to the fundamental work of Friedrich [5, 7], the set of asymptotically simple solutions to (2.1), with $\lambda > 0$, is open in the set of all maximal globally hyperbolic solutions with compact spatial sections. Thus, by Theorem 2.1, in conjunction with the work of Friedrich, a sufficiently small perturbation of the Cauchy data on a fixed Cauchy hypersurface in de Sitter space will in general destroy all the null lines of de Sitter space, i.e., the resulting space-time that develops from the perturbed Cauchy data will not contain any null lines. While one would expect many of the null lines to be destroyed, it is somewhat surprising that *none* of the null lines persist. One is tempted to draw a similar conclusion in the $\lambda = 0$ case. However, the nonlinear stability of asymptotic simplicity in this case has not been established, and, indeed may not hold, cf., [8, 9] for further discussion. In fact at present, Minkowski space is the only known asymptotically simple solution to (2.1) with $\lambda = 0$ (though evidence is mounting that there are solutions distinct from, but in a suitable sense, close to Minkowski space; by Theorem 2.1, such solutions would have no null lines). Finally, we note that an asymptotically simple and de Sitter space-time M, which does not contain any null lines, will not have any particle horizons, i.e., the "past null cones" $\partial J^-(p)$ will be compact for all $p \in M$ sufficiently close to $\mathscr{I}^+$. As all such null cones in de Sitter space are non-compact, this further serves to illustrate the delicate nature of the causal structure of de Sitter space (see also [13, Corollary 1]).

Theorem 2.1 is a consequence of the *null splitting theorem* obtained in [10]. This latter result establishes the rigidity in general of null geodesically complete space-times which obey the null energy condition. We discuss this result in the next section. In Sect. 2.3 we present the proof of Theorem 2.1. For results of related interest concerning the rigidity of asymptotically simple space-times, see, for example, [6, 15].

2.2 The Null Splitting Theorem

The statement of the null splitting theorem involves null hypersurfaces. We recall briefly some facts about the geometry of null hypersurfaces; see e.g., [10, 14, 19] for more detailed discussions from slightly varying points of view.

A smooth null hypersurface in a space-time (M, g) is a smooth co-dimension one submanifold S along which the Lorentz metric g is degenerate. Hence, S admits a smooth future directed null vector field K, which is unique up to a positive scale factor. As is well-known, the integral curves of K, when suitably parameterized, are null geodesics, and are referred to as the null generators of S.

The *null expansion tensor* (or *null second fundamental form*) Θ of S measures the variations in the spatial separation of the null generators of S. Let TS/K denote the tangent vectors to S mod K; for each $p \in S$, $T_pS/K = \{\bar{X} : X \in T_pS\}$, where $\bar{X} = \{Y \in T_pS : Y = X \bmod K\}$. TS/K is an $n-2$ dimensional vector bundle over S ($n = \dim M$). For $\bar{X}$, $\bar{Y} \in T_pS/K$, define $h(\bar{X}, \bar{Y}) = g(X, Y)$; h is a well-defined smooth Riemannian metric on TS/K. Θ is defined as, $\Theta(\bar{X}, \bar{Y}) = g(\nabla_X K, Y)$, where ∇ is the Levi-Civita connection of (M, g); Θ is a well-defined symmetric bilinear form on TS/K, unique up to the scaling of K. By tracing Θ with respect to h, we obtain the null expansion scalar $\theta = h^{ab}\theta_{ab}$, which measures the divergence of the null generators towards the future. Along an affinely parameterized null generator, $s \to \eta(s)$, the null expansion $\theta = \theta(s)$ satisfies (provided K is appropriately scaled) the Raychaudhuri equation for an irrotational null geodesic congruence,

$$\frac{d\theta}{ds} = -\mathrm{Ric}\,(\eta', \eta') - \sigma^2 - \frac{1}{n-2}\theta^2 , \tag{2.2}$$

where $\sigma^2 = \sigma_{ab}\sigma^{ab}$, and σ_{ab} is the shear tensor, $\sigma_{ab} = \theta_{ab} - \frac{1}{n-2}\theta$.

We say that S is *totally geodesic* if and only if the expansion tensor vanishes, $\theta_{ab} \equiv 0$, or, equivalently, if and only if the expansion scalar and shear vanish, $\theta \equiv 0$, $\sigma_{ab} \equiv 0$. This has the usual geometric meaning: A geodesic in M starting tangent to a totally geodesic null hypersurface S remains in S. Null hyperplanes in Minkowski space are totally geodesic, as is the event horizon in Schwarzschild space-time.

We now state the null splitting theorem.

Theorem 2.2. *Let M be a null geodesically complete space-time which obeys the null energy condition,* $\mathrm{Ric}(X,X) = R_{ij}X^iX^j \geq 0$*, for all null vectors X. If M admits a null line η then η is contained in a smooth achronal edgeless* totally geodesic *null hypersurface S.*

The simplest illustration of Theorem 2.2 is Minkowski space: Each null line ℓ in Minkowski space is contained in a unique null hyperplane Π.

We make some comments about the proof, which is based on a maximum principle for C^0 null hypersurfaces, see [10] for details. First, by way of motivation, note that the null plane Π above can be realized as the limit of the future null cone $\partial I^+(x)$ as x goes to past null infinity along the null line ℓ. Π can also be realized as the limit of the past null cone $\partial I^-(x)$ as x goes to future null infinity along the null line ℓ. In fact, one sees that $\Pi = \partial I^+(\ell) = \partial I^-(\ell)$.

Thus, in the setting of Theorem 2.2, consider the *achronal boundaries* $S_+ = \partial I^+(\eta)$ and $S_- = \partial I^-(\eta)$. By standard causal theoretic results [19], S_+ and S_- are achronal, edgeless, C^0 (but in general not smooth) hypersurfaces in M. Since η is a achronal, it follows that S_+ and S_- both contain η. For simplicity, assume S_+ and S_- are connected (otherwise restrict attention to the component of each containing η). The proof then consists of showing that S_+ and S_- agree and form a smooth totally geodesic null hypersurface. We give some indication as to how this works.

By further properties of achronal boundaries (see especially [19, Lemma 3.19]), each point $p \in S_-$ is the past end point of a null geodesic contained in S_- which is future inextendible in M, and hence future complete, i.e., S_- is a C^0 null hypersurface ruled by future complete null geodesics. Similarly, S_+ is a C^0 null hypersurface ruled by past complete null geodesics. Now suppose S_+ and S_- are actually *smooth* null hypersurfaces. Then by standard arguments, based on Raychaudhuri's equation (eq. (2.2)), S_- must have null expansion $\theta_- \geq 0$. Indeed, if $\theta_- < 0$ at some point p then the null generator through p would encounter a null focal point in a finite affine parameter time to the future, forcing the generator to leave S_-, which is impossible. (This is the basis of the proof of the black hole area theorem.) Time-dually, S_+ has null expansion $\theta_+ \leq 0$. Now, let q be a point on both S_+ and S_-. By simple causal considerations one observes that in the vicinity of q, S_+ lies to the future side of S_-. Using the inequalities $\theta_+ \leq 0 \leq \theta_-$, and the fact that the null expansion scalar can be expressed as a second order quasi-linear elliptic operator acting on some function related to the graph of the null hypersurface, one can apply the strong maximum principle to conclude that S_+ and S_- agree near q, c.f., [10, Theorem II.1]. Thus, the nonempty set $S_- \cap S_+$ is both open and closed in S_-, and in S_+. Hence $S_- = S_+$, and this common null hypersurface, call it S, must have vanishing null expansion, $\theta \equiv 0$. Equation (2.2) then implies that the shear of the generators also vanishes, and hence S is totally geodesic.

The principal difficulty in proving Theorem 2.2 is in showing that the arguments of the preceding paragraph extend to the C^0 setting. In fact it is possible to show that S_+ and S_- satisfy, $\theta_+ \leq 0 \leq \theta_-$, in a certain weak sense, namely, in the sense of smooth *support* null hypersurfaces. Then, as an application of the weak version of the strong maximum principle obtained in [2] (which includes a regularity assertion), it is again possible to show that S_+ and S_- agree and form a smooth null hypersurface with vanishing null expansion scalar, cf., [10, Theorem III.4].

2.3 Proof of Theorem 2.1

A proof of Theorem 2.1 in the case $\lambda = 0$ was given in [10]. Here we give a proof of Theorem 2.1, different in a number of respects, which accommodates the case $\lambda > 0$. We also correct a mistake in the proof given in [10]. The proof made use of the erroneous assertion that $M^- = M \cup \mathscr{I}^-$ is causally simple (i.e., that the sets of the form $J^\pm(K, M^-)$ are closed for all compact $K \subset M^-$). As pointed out in [16], without some further assumption on $\mathscr{I}$, this in principle need not hold. In the proof presented here we circumvent the use of causal simplicity.

The first, and main step in proving Theorem 2.1, is to show that (M, g) has constant curvature. Let η be the assumed null line in M, and let S be the component of $\partial I^+(\eta, M)$ containing η. By Theorem 2.2, S is a smooth totally geodesic null hypersurface in M. The null line η acquires a past end point p on $\mathscr{I}^-$ and a future end point q on $\mathscr{I}^+$. We examine the structure of S near p. (A similar structure will hold near q.) For this purpose it is convenient to extend $M^- = M \cup \mathscr{I}^-$ slightly beyond its boundary. Thus, without loss of generality, we may assume that M^- is contained in a space-time (without boundary) P such that $\mathscr{I}^-$ separates P. It follows that $\mathscr{I}^-$ is a globally achronal null hypersurface in P.

Observe that $I^+(\eta, M) = I^+(p, P)$, from which it follows that $\partial I^+(\eta, M) = \partial I^+(p, P) \cap M$ (where $\partial I^+(A, X)$ refers to the boundary in X). Asymptotic simplicity then implies that the generators of $\partial I^+(\eta, M)$ must have past end points on $\mathscr{I}^-$ at p. (A generator γ of $\partial I^+(p, P)$ starting in M either extends to p, or else is past inextendible in P, while remaining in $\partial I^+(p, P)$. In the latter case, γ meets $\mathscr{I}^-$ transversely and enters $I^-(\mathscr{I}^-, P)$, violating the achronality of $\mathscr{I}^-$.) The generators of $S \subset \partial I^+(\eta, M)$ must coincide with those of $\partial I^+(\eta, M)$, and hence have past end points on $\mathscr{I}^-$ at p, as well. Let U be a convex normal neighborhood of p, and let A be the "null cone" in U generated by the future directed null geodesics in U emanating from p. A is a smooth null hypersurface in U, except for the conical singularity at p. We can choose a smooth space-like hypersurface in U, passing slightly to the future of p, which meets A in a 2-sphere Σ. By invariance of domain, and the fact that S is closed in M, it is easily seen that the subset $\Sigma \cap S$ of $\Sigma \cap M$ is both open and closed in $\Sigma \cap M$. It follows that all the null geodesics forming A correspond precisely to the generators of S, except, in the asymptotically flat case, for the generator γ_p of $\mathscr{I}^-$ with past end point p. Note in particular, this implies that $S = \partial I^+(\eta, M)$.

Now, set $N_p = S \cup \gamma_p$ in the asymptotically flat case, and $N_p = S \cup \{p\}$ in the asymptotically de Sitter case. From the above, N_p is made up of all the future inextendible null geodesics in P emanating from p. In the asymptotically de Sitter case, $\mathscr{I}^-$ is space-like, so $\partial I^+(p, P)$ meets $\mathscr{I}^-$ only at p. It follows from statements above that $N_p = \partial I^+(p,P)$. Suppose, in the asymptotically flat case, that $\partial I^+(p,P)$ meets $\mathscr{I}^-$ at a point x, say. Then there exists a null geodesic $\gamma \subset \partial I^+(p,P)$ with future end point x which either extends in the past to p or is past inextendible in $M \cup \mathscr{I}^-$. In the latter case, Lemma 4.2 in [16] implies that $M \subset I^+(\gamma)$, from which it follows that $M \subset I^+(p,P)$. But this contradicts the achronality of η. It follows that $\partial I^+(p,P) \cap \mathscr{I}^- = \gamma_p$, and again we have $N_p = \partial I^+(p,P)$. Thus, N_p is an achronal boundary, and hence is an achronal edgeless C^0 hypersurface in P. Moreover, since the future directed null geodesics in P emanating from p do not cross and, being achronal, do not have points conjugate to p, $N_p \setminus \{p\}$ is the diffeomorphic image under the exponential map $\exp : \mathcal{O} \subset T_pP \to P$ of $(\Lambda_p^+ \setminus \{0\}) \cap \mathcal{O}$, where Λ_p^+ is the future null cone in T_pP and $\mathcal{O}$ is the maximal open set on which exp is defined. Thus, apart from the conical singularity at p, N_p is a smooth null hypersurface. We are justified in thinking of N_p as the future null cone in P at p.

The shear tensor σ_{ab} of S in the physical metric g vanishes, and so, since the shear is a conformal invariant, the shear tensor $\tilde{\sigma}_{ab}$ of $N_p \setminus \{p\}$ vanishes in the unphysical metric $\tilde{g}$. Then by the well-known propagation equation for the shear tensor [14, Eq. 4.36], the components $\tilde{C}_{a0b0}$ (with respect to an appropriately chosen frame in which e_0 is aligned along the generators) of the conformal tensor of $\tilde{g}$ vanish on $N_p \setminus \{p\}$. We can now invoke an argument of Friedrich [6] used in essentially the same situation to prove a cosmic no-hair theorem. By use of the regular conformal field equations, specifically the divergencelessness of the rescaled conformal tensor,

$$\tilde{\nabla}_i d^i_{jkl} = 0\,, \qquad d^i_{jkl} = \Omega^{-1}\tilde{C}^i_{jkl}\,,$$

Friedrich shows that the full rescaled conformal tensor d^i_{jkl} vanishes on the future domain of dependence $D^+(N_p,P)$ of N_p. Hence, the conformal tensor with respect to the physical metric vanishes on $D^+(N_p,P) \cap M$. Together with equation (2.2), this implies that (M,g) has constant curvature on $D^+(N_p,P) \cap M$. In a precisely time-dual fashion (M,g) has constant curvature on $D^-(N_q,Q) \cap M$, where Q is an extension of $M \cup \mathscr{I}^+$ analogous to P, and N_q is the past null cone in Q at q.

To conclude that M is everywhere of constant curvature we need to show that $M \subset D^+(N_p,P) \cup D^-(N_q,Q)$. Since $S = \partial I^+(\eta,M) = \partial I^-(\eta,M)$, M can be expressed as the disjoint union,

$$M = I^-(S,M) \cup S \cup I^+(S,M)\,. \tag{2.3}$$

This decomposition follows from [19, Proposition 3.15], but can be easily shown directly as follows. Let $x \in M \setminus S$. There exists a curve σ in M (not necessarily causal) from x to $y \in S$ that meets S only at y. Either $\sigma \cap I^-(S,M) \neq \emptyset$ or $\sigma \cap I^+(S,M) \neq \emptyset$. If the latter holds and σ is not contained in $I^+(S,M)$, then σ meets $\partial I^+(S,M)$ at some point $z \notin S$. But since $I^+(S,M) = I^+(\partial I^+(\eta,M)) = I^+(\eta,M)$, $z \in \partial I^+(\eta,M) = S$, which is a contradiction. Hence, $\sigma \subset I^+(S,M)$, and so $x \in I^+(S,M)$. Similarly, $\sigma \cap I^-(S,M) \neq \emptyset$ leads to $x \in I^-(S,M)$, which establishes (2.3).

We now show that each term in (2.3) is a subset of $D^+(N_p, P) \cup D^-(N_q, Q)$. Trivially, $S \subset N_p \subset D^+(N_p, P)$. Consider $I^+(S, M) \subset J^+(N_p, P) \cap M$. We claim that $J^+(N_p, P) \cap M \subset D^+(N_p, P) \cap M$. If not, then $H^+(N_p, P) \cap M \neq \emptyset$. Choose a point $x \in H^+(N_p, P) \cap M$, and let ν be a null generator of $H^+(N_p, P)$ with future end point x. Since N_p is edgeless, ν remains in $H^+(N_p, P)$ as it is extended into the past. By asymptotic simplicity, ν must meet $\mathscr{I}^-$. In fact, ν must meet $\mathscr{I}^-$ transversely (even in the asymptotically flat case, since ν starts in M) and then enters $I^-(\mathscr{I}^-, P)$. But this means that ν has left $J^+(N_p, P)$, which is a contradiction. Hence, $I^+(S, M) \subset D^+(N_p, P)$, and by the time-dual argument, $I^-(S, M) \subset D^-(N_q, Q)$.

Thus, M is globally of constant curvature. By the uniqueness of simply connected Lorentzian space forms, Theorem 2.1 will follow once we show that M is simply connected and geodesically complete. In the asymptotically flat case, it is shown in [16] that M is simply connected (in fact, is homeomorphic to $\mathbb{R}^4$). Simple connectivity in the anti-de Sitter case may be established as follows. Each generator of S has a past end point at p and a future end point at q. By the structure of S near p and q established above, it follows that $S \cup \{p, q\}$ is homeomorphic to the 3-sphere. Perturbing $S \cup \{p, q\}$ slightly near p and q, we obtain a smooth achronal 3-sphere S_0 in M. By an argument similar to one given above, one can show that $H^+(S_0, P) \cap M = H^+(S_0, Q) \cap M = \emptyset$, from which it follows that S_0 is a Cauchy surface for M. (Alternatively, it follows from Proposition 2.1 in [1] that M is globally hyperbolic. Then, since S_0 is compact and achronal it must be a Cauchy surface.) Thus M has topology $\mathbb{R} \times S_0$, and so is simply connected.

It remains to show that M is geodesically complete. Being asymptotically simple, we know that M is null geodesically complete, which is essential to the proof of full completeness. Let $\bar{M}$ denote Minkowski space in the case $\lambda = 0$, and de Sitter space in the case $\lambda > 0$. By standard results, for each $x \in M$ and $\bar{x} \in \bar{M}$, there is a neighborhood of x isometric to a neighborhood of $\bar{x}$. Since M is simply connected, these local isometries can be pieced together, by an analytic continuation type argument, to produce a local isometry $\phi : M \to \bar{M}$, see, for example, Theorem 8.17 in [17]. (The assumption in [17, Th. 8.17] that M is complete is not needed to produce the local isometry; it is used only to conclude that ϕ is a covering map.)

Let x be any point in M, and let $\bar{x} = \phi(x)$. Then, from the fact that ϕ is a local isometry and M is null geodesically complete, it follows that any (broken) null geodesic segment in $\bar{M}$ starting at $\bar{x}$ can be lifted uniquely via ϕ to a (broken) null geodesic segment in M starting at x. From this it follows that ϕ is onto: Fix $x \in M$ and let $\bar{x} = \phi(x)$. Let $\bar{\nu}$ be a broken null geodesic in $\bar{M}$ from $\bar{x}$ to any other point $\bar{y}$. Let ν be the lift of $\bar{\nu}$ starting at x, and let y be its final end point; since ν covers $\bar{\nu}$, we must have $\phi(y) = \bar{y}$.

To establish geodesic completeness, it is sufficient to show that any unit speed time-like or space-like geodesic $\gamma : [0, a) \to M$, $t \to \gamma(t)$, continuously extends to $t = a$. Let $\bar{\gamma} = \phi \circ \gamma$; $\bar{\gamma}$ can be extended to a complete geodesic in $\bar{M}$ which we still refer to as $\bar{\gamma}$. Fixing $u_0 \in [0, a)$ sufficiently close to a, one easily constructs a C^0 homotopy of curves $\{\bar{\sigma}_u\}$, $u_0 \le u < a$, from $\bar{\gamma}(0)$ to $\bar{\gamma}(a)$, where, for each $u \in [u_0, a)$, $\bar{\sigma}_u$ consists of the segment $\bar{\gamma}|_{[0,u]}$, followed by a suitably chosen two-segment broken null geodesic $\bar{\nu}_u$ from $\bar{\gamma}(u)$ to $\bar{\gamma}(a)$, such that the length of $\bar{\nu}_u$ (in some background Riemannian metric) goes to zero as $u \to a$. The broken null segments $\bar{\nu}_u$ can be constructed from the two families of null geodesics foliating a totally geodesic time-like 2-surface containing $\bar{\gamma}|_{[u_0,a]}$. By the nature of the lifting procedure, the homotopy $\{\bar{\sigma}_u\}$, $u_0 \le u < a$ can be lifted to a C^0 homotopy $\{\sigma_u\}$, $u_0 \le u < a$, where, for each u, σ_u consists of the segment $\gamma|_{[0,u]}$, followed by a two-segment broken geodesic ν_u covering $\bar{\nu}_u$. The curves σ_u must have a common final end point q, say. Since the length of ν_u goes to zero (in the lifted Riemannian metric) as $u \to a$, it follows that $\gamma(u) \to q$ as $u \to a$, i.e., γ is extendible to q. This completes the proof of Theorem 2.1.

2.4 Concluding Remarks

It is possible to formulate a version of Theorem 2.1 applicable to anti-de Sitter space. However, since in this case $\mathscr{I}$ is time-like, the characteristic initial value problem encountered in the proof of Theorem 2.1 would become an initial-boundary value problem, and it would be necessary to impose boundary data (namely the vanishing of the conformal tensor) on $\mathscr{I}$. In this case one might expect to establish the uniqueness of anti-de Sitter space without the assumption of a null line.

We believe that one should be able to generalize Theorem 2.1 in certain directions. For example, in the asymptotically flat case, we expect that the vacuum assumption could be weakened to allow for the presence of matter. For instance, if we assume space-time M satisfies the null energy condition, and is vacuum in a neighborhood of $\mathscr{I}$, then the proof of Theorem 2.1 implies the existence of open sets near scri which are flat. If the vacuum region is analytic then it follows that a neighborhood of scri is flat, and so M should have vanishing mass. By a suitable version of the positive mass theorem, M should be isometric to Minkowski space. Thus, we conjecture that, in the asymptotically flat case, the vacuum assumption can be replaced by the null energy condition and a requirement that the space-time Ricci tensor falls off at an appropriate rate on approach to $\mathscr{I}$. It may also be possible, in the asymptotically flat case, to prove a version of Theorem 2.1 for *weakly* asymptotically simple space-times, which allows for the occurrence of black holes. By imposing suitable conditions on the domain of outer communications $I^-(\mathscr{I}^+) \cap I^+(\mathscr{I}^-)$ (e.g., that the DOC be globally hyperbolic) one might expect to be able to show that the DOC is flat. To accomplish this, one may be able to exploit the fact that the proof of the null splitting theorem does not actually require the full null completeness of space-time. As the discussion of Theorem (2.2) shows, if η is the null line, it is sufficient to require that the generators of $\partial I^-(\eta)$ (respectively, $\partial I^+(\eta)$) be future (resp., past) complete.

We mention in closing that the null splitting theorem has recently been used [1] to obtain restrictions on the topology of asymptotically simple and de Sitter space-times obeying the null energy condition. It is shown that the Cauchy surfaces of such space-times must be compact and have finite fundamental group.

Acknowledgments

This paper is based on a talk given at the Workshop on the Conformal Structure of Space-Times, held in Tübingen, April 2–4, 2001. We wish to thank the organizers Jörg Frauendiener and Helmut Friedrich for their gracious hospitality and support. We also wish to thank Piotr Chruściel for comments on an earlier draft.

References

1. L. Andersson and G. J. Galloway, *dS/CFT and spacetime topology*, preprint, hep-th/0202161.
2. L. Andersson, G. J. Galloway, and R. Howard, *A strong maximum principle for weak solutions of quasi-linear elliptic equations with applications to Lorentzian and Riemannian geometry*, Comm. Pure Appl. Math. **51**, 581–624 (1998).
3. P. T. Chruściel, E. Delay, G. J. Galloway, and R. Howard, *Regularity of horizons and the area theorem*, Ann. Henri Poincaré **2**, 109–178 (2001).
4. J. L. Friedman, K. Schleich, and D. M. Witt, *Topological censorship*, Phys. Rev. Lett. **71**, 1486–1489 (1993).
5. H. Friedrich, *On the existence of n-geodesically complete or future complete solutions of Einstein's field equations with smooth asymptotic structure.*, Comm. Math. Phys. **107**, 587–609 (1986).
6. H. Friedrich, *Existence and structure of past asymptotically simple solutions of Einstein's field equations with positive cosmological constant*, J. Geom. Phys. **3**, 101–117 (1986).
7. H. Friedrich, *On the global existence and the asymptotic behaviour of solutions to the Einstein-Maxwell-Yang-Mills equations*, J. Diff. Geom. **3**, 275–345 (1991).
8. H. Friedrich, *Einstein's equation and conformal structure*, in *The Geometric universe: Science, geometry and the work of Roger Penrose*, eds. S. A. Huggett et al., Oxford University Press, Oxford, 1996.
9. H. Friedrich, *Einstein's equations and geometric asymptotics*, in *Gravitation and relativity: At the turn of the millennium.* Proceedings of the GR-15 Conference at Pune, India, in June 1997, eds. N. Dadhich and J. Narlikar, IUCAA, Pune, India.
10. G. J. Galloway, *Maximum Principles for null hypersurfaces and null splitting theorems*, Ann. Henri Poincaré **1**, 543–567 (2000).
11. G. Galloway, K. Schleich, D. Witt, and E. Woolgar, *Topological Censorship and Higher Genus Black Holes*, Phys. Rev. D **60**, 104039 (1999).
12. G. J. Galloway, S. Surya and E. Woolgar, *A uniqueness theorem for the AdS soliton*, Phys. Rev. Lett. **88**, 101102 (2002)
13. S. Gao and R. M. Wald, *Theorems on gravitational time delay and related issues*, Class. Quant. Grav. **17**, 4999-5008 (2000).
14. S.W. Hawking and G.F.R. Ellis, *The large scale structure of space-time*, Cambridge University Press, Cambridge, 1973.
15. L. J. Mason, *The asymptotic structure of algebraically special spacetimes*, Class. Quantum Grav. **15**, 1019–1030 (1998).

16. R.P.A.C. Newman, *The global structure of simple space-times*, Commun. Math. Phys. **123**, 17–52 (1989).
17. B. O'Neill, *Semi–Riemannian geometry*, Academic Press, New York, 1983.
18. R. Penrose, *Zero rest-mass fields including gravitation: asymptotic behavior*, Proc. Roy. Soc. Lond. A **284**, 159–203 (1965).
19. R. Penrose, *Techniques of differential topology in relativity*, SIAM, Philadelphia, 1972, (Regional Conf. Series in Appl. Math., vol. 7).
20. R. Penrose and W. Rindler, *Spinors and Spacetime*, vol. 2, chapter 9, Cambridge University Press, Cambridge, 1986.
21. R. Penrose, R. D. Sorkin and E. Woolgar, *A positive mass theorem based on the focusing and retardation of null geodesics*, preprint, gr-qc/9301015.
22. R. M. Wald, *General relativity*, University of Chicago Press, Chicago, 1984.
23. E. Woolgar, *The positivity of energy for asymptotically anti-de Sitter spacetimes*, Class. Quant. Grav. **11**, 1881–1900 (1994).

3 Black Holes

Piotr T. Chruściel*

Département de Mathématiques, Faculté des Sciences,
Parc de Grandmont, F 37200 Tours, France

Abstract. This paper is concerned with several not-quantum aspects of black holes, with emphasis on theoretical and mathematical issues related to numerical modeling of black hole space-times. Part of the material has a review character, but some new results or proposals are also presented. We review the experimental evidence for existence of black holes. We propose a definition of black hole region for any theory governed by a symmetric hyperbolic system of equations. Our definition reproduces the usual one for gravity, and leads to the one associated with the Unruh metric in the case of Euler equations. We review the global conditions which have been used in the scri-based definition of a black hole and point out the deficiencies of the scri approach. Various results on the structure of horizons and apparent horizons are presented, and a new proof of semi-convexity of horizons based on a variational principle is given. Recent results on classification of stationary singularity-free vacuum solutions are reviewed. Two new frameworks for discussing black holes are proposed: a "naive approach", based on coordinate systems, and a "quasi-local approach", based on timelike boundaries satisfying a null convexity condition. Some properties of the resulting black holes are established, including an area theorem, topology theorems, and an approximation theorem for the location of the horizon.

3.1 Introduction

Black holes belong to the most fascinating objects predicted by Einstein's theory of gravitation. Although they have been studied for years,[1] they still attract quite a lot of attention in the physics and astrophysics literature and, in fact, an exponential growth of the number of related papers can be observed.[2] It turns out that several field theories are known to possess solutions which exhibit black hole properties:

- The "standard" gravitational ones which, according to our current postulates, are black holes for all classical fields.
- The "dumb holes", which are the sonic counterparts of black holes, first discussed by Unruh [127].

---

* Partially supported by a Polish Research Council grant # KBN 2 P03B 130 16.

[1] The reader is referred to the introduction to [27] for an excellent concise review of the history of the concept of a black hole, and to [26, 82] for more detailed ones.

[2] A search on the key word "black holes" on 2.11.2001 reveals 213 papers on gr-qc for the current year and 773 papers from 1991, date of the beginning of the archive; the figures on astro-ph and hep-th are very similar.

- The "optical" ones – the black-hole counterparts arising in the theory of moving dielectric media, or in non-linear electrodynamics [95, 111].
- The "numerical black holes" – objects constructed by numerical general relativists. We shall argue below that this leads to the need of introducing, and studying, new frameworks for the notion of black holes; two such frameworks ("naive black holes" and "quasi-local black holes") will be introduced and studied in Sects. 3.8.1 and 3.8.2 below.

(An even longer list of models and submodels can be found in [9].) In this work we shall discuss various aspects of the above. The reader is referred to [21, 51, 77, 83, 113, 133] and references therein for a review of quantum aspects of black holes. Let us start with a short review of the observational status of black holes in astrophysics.

3.2 Experimental Evidence

While there is growing evidence that black holes do indeed exist in astrophysical objects, and that alternative explanations for the observations discussed below seem less convincing, it should be borne in mind that no undisputed evidence of occurrence of black holes has been presented so far. The flagship black hole candidate used to be Cygnus X-1, known and studied for years (cf., e.g., [27]). Its published mass has been going up and down over time: an optimistic interpretation of this phenomenon is that there has been mass accretion during the ascending periods, and Hawking radiation during the descending ones; a more realistic one is that there is still considerable uncertainty in the determination of this mass. Table 3.1[3] lists a series of very strong black hole candidates in X-ray binary systems; M_c is mass of the compact object and M_* is that of its optical companion; some other candidates, as well as references, can be found in [103, 109]. The binaries have been divided into two families: the High Mass X-ray Binaries (HMXB), where the companion star is of (relatively) high mass, and the Low Mass X-ray Binaries (LMXB), where the companion is typically below a solar mass. The LMXB's include the "X-ray transients", so-called because of flaring-up behaviour. This particularity allows to make detailed studies of their optical properties during the quiescent periods, which would be impossible during the periods of intense X-ray activity. The stellar systems listed have X-ray spectra which are neither periodic (that would correspond to a rotating neutron star), nor recurrent (which is interpreted as thermonuclear explosions on a neutron star's hard surface). The final selection criterion is that of the mass M_c exceeding the Chandrasekhar limit $M_C \approx 3$ solar masses $M_\odot$.[4] According to the authors of [27], the strongest black hole candidate in 1999 was V404 Cygni, which belongs to the LMXB class.

Table 3.1 should be put into perspective by realizing that, by some estimates [99], a typical galaxy – such as ours – should harbour $10^7 - 10^8$ stellar

[3] The recent review [109] lists thirteen black hole candidates.

[4] See [103] for a discussion and references concerning the value of M_C.

Table 3.1. Stellar mass black hole candidates (from [99])

Type	Binary system	$M_c/M_\odot$	$M_*/M_\odot$
HMXB:	Cygnus X-1	11–21	24–42
	LMC X-3	5.6–7.8	20
	LMC X-1	≥ 4	4–8
LMXB:	V 404 Cyg	10–15	≈ 0.6
	A 0620-00	5–17	0.2–0.7
	GS 1124-68 (Nova Musc)	4.2–6.5	0.5–0.8
	GS 2000+25 (Nova Vul 88)	6–14	≈ 0.7
	GRO J 1655-40	4.5–6.5	≈ 1.2
	H 1705-25 (Nova Oph 77)	5–9	≈ 0.4
	J 04224+32	6–14	≈ 0.3–0.6

black mass holes. We note an interesting proposal, put forward in [28], to carry out observations by gravitational microlensing of some 20 000 stellar mass black holes that are predicted [105] to cluster within 0.7 pc of Sgr A* (the centre of our galaxy).

It is now widely accepted that quasars and active galactic nuclei are powered by accretion onto massive black holes [101, 135]. Further, over the last few years there has been increasing evidence that massive dark objects may reside at the centres of most, if not all, galaxies [100, 120]. In several cases the best explanation for the nature of those objects is that they are "heavyweight" black holes, with masses ranging from 10^6 to 10^{10} solar masses. Table 3.2[5] lists some supermassive black hole candidates; some other candidates, as well as precise references, can be found in [87, 103, 104, 119]. The main criterion for finding candidates for such black holes is the presence of a large mass within a small region; this is determined by maser line spectroscopy, gas spectroscopy, or by measuring the motion of stars orbiting around the galactic nucleus. The reader is referred to [108] for a discussion of the maser emission lines and their analysis for the supermassive black hole candidate NGC 4258. An example of measurements via gas spectrography is given by the analysis of the Hubble Space Telescope (HST) observations of the radio galaxy M 87 [126] (compare [101]): A spectral analysis shows the presence of a disk-like structure of ionized gas in the innermost few arc seconds in the vicinity of the nucleus of M 87. The velocity of the gas measured by spectroscopy (cf. Fig. 3.1) at a distance from the nucleus of the order of 6×10^{17} m, shows that the gas recedes from us on one side, and approaches us on

[5] The table lists those galaxies which are listed both in [104] and [87]; we note that some candidates from earlier lists [119] do not occur any more in [87, 104]. Nineteen of the observations listed have been published in 2000 or 2001.

Table 3.2. Twenty-nine supermassive black hole candidates (from [87, 104])

dynamics of	host galaxy	$M_h/M_\odot$	host galaxy	$M_h/M_\odot$
water maser discs:	NGC 4258	4×10^7		
gas discs:	IC 1459	2×10^8	M 87	3×10^9
	NGC 2787	4×10^7	NGC 3245	2×10^8
	NGC 4261	5×10^8	NGC 4374	4×10^8
	NGC 5128	2×10^8	NGC 6251	6×10^8
	NGC 7052	3×10^8		
stars:	NGC 821	4×10^7	NGC 1023	4×10^7
	NGC 2778	1×10^7	NGC 3115	1×10^9
	NGC 3377	1×10^8	NGC 3379	1×10^8
	NGC 3384	1×10^7	NGC 3608	1×10^8
	NGC 4291	2×10^8	NGC 4342	3×10^8
	NGC 4473	1×10^8	NGC 4486B	5×10^8
	NGC 4564	6×10^7	NGC 4649	2×10^9
	NGC 4697	2×10^8	NGC 4742	1×10^7
	NGC 5845	3×10^8	NGC 7457	4×10^6
	Milky Way	2.5×10^6		

the other, with a velocity difference of about 920 km s^{-1} . This leads to a mass of the central object of $\sim 3 \times 10^9\ M_\odot$, and no form of matter can occupy such a small region except for a black hole. Figure 3.2 shows another image, reconstructed out of HST observations, of a recent candidate for a supermassive black hole – the (active) galactic nucleus of NGC 4438 [85]. To close the discussion of Table 3.2, we note that the determination of mass of the galactic nuclei via direct measurements of star motions has been made possible by the unprecedentedly high angular resolution and sensitivity of the HST. There seems to be consensus [87, 104, 120] that the two most convincing supermassive black hole candidates are the galactic nuclei of NGC 4258 and of our own Milky Way.

There is a huge gap between the masses of stellar mass black holes and those of the supermassive black holes and, in Bromley's terminology [20], "middleweight" candidates would be welcome. Such a tentative black hole with M of the order of $460 M_\odot$ has been identified in M 82 by A. Ptak and R. Griffiths [117]. E. Colbert and R. Mushotzky [48] give a list of compact X-ray sources which could include some further black holes with masses in the $10^2 - 10^4 M_\odot$ range.

Let us close this section by pointing out the review paper [25] which discusses both theoretical and experimental issues concerning *primordial* black holes.

3.3 Causality for Symmetric Hyperbolic Systems

The usual, Scri based, definition of a black hole proceeds as follows: one starts by introducing a preferred asymptotic region – this is usually taken to be a conformal boundary. Then the black hole region is defined as the set of points which cannot send information to the asymptotic region; we will return to the details of this construction in Sect. 3.4 below. The key ingredient here is the notion of "not being able to send information" – this is usually defined using the metric, together with the postulate that the propagation speed for objects carrying information is bounded from above by that of null geodesics. The aim of this section is to provide a precise mathematical version of this notion, via Proposition 3.1 below, for any symmetric hyperbolic system of differential equations, without introducing any supplementary postulates. The point is that every physical theory defines its own causality via the corresponding system of equations. We will then show in Sect. 3.3.1 how the "Unruh metric" arises in the application of our construction to the Euler equations.

Let, thus, L be a quasi-linear, symmetric hyperbolic, first order partial differential operator acting on sections φ of a bundle E, with scalar product $\langle \cdot, \cdot \rangle$, over a manifold $\mathscr{M}$; in a local trivialization over a coordinate patch $\mathscr{U} \ni x^\mu$ we

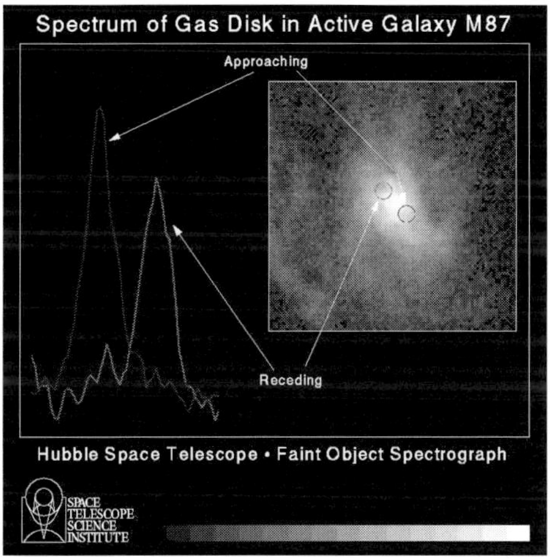

Fig. 3.1. Hubble Space Telescope observations of spectra of gas in the vicinity of the nucleus of the radio galaxy M 87, NASA and H. Ford (STScI/JHU) [123]

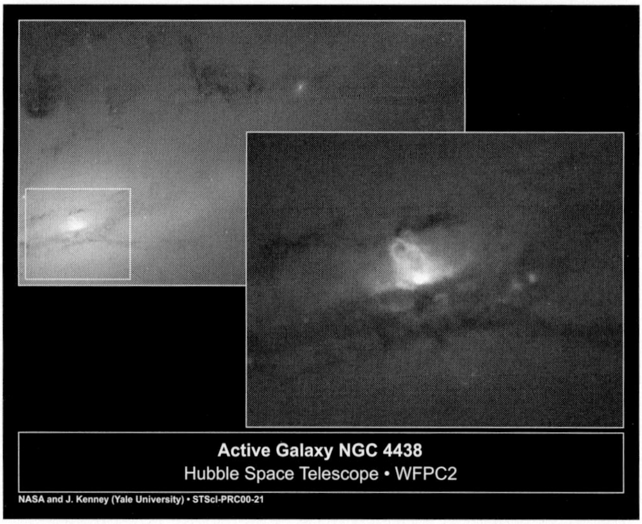

Fig. 3.2. Hubble Space Telescope observations [85] of the nucleus of the galaxy NGC 4438, from the STScI Public Archive [123]

have

$$L[\varphi] := A^\mu(\varphi, x^\beta)\partial_\mu\varphi + B(\varphi) \,, \tag{3.1}$$

where the A^μ's are endomorphisms of the fibers of E, symmetric with respect to $\langle \cdot, \cdot \rangle$. Given a section φ of E (defined perhaps on a subset of $\mathscr{M}$), a hypersurface $\mathscr{S}$ is said to be *space-like*[6] if one of its fields of co-normals n_μ satisfies

$$A^\mu\left(\varphi\left(p\right), p\right) n_\mu(p) > 0 \,, \quad p \in \mathscr{S} \,;$$

this is understood in the sense of strict positive definiteness of endomorphisms. By definition of a symmetric hyperbolic system, every point of the domain of definition of φ lies on some space-like hypersurface. For $p \in \mathscr{M}$ we set

$$\mathscr{T}_p^\flat := \{\alpha \in T_p^*\mathscr{M} \mid A^\mu\left(\varphi\left(p\right), p\right) \alpha_\mu \text{ is positive definite}\} \subset T_p^*\mathscr{M} \,, \tag{3.2a}$$

$$\mathscr{C}_p^\flat := \overline{\mathscr{T}_p^\flat} \subset \{\alpha \in T_p^*\mathscr{M} \mid A^\mu\left(\varphi\left(p\right), p\right) \alpha_\mu \text{ is non-negative definite}\} \subset T_p^*\mathscr{M} \,, \tag{3.2b}$$

and

$$\mathscr{T}^\flat := \cup_{p\in\mathscr{M}} \mathscr{T}_p^\flat \subset T^*\mathscr{M} \,, \qquad \mathscr{C}^\flat := \cup_{p\in\mathscr{M}} \mathscr{C}_p^\flat = \overline{\mathscr{T}^\flat} \,. \tag{3.3}$$

The bundle $\mathscr{T}^\flat$ is thus the bundle of covectors normal to space-like hypersurfaces. Because positive definite matrices form a convex set, each set $\mathscr{T}_p^\flat$ is a convex

[6] In quasi-linear theories, in which the coefficients of the highest derivatives depend upon the fields, the notion of space-likeness will of course depend upon the given set of fields.

cone in $T_p^*\mathscr{M}$. The bundles $\mathscr{T}^\sharp \subset T\mathscr{M}$, respectively $\mathscr{C}^\sharp \subset T\mathscr{M}$, of *time-like*, respectively *causal*, vectors are defined by duality:

$$\mathscr{T}_p^\sharp := \{X \in T_p\mathscr{M} \,|\forall\, \alpha \in \mathscr{T}_p^\flat \; X^\mu \alpha_\mu > 0\} \,, \tag{3.4a}$$

$$\mathscr{C}_p^\sharp := \overline{\mathscr{T}_p^\sharp} = \{X \in T_p\mathscr{M} \,|\forall\, \alpha \in \mathscr{T}_p^\flat \; X^\mu \alpha_\mu \geq 0\} \,. \tag{3.4b}$$

Recall that all standard classical field equations, i.e., Euler's equations, the scalar wave equation, the wave map equation, Maxwell's equations, Yang-Mills equations, or Einstein's equations, can be written as a symmetric hyperbolic system (cf., e.g., [118] or [125, Vol. IV]). For example, the wave equation for a scalar field u on a Lorentzian manifold $(\mathscr{M}, g)$ can be written in the form (3.1) by introducing a new set of variables

$$\varphi := (u, e_\mu(u)) \,,$$

where the e_μ's, $\mu = 0, \ldots, n$, form an ON-frame for g. In this case $\mathscr{T}^\sharp$ coincides with the bundle of time-like future pointing vectors of the metric g (where the future direction is determined by that of e_0), $\mathscr{C}^\sharp$ coincides with the bundle of causal future pointing or vanishing vectors of g, while $\mathscr{T}^\flat$ and $\mathscr{C}^\flat$ are, modulo sign (we are using the signature $(-,+,\ldots,+)$), "$\mathscr{T}^\sharp$ and $\mathscr{C}^\sharp$ with indices lowered using the metric". The same remains true for wave maps, for linear electrodynamics (considered as a first order system for $\boldsymbol{E}$ and $\boldsymbol{B}$), for the Yang-Mills equations in the Lorentz gauge, and for the usual symmetric hyperbolic reduction of Einstein's equations based on harmonic coordinates. (See the end of this section for more comments about the Einstein case.)

We emphasize that the bundles $\mathscr{T}^\sharp$, etc., are defined solely by the system of equations under consideration, independently of any metric. They can be used to determine the causality properties of a given symmetric hyperbolic system by mimicking the usual metric constructions [14, 73, 112, 115]: *time-like future directed paths* are defined as piecewise differentiable maps γ from I to $\mathscr{M}$, where $I \subset \mathbb{R}$ is an interval, with the property that the tangent $\dot\gamma$ is in $\mathscr{T}^\sharp$ wherever defined; *causal future directed paths* are defined as Lipschitz continuous paths such that $\dot\gamma$ is in $\mathscr{C}^\sharp$ wherever defined. *Past directed paths* are paths which are obtained from future directed paths by a reversal of parameterization. The *time-like future* $I^+(\Omega)$, respectively the *causal future* $J^+(\Omega)$, of a set $\Omega \subset \mathscr{M}$ is defined as the set of points which can be reached form Ω by following a future directed time-like, respectively causal, path. Causal and time-like futures J^- and I^- are obtained form the above definition by replacing "future" by "past". Notions such as global hyperbolicity, and so on, can be defined in the usual way. It might be convenient to write $I^\pm(\Omega, L)$, etc., to emphasize the dependence upon the system of equations L, if ambiguities can arise. We note that for semi-linear systems – by definition, these are the systems in which the coefficients A^μ in (3.1) do not depend upon φ – the resulting causal constructs are independent of the solution φ under consideration; however, this will not be the case in general.

We are ready now to address the issue of main interest in this work – the notion of a black hole. Given a set $\Omega \subset \mathscr{M}$ and a symmetric hyperbolic system

L we define the *black hole region* $\mathscr{B}_\Omega(L)$ *of* Ω as

$$\mathscr{B}_\Omega(L) := \mathscr{M} \setminus J^-(\Omega, L) \,. \tag{3.5}$$

The object so defined depends upon the set Ω, and acquires its full meaning when Ω is naturally distinguished by the problem at hand.

In several cases, some of which have already been mentioned, the causality theory constructed above arises out of a metric; we will show in Sect. 3.3.1 below that this holds also for the Euler equations. The intuitive meaning associated to the notion of a black hole region is that no information from $\mathscr{B}_\Omega(L)$ can reach Ω. This is made precise by the following result, the proof of which proceeds by causality arguments which are known in principle:

Proposition 3.1. *Let* φ *be a solution of*

$$L[\varphi] = J \,, \tag{3.6}$$

where L *is symmetric hyperbolic operator (3.1) on a manifold* $\mathscr{M}$*, and suppose that there exists a Lorentzian metric* g *such that the sets* $\mathscr{C}^\sharp$ *defined in (3.4b) are future light cones of* g*. Assume that* $(\mathscr{M}, g)$ *is globally hyperbolic with Cauchy surface* $\mathscr{S}$ *and consider a set* $\Omega \subset \mathscr{M}$ *such that* $\mathscr{B}_\Omega(L) \neq \emptyset$*. If* φ' *is a solution of*

$$L[\varphi'] = J' \quad \textit{with} \quad \varphi|_\mathscr{S} = \varphi'|_\mathscr{S} \,, \tag{3.7}$$

and if the difference of the source terms $J - J'$ *is supported in* $\mathscr{B}_\Omega(L)$*, then* $\varphi = \varphi'$ *on* $J^-(\Omega, L)$*.*

We believe that the conclusion above is true *without* the assumption that causality is determined by a metric, but we are not aware of any studies of this question in the current context; it would be of interest to settle this. The reader is referred to [97] for an analogous analysis of strictly hyperbolic systems, and to [30] for that of second order Lagrangean hyperbolic systems.

We close this section by mentioning that care has to be taken when interpreting the notions above for systems of equations which are not directly in first order symmetric hyperbolic form: it can happen that the physical equations can be then rewritten in several different symmetric hyperbolic forms, leading to different notions of pasts, futures, black hole regions, etc. This does indeed happen for the Einstein equations: in [57, Sect. 4.1] a symmetric hyperbolic system is presented which can be used to solve the Einstein equations and in which the causal cones, as defined above, differ from the metric ones; see the discussion after Equation (4.25) there. Another example of this kind arises if the T-vector of the system of equations given in Sect. 5.2 of [55] is chosen to be space-like. This is related to the acausal (in the usual metric sense) propagation of gauge degrees of freedom in the associated systems, and does of course not affect the propagation of any physically relevant quantities. Given a system of equations, say (S), a simple way out of this problem is to define the physical black hole region as the intersection, as L runs over the set of those symmetric hyperbolic systems L the solutions of which include solutions of (S), of the associated black hole regions $\mathscr{B}_\Omega(L)$. We note that the existence of exact solutions such as the pp-waves, where nonlinear perturbations propagate along null geodesics, guarantees that every symmetric hyperbolic system L which reproduces the Einstein's equations will necessary contain the metric light cone as its causal light cone associated to L. It follows that the above prescription does indeed reproduce the usual metric definition of a black hole region for the vacuum Einstein equations.

3.3.1 Dumb Holes

Let us illustrate the considerations of the previous section in the case of the Euler equations

$$\begin{aligned} \rho\,(\dot{\mathbf{v}} + \mathbf{v}\cdot\nabla\mathbf{v}) + \nabla p(\rho) &= 0\;, \\ \dot{\rho} + \nabla(\rho\mathbf{v}) &= 0\;. \end{aligned} \tag{3.8}$$

For strictly positive ρ's (3.8) can be rewritten as a symmetric hyperbolic system of the form (3.1) with $B = 0$, for a field $\varphi = (v^i, \rho)$, by introducing the matrices

$$A^0 = \begin{pmatrix} \rho\mathrm{id}_{\mathbb{R}^3} & 0 \\ 0 & \frac{p'}{\rho} \end{pmatrix}\;, \quad A^i = \begin{pmatrix} \rho v^i\mathrm{id}_{\mathbb{R}^3} & p'\delta^i_j \\ p'\delta^i_j & \frac{p'}{\rho}v^i \end{pmatrix}\;, \tag{3.9}$$

which are clearly symmetric with respect to the standard scalar product on $\mathbb{R}^3 \times \mathbb{R} \ni (\boldsymbol{v}, \rho)$. Straightforward algebra shows that $A^\mu n_\mu$ is strictly positive if and only if $\rho > 0$, $dp/d\rho > 0$ and

$$-(n_0 + v^i n_i)^2 + p'\sum_i (n_i)^2 < 0 \quad \Longleftrightarrow \quad g^{\mu\nu}n_\mu n_\nu < 0\;, \tag{3.10}$$

where the *Unruh metric* g is defined as [127, 129]

$$g^{\mu\nu} = \frac{1}{c_s\rho}\begin{pmatrix} cc - 1 & -v^i \\ -v^j & c_s^2\delta^{ij} - v^i v^j \end{pmatrix} \Longleftrightarrow g_{\mu\nu} = \frac{\rho}{c_s}\begin{pmatrix} -c_s^2 + \boldsymbol{v}^2 & -v^i \\ -v^j & \delta_{ij} \end{pmatrix}\;, \tag{3.11}$$

$$c_s^2 = \frac{dp}{d\rho}\;.$$

This metric exhibits a typical black hole behaviour for stationary solutions in which the speed $\boldsymbol{v}$ of the fluid velocity meets the speed of sound c_s across a smooth hypersurface.

The original argument of Unruh leading to (3.11) was a perturbational one, for irrotational flows:

$$\boldsymbol{v} = \nabla\Phi\;. \tag{3.12}$$

Small perturbations $\phi \equiv \delta\Phi$ of Φ satisfy then a scalar wave equation [127],

$$\partial_\mu(\sqrt{-g}g^{\mu\nu}\partial_\nu\phi) = 0\;,$$

in the metric (3.11). Our discussion above shows that neither the irrotationality of the flow, nor the perturbative character of the argument are essential for the problem at hand. The fact that causality for the full non-linear Euler's equations is governed by a metric can essentially be found in Courant and Hilbert, and is certainly well known to some researchers (H. Friedrich, private communication).

There was renewed interest in the above because of the recent experiments with the Bose-Einstein condensates. Recall that in the Madelung formulation, the Gross-Pitaevskii equation governing the dynamics of the condensates can be rewritten in Euler form. There have been suggestions that some of the effects associated to black-hole type causality, including the analogue of the Hawking radiation, could be observed in such systems, see [10, 83, 131] and references therein. While the resulting black holes are still referred to as *sonic*, this is rather misleading as the underlying Gross-Pitaevskii equation does not have anything to do with sound propagation in liquid or gaseous physical media.

3.3.2 Optical Holes

According to M. Visser [128], in a dielectric fluid "the refractive index n, the fluid velocity v, and the background Minkowski metric η can be combined algebraically to provide an effective metric g"

$$g_{\mu\nu} = \eta_{\mu\nu} - (n^2 - 1)v_\mu v_\nu \ .$$

This leads to black-hole effective geometries for refractive indices which exceed 1 in some regions. Phenomenological models for this behaviour have been proposed by U. Leonhardt and P. Piwnicki, cf. [95] and references therein[7]. Similarly, a (different) effective metric is obtained in nonlinear electrodynamics, leading – according to [111] – to models which sometimes contain closed time-like curves. The status of these "analogous models" seems to be somewhat less clear than that of the sonic ones.

3.3.3 Trapped Surfaces

We shall close this section by noting the discussion in [96, 130] concerning the notion of trapped surfaces[8] for black hole geometries – such surfaces are useful detectors of black hole regions in GR with matter fields carying positive energy. It should be borne in mind that while causality concepts can be introduced for any symmetric hyperbolic system, the existence of a trapped surface signals the presence of a black hole region *only* when causality is governed by *a metric satisfying an energy positivity condition* (and, if in a Scri context, when Scri satisfies various regularity conditions, see Sect. 3.4 below). In the non-gravitational black hole models the metric is the one arising out of the causality structure of the theory, and there are no reasons in general to believe that its Ricci tensor should have any preferred properties; to start with, it is defined only up to a conformal factor anyway. In addition, the "trapped surface" terminology is used in [130] for objects which are completely unrelated to the usual differential geometric context, which is extremely misleading and obscures the problems at hand.

[7] See also URL `http://www.st-and.ac.uk/~www_pa/group/quantumoptics`.

[8] Cf. Sect 3.6 for the differential-geometric definition of this notion.

3.4 Standard Black Holes

The standard way of defining black holes is by using conformal completions: A pair $(\widetilde{\mathscr{M}}, \tilde{g})$ is called a *conformal completion* of $(\mathscr{M}, g)$ if $\widetilde{\mathscr{M}}$ is a manifold with boundary such that:

1. $\mathscr{M}$ is the interior of $\widetilde{\mathscr{M}}$.
2. There exists a function Ω, with the property that the metric $\tilde{g}$, defined to be $\Omega^2 g$ on $\mathscr{M}$, extends by continuity to the boundary of $\widetilde{\mathscr{M}}$, with the extended metric still being non-degenerate throughout.
3. Ω is positive on $\mathscr{M}$, differentiable on $\widetilde{\mathscr{M}}$, vanishes on $\mathscr{I}$, with $d\Omega$ *nowhere vanishing* on $\mathscr{I}$.

We emphasize that no assumptions about the causal nature of Scri are made so far. The boundary of $\widetilde{\mathscr{M}}$ will be called Scri, denoted $\mathscr{I}$. In the usual textbooks [73, 132] smoothness of both the conformal completion and the metric $\tilde{g}$ is imposed, though this can be weakened for many purposes. We set

$$\mathscr{I}^+ = \{p \in \mathscr{I} \mid I^-(p; \widetilde{\mathscr{M}}) \cap \mathscr{M} \neq \emptyset\} .$$

Assuming various global regularity conditions on $\widetilde{\mathscr{M}}$, to which we shall return in Sect. 3.4.1, one then defines the black hole region $\mathscr{B}$ as

$$\mathscr{B} := \mathscr{M} \setminus J^-(\mathscr{I}^+) . \tag{3.13}$$

Let us point out some drawbacks of this approach:

- **Non-equivalent Scri's:** Conformal completions at null infinity do not have to be unique, an example can be constructed as follows: the Taub-NUT metrics can be locally written in the form [106]

$$-U^{-1}dt^2 + (2L)^2 U\sigma_1^2 + (t^2 + L^2)(\sigma_2^2 + \sigma_3^2) , \tag{3.14}$$

$$U(t) = -1 + \tfrac{2(mt+L^2)}{t^2+L^2} . \tag{3.15}$$

  where σ_1, σ_2 and σ_3 are left invariant one-forms on $SU(2) \approx S^3$. The constants L and m are real numbers with $L > 0$. Parameterizing S^3 with Euler angles (μ, θ, φ) one is led to the following form of the metric

$$g = -U^{-1}dt^2 + (2L)^2 U(d\mu + \cos\theta d\varphi)^2 + (t^2 + L^2)(d\theta^2 + \sin^2\theta d\varphi^2) .$$

  The standard way of performing extensions across the Cauchy horizons $t_\pm := M \pm \sqrt{M^2 + L^2}$ is to introduce new coordinates

$$(t, \mu, \theta, \varphi) \to (t, \mu \pm \int_{t_0}^{t} [2LU(s)]^{-1} ds, \theta, \varphi) , \tag{3.16}$$

  which gives

$$\begin{aligned} g_\pm = {} & \pm 4L(d\mu + \cos\theta d\varphi)dt \\ & + (2L)^2 U(d\mu + \cos\theta d\varphi)^2 + (t^2 + L^2)(d\theta^2 + \sin^2\theta d\varphi^2) . \end{aligned} \tag{3.17}$$

Each of the metrics $g_\pm$ can be smoothly conformally extended to the boundary at infinity "$t=\infty$" by introducing

$$x = 1/t\ ,$$

so that (3.17) becomes

$$\begin{aligned} g_\pm = x^{-2}\Big(&\mp 4L(d\mu + \cos\theta d\varphi)dx \\ &+(2L)^2x^2U(d\mu + \cos\theta d\varphi)^2 + (1+L^2x^2)(d\theta^2 + \sin^2\theta d\varphi^2)\Big) \end{aligned} \quad (3.18)$$

In each case this leads to a Scri diffeomorphic to S^3. There is a simple isometry between g_+ and g_- given by

$$(x,\mu,\theta,\varphi) \to (x,-\mu,\theta,-\varphi)$$

(this does correspond to a smooth map of the region $t\in(t_+,\infty)$ into itself, cf. [42]), so that the two Scri's so obtained are isometric. However, in addition to the two ways of attaching Scri to the region $t\in(t_+,\infty)$ there are the two corresponding ways of extending this region across the Cauchy horizon $t=t_+$, leading to four possible manifolds with boundary. It can then be seen, using ,e.g., the arguments of [42], that the four possible manifolds split into two pairs, each of the manifolds from one pair *not* being isometric to one from the other. Taking into account the corresponding completion at "$t=-\infty$", and the two extensions across the Cauchy horizon $t=t_-$, one is led to four inequivalent conformal completions of each of the two inequivalent [42] time-oriented, maximally extended, standard Taub-NUT space-times.

Under mild completeness conditions in the spirit of [66], uniqueness of $\mathscr{I}^+$ as a point set in *past-distinguishing* space-times should follow from the TIP and TIF construction of [67]; however, uniqueness of Scri's differentiable structure within that framework is far from being clear.

Yet another approach to uniqueness has been proposed by Geroch [65]: A completion as defined at the beginning of this section is said to be *a regular asymptote* if, given any point $p\in\mathscr{I}$, and any non-zero null vector $\ell\in T_q\mathscr{M}$ such that the maximally extended null geodesic with tangent ℓ does not meet p, then there are neighbourhoods $\mathscr{U}\subset\tilde{\mathscr{M}}$ of p and $\mathscr{V}\subset T\tilde{\mathscr{M}}$ of ℓ such that no maximally extended null geodesic with tangent in $\mathscr{V}$ meets $\mathscr{U}$. This regularity condition is not satisfied by the Taub-NUT completions described above. Supposing that the set of regular asymptotes is non-empty, Geroch [65, Theorem 2, p. 14] gives an argument for existence of a maximal regular asymptote, unique up to well behaved conformal transformations. Now, we have not been able to fill in the details[9] of some of the arguments

[9] It is not completely clear that the map φ defined by Geroch in his proof is simultaneously differentiable for all regular asymptotes, as asserted in [65, p. 14]. The regularity condition guarantees Hausdorffness of the maximal regular asymptote constructed in [65, p. 14], but does not seem to guarantee its differentiability in any obvious way.

suggested in [65], but the construction given is plausible enough so that we expect that it might actually be correct. Whatever the case, Geroch's regularity requirement is a global condition which seems to be difficult to control in general situations.[10] In particular it could happen that many space-times of interest admitting conformal completions do not admit any regular ones. It should be borne in mind that one of the main applications of the $\mathscr{I}$ framework is the possibility of *uniquely* defining global charges such as mass, angular momentum, etc. (cf. [43] for a new approach to this question), in which issues about global behaviour of *all* geodesics seem completely irrelevant. The fact that there is no suitable uniqueness result for the differentiable structure of conformal completions is a serious gap in our understanding of objects such as the Trautman-Bondi mass.

We note that uniqueness of a class of Riemannian conformal completions at infinity has been established in [41, Section 6]; this result can probably be used to obtain uniqueness of differentiable structure of Lorentzian conformal completions for Scri's admitting cross-sections. Further partial results on the problem at hand can be found in [121].

- **Poorly differentiable Scri's:** In all standard treatments [65, 73, 132] it is assumed that both the conformal completion $\widetilde{\mathscr{M}} = \mathscr{M} \cup \mathscr{I}$ and the extended metric $\tilde{g}$ are smooth, or have a high degree of differentiability [114]. This is a restriction which excludes most space-times which are asymptotically Minkowskian in light-like directions, see [5, 90] and references therein. Poor differentiability properties of $\mathscr{I}$ change the peeling properties of the gravitational field [44], but most – if not all – essential properties of black holes should be unaffected by conformal completions with, e.g., polyhomogeneous differentiability properties as considered in [6, 44]. It should, however, be borne in mind that the hypothesis of smoothness has been done in the standard treatments, so that in a complete theory the validity of various claims should be reexamined. Some new results concerning existence of space-times with a poorly differentiable Scri can be found in [94].
- **The structure of i^+:** The current theory of black holes is entirely based on intuitions originating in the Kerr and Schwarzschild geometries. In those space-times we have a family of preferred "stationary" observers which follow the orbits of the Killing vector field ∂_t in the asymptotic region, and their past coincides with that of $\mathscr{I}^+$. It is customary to denote by i^+ the set consisting of the points $t = \infty$, where t is the Killing time parameter for those observers. Now, the usual conformal diagrams for those space-times [73, 107] leave the highly misleading impression that i^+ is a regular point in the conformally rescaled manifold, which, to the best of our knowledge, is not the

[10] Globally hyperbolic conformal completions (in the sense of manifolds with boundary) should satisfy Geroch's condition. Nevertheless, it should be borne in mind that good causal properties of a space-time might fail to survive the process of adding a conformal boundary. For example, adding a Cauchy horizon to a maximal globally hyperbolic space-time sometimes leads to space-times which are not globally hyperbolic in the sense of manifolds with boundary.

case. In dynamical cases the situation is likely to become worse. For example, one can imagine black hole space-times with a conformal diagram which, to the future of a Cauchy hypersurface $t = 0$, looks as in Fig. 3.3. In that

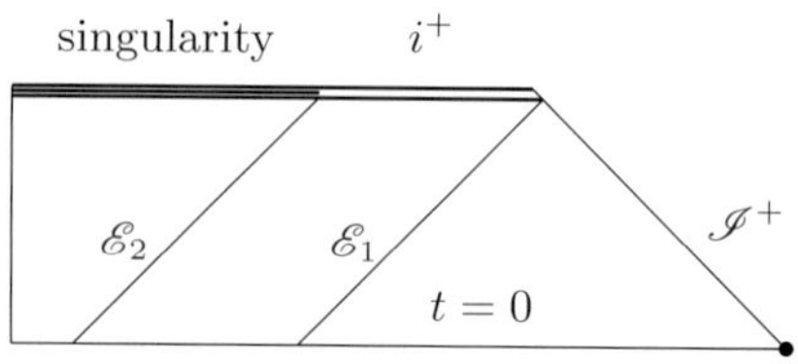

Fig. 3.3. An asymptotically flat space-time with an unusual i^+.

diagram the set i^+ should be thought of as the addition to the space-time manifold $\mathscr{M}$ of a set of points "$\{t = \infty, q \in \mathscr{O}\}$", where $t \in [0, \infty)$ is the proper time for a family of observers $\mathscr{O}$. The part of the boundary of $\widetilde{\mathscr{M}}$ corresponding to i^+ is a singularity of the conformally rescaled metric, but we assume that it does *not* correspond to singular behaviour in the physical space-time. In this space-time there is the usual event horizon $\mathscr{E}_1$ corresponding to the boundary of the past of $\mathscr{I}^+$, which is completely irrelevant for the family of observers $\mathscr{O}$, and an event horizon $\mathscr{E}_2$ which is the boundary of the *true* black hole region for the family $\mathscr{O}$, i.e., the region that is not accessible to observations for the family $\mathscr{O}$. Clearly the usual black hole definition carries no physically interesting information in such a setting.

- **Causal regularity of Scri:** As already pointed out, in order to be able to prove interesting results the definition (3.13) should be complemented by causal conditions on $\widetilde{\mathscr{M}}$. The various approaches to this question, discussed in Sect.3.4.1, are aesthetically highly unsatisfactory: it appears reasonable to impose causal regularity conditions on a space-time, but why should some unphysical completion have any such properties? Clearly, the physical properties of a black hole should not depend upon the causal regularity – or lack thereof – of some artificial boundary which is being attached to the space-time. While it seems reasonable and justified to restrict attention to space-times which possess good causal properties, it is not clear why the addition of artificial boundaries should preserve those properties, or even be consistent with them. Physically motivated restrictions are relevant when dealing with physical objects, they are not when non-physical constructs are considered.

- **Inadequacy for numerical purposes:** Several[11] numerical studies of black holes have been performed on numerical grids which cover finite space-time regions [1, 2, 19, 86]. Clearly, it would be convenient to have a set-up which is more compatible in spirit with such calculations than the Scri one.

We will suggest, in Sects. 3.8.1 and 3.8.2 below, two approaches in which the above listed problems are avoided. Before doing this, let us complete the presentation of the usual, Scri based, approach to black holes.

3.4.1 Scri Regularity Conditions, and the Area Theorem

It is easily seen that the definition (3.13) is not very useful without some *completeness conditions* on $\mathscr{I}^+$. For instance, denote by $\widetilde{\mathscr{M}}$ the standard conformal completion[12] of the Minkowski space-time $\mathbb{R}^{1,3}$, with $\mathscr{I}^+ \approx \mathbb{R} \times S^2$, where the $\mathbb{R}$ factor is parameterized by a Bondi coordinate u, with $u = t - r$ on $\mathbb{R}^{1,3}$. One can obtain a new completion $\widetilde{\mathscr{M}}_2$ by replacing $\mathscr{I}^+$ by an open subset thereof, e.g. restricting the range of u's to $(-\infty, 0)$. This will give a non-empty black hole region $\mathscr{B}$ in $\mathbb{R}^{1,3}$,

$$\mathscr{B} = J^+(0) ,$$

which is clearly a completely uninteresting statement. A way out of this problem (as well as of some other related ones, discussed in [66]) has been proposed[13] in [66], for space-times which are vacuum near the conformal boundary, assuming that the cosmological constant is zero: R. Geroch and G. Horowitz introduce a preferred family of conformal factors Ω such that the matrix Hess Ω of second covariant derivatives of Ω vanishes at $\mathscr{I}$. It is then required that both $\mathscr{I}^+$ and $\mathscr{I}^-$ be diffeomorphic to $\mathbb{R} \times S^2$, with the $\mathbb{R}$ factor corresponding to integral curves of $\widetilde{\nabla}\Omega$, which are assumed to be complete.

Let us pass to a presentation of the causal regularity conditions which are imposed in Hawking and Ellis's approach [73]. Now, a set of conditions is only useful insofar as it allows to prove something. Since the "area theorem"[14] is one of the landmark theorems in the theory of black holes, we shall concentrate on the set of conditions from Hawking and Ellis as used in their treatment of the area theorem; we follow the presentation of [38, Appendix B]. One of the conditions of the Hawking–Ellis area theorem [73, Proposition 9.2.7, p. 318] is that space-time $(\mathscr{M}, g)$ is *weakly asymptotically simple and empty* ("WASE", [73, p. 225]). This means that there exists an open set $\mathscr{U} \subset \mathscr{M}$ which is isometric

[11] The numerical simulations in [12, 16, 70] cover regions extending all the way to infinity, within frameworks which seem to be closely related to the "naive" framework of Sect. 3.8.1 below.

[12] Introduce coordinates $u = t - r$, $x = 1/r$, θ, φ on $R^4 \setminus \{r = 0\}$, with (r, θ, φ) – the usual spherical coordinates on $\mathbb{R}^3$, and set $\mathscr{I}^+ := \{x = 0\}$ in those coordinates.

[13] The "WASE setup" of [73], presented below, does not solve the problem, cf. [66].

[14] Recall that the area theorem is the statement that the area of cross-sections of event horizons is non-increasing towards the future. One possible precise version of this theorem can be found in Theorem 3.6 below.

to $\mathscr{U}' \cap \mathscr{M}'$, where $\mathscr{U}'$ is a neighborhood of null infinity in an asymptotically simple and empty ("ASE") space-time $(\mathscr{M}', g')$ [73, p. 222]. It is further assumed that $\mathscr{M}$ admits a partial Cauchy surface $\mathscr{S}$ with respect to which $\mathscr{M}$ is *future asymptotically predictable* ([73], p. 310). This is defined by the requirement that $\mathscr{I}^+$ is contained in the closure of the future domain of dependence $\mathcal{D}^+(\mathscr{S};\mathscr{M})$ of $\mathscr{S}$, where the closure is taken in the conformally completed manifold $\overline{\mathscr{M}} = \mathscr{M} \cup \mathscr{I}^+ \cup \mathscr{I}^-$, with both $\mathscr{I}^+$ and $\mathscr{I}^-$ being null hypersurfaces. Next, one says that $(\mathscr{M}, g)$ is *strongly future asymptotically predictable* ([73], p. 313) if it is future asymptotically predictable and if $J^+(\mathscr{S}) \cap \bar{J}^-(\mathscr{I}^+;\overline{\mathscr{M}})$ is contained in $\mathcal{D}^+(\mathscr{S};\mathscr{M})$. Finally ([73], p. 318), $(\mathscr{M}, g)$ is said to be a *regular predictable space* if $(\mathscr{M}, g)$ is strongly future asymptotically predictable and if the following three conditions hold:

(α) $\mathscr{S} \cap \bar{J}^-(\mathscr{I}^+;\overline{\mathscr{M}})$ is homeomorphic to $\mathbb{R}^3 \backslash$(an open set with compact closure).
(β) $\mathscr{S}$ is simply connected.
(γ) The family of hypersurfaces $\mathscr{S}(\tau)$ constructed in [73, Proposition 9.2.3,p. 313] has the property that for sufficiently large τ the sets $\mathscr{S}(\tau) \cap \bar{J}^-(\mathscr{I}^+;\overline{\mathscr{M}})$ are contained in $\bar{J}^+(\mathscr{I}^-;\overline{\mathscr{M}})$.

It should be clear at this stage that this set of conditions is rather intricate. As it turns out, it is not clear whether or not it does suffice for a proof of the area theorem, as asserted in [73, Proposition 9.2.7, p. 318]: indeed, in the proof of [73, Proposition 9.2.1, p. 311] (which is one of the results used in the proof of [73, Proposition 9.2.7, p. 318]) Hawking and Ellis write: "This shows that if $\mathscr{W}$ is any compact set of $\mathscr{S}$, every generator of $\mathscr{I}^+$ leaves $J^+(\mathscr{W};\overline{\mathscr{M}})$." The justification of this given in [73] is wrong. If one is willing to impose the sentence in quotation marks as *yet one more regularity hypothesis* on $\mathscr{I}^+$, then the arguments given in [73] apply to prove the area theorem for black holes with a piecewise smooth event horizon; the results in [38] show that this remains true with no supplementary conditions on the differentiability of the horizon.

Not only is the above set of hypotheses aesthetically unappealing, it is far from being unique. An alternative way to guarantee that the area theorem will hold in the "future asymptotically predictable WASE" set-up of [73] is to impose some mild additional conditions on $\mathscr{U}$ and $\mathscr{S}$. There are quite a few possibilities, one such set of conditions has been described in [38, Appendix B]: Let $\psi : \mathscr{U} \to \mathscr{U}' \cap \mathscr{M}'$ denote the isometry arising in the definition of the WASE space-time $\mathscr{M}$. First, one requires that ψ can be extended by continuity to a continuous map, still denoted by ψ, defined on $\overline{\mathscr{U}}$. Next, one demands that there exists a compact set $K \subset \mathscr{M}'$ such that,

$$\psi(J^+(\mathscr{S};\mathscr{M}) \cap \partial\mathscr{U}) \subset J^+(K;\mathscr{M}') \,, \tag{3.19}$$

see Fig. 3.4. Under those conditions the area theorem does again hold.

Several other proposals how to modify the WASE conditions of [73] to obtain sufficient control of the space-times at hand have been proposed in [47, 89, 110].

A completely different, and considerably simpler, treatment of the question of regularity of Scri needed for the area theorem, has been proposed in [38]: A point

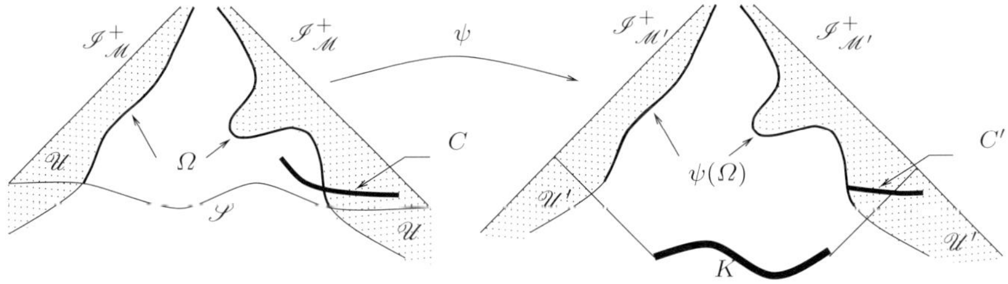

Fig. 3.4. The set $\Omega \equiv J^+(\mathscr{S};\mathscr{M}) \cap \partial\mathscr{U}$ and its image under ψ.

q in a set $A \subset B$ is said to be a *past* point of A with respect to B if $J^-(q;B) \cap A = \{q\}$. One then says that $\mathscr{I}^+$ is $\mathscr{H}$*–regular* if there exists a neighborhood $\mathcal{O}$ of $\mathscr{H}$ such that for every compact set $C \subset \mathcal{O}$ satisfying $I^+(C;\bar{M}) \cap \mathscr{I}^+ \neq \emptyset$ there exists a past point *with respect to* $\mathscr{I}^+$ in $\partial I^+(C;\bar{M}) \cap \mathscr{I}^+$. The condition of $\mathscr{H}$–regularity of $\mathscr{I}^+$ is precisely what is needed for the arguments of the area theorem to go through. This condition is somewhat related to the i^0*-avoidance condition* used by Galloway and Woolgar [63].

Yet another approach has been advocated by Wald [132]: there one considers globally hyperbolic completions in a set-up that includes i^0. While the conditions of [132] do lead to a coherent set-up for the validity of the area theorem, they introduce several new difficulties related to the low differentiability of the conformally rescaled metric at i^0, not to mention the question of mere existence of such completions. This can actually be relaxed to the related weaker requirement of global hyperbolicity of $\mathscr{M} \cup \mathscr{I}^+$ (in the sense of a manifold with boundary) which enforces $\mathscr{H}$–regularity of $\mathscr{I}^+$ [38], so that the area theorem again holds.

3.5 Horizons

A key notion related to the concept of a black hole is that of an *event horizon*,

$$\mathscr{E} := \partial\mathscr{B} . \tag{3.20}$$

(Actually, when $\mathscr{E}$ is not connected some care is required for a useful definition, but this does not affect the discussion that follows.) Event horizons are a special case of a family of objects called *future horizons*: by definition, these are closed topological hypersurfaces $\mathscr{H}$ threaded by null geodesics, called *generators*, with no *future end* points, and possibly with *past end* points. At the latter, differentiability of $\mathscr{H}$ breaks down in general; a necessary and sufficient condition for this breakdown has been given in [13]. It seems that most authors have been taking for granted that horizons are nice, piecewise smooth hypersurfaces. This is, however, not the case, and examples of *nowhere* C^1 horizons have been constructed in [40]. Further, nowhere C^1 horizons are generic in the class of convex horizons in Minkowski space-time [23]. This leads to various difficulties when trying to study their structure, e.g., attempting to prove results such as the area theorem discussed in Sect. 3.4.1, compare Theorem 3.6 below. Horizons are always *locally Lipschitz-continuous* topological hypersurfaces [115] (compare Corollary A.1 below), i.e., they are locally graphs of functions satisfying

$$|\phi(x) - \phi(y)| \leq C|x - y| \ .$$

In spite of the potential low differentiability, the usual optical scalars can be defined almost everywhere in an *Alexandrov sense*, as follows: A function is said to be *semi-convex* if it is the sum of a C^2 function and of a convex function. It has been shown in [38][15] that future horizons are semi-convex. We give an alternative proof of this result in an appendix: the new argument has some interest on its own, as it proceeds via a variational principle for horizons, which could be useful for numerical applications. The interest of the semi-convexity property relies on the following theorem of Alexandrov:

Theorem 3.1. (Alexandrov [52, Appendix E]) *Semi-convex functions $f : B \to \mathbb{R}$ are "twice-differentiable"* almost everywhere *in the following sense: there exists a set $B_{\mathcal{A}l}$ with full measure in B such that*

$$\forall\ x \in B_{\mathcal{A}l}\ \exists\ Q \in (\mathbb{R}^p)^* \otimes (\mathbb{R}^p)^* \text{ such that } \forall\ y \in B$$
$$f(y) - f(x) - df(x)(y - x) = Q(x - y, x - y) + r_2(x, y) \ ,$$

with $r_2(x,y) = o(|x-y|^2)$. The symmetric quadratic form Q above will be denoted by $\frac{1}{2}D^2 f(x)$, and will be called the second Alexandrov derivative *of f at x.*

Points in $B_{\mathcal{A}l}$ will be referred to as *Alexandrov points* of B.

One can now use Theorem 3.1 to define the equivalent to the usual divergence θ of the horizon, by writing $\mathscr{H}$ locally as the graph of a function f, and using the second Alexandrov derivatives of f to define (almost everywhere on $\mathscr{H}$) the divergence $\theta_{\mathcal{A}l}$ of the horizon. More precisely, let $p = (t = f(q), q)$ be an Alexandrov point of $\mathscr{H}$ and let e_i, $i = 1, 2, 3$ a basis of $T_p\mathscr{H}$ such that

$$e_3 = K_\mu(p)dx^\mu = -dt + df(q) \ , \ g(e_1, e_1) = (e_2, e_2) = 1 \ , \ g(e_1, e_2) = 0 \ .$$

One then sets

$$\nabla_i K_j = D^2_{ij} f - \Gamma^\mu_{ij} K_\mu \ , \qquad \theta_{\mathcal{A}l} = (e_1^i e_1^j + e_2^i e_2^j)\nabla_i K_j \ .$$

where the Hessian matrix $D^2_{ij} f$ above is that of second Alexandrov derivatives. This reduces to the standard definition of θ when $f \in C^2$.

Several results of causality theory go through with θ replaced by $\theta_{\mathcal{A}l}$, though the standard arguments sometimes have to be replaced by completely different ones, the reader is referred to [38] for details. In particular the Raychaudhuri equation, as well as the remaining optical equations, hold on almost all generators of $\mathscr{H}$.

Somewhat surprisingly, event horizons in smooth, stationary, globally hyperbolic, asymptotically flat space-times satisfying the null energy condition are always smooth null hypersurfaces, analytic if the metric is analytic; this is a corollary of the area theorem of [38].

[15] Some further results concerning differentiability properties of horizons can be found in [13, 35, 39].

3.6 Apparent Horizons

In spite of their name, apparent horizons are *not* horizons. They are usually [73] defined on space-like hypersurfaces $\mathscr{S} \subset \mathscr{M}$, as follows: let $\Omega \subset \mathscr{M}$ be the set covered by *(future) trapped surfaces*: by definition, those are compact, boundaryless, smooth surfaces $S \subset \mathscr{S}$ with the property that

$$\theta(S) := \lambda - (g^{ij} - n^i n^j)K_{ij} < 0 \ , \tag{3.21}$$

where λ is the (outwards) mean extrinsic curvature of S in $\mathscr{S}$, n^i is the outwards pointing unit normal to S in $\mathscr{S}$, and K_{ij} is the extrinsic curvature of $\mathscr{S}$ in $\mathscr{M}$. There is no reason for Ω to be nonempty in general; on the other hand, in appropriately censored space-times, a non-empty Ω implies the existence of a black hole region. Hawking and Ellis define the apparent horizon $\mathscr{A}$ as

$$\mathscr{A} := \partial\Omega \ , \tag{3.22}$$

and argue that

$$\theta(\mathscr{A}) = 0 \ . \tag{3.23}$$

Their argument is correct if one assumes that $\mathscr{A}$ is C^2. However, $\mathscr{A}$ could be a priori a very rough set, with θ not even defined, in which case the arguments of [73] do not apply. Probably the simplest way out is to use (3.23) as the definition of apparent horizon, and forget about Ω. The existence of a C^2 compact, boundaryless surface satisfying (3.23) does again imply the existence of a black hole in a Scri framework, assuming appropriate causal properties of $\mathscr{I}^+$.

We note that if $K_{ij} = 0$, then (3.22) is the equation for a minimal surface. Now, in the course of their proof of the Penrose inequality, G. Huisken and T. Ilmanen [80] prove that the outermost minimal (in the sense of calculus of variations) surface is always smooth, which supports the validity of (3.23) at least for the outermost component of $\mathscr{A}$. However, they provide examples which show that this outermost minimal surface might not be the boundary of Ω in general. This illustrates another deficiency of the definition (3.22). Some partial results concerning the differentiability properties of $\mathscr{A}$ have been obtained by M. Kriele and S. Hayward in [88]. R. Howard and J. Fu have recently shown [79] that $\partial\Omega$ satisfies (3.22) in a *viscosity sense* – this is defined as follows: Let D be an open set, and let $p \in \partial D$. Then U is an inner (outer) support domain if U is an open subset of D (of $\complement D$), ∂U is a C^2 hypersurface and $p \in \partial U$. One says that $\theta(\partial D) \le 0$ (respectively $\theta(\partial D) \ge 0$) in the viscosity sense if for any point $p \in \partial D$ and for any inner support (outer support) domain we have

$$\theta(\partial U)(p) \le 0 \qquad \text{(respectively } \theta(\partial U)(p) \ge 0) \ .$$

This coincides with the usual inequality $\theta \le 0$ ($\theta \ge 0$) on any subset of ∂D which is C^2. Finally $\theta(\partial D) = 0$ in the viscosity sense if for any point $p \in \partial D$ we have both $\theta(\partial D) \le 0$ in the viscosity sense, and $\theta(\partial D) \ge 0$ in the viscosity sense.

It is tempting to conjecture the following: if $\widehat{\Omega}$ is the set covered by smooth compact S's satisfying

$$\theta(S) := \lambda + (g^{ij} - n^i n^j)K_{ij} \le 0 \ ,$$

then any outermost connected component of $\partial\widehat{\Omega}$ is smooth, separating, and satisfies (3.22). When $K_{ij} = 0$ this statement is known to be true by the already mentioned result of G. Huisken and T. Ilmanen.

We refer the reader to [3, 72] and references therein for a numerical treatment of apparent horizons.

3.7 Classification of Stationary Solutions ("No Hair Theorems")

Uniqueness theorems for stationary solutions have been one of the central fields of research in the theory of black holes, the reader is referred to [33, 34] and references therein for a review, as well as lists of open problems; cf. also [75, 76]. Actually the perspective here is somewhat larger, as one is interested in classifying all solutions, not only those that contain black holes. Since the time of writing of [33, 34] some progress has been made, both for zero and non-zero cosmological constant; it is convenient to discuss various cases separately. Throughout this section, *stationary* means that there exists an action of $\mathbb{R}$ on $\mathscr{M}$ by isometries, with the orbits of the group being time-like sufficiently far away in an asymptotic region; *strictly stationary* means that the orbits are time-like everywhere. Stationary black hole solutions are never strictly stationary.

$\Lambda = 0$: The following fundamental theorem has been recently proved by M. Anderson:

Theorem 3.2 (Anderson [4]). *The only geodesically complete strictly stationary solutions of the vacuum Einstein equations which do not contain closed timelike curves are the Minkowski space-time and quotients thereof.*

This is a milestone result in general relativity, involving only the natural geometric conditions of completeness together with absence of closed time-like curves. This should be contrasted with the version of this result originating in Lichnerowicz's work, the state-of-the art form of which, as based on previous techniques, assumes *in addition* global hyperbolicity, *and* asymptotic flatness, *and* existence of a compact interior. Last but not least, the paper introduces techniques which have not been used in mathematical general relativity so far, and which are likely to be useful tools in future work.

In the black hole case, the only significant progress made since the review paper [34] is for *static* black holes: In [36] the classification of static vacuum black holes which contain an asymptotically flat space-like hypersurface with compact interior has been, essentially,[16] finished. This problem has a long history, starting with the pioneering work of Israel [81]. The most complete result existing

[16] Some comments about the qualification "essentially" are in order: uniqueness theorems usually proceed in two steps, the first being the reduction of the problem to a Riemannian one, the second consisting of a uniqueness theorem for the Riemannian problem. The latter part of the proof in [36] seems to be optimal; one could consider improving the former, see [36] for a detailed discussion.

in the literature prior to [36] was that of Bunting and Masood-ul-Alam [24] who showed, roughly speaking, that all appropriately regular such black holes which *do not contain degenerate horizons* belong to the Schwarzschild family. In [36] the condition of non-degeneracy of the event horizon is removed, showing that the Schwarzschild black holes exhaust the family of all appropriately regular black hole space-times. The proof requires an extension of the positive mass theorem which applies to asymptotically flat *complete* Riemannian manifolds, proved in [11]. The paper [37] contains an improvement of similar previous results concerning the electro-vacuum black holes, under the restrictive condition that all degenerate components of the black hole carry charges of the same sign. This seems a reasonable condition from a physical point of view: opposite charges would attract, which added to the gravitational attraction should prevent the configuration from being static. However, a proper mathematical treatment which would transform this argument into a proof is still missing.

$\Lambda > 0$: The natural boundary conditions for stationary solutions with $\Lambda > 0$ are the "no boundary ones": one considers globally hyperbolic space-times containing compact, boundaryless spatial surfaces. In this case there is no asymptotic region, and *stationarity* is defined by the requirement that the set of points at which the Killing vector field is time-like is non-empty. In [92] J. Lafontaine and L. Rozoy have announced a complete classification of static space-times with a compact (boundaryless) totally geodesic space-like hypersurface $\mathscr{S}$, under the assumption that the Lorentzian norm $\sqrt{-g_{\mu\nu}X^\mu X^\nu}$ of the Killing vector field is a Morse-Bott function, and that of analyticity[17] of the metric on $\mathscr{S}$. The hypothesis that $\mathscr{S}$ is totally geodesic is the usual hypothesis of staticity. This is the first result of such generality in this context. The proof is an extension of an argument of R. Beig and W. Simon [15] done in a related context, and proceeds by proving that the metric on $\mathscr{S}$ must be conformally flat; one can then conclude using [91].

Uniqueness results under the completely different hypothesis of existence of a conformal completion have been previously established by H. Friedrich [54] and by Boucher [17].

$\Lambda < 0$: Our understanding of the class of stationary space-times with a negative cosmological constant is much poorer than that of the $\Lambda = 0$ ones. It seems that the only results available so far concern static space-times. To start with, the question of boundary conditions which should be imposed in the asymptotic region are far from being understood, a beginning of a systematic analysis of this question can be found in [45]. A uniqueness argument for the de Sitter metric in the strictly stationary case has been proposed in [18]. It is only recently that complete proofs of the required positive energy theorem have been given [41, 134,

[17] This is a condition on the analyticity of the metric at the set of points at which the Killing vector vanishes or becomes null, since the metric is necessarily analytic elsewhere.

136]; the results in [41] do actually lead to a uniqueness result under much weaker asymptotic conditions than originally suggested [18]. It should be pointed out that the nature of the corresponding Riemannian problem is completely different from that of the asymptotically flat case: here one needs to analyze the set of solutions of the equations for a Riemannian metric g and a function V on a three dimensional manifold $\mathscr{S}$,

$$\Delta V = -\Lambda V \ , \tag{3.24}$$
$$R_{ij} = V^{-1} D_i D_j V + \Lambda g_{ij} \ , \tag{3.25}$$

and it is customary to assume that $(\mathscr{S}, g)$ can be conformally compactified. Now, the uniqueness theorems mentioned require that $V^{-2}g$ extends by continuity to a smooth metric on the conformal boundary, with constant scalar curvature there. No reasons are known why this should be the case, and it is conceivable that more general static solutions would exist which do not satisfy this condition. It would be of interest to construct such solutions, or to prove their non-existence.

The topology of the boundary of the conformally compactifiable solutions of (3.24)-(3.25) does not have to be spherical, as is the case for the anti-de Sitter metric, while the uniqueness result mentioned above assumes that the boundary at infinity $\partial_\infty \mathscr{S}$ is a sphere. When $\partial_\infty \mathscr{S}$ is a torus, a strictly static solution has been found by Horowitz and Myers [78]. The metric in $n+1 \geq 4$ space-time dimensions is

$$ds^2 = -r^2 dt^2 + \frac{1}{V(r)} dr^2 + V(r) d\phi^2 + r^2 \sum_{i=1}^{n-2} (dy^i)^2 \ , \tag{3.26}$$

where $V(r) = \frac{r^2}{\ell^2}\left(1 - \frac{\rho^n}{r^n}\right)$, $\ell^2 = -\frac{n(n-1)}{2\Lambda}$, and ρ is a constant. Regularity demands that ϕ be identified with period $\beta_0 = \frac{4\pi\ell^2}{n\rho}$. The periods of the y^i's are arbitrary. The space-time metric is "asymptotically locally anti-de Sitter" with boundary at conformal infinity foliated by space-like $(n-1)$-tori. The time slices of space-time itself, when conformally completed, have topology $D^2 \times T^{n-2}$, which in dimension 3+1 is a solid torus. Under a convexity condition on the conformal boundary, G. Galloway, S. Surya and E. Woolgar have proved a uniqueness theorem for the corresponding metrics [62].

The above mentioned results concern the strictly static case, and the question of classification of black holes with a Killing vector which is not time-like everywhere is essentially open. In [45] an argument has been presented which shifts the problem of proving uniqueness of a class of such black holes to that of proving a Penrose-type inequality[18] for conformally compactifiable Riemannian manifolds with minimal boundary, and with scalar curvature bounded from below by a negative constant. While there is some hope that some such results could be proved by extending the arguments of [80], this remains to be seen.

---

[18] A version of such an inequality has been originally proposed by Gibbons [69]; the inequality proposed there is not quite correct, see [45].

3.8 Black Holes Without Scri

There has been considerable progress in the numerical analysis of black hole solutions of Einstein's equations; here one of the objectives is to write a stable code which would solve the full four dimensional Einstein equations, with initial data containing a non-connected black hole region that eventually merges into a connected one. One wishes to be able to consider initial data which do not possess any symmetries, and which have various parameters – such as the masses of the individual black holes, their angular momenta, as well as distances between them – which can be varied in significant ranges. Finally one wishes the code to run to a stage where the solution settles to a state close to equilibrium. The challenge then is to calculate the gravitational wave forms for each set of parameters, which could then be used in the gravitational wave observatories to determine the parameters of the collapsing black holes out of the observations made. This program has been being undertaken for years by several groups of researchers, with steady progress being made [1, 2, 12, 16, 19, 70, 71, 86, 93].[19]

There is a fundamental difficulty above, of deciding whether or not one is dealing indeed with the desired black hole initial data: the definition (3.13) of a black hole requires a conformal boundary $\mathscr{I}$ satisfying some – if not all – properties discussed in Sect. 3.4.1. Clearly there is no way of ensuring those requirements in a calculation performed on a finite space-time grid.[20]

In practice what one does is to set up initial data on a finite grid so that the region near the boundary is close to flat (in the conformal approach the whole asymptotically flat region is covered by the numerical grid, and does not need to be near the boundary of the numerical grid; this distinction does not affect the discussion here). Then one evolves the initial data as long as the code allows. The gravitational waves emitted by the system are then extracted out of the metric near the boundary of the grid. Now, our understanding of energy emitted by gravitational radiation is essentially based on an analysis of the metric in an asymptotic region where g is nearly flat. In order to recover useful information out of the numerical data it is thus necessary for the metric near the boundary of the grid to remain close to a flat one. If we want to be sure that the information extracted contains all the essential dynamical information about the system, the metric near the boundary of the grid should quiet down to an almost stationary state as time evolves. Now, it is straightforward to set-up

[19] Some spectacular visualizations of the calculations performed can be found at the URL `http://jean-luc.aei.mpg.de/NCSA1999/GrazingBlackHoles`.

[20] The conformal approach developed by Friedrich (cf., e.g., [56] and references therein, as well as S. Husa's and J. Frauendiener's contributions to this volume) provides an ideal numerical framework for studying gravitational radiation in situations where the extended space-time is smoothly conformally compactifiable across i^+, since then one can hope that the code will be able to "calculate Scri" globally to the future of the initial hyperboloidal hypersurface. It is not clear whether a conformal approach could provide more information than the non-conformal ones when i^+ is itself a singularity of the conformally rescaled equations, as is the case for black holes.

a mathematical framework to describe such situations without having to invoke conformal completions, this is done in the next section.

3.8.1 Naive Black Holes

Consider a *globally hyperbolic* space-time $\mathscr{M}$ which contains a region covered by coordinates (t, x^i) with ranges

$$r := \sqrt{\sum_i (x^i)^2} \geq R_0 , \qquad T_0 - R_0 + r \leq t < \infty , \tag{3.27}$$

such that the metric g satisfies there

$$|g_{\mu\nu} - \eta_{\mu\nu}| \leq C_1 r^{-\alpha} \leq C_2 , \quad \alpha > 0 , \tag{3.28}$$

for some positive constants C_1, C_2, α; clearly C_2 can be chosen to be less than or equal to $C_1 R_0^{-\alpha}$. Making R_0 larger one can thus make C_2 as small as desired, e.g.

$$C_2 = 10^{-2} , \tag{3.29}$$

which is a convenient number in dimension 3 + 1 to guarantee that objects algebraically constructed out of g (such as $g^{\mu\nu}$, $\sqrt{\det g}$) are well controlled; (3.29) is certainly not optimal, and any other number suitable for the purposes at hand would do. To be able to prove theorems about such space-times one would need to impose some further, perhaps not necessarily uniform, decay conditions on a finite number of derivatives of g; there are various possibilities here, but we shall ignore this for the moment. Then one can define the *exterior region* $\mathscr{M}_{\text{ext}}$, the *black hole region* $\mathscr{B}$ and the *future event horizon* $\mathscr{E}$ as

$$\mathscr{M}_{\text{ext}} := \cup_{\tau \geq T_0} J^-(\mathcal{S}_{\tau,R_0}) = J^-(\cup_{\tau \geq T_0} \mathcal{S}_{\tau,R_0}) , \tag{3.30}$$

$$\mathscr{B} := \mathscr{M} \setminus \mathscr{M}_{\text{ext}} , \quad \mathscr{E} := \partial \mathscr{B} , \tag{3.31}$$

where

$$\mathcal{S}_{\tau,R_0} := \{t = \tau, r = R_0\} . \tag{3.32}$$

We will refer to the definition (3.27)-(3.32) as that of a *naive* black hole.

In the setup of (3.27)-(3.32) an arbitrarily chosen R_0 has been used; for this definition to make sense $\mathscr{B}$ so defined should *not* depend upon this choice. This is indeed the case, as can be seen as follows:

Proposition 3.2. *Let $\mathscr{O}_a \subset \mathbb{R}^3 \setminus B(0, R_0)$, $a = 1, 2$, and let $\mathscr{U}_a \subset \mathscr{M}$ be of the form $\{(t \geq T_0 - R_0 + r(\boldsymbol{x}), \boldsymbol{x}) , \boldsymbol{x} \in \mathscr{O}_a\}$ in the coordinate system of (3.28). Then*

$$I^-(\mathscr{U}_1) = J^-(\mathscr{U}_1) = I^-(\mathscr{U}_2) = J^-(\mathscr{U}_2) .$$

Proof. If Γ is a future directed causal path from $p \in \mathscr{M}$ to $q = (t, \boldsymbol{x}) \in \mathscr{U}_1$, then the path obtained by concatenating Γ with the path $[0, 1] \ni s \to (t(s) := t + s, \boldsymbol{x}(s) := \boldsymbol{x})$ is a causal path which is not a null geodesic, hence can be deformed to a time-like path from p to $(t+1, \boldsymbol{x}) \in \mathscr{U}_1$. It follows that $I^-(\mathscr{U}_1) = J^-(\mathscr{U}_1)$; clearly the same holds for $\mathscr{U}_2$. Next, let $\boldsymbol{x}_a \in \mathscr{O}_a$, and let $\gamma : [0,1] \to \mathbb{R}^3 \setminus B(0, R_0)$ be any differentiable path such that $\gamma(0) = \boldsymbol{x}_1$ and $\gamma(1) = \boldsymbol{x}_2$. Then for any $t_0 \geq T_0 - R_0 + r(\boldsymbol{x}_1)$ the causal curve $[0,1] \ni s \to \Gamma(s) = (t := Cs + t_0, \boldsymbol{x}(s) := \gamma(s))$ will be causal for the metric g by (3.28) if the constant C is chosen large enough, with a similar result holding when $\boldsymbol{x}_1$ is interchanged with $\boldsymbol{x}_2$. The equality $I^-(\mathscr{U}_1) = I^-(\mathscr{U}_2)$ easily follows from this observation. □

Summarizing, Prop. 3.2 shows that there are many possible equivalent definitions of $\mathscr{M}_{\mathrm{ext}}$: in (3.30) one can replace $J^-(\mathcal{S}_{\tau,R_0})$ by $J^-(\mathcal{S}_{\tau,R_1})$ for any $R_1 \geq R_0$, but also simply by $J^-((t+\tau, q))$, for any $p = (t,q) \in \mathscr{M}$ which belongs to the region covered by the coordinate system (t, x^i).

The following remarks concerning the definition (3.30)-(3.31) are in order:

- For vacuum, stationary, asymptotically flat space-times the definition is equivalent to the usual one with $\mathscr{I}$ [46, Footnote 7, p. 572]; here the results of [49, 50] are used. However, one does not expect the existence of a smooth $\mathscr{I}^+$ to follow from (3.27)-(3.28) in general.
- Suppose that $\mathscr{M}$ admits a conformal completion in the sense defined at the beginning of Sect. 3.4, and that $\mathscr{I}$ is *semi-complete to the future*, in the sense that the Geroch-Horowitz condition of the beginning of Sect. 3.4.1 holds with $\mathscr{I} \approx \mathbb{R} \times S^2$ there replaced by $\mathscr{I} \supset \mathbb{R}^+ \times S^2$. Then for any finite interval $[T_0, T_1]$ there exists $R_0(T_0, T_1)$ and a coordinate system satisfying (3.28) and covering a set $r \geq R_0(T_0,T_1)$, $T_0 - R_0 \leq t \leq T_1 - R_0$. This follows from the Tamburino-Winicour construction of Bondi coordinates (u, r, θ, φ) near $\mathscr{I}^+$ [124], followed by the introduction of the usual Minkowskian coordinates $t = u + r$, $x = r \sin\theta \cos\varphi$, etc. The problem is that $R(T_1, T_2)$ could shrink to zero as T_2 goes to infinity. Thus, when $\mathscr{I}^+$ exists, conditions (3.27)-(3.28) are *uniformity* conditions on $\mathscr{I}^+$ to the future – the metric remains uniformly controlled on a uniform neighbourhood of $\mathscr{I}^+$ as the retarded time goes to infinity.
- It should not be too difficult to check whether or not the future geodesically complete space-times of Friedrich [53, 58] and of Christodoulou and Klainerman [31], or the Robinson-Trautman black-holes [32], admit coordinate systems satisfying (3.27)-(3.28).

It is not clear if asymptotically flat space-times in which no such control is available do exist at all, in fact, it is tempting to formulate the following version of the *(weak) cosmic censorship* conjecture:

> The maximal globally hyperbolic development of generic[21], asymptotically flat, vacuum initial data contains a region with coordinates satisfying (3.27)-(3.28).

[21] The examples constructed by Christodoulou [29] with spherically symmetric gravitating scalar fields suggest that the genericity condition is unavoidable, though no corresponding vacuum examples are known.

Whatever the status of this conjecture, one can hardly envisage numerical simulations leading to the calculation of an essential fraction of the total energy radiated away in space-times in which some uniformity conditions do not hold.

3.8.2 Quasi-local Black Holes

As already argued, the naive approach of the previous section should be more convenient for numerical simulations of black hole space-times, as compared to the usual one based on Scri. It appears to be even more convenient to have a framework in which all the issues are localized in space; we wish to suggest such a framework here. When numerically modeling an asymptotically flat space-time, whether in a conformal or a direct approach, a typical numerical grid will contain large spheres $S(R)$ on which the metric is nearly flat, so that an inequality such as (3.28)-(3.29) will hold in a neighbourhood of $S(R)$. On slices $t = \text{const}$ the condition (3.28) is usually complemented with a fall-off condition on the derivatives of the metric

$$|\partial_\sigma g_{\mu\nu}| \leq C r^{-\alpha-1} , \tag{3.33}$$

however condition (3.33) is inadequate in the radiation regime, where retarded time derivatives of the metric are not expected to fall off faster than r^{-1}. It turns out that there is a condition on derivatives of the metric in null directions which is fulfilled at large distance both in space-like and in null directions: let K_a, $a = 1, 2$, be null future pointing vector fields on $S(R)$ orthogonal to $S(R)$, with K_1 inwards pointing and K_2 outwards pointing; these vector fields are unique up to scaling. Let θ_a denote the associated null second fundamental forms defined as

$$\forall\, X, Y \in TS(R) \quad \chi_a(X, Y) := g(\nabla_X K_a, Y) . \tag{3.34}$$

It can be checked, e.g. using the asymptotic expansions for the connection coefficients near $\mathscr{I}^+$ from [43, Appendix C], that χ_1 is negative definite and χ_2 is positive definite for Bondi spheres $S(R)$ sufficiently close to $\mathscr{I}^+$; similarly for $\mathscr{I}^-$. This property is not affected by the rescaling freedom at hand. Following G. Galloway [60], a two-dimensional space-like submanifold of a four-dimensional space-time will be called *weakly null convex* if χ_1 is semi-positive definite, with the trace of χ_2 negative.[22] The null convexity condition is easily verified for sufficiently large spheres in a region asymptotically flat in the sense of (3.33). It does also hold for large spheres in a large class of space-times with negative cosmological constant. The null convexity condition is then the condition which we propose as a starting point to defining "quasi-local" black holes and horizons. The point is that several of the usual properties of black holes carry over to the weakly null convex setting. In retrospect, the situation can be summarized as follows: the usual theory of Scri based black holes exploits the existence of conjugate points on appropriate null geodesics whenever those are complete to the

[22] Galloway defines *null convexity* through the requirement of positive definiteness of χ_1 and negative definiteness of χ_2. However, he points out himself [60, p. 1472] that the weak null convexity as defined above suffices for his arguments to go through.

future; this completeness is guaranteed by the fact that the conformal factor goes to zero at the conformal boundary at an appropriate rate. Galloway's discovery in [60] is that weak null convexity of large spheres near Scri provides a second, in principle completely independent, mechanism to produce the needed focusing behaviour.

Throughout this section we will consider a globally hyperbolic space-time $(\mathscr{M}, g)$ with time function t. Let $\mathscr{T} \subset \mathscr{M}$ be a finite union of connected time-like hypersurfaces $\mathscr{T}_\alpha$ in $\mathscr{M}$, we set

$$\mathscr{S}_\tau := \{t = \tau\} \;, \quad \mathscr{T}(\tau) := \mathscr{T} \cap \mathscr{S}_\tau \;, \quad \mathscr{T}_\alpha(\tau) := \mathscr{T}_\alpha \cap \mathscr{S}_\tau \;. \tag{3.35}$$

For further purposes anything that happens on the exterior side of $\mathscr{T}$ is completely irrelevant, so it is convenient to think of $\mathscr{T}$ as a boundary of $\mathscr{M}$; global hyperbolicity should then be understood in the sense that $(\overline{\mathscr{M}} := \mathscr{M} \cup \mathscr{T}, g)$ is strongly causal, and that $J^+(p; \overline{\mathscr{M}}) \cap J^-(q; \overline{\mathscr{M}})$ is compact in $\overline{\mathscr{M}}$ for all $p, q \in \overline{\mathscr{M}}$. Recall that the null convergence condition is the requirement that

$$\mathrm{Ric}(X, X) \geq 0 \quad \text{for all } X \in TM \;. \tag{3.36}$$

We have the following *topological censorship* theorem for weakly null convex time-like boundaries:

Theorem 3.3 (Galloway [60]). *Suppose that a globally hyperbolic space-time $(\overline{\mathscr{M}}, g)$ satisfying the null convergence condition (3.36) has a time-like boundary $\mathscr{T} = \cup_{\alpha=1}^{I} \mathscr{T}_\alpha$ and a time function t such that the level sets of t are Cauchy surfaces, with each section $\mathscr{T}(\tau)$ of $\mathscr{T}$ being null convex. Then distinct $\mathscr{T}_\alpha$'s cannot communicate with each other:*

$$\alpha \neq \beta, \qquad J^+(\mathscr{T}_\alpha) \cap J^-(\mathscr{T}_\beta) = \emptyset \;.$$

As is well known, topological censorship implies constraints on the topology:

Theorem 3.4 (Galloway [60]). *Under the hypotheses of Theorem 3.3 suppose further that the cross-sections $\mathscr{T}_\alpha(\tau)$ of $\mathscr{T}_\alpha$ have spherical topology.*[23] *Then the* α-domain of outer communication

$$\langle\langle \mathscr{T}_\alpha \rangle\rangle := J^+(\mathscr{T}_\alpha) \cap J^-(\mathscr{T}_\alpha) \tag{3.37}$$

is simply connected.

It follows in particular from Theorem 3.4 that $\overline{\mathscr{M}}$ can be replaced by a subset thereof such that $\mathscr{T}$ is connected in the new space-time, with all essential properties relevant for the discussion in the remainder of this section being unaffected by that replacement. We shall not do that, to avoid a lengthy discussion of which properties are relevant and which are not, but the reader should keep

[23] The reader is referred to [61] and references therein for results without the hypothesis of spherical topology. The results there, presented in a Scri setting, generalize immediately to the weakly null convex one.

in mind that the hypothesis of connectedness of $\mathscr{T}$ can indeed be done without any loss of generality for most purposes.

We define the *quasi-local black hole region* $\mathscr{B}_{\mathscr{T}_\alpha}$ and the *quasi-local event horizon* $\mathscr{E}_{\mathscr{T}_\alpha}$ associated with the hypersurface $\mathscr{T}_\alpha$ by

$$\mathscr{B}_{\mathscr{T}_\alpha} := \mathscr{M} \setminus J^-(\mathscr{T}_\alpha) \,, \qquad \mathscr{E}_{\mathscr{T}_\alpha} := \partial \mathscr{B}_{\mathscr{T}_\alpha} \,. \tag{3.38}$$

If $\mathscr{T}$ is the hypersurface $\cup_{\tau \ge T_0} \mathcal{S}_{\tau, R_0}$ of Sect. 3.8.1 then the resulting black hole region coincides with that defined in (3.31), hence does not depend upon the choice of R_0 by Proposition 3.2; however, $\mathscr{B}_{\mathscr{T}_\alpha}$ might depend upon the chosen family of observers $\mathscr{T}_\alpha$ in general. It is certainly necessary to impose some further conditions on $\mathscr{T}$ to reduce this dependence. A possible condition, suggested by the geometry of the large coordinate spheres considered in the previous section, could be that the light-cones of the induced metric on $\mathscr{T}$ are uniformly controlled both from outside and inside by those of two static, future complete reference metrics on $\mathscr{T}$. However, neither the results above, nor the results that follow, do require that condition.

The Scri-equivalents of Theorem 3.4 [22, 46, 59, 61, 63, 84] allow one to control the topology of "good" sections of the horizon, and for the standard stationary black-holes this does lead to the usual $S^2 \times \mathbb{R}$ topology of the horizon [46, 73]. In particular, in stationary, asymptotically flat, appropriately regular space-times the intersection of a partial Cauchy hypersurface with an event horizon will necessary be a finite union of spheres. In general space-times such intersections do not even need to be manifolds: for example, in the usual spherically symmetric collapsing star the intersection of the event horizon with level sets of a time function will be a point at the time of appearance of the event horizon. We refer the reader to [38, Section 3] for other such examples, including one in which the topology of sections of horizon changes type from toroidal to spherical as time evolves. This behaviour can be traced back to the existence of past end points of the generators of the horizon. Nevertheless, some sections of the horizon have controlled topology – for instance, we have the following:

Theorem 3.5. *Under the hypotheses of Theorem 3.3, consider a connected component* $\mathscr{T}_\alpha$ *of* $\mathscr{T}$ *such that* $\mathscr{E}_{\mathscr{T}_\alpha} \neq \emptyset$. *Let*

$$\mathscr{C}_\alpha(\tau) := \partial J^+(\mathscr{T}_\alpha(\tau)) \,.$$

If $\mathscr{C}_\alpha(\tau) \cap \mathscr{E}_{\mathscr{T}_\alpha}$ *is a topological manifold, then each connected component thereof has spherical topology.*

Proof. Consider the open subset $\mathscr{M}_\tau$ of $\overline{\mathscr{M}}$ defined as

$$\mathscr{M}_\tau := I^+(\mathscr{C}_\alpha(\tau); \overline{\mathscr{M}}) \cap I^-(\mathscr{T}_\alpha; \overline{\mathscr{M}}) \subset \langle\langle \mathscr{T}_\alpha \rangle\rangle \,.$$

We claim that $(\mathscr{M}_\tau, g|_{\mathscr{M}_\tau})$ is globally hyperbolic: indeed, let $p, q \in \mathscr{M}_\tau$; global hyperbolicity of $\overline{\mathscr{M}}$ shows that $J^-(p; \overline{\mathscr{M}}) \cap J^+(q; \overline{\mathscr{M}})$ is a compact subset of $\overline{\mathscr{M}}$, which is easily seen to be included in $\mathscr{M}_\tau$. It follows that $J^-(p; \mathscr{M}_\tau) \cap J^+(q; \mathscr{M}_\tau)$ is compact, as desired. By the usual decomposition we thus have

$$\mathscr{M}_\tau \approx \mathbb{R} \times \mathcal{S} \,,$$

where $\mathcal{S}$ is a Cauchy hypersurface for $\mathscr{M}_\tau$. Applying Theorem 3.4 to the globally hyperbolic space-time $\mathscr{M}_\tau$ (which has a weakly null convex boundary $\mathscr{T}_\alpha \cap \{t > \tau\}$) one finds that $\mathscr{M}_\tau$ is simply connected, and thus so is $\mathcal{S}$. Since $\mathscr{C}_\alpha(\tau)$ and $\mathscr{E}_\alpha$ are null hypersurfaces in $\mathscr{M}$, it is easily seen that the closure in $\overline{\mathscr{M}}$ of the Cauchy surface $\{0\} \times \mathcal{S}$ intersects $\mathscr{E}_\alpha$ precisely at $\mathscr{C}_\alpha(\tau) \cap \mathscr{E}_{\mathscr{T}_\alpha}$. It follows that $\mathcal{S}$ is a compact, simply connected, three dimensional topological manifold with boundary, and a classical result [74, Lemma 4.9] shows that each connected component of $\partial\mathcal{S}$ is a sphere. The result follows now from $\partial\mathcal{S} \approx \mathscr{C}_\alpha(\tau) \cap \mathscr{E}_{\mathscr{T}_\alpha}$. □

Yet another class of "good sections" of $\mathscr{E}_{\mathscr{T}}$ can be characterized[24] as follows: suppose that $\langle\langle \mathscr{T}_\alpha \rangle\rangle \cap \mathscr{S}_\tau$ is a submanifold with boundary of $\mathscr{M}$ which is, moreover, a retract of $\langle\langle \mathscr{T}_\alpha \rangle\rangle$. Then $\langle\langle \mathscr{T}_\alpha \rangle\rangle \cap \mathscr{S}_\tau$ is simply connected by Theorem 3.4, and spherical topology of all boundary components of $\langle\langle \mathscr{T}_\alpha \rangle\rangle \cap \mathscr{S}_\tau$ follows again from [74, Lemma 4.9]. It is not clear whether there always exist time functions t such that the retract condition is satisfied; similarly it is not clear that there always exist τ's for which the conditions of Theorem 3.5 are met for metrics which are not stationary (one would actually want "a lot of τ's"). It would be of interest to understand this better.

We have an area theorem for $\mathscr{E}_{\mathscr{T}}$:

Theorem 3.6. *Under the hypotheses of Theorem 3.3, suppose further that $\mathscr{E}_{\mathscr{T}} \neq \emptyset$. Let $\mathscr{S}_a$, $a = 1, 2$ be two achronal space-like embedded hypersurfaces of C^2 differentiability class, set $S_a = \mathscr{S}_a \cap \mathscr{E}_{\mathscr{T}}$. Then:*

1. *The area of S_a is well defined.*
2. *If*
$$S_1 \subset J^-(S_2) \ ,$$
then the area of S_2 is larger than or equal to that of S_1. (Moreover, this is true even if the area of S_1 is counted with multiplicity[25] of generators provided that $S_1 \cap S_2 = \emptyset$.)

We note that point 1 is less trivial as it appears, because horizons can be rather rough sets, and it requires a certain amount of work to establish that claim.

Proof. The result is obtained by a mixture of methods of [38] and of [60], and proceeds by contradiction: assume that the Alexandrov divergence $\theta_{\mathcal{A}l}$ of $\mathscr{E}_{\mathscr{T}}$ is negative, and consider the $S_{\epsilon,\eta,\delta}$ deformation of the horizon as constructed in Proposition 4.1 of [38], with parameters chosen so that $\theta_{\epsilon,\eta,\delta} < 0$. Global hyperbolicity implies the existence of an achronal null geodesic from $S_{\epsilon,\eta,\delta}$ to some cut $\mathscr{T}(\tau)$ of $\mathscr{T}$. The geodesic can further be chosen to be "extremal", in the sense that it meets $\mathscr{T}(t)$ for the smallest possible value of t among all generators of the boundary of $J^+(S_{\epsilon,\eta,\delta})$ meeting $\mathscr{T}$. The argument of the proof of Theorem 1

[24] I am grateful to G. Galloway for useful discussions concerning this question, as well as many other points presented in this section.

[25] See [38] for details.

of [60] shows that this is incompatible with the null energy condition and with weak null convexity of $\mathscr{T}(\tau)$. It follows that $\theta_{Al} \geq 0$, and the result follows from [38, Proposition 3.3 and Theorem 6.1]. □

It immediately follows from the proof above that, under the hypotheses of Theorem 3.3, the occurrence of twice differentiable future trapped (compact) surfaces implies the presence of a black hole region. The same result holds for semi-convex compact surfaces which are trapped in an Alexandrov sense, that is, (3.21) holds with λ there defined in a way which should be clear from the discussion following Theorem 3.1. It is, however, not known if the existence of marginally trapped surfaces – whether defined in a classical, or Alexandrov, or a viscosity sense – does signal the occurrence of black hole; it would be of interest to settle that.

In summary, we have shown that the quasi-local black holes, defined using weakly null convex time-like hypersurfaces, or boundaries, possess several properties usually associated with the Scri-based black holes, without the associated problems. We believe they provide a reasonable alternative, well suited for numerical calculations. In the next section we address the question of numerically finding the resulting horizons.

3.8.3 Finding Horizons

Consider the large coordinate spheres $\mathcal{S}_{\tau,R_1}$ of Sect. 3.8.1; under reasonable restrictions on $\mathscr{M}$ one would expect that $\partial J^-(\mathcal{S}_{\tau,R_1})$ converges as $\tau \to \infty$ – *e.g.* in Hausdorff topology – to the event horizon $\mathscr{E}$ of (3.31). Similarly, the achronal boundaries $\partial J^-(\mathscr{T}(\tau))$ should sometimes converge to the event horizon $\mathscr{E}_{\mathscr{T}}$. Whenever that is the case, the $\partial J^-(\mathcal{S}_{\tau,R_1})$'s or the $\partial J^-(\mathscr{T}(\tau))$'s can be taken as good approximations of the event horizon when τ is chosen to be large enough. This is of practical significance, as it provides a way of approximately locating the horizon in numerical simulations. Let us establish one such approximation result, for the $\partial J^-(\mathscr{T}(\tau))$'s; clearly the corresponding result for the $J^-(\mathcal{S}_{\tau,R_1})$'s follows by specialization. We assume that $\overline{\mathscr{M}}$ is spatially compact and globally hyperbolic, which seem rather natural assumptions in the context of numerical simulations of space-times:

Theorem 3.7. *Consider a globally hyperbolic space-time $(\overline{\mathscr{M}}, g)$ with compact Cauchy hypersurfaces $\mathscr{S}_\tau$ and time-like boundary $\mathscr{T}$, and suppose that*

$$\mathscr{B}_{\mathscr{T}} \neq \emptyset \,. \tag{3.39}$$

Then for any τ we have

$$\partial J^-(\mathscr{T}(\sigma)) \cap \mathscr{S}_\tau \to_{\sigma\to\infty} \mathscr{E}_{\mathscr{T}} \cap \mathscr{S}_\tau \,,$$

where the convergence is in Hausdorff distance.

Proof. Under the usual identification of $\mathscr{M}$ with $\mathbb{R} \times \mathscr{S}_0$, let f be the graphing function of $\partial J^-(\mathscr{T}(\sigma))$ over the projection $\mathrm{pr}_2\mathscr{E}$ of $\mathscr{E}$ on $\mathscr{S}_0$; $\mathrm{pr}_2\mathscr{E}$ is an open subset of $\mathscr{S}_0$ by the invariance-of-the-domain theorem. If $p = (t, \boldsymbol{x})$ is such that $t < f(\boldsymbol{x})$, then there exists a causal curve from p to $\mathscr{T}$, hence $p \in J^-(\mathscr{T}(\sigma))$ for some σ. It follows that

$$\mathrm{pr}_2 \mathscr{E} = \cup_\sigma \mathrm{pr}_2 J^-(\mathscr{T}(\sigma)) \, .$$

Let f_σ be the graphing function of $\partial J^-(\mathscr{T}(\sigma))$ over $\mathrm{pr}_2 J^-(\mathscr{T}(\sigma)) \subset \mathscr{S}_0$; clearly

$$\mathrm{pr}_2 J^-(\mathscr{T}(\sigma)) \subset \mathrm{pr}_2 J^-(\mathscr{T}(\sigma')) \qquad \sigma < \sigma' \, ,$$

which shows that f and all the $f_{\sigma'}$'s are defined on $\mathrm{pr}_2 J^-(\mathscr{T}(\sigma))$ for $\sigma' > \sigma$. The f_τ's are uniformly Lipschitz and bounded over any compact subset $K \subset \mathrm{pr}_2 \mathscr{E}$ of $\mathscr{S}_0$, and hence for any sequence τ_i there exists a subsequence such that the $f_{\tau_{i_j}}$'s converge to a Lipschitz function h. Since $f_\tau \le f$ for all τ we have $h \le f|_K$. Suppose that there exists $\boldsymbol{x}$ such that $h(\boldsymbol{x}) < f(\boldsymbol{x})$, then $p = ((h(\boldsymbol{x}) + f(\boldsymbol{x}))/2, \boldsymbol{x})$ is in $J^-(\mathscr{T})$, hence in $J^-(\mathscr{T}(\sigma))$ for some σ, so that $(h(\boldsymbol{x}) + f(\boldsymbol{x}))/2 \le f_\sigma(\boldsymbol{x})$ which is not possible; we thus have $h = f|_K$. It follows that the f_τ's converge pointwise to f. The f_τ's are monotonously increasing as τ increases, and Dini's theorem implies that the f_τ's converge uniformly to f over any compact set. The result follows now by elementary considerations using the fact that $\mathscr{E}_{\mathscr{I}} \cap \mathscr{S}_\tau$ is compact, being the intersection of a closed set with a compact one. □

Theorem 3.7 gives partial justification for the numerical analysis of Anninos [7], Libson et al. [98], or Massó et al. [102],[26] where the event horizon $\mathscr{E}$ is taken to be

$$\mathscr{E} \approx \partial J^-(\mathcal{S}_{\tau,R_1}) \, , \quad \tau \text{ as big as the computer simulation allows} \, .$$

In those works $\partial J^-(\mathcal{S}_{\tau,R_1})$ is further numerically approximated as the solution of an eikonal equation, which leads to difficulties at past end points of the generators of $\partial J^-(\mathcal{S}_{\tau,R_1})$. We conjecture that a more stable method of locating objects such as $\partial J^-(\mathcal{S}_{\tau,R_1})$ is provided by the following straightforward modification of the Fermat-type variational principle presented in the Appendix below, see (3.43): let $\mathcal{P}(\boldsymbol{x})(x)$ be defined as in the paragraph preceding (3.43), with $\mathscr{H}_\sigma$ there replaced by $\mathcal{S}_{\tau,R_1}$, and let $\hat\gamma$ be the null lifts of paths γ as defined in the paragraph preceding (3.45). An argument similar to that of the proof of Corollary A.1 shows that in globally hyperbolic space-times the set $\partial J^-(\mathcal{S}_{\tau,R_1})$ is a graph of a function f such that

$$f(\boldsymbol{x}) = \sup_{\hat\gamma \in \mathcal{P}(\boldsymbol{x})(x)} t(\hat\gamma(a)) \, . \tag{3.40}$$

Such a maximization over null lifts $\hat\gamma$ of paths γ which have image in a fixed space-like hypersurface should be easy to perform numerically by, e.g., Newton's method. The variational principle above automatically takes care of the past end-points of the horizon.

---

[26] We note that the claim in [7, 98], that the horizon acts as an attractor for backwards-directed null geodesics, does not seem to be justified. In fact, such a statement is coordinate-dependent: it is easy to introduce coordinate systems in the Schwarzschild space-time where backwards-directed null geodesics will be actually repelled by the horizon.

Acknowledgments

The author is grateful to Y. Choquet-Bruhat, G. Horowitz, S. Husa, R. Narayan, R. Parentani, D. Polarski and H. Reeves for bibliographical advice. Thanks are due to H. Friedrich and G. Galloway for suggesting many improvements to a previous version of this manuscript, and to R. Geroch and J. Jezierski for useful comments and discussions.

Appendix A: Future Horizons Are Semi-convex

A hypersurface $\mathscr{H} \subset M$ will be said to be *future null geodesically ruled* if every point $p \in \mathscr{H}$ belongs to a future inextensible null geodesic $\Gamma \subset \mathscr{H}$; those geodesics will be called *the generators* of $\mathscr{H}$. We emphasize that the generators are allowed to have past endpoints on $\mathscr{H}$, but no future endpoints. *Past null geodesically ruled* hypersurfaces are defined by changing the time orientation. Examples of future geodesically ruled hypersurfaces include past Cauchy horizons $\mathcal{D}^-(\Sigma)$ of achronal sets Σ [115, Theorem 5.12] and black hole event horizons $\partial J^-(\mathscr{I}^+)$ [73, p. 312].

Let $\dim M = n+1$ and suppose that $\mathcal{O}$ is a domain in $\mathbb{R}^n$. Recall that a continuous function $f : \mathcal{O} \to \mathbb{R}$ is called semi-convex if there exists a C^2 function $\phi : \mathcal{O} \to \mathbb{R}$ such that $f + \phi$ is convex. We shall say that the graph of f is a semi-convex hypersurface if f is semi-convex. A hypersurface $\mathscr{H}$ in a manifold M will be said semi-convex if $\mathscr{H}$ can be covered by coordinate patches $\mathcal{U}_\alpha$ such that $\mathscr{H} \cap \mathcal{U}_\alpha$ is a semi-convex graph for each α.

Consider an achronal hypersurface $\mathscr{H} \neq \emptyset$ in a globally hyperbolic space-time (M,g). Let t be a time function on M which induces a diffeomorphism of M with $\mathbb{R} \times \Sigma$ in the standard way [64, 122], with the level sets $\Sigma_\tau \equiv \{p | t(p) = \tau\}$ of t being Cauchy surfaces. As usual we identify Σ_0 with Σ, and in the identification above the curves $\mathbb{R} \times \{q\}$, $q \in \Sigma$, are integral curves of ∇t. Define

$$\Sigma_{\mathscr{H}} = \{q \in \Sigma | \mathbb{R} \times \{q\} \text{ intersects } \mathscr{H}\} . \tag{3.41}$$

For $q \in \Sigma_{\mathscr{H}}$ the set $(I \times q) \cap H$ is a point by achronality of $\mathscr{H}$, we shall denote this point by $(f(q), q)$. Thus an achronal hypersurface $\mathscr{H}$ in a globally hyperbolic space-time is a graph over $\Sigma_{\mathscr{H}}$ of a function f. The invariance-of-the-domain theorem shows that $\Sigma_{\mathscr{H}}$ is an open subset of Σ. We have the following:

Theorem A.1. *Let $\mathscr{H} \neq \emptyset$ be an achronal future null geodesically ruled hypersurface in a globally hyperbolic space-time $(M = \mathbb{R} \times \Sigma, g)$. Then $\mathscr{H}$ is the graph of a semi-convex function f defined on an open subset $\Sigma_{\mathscr{H}}$ of Σ, in particular $\mathscr{H}$ is semi-convex.*

Proof. As discussed above, $\mathscr{H}$ is the graph of a function f. The idea of the proof is to show that f satisfies a variational principle, the semi-concavity of f follows then by a standard argument. Let $p \in \mathscr{H}$ and let $\mathcal{O}$ be a coordinate patch in a neighborhood of p such that $x^0 = t$, with $\mathcal{O}$ of the form $I \times B(3R)$, where $B(R)$ denotes a coordinate ball centered at 0 of radius R in $\mathbb{R}^3$, with $p = (t(p), 0)$.

Here I is the range of the coordinate x^0, we require it to be a bounded interval the size of which will be determined later on. We further assume that the curves $I \times \{\boldsymbol{x}\}$, $\boldsymbol{x} \in B(3R)$, are integral curves of ∇t. Define

$$\mathcal{U}_0 = \{\boldsymbol{x} \in B(3R)|\text{ the causal path } I \ni t \to (t, \boldsymbol{x}) \text{ intersects } \mathscr{H}\} .$$

We note that $\mathcal{U}_0$ is non-empty, since $0 \in \mathcal{U}_0$. Set

$$\mathscr{H}_\sigma = \mathscr{H} \cap \Sigma_\sigma , \tag{3.42}$$

and choose σ large enough so that $\mathcal{O} \subset I^-(\Sigma_\sigma)$. Now p lies on a future inextensible generator Γ of $\mathscr{H}$, and global hyperbolicity of (M, g) implies that $\Gamma \cap \Sigma_\sigma$ is nonempty, hence $\mathscr{H}_\sigma$ is nonempty.

For $\boldsymbol{x} \in B(3R)$ let $\mathcal{P}(\boldsymbol{x})(x)$ denote the collection of piecewise differentiable future directed null curves $\Gamma : [a, b] \to M$ with $\Gamma(a) \in \mathbb{R} \times \{\boldsymbol{x}\}$ and $\Gamma(b) \in \mathscr{H}_\sigma$. We define

$$\tau(\boldsymbol{x}) = \sup_{\Gamma \in \mathcal{P}(\boldsymbol{x})(x)} t(\Gamma(a)) . \tag{3.43}$$

We emphasize that we allow the domain of definition $[a, b]$ to depend upon Γ, and that the "a" occurring in $t(\Gamma(a))$ in (3.43) is the lower bound for the domain of definition of the curve Γ under consideration. We have the following result (compare [8, 68, 116])·

Proposition A.1 (Fermat principle). *For $\boldsymbol{x} \in \mathcal{U}_0$ we have*

$$\tau(\boldsymbol{x}) = f(\boldsymbol{x}) .$$

Proof. Let Γ be any generator of $\mathscr{H}$ such that $\Gamma(0) = (f(\boldsymbol{x}), \boldsymbol{x})$, clearly $\Gamma \in \mathcal{P}(\boldsymbol{x})(x)$ so that $\tau(\boldsymbol{x}) \geq f(\boldsymbol{x})$. To show that this inequality has to be an equality, suppose for contradiction that $\tau(\boldsymbol{x}) > f(\boldsymbol{x})$, thus there exists a null future directed curve Γ such that $t(\Gamma(0)) > f(\boldsymbol{x})$ and $\Gamma(1) \in \mathscr{H}_\sigma \subset \mathscr{H}$. Then the curve $\tilde{\Gamma}$ obtained by following $\mathbb{R} \times \{\boldsymbol{x}\}$ from $(f(\boldsymbol{x}), \boldsymbol{x})$ to $(t(\Gamma(0)), \boldsymbol{x})$ and following Γ from there on is a causal curve with endpoints on $\mathscr{H}$ which is not a null geodesic. By [73, Proposition 4.5.10] $\tilde{\Gamma}$ can be deformed to a time-like curve with the same endpoints, which is impossible by achronality of $\mathscr{H}$. □

The Fermat principle, Proposition A.1, shows that f is a solution of the variational principle (3.43). Now this variational principle can be rewritten in a somewhat more convenient form as follows: The identification of M with $\mathbb{R} \times \Sigma$ by flowing from $\Sigma_0 \equiv \Sigma$ along the gradient of t leads to a global decomposition of the metric of the form

$$g = \alpha(-dt^2 + h_t) ,$$

where h_t denotes a t–dependent family of Riemannian metrics on Σ. Any future directed differentiable null curve $\Gamma(s) = (t(s), \gamma(s))$ satisfies

$$\frac{dt(s)}{ds} = \sqrt{h_{t(s)}(\dot{\gamma}, \dot{\gamma})} ,$$

where $\dot\gamma$ is a shorthand for $d\gamma(s)/ds$. It follows that for any $\Gamma \in \mathcal{P}(\boldsymbol{x})(x)$ it holds that

$$t(\Gamma(a)) = t(\Gamma(b)) - \int_a^b \frac{dt}{ds} ds$$
$$= \sigma - \int_a^b \sqrt{h_{t(s)}(\dot\gamma,\dot\gamma)} ds \ .$$

This allows us to rewrite (3.43) as

$$\tau(\boldsymbol{x}) = \sigma - \mu(\boldsymbol{x}) \ , \qquad \mu(\boldsymbol{x}) \equiv \inf_{\Gamma\in\mathcal{P}(\boldsymbol{x})(x)} \int_a^b \sqrt{h_{t(s)}(\dot\gamma,\dot\gamma)} ds \ . \tag{3.44}$$

We note that in static space-times $\mu(\boldsymbol{x})$ is the Riemannian distance from $\boldsymbol{x}$ to $\mathscr{H}_\sigma$. In particular Equation (3.44) implies the well known fact, that in globally hyperbolic static space-times Cauchy horizons of open subsets of level sets of t are graphs of the distance function from the boundary of those sets.

Let $\gamma : [a,b] \to \Sigma$ be a piecewise differentiable path, for any $p \in \mathbb{R}\times\{\gamma(b)\}$ we can find a null future directed curve $\hat\gamma : [a,b] \to M$ of the form $\hat\gamma(s) = (\phi(s),\gamma(s))$ with future end point p by solving the problem

$$\begin{cases} \phi(b) = t(p) \ , \\ \dfrac{d\phi(s)}{ds} = \sqrt{h_{\phi(s)}(\dot\gamma(s),\dot\gamma(s))} \ . \end{cases} \tag{3.45}$$

The path $\hat\gamma$ will be called the *null lift of γ with endpoint p.*

As an example of application of Proposition A.1 we recover the following well known result [115]:

Corollary A.1. *f is Lipschitz continuous on any compact subset of its domain of definition.*

Proof. For $\boldsymbol{y},\boldsymbol{z} \in B(2R)$ let $K \subset \mathbb{R}\times B(2R)$ be a compact set which contains all the null lifts $\Gamma_{\boldsymbol{y},\boldsymbol{z}}$ of the coordinate segments $[\boldsymbol{y},\boldsymbol{z}] := \{\lambda\boldsymbol{y}+(1-\lambda)\boldsymbol{z} \ , \ \lambda\in[0,1]\}$ with endpoints $(\tau(\boldsymbol{z}),\boldsymbol{z})$. Define

$$\hat C = \sup\{\sqrt{h_p(n,n)} | p \in K, |n|_\delta = 1\} \ , \tag{3.46}$$

where the supremum is taken over all points $p \in K$ and over all vectors $n \in T_pM$ the coordinate components n^i of which have Euclidean length $|n|_\delta$ equal to one. Choose I to be a bounded interval large enough so that $K \subset I \times B(2R)$ and, as before, choose σ large enough so that $I\times B(2R)$ lies to the past of Σ_σ. Let $\boldsymbol{y},\boldsymbol{z} \in B(2R)$ and consider the causal curve $\Gamma = (t(s),\gamma(s))$ obtained by following the null lift $\Gamma_{\boldsymbol{y},\boldsymbol{z}}$ in the parameter interval $s \in [0,1]$, and then a generator of $\mathscr{H}$ from $(\tau(\boldsymbol{z}),\boldsymbol{z})$ until $\mathscr{H}_\sigma$ in the parameter interval $s\in[1,2]$. Then we have

$$\mu(\boldsymbol{z}) = \int_1^2 \sqrt{h_{t(s)}(\dot\gamma,\dot\Gamma)} ds \ .$$

Further $\Gamma \in \mathcal{P}(\boldsymbol{x})(x)$ so that

$$\begin{aligned}\mu(\boldsymbol{y}) &\leq \int_0^2 \sqrt{h_{t(s)}(\dot{\gamma},\dot{\gamma})}ds \\ &= \int_0^1 \sqrt{h_{t(s)}(\dot{\gamma},\dot{\gamma})}ds + \int_1^2 \sqrt{h_{t(s)}(\dot{\gamma},\dot{\gamma})}ds \\ &\leq \hat{C}|\boldsymbol{y}-\boldsymbol{z}|_\delta + \mu(\boldsymbol{z}) \ , \end{aligned} \tag{3.47}$$

where $|\cdot|_\delta$ denotes the Euclidean norm of a vector, and with $\hat{C}$ defined in (3.46). Setting first $\boldsymbol{y} = \boldsymbol{x}$, $\boldsymbol{z} = \boldsymbol{x}+\boldsymbol{h}$ in (3.47) and then $\boldsymbol{z} = \boldsymbol{x}$, $\boldsymbol{y} = \boldsymbol{x}+\boldsymbol{h}$, the Lipschitz continuity of f on $B(2R)$ follows. The general result is obtained now by a standard covering argument. □

Returning to the proof of Theorem A.1, for $\boldsymbol{x} \in B(R)$ let $\Gamma_{\boldsymbol{x}}$ be a generator of $\mathscr{H}$ such that $\Gamma_{\boldsymbol{x}}(0) = (\tau(\boldsymbol{x}),\boldsymbol{x})$, and, if we write $\Gamma_{\boldsymbol{x}}(s) = (\phi_{\boldsymbol{x}}(s),\gamma_{\boldsymbol{x}}(s))$, then we require that $\gamma_{\boldsymbol{x}}(s) \in B(2R)$ for $s \in [0,1]$. For $s \in [0,1]$ and $\boldsymbol{h} \in B(R)$ let $\gamma_{\boldsymbol{x},\pm}(s) \in \Sigma$ be defined by

$$\gamma_{\boldsymbol{x},\pm}(s) = \gamma_{\boldsymbol{x}}(s) \pm (1-s)\boldsymbol{h} = s\gamma_{\boldsymbol{x}}(s) + (1-s)(\gamma_{\boldsymbol{x}}(s)\pm\boldsymbol{h}) \in B(2R) \ .$$

We note that

$$\gamma_{\boldsymbol{x},\pm}(0) = \boldsymbol{x}\pm\boldsymbol{h} \ , \qquad \gamma_{\boldsymbol{x},\pm}(1) = \gamma_{\boldsymbol{x}}(1) \ , \qquad \dot{\gamma}_{\boldsymbol{x},\pm} - \dot{\gamma}_{\boldsymbol{x}} = +\boldsymbol{h} \ .$$

Let $\Gamma_{\boldsymbol{x},\pm} = (\phi_\pm(s),\gamma_{\boldsymbol{x},\pm})$ be the null lifts of the paths $\gamma_{\boldsymbol{x},\pm}$ with endpoints $\Gamma_{\boldsymbol{x}}(1)$. Let K be a compact set containing all the $\Gamma_{\boldsymbol{x},\pm}$'s, where $\boldsymbol{x}$ and $\boldsymbol{h}$ run through $B(R)$. Let I be any bounded interval such that $I\times B(2R)$ contains K. As before, choose σ so that $I\times B(2R)$ lies to the past of Σ_σ, and let b be such that $\Gamma_{\boldsymbol{x}}(b) \in \mathscr{H}_\sigma$. (The value of the parameter b will of course depend upon $\boldsymbol{x}$). Let $\Gamma_\pm$ be the null curve obtained by following $\Gamma_{\boldsymbol{x},\pm}$ for parameter values $s\in[0,1]$, and then $\Gamma_{\boldsymbol{x}}$ for parameter values $s\in[1,b]$. Then $\Gamma_\pm \in \mathcal{P}(\boldsymbol{x})(\boldsymbol{x}\pm\boldsymbol{h})$ so that we have

$$\mu(\boldsymbol{x}\pm\boldsymbol{h}) \leq \int_0^1 \sqrt{h_{\phi_\pm(s)}(\dot{\gamma}_{\boldsymbol{x},\pm},\dot{\gamma}_{\boldsymbol{x},\pm})}ds + \int_1^b \sqrt{h_{\phi_\pm(s)}(\dot{\gamma}_{\boldsymbol{x}},\dot{\gamma}_{\boldsymbol{x}})}ds \ .$$

Further

$$\mu(\boldsymbol{x}) = \int_0^b \sqrt{h_{\phi_{\boldsymbol{x}}(s)}(\dot{\gamma}_{\boldsymbol{x}},\dot{\gamma}_{\boldsymbol{x}})}ds \ ,$$

hence

$$\begin{aligned}&\frac{\mu(\boldsymbol{x}+\boldsymbol{h})+\mu(\boldsymbol{x}-\boldsymbol{h})}{2} - \mu(\boldsymbol{x}) \leq \\ &\int_0^1 \left(\frac{\sqrt{h_{\phi_\pm(s)}(\dot{\gamma}_{\boldsymbol{x},+},\dot{\gamma}_{\boldsymbol{x},+})}+\sqrt{h_{\phi_\pm(s)}(\dot{\gamma}_{\boldsymbol{x},-},\dot{\gamma}_{\boldsymbol{x},-})}}{2} - \sqrt{h_{\phi_{\boldsymbol{x}}(s)}(\dot{\gamma}_{\boldsymbol{x}},\dot{\gamma}_{\boldsymbol{x}})}\right) ds \ .\end{aligned} \tag{3.48}$$

Since solutions of ODE's with parameters are differentiable functions of those, we can write

$$\phi_\pm(s) = \phi_{\boldsymbol{x}}(s) + \psi_i(s)h^i + r(s,h), \qquad |r(s,h)| \le C|h|^2_\delta \ , \tag{3.49}$$

for some functions ψ_i, with a constant C which is independent of $\boldsymbol{x}, \boldsymbol{h} \in B(R)$ and $s \in [0,1]$. Inserting (3.49) in (3.48), second order Taylor expanding the function $\sqrt{h_{\phi_\pm(s)}(\dot\gamma_{\boldsymbol{x},\pm}, \dot\gamma_{\boldsymbol{x},\pm})(s)}$ in all its arguments around $(\phi_{\boldsymbol{x}}(s), \gamma_{\boldsymbol{x}}(s), \dot\gamma_{\boldsymbol{x}}(s))$ and using compactness of K one obtains

$$\frac{\mu(\boldsymbol{x}+\boldsymbol{h}) + \mu(\boldsymbol{x}-\boldsymbol{h})}{2} - \mu(\boldsymbol{x}) \le C|\boldsymbol{h}|^2_\delta \ , \tag{3.50}$$

for some constant C. Set

$$\psi(\boldsymbol{x}) = \mu(\boldsymbol{x}) - C|\boldsymbol{x}|^2_\delta \ .$$

Equation (3.50) shows that

$$\forall \boldsymbol{x}, \boldsymbol{h} \in B(R) \qquad \psi(\boldsymbol{x}) \ge \frac{\psi(\boldsymbol{x}+\boldsymbol{h}) + \psi(\boldsymbol{x}-\boldsymbol{h})}{2} \ .$$

A standard argument implies that ψ is concave. It follows that

$$f(\boldsymbol{x}) + C|\boldsymbol{x}|^2_\delta = \tau(\boldsymbol{x}) + C|\boldsymbol{x}|^2_\delta = \sigma - \mu(\boldsymbol{x}) + C|\boldsymbol{x}|^2_\delta = \sigma - \psi(\boldsymbol{x})$$

is convex, which is what had to be established. □

References

1. M. Alcubierre, B. Brugmann, et al. *Black hole excision for dynamic black holes.* Phys. Rev., **D64**, 061501 (5pp.), 2001. Gr-qc/0104020.
2. M. Alcubierre et al. *The 3d grazing collision of two black holes.* 2000. Gr-qc/0012079.
3. M. Alcubierre et al. *Test-beds and applications for apparent horizon finders in numerical relativity.* Class. Quant. Grav., **17**, 2159–2190, 2000. Gr-qc/9809004.
4. M. Anderson. *On the structure of solutions to the static vacuum Einstein equations.* Annales H. Poincaré, **1**, 995–1042, 2000. Gr-qc/0001018.
5. L. Andersson and P. Chruściel. *On "hyperboloidal" Cauchy data for vacuum Einstein equations and obstructions to smoothness of null infinity.* Phys. Rev. Lett., **70**, 2829–2832, 1993.
6. L. Andersson and P. Chruściel. *On asymptotic behaviour of solutions of the constraint equations in general relativity with "hyperboloidal boundary conditions".* Dissert. Math., **355**, 1–100, 1996.
7. P. Anninos, D. Bernstein, et al. *Dynamics of apparent and event horizons.* Phys. Rev. Lett., **74**, 630–633, 1995. Gr-qc/9403011.
8. F. Antonacci and P. Piccione. *A Fermat principle on Lorentzian manifolds and applications.* Appl. Math. Lett., **9**, 91–95, 1996.
9. C. Barceló, S. Liberati, and M. Visser. *Analog gravity from field theory normal modes?* Class. Quant. Grav., **18**, 3595–3610, 2001. Gr-qc/0104001.

10. C. Barceló, S. Liberati, and M. Visser. *Towards the observation of Hawking radiation in Bose-Einstein condensates.* 2001. Gr-qc/0110036.
11. R. Bartnik and P. Chruściel. *Spectral boundary conditions for Dirac–type equations.* in preparation.
12. R. Bartnik and A. Norton. *NQS Einstein solver.* `http://beth.ise.canberra.edu.au/mathstat/StaffPages/Robert2.htm`.
13. J. Beem and A. Królak. *Cauchy horizon endpoints and differentiability.* Jour. Math. Phys., **39**, 6001–6010, 1998. Gr-qc/9709046.
14. J. K. Beem and P. E. Ehrlich. *Global Lorentzian Geometry*, vol. 67 of *Pure and Applied Mathematics* (Marcel Dekker, New York, 1981).
15. R. Beig and W. Simon. *On the uniqueness of static perfect–fluid solutions in general relativity.* Commun. Math. Phys., **144**, 373–390, 1992.
16. N. Bishop, R. Gomez, et al. *High-powered gravitational news.* Phys. Rev., **D56**, 6298–6309, 1997. Gr-qc/9708065.
17. W. Boucher. *Cosmic no hair theorems.* In W. Bonnor and M. MacCallum, eds., *Classical general relativity*, 43–52 (Cambridge University Press, Cambridge, 1984).
18. W. Boucher, G. Gibbons, and G. Horowitz. *Uniqueness theorem for anti–de Sitter spacetime.* Phys. Rev. D, **30**, 2447–2451, 1984.
19. S. Brandt et al. *Grazing collisions of black holes via the excision of singularities.* Phys. Rev. Lett., **85**, 5496–5499, 2000. Gr-qc/0009047.
20. B. Bromley. *A lightweight review of middleweight black holes.* Matters of gravity, **14**, 6–7, 1999. Gr-qc/9909022.
21. R. Brout, S. Massar, R. Parentani, and P. Spindel. *A primer for black hole quantum physics.* Phys. Rept., **260**, 329–454, 1995.
22. S. Browdy and G. Galloway. *Topological censorship and the topology of black holes.* Jour. Math. Phys., **36**, 4952–4961, 1995.
23. R. Budzyński, W. Kondracki, and A. Królak. *On the differentiability of Cauchy horizons.* Jour. Math. Phys., **40**, 5138–5142, 1999.
24. G. Bunting and A. Masood–ul–Alam. *Nonexistence of multiple black holes in asymptotically euclidean static vacuum space-time.* Gen. Rel. Grav., **19**, 147–154, 1987.
25. B. Carr. *Primordial black holes as a probe of the early universe and a varying gravitational constant.* Astro-ph/0102390.
26. B. Carter. *Has the black hole equilibrium problem been solved?* (in Proc. of the 8th Marcel Grossmann Meeting on Relativistic Astrophysics - MG 8, Jerusalem, Israel, 22 - 27 June 1997, T. Piran, Ed., World Scientific, Singapore, 1999). Gr-qc/9712038.
27. A. Celotti, J. Miller, and D. Sciama. *Astrophysical evidence for the existence of black holes.* Class. Quantum Grav., **16**, A3–A21, 1999. Astro-ph/9912186.
28. J. Chaname, A. Gould, and J. Miralda-Escude. *Microlensing by the cluster of black holes around Sgr A^*.* 2001. Astro-ph/0102481.
29. D. Christodoulou. *Examples of naked singularity formation in the gravitational collapse of a scalar field.* Ann. Math., **140**, 607–653, 1994.
30. D. Christodoulou. *The action principle and partial differential equations*, vol. 146 of *Annals of Mathematics Studies* (Princeton Univ. Press, Princeton, 2000).
31. D. Christodoulou and S. Klainermann. *Nonlinear Stability of Minkowski Space* (Princeton University Press, Princeton, 1993).
32. P. Chruściel. *On the global structure of Robinson–Trautman space–times.* Proc. Roy. Soc. London, **A 436**, 299–316, 1992.

33. P. Chruściel. *"No Hair" Theorems – folklore, conjectures, results.* In J. Beem and K. Duggal, eds., *Differential Geometry and Mathematical Physics*, vol. 170 of *Cont. Math.*, 23–49, gr–qc/9402032 (AMS, Providence, 1994).
34. P. Chruściel. *Uniqueness of black holes revisited.* vol. 69, 529–552 (Helv. Phys. Acta, 1996). Proceedings of Journés Relativistes 1996, gr-qc/9610010.
35. P. Chruściel. *A remark on differentiability of Cauchy horizons.* Class. Quantum Grav., 3845–3848, 1998. Gr-qc/9807059.
36. P. Chruściel. *The classification of static vacuum space–times containing an asymptotically flat spacelike hypersurface with compact interior.* Class. Quantum Grav., **16**, 661–687, 1999. Gr-qc/9809088.
37. P. Chruściel. *Towards the classification of static electro–vacuum space–times containing an asymptotically flat spacelike hypersurface with compact interior.* Class. Quantum Grav., **16**, 689–704, 1999. Gr-qc/9810022.
38. P. Chruściel, E. Delay, G. Galloway, and R. Howard. *Regularity of horizons and the area theorem.* Annales Henri Poincaré, **2**, 109–178, 2001. URL{URL}\url{http://www.phys.univ-tours.fr/~piotr/papers/area}. Gr-qc/0001003.
39. P. Chruściel, J. Fu, G. Galloway, and R. Howard. *On fine differentiability properties of horizons and applications to Riemannian geometry.* Jour. Geom. Phys., 2001. URL{URL}\url{http://www.phys.univ-tours.fr/~piotr/papers/area}. In press, gr-qc/0001003.
40. P. Chruściel and G. Galloway. *Horizons non–differentiable on dense sets.* Commun. Math. Phys., **193**, 449–470, 1998. Gr-qc/9611032.
41. P. Chruściel and M. Herzlich. *The mass of asymptotically hyperbolic Riemannian manifolds.* 2001. Dg-ga/0110035.
42. P. Chruściel and J. Isenberg. *Non–isometric vacuum extensions of vacuum maximal globally hyperbolic space–times.* Phys. Rev., **D48**, 1616–1628, 1993.
43. P. Chruściel, J. Jezierski, and J. Kijowski. *Hamiltonian field theory in the radiating regime*, vol. m70 of *Lect. Notes in Physics* (Springer, Berlin, Heidelberg, New York, 2001). URL `http://www.phys.univ-tours.fr/~piotr/papers/hamiltonian_structure`.
44. P. Chruściel, M. MacCallum, and D. Singleton. *Gravitational waves in general relativity. XIV: Bondi expansions and the "polyhomogeneity" of Scri.* Phil. Trans. Roy. Soc. London A, **350**, 113–141, 1995.
45. P. Chruściel and W. Simon. *Towards the classification of static vacuum spacetimes with negative cosmological constant.* Jour. Math. Phys., **42**, 1779–1817, 2001. Gr-qc/0004032.
46. P. Chruściel and R. Wald. *On the topology of stationary black holes.* Class. Quantum Grav., **11**, L147–152, 1994.
47. C. Clarke and F. de Felice. *Globally noncausal space-times.* Jour. Phys. A, **15**, 2415–2417, 1982.
48. E. Colbert and R. Mushotzky. *The nature of accreting black holes in nearby galaxy nuclei.* Astroph. Jour., **519**, 89–107, 1999. Astro-ph/9901023.
49. S. Dain. *Initial data for stationary space-times near space-like infinity.* Class. Quantum Grav., **18**, 4329–4338, 2001. Gr-qc/0107018.
50. T. Damour and B. Schmidt. *Reliability of perturbation theory in general relativity.* Jour. Math. Phys., **31**, 2441–2453, 1990.
51. S. Das and S. Mathur. *The quantum physics of black holes: Results from string theory.* Ann. Rev. Nucl. Part. Sci., **50**, 153–206, 2000.
52. W. Fleming and H. M. Soner. *Controlled Markov processes and viscosity solutions*, vol. 25 of *Applications of Mathematics* (Springer–Verlag, New York, Heidelberg, 1993).

53. H. Friedrich. *Cauchy problem for the conformal vacuum field equations in general relativity.* Commun. Math. Phys., **91**, 445–472, 1983.
54. H. Friedrich. *Existence and structure of past asymptotically simple solutions of Einstein's field equations with positive cosmological constant.* Jour. Geom. Phys., **3**, 101–117, 1986.
55. H. Friedrich. *Hyperbolic reductions for Einstein's equations.* Class. and Quantum Grav., **13**, 1451–1469, 1996.
56. H. Friedrich. *Einstein's equation and geometric asymptotics.* In *Gravitation and Relativity: At the turn of the Millennium, N. Dahdich and J. Narlikar (eds.)*, 153–176 (IUCAA, Pune, 1998). Proceedings of GR15.
57. H. Friedrich and A. Rendall. *The Cauchy problem for the Einstein equations.* Lect. Notes Phys., **540**, 127–224, 2000. Gr-qc/0002074.
58. H. Friedrich and B. Schmidt. *Conformal geodesics in general relativity.* Proc. Roy. Soc. London Ser. A, **414**, 171–195, 1987.
59. G. Galloway. *On the topology of the domain of outer communication.* Class. Quantum Grav., **12**, L99–L101, 1995.
60. G. Galloway. *A "finite infinity" version of the FSW topological censorship.* Class. Quantum Grav., **13**, 1471–1478, 1996.
61. G. Galloway, K. Schleich, D. Witt, and E. Woolgar. *The AdS/CFT correspondence conjecture and topological censorship.* Phys. Lett., **B505**, 255–262, 2001. Hep-th/9912119.
62. G. Galloway, S. Surya, and E. Woolgar. *A uniqueness theorem for the adS soliton.* 2001. Hep-th/0108170.
63. G. Galloway and E. Woolgar. *The cosmic censor forbids naked topology.* Class. Quantum Grav., **14**, L1–L7, 1996. Gr-qc/9609007.
64. R. Geroch. *Domain of dependence.* Jour. Math. Phys., **11**, 437–449, 1970.
65. R. Geroch. *Asymptotic structure of space-time.* In F. Esposito and L. Witten, eds., *Asymptotic Structure of Space-Time*, 1–106 (Plenum Press, New York, 1977).
66. R. Geroch and G. Horowitz. *Asymptotically simple does not imply asymptotically Minkowskian.* Phys. Rev. Lett., **40**, 203–206, 1978.
67. R. Geroch, E. Kronheimer, and R. Penrose. *Ideal points in space-time.* Proc. Roy. Soc. London, **A 327**, 545–567.
68. F. Giannoni, A. Masiello, and P. Piccione. *A variational theory for light rays in stably causal Lorentzian manifolds: regularity and multiplicity results.* Commun. Math. Phys., **187**, 375–415, 1997.
69. G. Gibbons. *Gravitational entropy and the inverse mean curvature flow.* Class. Quantum Grav., **16**, 1677–1687, 1999.
70. R. Gomez et al. *Stable characteristic evolution of generic 3-dimensional single-black-hole spacetimes.* Phys. Rev. Lett., **80**, 3915–3918, 1998. Gr-qc/9801069.
71. P. Grandclément, E. Gourgoulhon, and S. Bonazzola. *Binary black holes in circular orbits. II. Numerical methods and first results.* 2001. Gr-qc/0106016.
72. C. Gundlach. *Pseudo-spectral apparent horizon finders: An efficient new algorithm.* Phys. Rev., **D57**, 863–875, 1998. Gr-qc/9707050.
73. S. Hawking and G. Ellis. *The large scale structure of space-time* (Cambridge University Press, Cambridge, 1973).
74. J. Hempel. *3–manifolds* (Princeton University Press, Princeton, 1976). Annals of Mathematics Studies No 86.
75. M. Heusler. *Black hole uniqueness theorems* (Cambridge University Press, Cambridge, 1996).
76. M. Heusler. *Stationary black holes: uniqueness and beyond.* Living Reviews, **1**, 1998. `http://www.livingreviews.org`.

77. G. Horowitz. *Quantum states of black holes.* In *R.M. Wald (ed.), Black holes and relativistic stars. Symposium held in Chicago, IL, USA, December 14-15, 1996. Univ. of Chicago Press, Chicago, Ill., pp. 241–266.*
78. G. Horowitz and R. Myers. *The AdS/CFT correspondence and a new positive energy conjecture for general relativity.* Phys. Rev., **D59**, 026005 (12 pp.), 1999.
79. R. Howard and J. Fu. *Hypersurfaces with prescribed mean curvature.* 2001. Preprint.
80. G. Huisken and T. Ilmanen. *The inverse mean curvature flow and the Riemannian Penrose inequality.* Jour. Diff. Geom., 2002. In press.
81. W. Israel. *Event horizons in static vacuum space-times.* Phys. Rev., **164**, 1776–1779, 1967.
82. W. Israel. *Dark stars: The evolution of an idea.* In *S.W. Hawking and W. Israel (eds.), Three hundred years of gravitation. Cambridge: Cambridge University Press, pp. 199-276* (1987).
83. T. Jacobson. *Trans-planckian redshifts and the substance of the space- time river.* Prog. Theor. Phys. Suppl., **136**, 1–17, 1999. Hep-th/0001085.
84. T. Jacobson and S. Venkatarami. *Topology of event horizons and topological censorship.* Class. Quantum Grav., **12**, 1055–1061, 1995.
85. J. Kenney and E. Yale. *Hubble space telescope imaging of bipolar nuclear shells in the disturbed Virgo Cluster Galaxy NGC 4438.* 2000. Astro-ph/0005052.
86. L. Kidder, M. Scheel, and S. Teukolsky. *Extending the lifetime of 3d black hole computations with a new hyperbolic system of evolution equations.* Phys. Rev., **D64**, 064017, 2001. Gr-qc/0105031.
87. J. Kormendy and K. Gebhardt. *Supermassive black holes in galactic nuclei.* 2001. Astro-ph/0105230.
88. M. Kriele and S. Hayward. *Outer trapped surfaces and their apparent horizon.* Jour. Math. Phys., **38**, 1593–1604.
89. A. Królak. *The existence of regular partially future asymptotically predictable space-times.* Jour. Math. Phys., **29**, 1786–1788, 1988.
90. J. Kroon. *Can one detect a non-smooth null infinity?* Class. Quantum Grav., **18**, 4311–4316, 2001.
91. J. Lafontaine. *Sur la géométrie d'une généralisation de l'équation différentielle d'Obata.* Jour. de Math. Pures et Appli., **62**, 63–72, 1983.
92. J. Lafontaine and L. Rozoy. *Courbure scalaire et trous noirs.* Séminaire de théorie spectrale et géométrie (Grenoble), **18**, 69–76, 2000.
93. L. Lehner. *Numerical relativity: A review.* Class. Quantum Grav., **18**, R25–R86, 2001. Gr-qc/0106072.
94. O. Lengard. *Solutions of the Einstein's equation, waves maps, and semilinear waves in the radiation regime.* Ph.D. thesis, Université de Tours, 2001. `http://www/phys.univ-tours.fr/~piotr/papers/batz`.
95. U. Leonhardt and P. Piwnicki. *Relativistic effects of light in moving media with extremely low group velocity.* Phys. Rev. Lett., **84**, 822–825, 2000.
96. U. Leonhardt and P. Piwnicki. *reply to the "Comment on 'Relativistic effects of light in moving media with extremely low group velocity'" by M. Visser.* Phys. Rev. Lett., **85**, 5253, 2000. Gr-qc/0003016.
97. J. Leray. *Hyperbolic differential equations.* mimeographed notes, 1953. Princeton.
98. J. Libson, J. Masso, et al. *Event horizons in numerical relativity. I:Methods and tests.* Phys. Rev., **D53**, 4335–4350, 1996. Gr-qc/9412068.
99. J.-P. Luminet. *Black holes: A general introduction.* In *F.W. Hehl, C. Kiefer and R.J.K. Metzler (eds.), Black holes: theory and observation. Proceedings of*

the 179th W.E. Heraeus seminar, Bad Honnef, Germany, August 18-22, 1997. Springer Lect. Notes Phys. 514, pp. 3-34. Astro-ph/9804006.
100. F. Macchetto. *Supermassive black holes and galaxy morphology.* 2001. Astro-ph/9910089.
101. G. Madejski. *Black holes in active galactic nuclei: observations.* 1999. Astro-ph/9903141.
102. J. Masso, E. Seidel, W.-M. Suen, and P. Walker. *Event horizons in numerical relativity II: Analyzing the horizon.* Phys. Rev., **D59**, 064015 (19 pp.), 1999. Gr-qc/9804059.
103. K. Menou, E. Quataert, and R. Narayan. *Astrophysical evidence for black hole event horizons.* In *Gravitation and Relativity: At the turn of the Millennium, N. Dahdich and J. Narlikar (eds.)*, 43–65 (IUCAA, Pune, 1998). Proceedings of GR15.
104. D. Merritt and L. Ferrarese. *Relationship of black holes to bulges.* 2001. Astro-ph/0107134.
105. J. Miralda-Escudé and A. Gould. *A cluster of black holes at the galactic center.* Astroph. Jour., **545**, 847–853, 2000. Astro-ph/0003269.
106. C. Misner. *Taub–NUT space as a counterexample to almost anything.* In *Relativity Theory and Astrophysics*, 160–169 (AMS, Providence, Rhode Island, 1967). Lectures in Appl. Math., vol. 8.
107. C. Misner, K. Thorne, and J. Wheeler. *Gravitation* (Freeman, San Fransisco, 1973).
108. J. Morgan, L. Greenhill, and J. Hernstein. *Observational evidence for massive black holes in the centers of active galaxies.* Astro-ph/0002085.
109. R. Narayan, M. Garcia, and J. McClintock. *X-ray Novae and the evidence for black hole event horizons.* 2001. Astro-ph/0107387.
110. R. Newman. *Cosmic censorship and curvature growth.* Gen. Rel. Grav., **15**, 641–353, 1983.
111. M. Novello, J. Salim, V. D. Lorenci, and E. Elbaz. *Nonlinear electrodynamics can generate a closed spacelike path for photons.* Phys. Rev., **D63**, 103516 (5 pp.), 2001.
112. B. O'Neill. *Semi–Riemannian Geometry* (Academic Press, New York, 1983).
113. A. Peet. *Tasi lectures on black holes in string theory.* 2000. Hep-th/0008241.
114. R. Penrose, *Zero rest-mass fields including gravitation.* Proc. Roy. Soc. London **A284** (1965), 159–203.
115. R. Penrose. *Techniques of Differential Topology in Relativity* (SIAM, Philadelphia, 1972). (Regional Conf. Series in Appl. Math., vol. 7).
116. V. Perlick. *On Fermat's principle in general relativity: I. The general case.* Class. Quantum Grav., **7**, 1319–1331, 1990.
117. A. Ptak and R. Griffiths. *Hard X-ray variability in M 82: Evidence for a nascent AGN?* Astroph. Jour., **517**, L85–L90, 1999. Astro-ph/9903372.
118. R. Racke. *Lectures on nonlinear evolution equations* (Friedr. Vieweg & Sohn, Braunschweig/Wiesbaden, 1992).
119. M. Rees. *Astrophysical evidence for black holes.* In *R.M. Wald (ed.), Black holes and relativistic stars. Symposium held in Chicago, IL, USA, December 14-15, 1996. Univ. of Chicago Press, Chicago, Ill., pp. 79-101.*
120. M. Rees. *Supermassive Black Holes: Their Formation, and Their Prospects as Probes of Relativistic Gravity.* Proceedings of ESO conference in honour of R. Giacconi on Black Holes in Binaries and Galactic Nuclei (2000). Astro-ph/9912346.
121. B. Schmidt. *On the uniqueness of boundaries at infinity of asymptotically flat spacetimes.* Class. Quantum Grav., **8**, 1491–1504, 1991.

122. H. Seifert. *Smoothing and extending cosmic time functions.* Gen. Rel. Grav., **8**, 815–831, 77.
123. T. H. site of the Space Telescope Science Institute. `http://hubble.stsci.edu`.
124. L. Tamburino and J. Winicour. *Gravitational fields in finite and conformal Bondi frames.* Phys. Rev., **150**, 1039–1053, 1966.
125. M. Taylor. *Partial Differential Equations* (Springer, New York, Berlin, Heidelberg, 1996).
126. Z. Tsvetanov, M. Allen, H. Ford, and R. Harms. *Morphology of the nuclear disk in M87* (1998). Proceedings of the M87 Workshop, Ringberg castle, Germany, 15-19 Sep 1997, astro-ph/9803178.
127. W. Unruh. *Dumb holes and the effects of high frequencies on black hole evaporation.* Phys. Rev., **D51**, 2827–2838, 1995. Gr-qc/9409008.
128. M. Visser. *Optical black holes?* Matters of gravity. Gr-qc/0002027.
129. M. Visser. *Acoustic black holes.* 1999. Gr-qc/9901047.
130. M. Visser. *Comment on 'Relativistic effects of light in moving media with extremely low group velocity'.* Phys. Rev. Lett., **85**, 5252, 2000. Gr-qc/0002011.
131. G. E. Volovik. *Superfluid analogies of cosmological phenomena.* Phys. Rept., **351**, 195–348, 2001. Gr-qc/0005091.
132. R. Wald. *General relativity* (University of Chicago Press, Chicago, Ill., 1984).
133. R. Wald. *The thermodynamics of black holes.* Living Reviews, **4**, 2001. Gr-qc/9912119, `http://www.livingreviews.org/`.
134. X. Wang. *Mass for asymptotically hyperbolic manifolds.* Jour. Diff. Geom., **57**, 273–299, 2001.
135. T. Zeeuw. *Evidence for massive black holes in nearby galactic nuclei.* 2000. Astro-ph/0009249.
136. X. Zhang. *A definition of total energy-momenta and the positive mass theorem on asymptotically hyperbolic 3 manifolds I.* 2001. Preprint.

4 Conformal Geometry, Differential Equations and Associated Transformations

Simonetta Frittelli[1], Niky Kamran[2], and Ezra T. Newman[3]

[1] Physics Dept., Duquesne University, Pittsburgh, PA, USA
[2] Math. Dept., McGill University, Montreal, Canada
[3] Dept. of Physics and Astronomy, Univ. of Pittsburgh, Pittsburgh, PA, USA

Abstract. By a series of examples, we illustrate some rather remarkable and surprising (at least to some) relationships between (1) all three-dimensional conformal Lorentzian manifolds and a wide class of 3rd order ODEs (all 3rd order ODEs with vanishing Wunschmann invariant) and (2) all four dimensional conformal Lorentzian manifolds and a wide class of pairs of 2nd order PDEs (all pairs of 2nd order PDEs with vanishing generalized Wunschmann invariant and satisfying a simple inequality). In addition, we study the relationship between the equivalence classes of differential equations under the group of contact transformations and the construction of null surfaces in the Lorentzian manifolds. General theorems are stated without proof.

4.1 Introduction

It is the purpose of this note to describe a rather remarkable connection between Lorentzian conformal geometry on both three and four-dimensional manifolds and large classes of differential equations. In particular we will discuss (1) the relationship between *all* three-dimensional conformal Lorentzian geometry and 3rd order ODEs of the form

$$\frac{d^3u}{ds^3} = F(u, \frac{du}{ds}, \frac{d^2u}{ds^2}, s) \tag{4.1}$$

where $F(u, u', u'', s)$ is a smooth function of its variables satisfying a differential condition (the vanishing of the Wunschmann invariant, $W[u] = 0$ [1–3]) and (2) the relationship between *all* four-dimensional conformal Lorentzian geometries and the pair of 2nd order PDEs of the form

$$\frac{\partial^2 u}{\partial s^2} = S(s, \bar{s}, \frac{\partial u}{\partial s}, \frac{\partial u}{\partial s}, \frac{\partial^2 u}{\partial s \partial \overline{s}}), \qquad \frac{\partial^2 u}{\partial \overline{s}^2} = \overline{S}(s, \bar{s}, \frac{\partial u}{\partial s}, \frac{\partial u}{\partial \overline{s}}, \frac{\partial^2 u}{\partial s \partial \overline{s}}), \tag{4.2}$$

where u is a real function of the complex conjugate pair $(s, \overline{s})$. The pair of complex conjugate functions, $(S, \overline{S})$ must be chosen (i) to satisfy integrability conditions, (ii) to satisfy the inequality

$$1 - S_{u_{s\overline{s}}} \overline{S}_{u_{s\overline{s}}} > 0, \tag{4.3}$$

which can be shown to imply that the solution space is defined by four independent parameters and (iii) to satisfy a certain differential condition (the vanishing of the generalized Wunschmann invariant, $M[S, \overline{S}]$ [4]).

The relationship between the differential equations and the Lorentzian manifolds in each case is established by first noting that for Eq.(4.1), the local coordinates of the Lorentzian manifold are given by the three constants of integration, x^i, $(i = 0, 1, 2)$, defining the solution space, $\mathfrak{M}_3$,, so that solutions are expressed as

$$u = z(x^i, s), \tag{4.4}$$

while for Eq.(4.2), the solutions space, $\mathfrak{M}_4$, is defined by four constants of integration, x^a, $(a = 0, 1, 2, 3)$, so solutions have the form

$$u = z(x^a, s, \bar{s}). \tag{4.5}$$

In addition, it has been shown [4] for both cases, that a conformal Lorentzian metric, $[g]$, can be defined on $\mathfrak{M}_3$ and $\mathfrak{M}_4$ by, respectively,

$$\begin{aligned} g^{ij}(x^k)\partial_i z \partial_j z &= 0, \\ g^{ab}(x^c)\partial_a z \partial_b z &= 0. \end{aligned} \tag{4.6}$$

In other words, the solutions, (4.4) and (4.5), of Eqs.(4.1) and (4.2), both define (conformal) Lorentzian metrics on respectively, $\mathfrak{M}_3$ and $\mathfrak{M}_4$, and *simultaneously* are interpretable as one and two parameter solutions to the eikonal equation associated with that metric, i.e., the solutions are complete integrals of the eikonal equations; the level surfaces, $u = constant$, are null or characteristic surfaces for each value of the independent parameters.

There is a further interesting issue that connects the theory of differential equations with the Lorentzian geometries, namely the connection between the theory of equivalence classes of differential equations under the variety of transformations, point transformations, fiber-preserving transformations and contact transformations. We will see that there is a simple but elegant geometric interpretation of these transformations in terms of geometric structures (the characteristic surfaces) on the associated Lorentzian manifolds.

Since many of the ideas and detailed proofs of the claims just referred to, have already appeared in the literature [5], [4], and involve lengthy and unattractive calculations, in this work we will, for the purpose of simplicity and clarity, restrict the discussion essentially to one example – namely flat three dimensional Minkowski space - which illustrates the entire theory - and then explain how it generalizes to arbitrary three and four dimensional Lorentzian manifolds. In the process, we will reverse the point of view just expressed; we will *begin* with the Lorentzian manifolds and their conformal metrics and derive the differential equations rather than start with the differential equations.

In particular, we will use, as an example, three-dimensional flat Lorentzian space-time. We introduce the "flat" coordinates, $x^i = (t, x, y)$ and conformal metric,

$$g = \Omega^2 \,\mathrm{diag}(1, -1, -1) = \Omega^2(dt^2 - dx^2 - dy^2)$$

and then show how the equation

$$\frac{d^3u}{ds^3} = 0,$$

and a family of similar equations naturally arise.

Before beginning the detailed discussion we want to point out that the idea of associating geometric structures to equivalence classes of families of differential equations goes back many years to the classical papers of Cartan and Chern. To a large extent the material reported here and in our earlier papers turns out to be a different formulation of some classical results of Cartan and Chern in the 3d case, and a generalization in the 4d case. Our motivation and point of view are, however, quite different from that of Cartan and Chern.

4.2 Example of Contact-Envelope Transformation

We begin with the flat three dimensional manifold $\mathfrak{M}_3$ with metric

$$g = \Omega^2 \operatorname{diag}(1, -1, -1) = \Omega^2 (dt^2 - dx^2 - dy^2) \tag{4.7}$$

and consider one-parameter solutions to the eikonal equation, (referred to as complete integrals)

$$g^{ij} \partial_i z \partial_j z = 0. \tag{4.8}$$

The simplest such solution is the one-parameter family of (level) surfaces given by

$$\widehat{u} = z(x, y, t, p) = (1 + p^2)t + 2px + (1 - p^2)y \tag{4.9}$$

where (t, x, y) are the Minkowski coordinates. For any fixed value of p, the level surfaces $\widehat{u} = constant$, are null surfaces or characteristic surfaces of the metric (4.7). This function, obtained as a solution of the eikonal equation on $\mathfrak{M}_3$ can alternatively be thought of as a solution of a third order ODE in the variable p.

By differentiating (4.9) three times with respect to p we obtain

$$\frac{d^3}{dp^3} \widehat{u} = 0. \tag{4.10}$$

If we had started with (4.10), instead of the Eikonal equation, (4.8), its general integral would have been (4.9), but now with (t, x, y) interpreted as constants of integration. By going further and substituting (4.9) into the Eikonal equation, the metric (4.7) could have been recovered. In other words the conformally flat Lorentzian geometry is contained or embodied in (4.10).

Though we have shown this for the simple case of (4.10), the result *can* be generalized to any Lorentzian $\mathfrak{M}_3$ or $\mathfrak{M}_4$.

An immediate question arises; what are the consequences of choosing different solutions (complete integrals) of the flat-space eikonal equation, solutions more general than (4.9). One simple and obvious way to obtain new complete integrals is to define a new independent variable and change the dependent variable by

$$\begin{aligned} s &= f(p), \\ u^* &= g(\widehat{u}, p), \end{aligned} \tag{4.11}$$

a change that is referred to as a fiber preserving transformation. An example is

$$u^* = \frac{\hat{u}}{1+p^2}, \qquad \cos s = \frac{2p}{1+p^2}, \tag{4.12}$$

so that the new solution of the eikonal equation becomes

$$u^* = t + x \cos s + y \sin s, \tag{4.13}$$

another representation of plane null waves. Either by taking three derivatives of (4.13) and eliminating the (t, x, y) or transforming the ODE (4.10) via (4.12), we obtain the new ODE

$$\frac{d^3u^*}{ds^3} = -\frac{du^*}{ds} \tag{4.14}$$

whose general 3-parameter solution is (4.13).

Another example is the simple rescaling of (4.13);

$$u = \frac{u^*}{\sin s} \tag{4.15}$$

or

$$u = x \cot s + y + \frac{t}{\sin s}, \tag{4.16}$$

again a representation of null waves but now satisfying a different third order ODE

$$\frac{d^3}{ds^3}u = 2\frac{d}{ds}u - 3\frac{d^2}{ds^2}u \cdot \cot s \tag{4.17}$$

when the (t, x, y) are eliminated.

What we have shown, by the three examples, is that the three very similar (geometrically identical) plane-wave complete solutions of the eikonal equation, (4.9),(4.13) and (4.16) are related to three different ODEs all obtained from each other by simple fiber preserving transformations, (4.11). The question then is whether there is a relationship between the ODEs when the solutions of the eikonal equation are geometrically completely different; e.g., what is the relationship between the ODEs when one goes from the plane wave null surfaces to light-cones?

If we define the two-parameter family of light cones by

$$(\overline{u} - y)^2 + (\overline{s} - x)^2 - t^2 = 0, \tag{4.18}$$

for fixed $(\overline{u}, \overline{s})$ and then treat this as an implicit one-parameter family of solutions (complete) of the eikonal equation,

$$\overline{u} = z(t, x, y, \overline{s}),$$

then it is easy to see that eliminating (t, x, y) from the 3rd derivative, we obtain the new 3rd order ODE

$$\frac{d^3}{d\overline{s}^3}\overline{u} = \frac{3}{1 + (\frac{d}{d\overline{s}}\overline{u})^2}\left(\frac{d}{d\overline{s}}\overline{u}\right)\left(\frac{d^2}{d\overline{s}^2}\overline{u}\right)^2. \tag{4.19}$$

Eq.(4.19) *can not* be obtained from any of our earlier ones by means of a fiber preserving transformation but it can be obtained by a beautiful generalization of them. This generalization has two apparently different statements or descriptions that on inspection turn out to yield exactly the same transformation. Though this generalization applies to the transform of any complete solution to any other complete solution, we will illustrate it with the example above — i.e., the transform from the plane waves to the null cones.

We begin with the classical procedure [6], (an envelope construction), applicable to either the Eikonal or the Hamilton-Jacobi equation, where from a complete solution one can construct any other solution.

This procedure, which we will use to go from one complete solution of the eikonal equation to another, is as follows: Assume that $u = z(t, x, y, s)$ is a complete solution of the Eikonal equation, then by taking an arbitrary function, $V(s, z(t, x, y, s), \overline{s})$ and then writing

$$\overline{u} = \overline{z}(t, x, y, \overline{s}) \equiv V(s, z(t, x, y, s), \overline{s}), \tag{4.20}$$

with $s = I(t, x, y, \overline{s})$, given by the implicit (the envelope) relationship,

$$\frac{dV}{ds}\Big|_{\overline{s}} \equiv \partial_s V + \partial_z V \cdot \partial_s z = 0. \tag{4.21}$$

It is easy to check that the new $\overline{z}(t, x, y, \overline{s})$ also satisfies the Eikonal equation for all $\overline{s}$.

If we begin with the plane waves, (4.16), $u = x \cot s + y + \frac{t}{\sin s}$, and choose

$$V(s, z(t, x, y, s), \overline{s}) = z - \overline{s} \cot s$$

or

$$\overline{z}(t, x, y, \overline{s}) = (x - \overline{s}) \cot s + y + \frac{t}{\sin s}$$

which, with the envelope condition, (4.21),

$$\cos s = \frac{(\overline{s} - x)}{t},$$

we obtain the family of null cones,

$$(\overline{u} - y)^2 + (\overline{s} - x)^2 - t^2 = 0.$$

As we saw, the $\overline{u}$ defined implicitly here, satisfies the ODE, (4.19).

Remark 4.1. This constitutes S. Lie's construction of three-dimensional Minkowski space as the space of all circles in the Euclidean plane.

An alternate means of performing this transformation is to concentrate on the 3rd order ODEs, (4.17) and (4.19), and completely forget about the space-time itself. Any third order ODE involves differential relations between the variables (s, u, u', u''), in the sense that

$$\frac{d^3u}{ds^3} = F(u, \frac{du}{ds}, \frac{d^2u}{ds^2}, s) \Leftrightarrow \tag{4.22}$$

$$\frac{du}{ds} = u', \quad \frac{d^2u}{ds^2} = u'', \quad \frac{d^3u}{ds^3} = F(u, u', u'', s). \tag{4.23}$$

We are interested in the transformation of these equations induced by changes in the variables. Though there are a wide variety of transformations that one could consider, we will consider only two classes:

a. Fiber preserving transformations, that we mentioned earlier (using a different notation),

$$\begin{aligned} \overline{s} &= f(s), \\ \overline{u} &= g(u, s). \end{aligned} \tag{4.24}$$

b. Contact transformations (i.e., transformations just among the variables (s, u, u'), technically, the first jet bundle over s) that are given in terms of a generating function, H, having the form $H(u, \overline{u}, s, \overline{s})$. The transformation,

$$(s, u, u') \Leftrightarrow (\overline{s}, \overline{u}, \overline{u}')$$

is given by the implicit relations [7],

$$H(u, \overline{u}, s, \overline{s}) = 0, \quad H_s + u'H_u = 0, \quad H_{\overline{s}} + \overline{u}'H_{\overline{u}} = 0, \tag{4.25}$$

with the assumption that the contact transformation is not a prolonged point transformation.

Remark 4.2. This class of transformations was first studied and applied, by Sophus Lie, to the problem of the equivalence classes of differential equations under a variety of transformations.

Returning to our example, by choosing

$$H(u, \overline{u}, s, \overline{s}) = (\overline{u} - u)\sin s + \overline{s}\cos s, \tag{4.26}$$

and using (4.25) we obtain the explicit contact transformation

$$\begin{aligned} \overline{u} \quad &= u - \overline{s}\cot s, \\ \overline{s} \quad &= -\sin^2 s \cdot du/ds, \\ d\overline{u}/d\overline{s} &= -\cot s, \end{aligned} \tag{4.27}$$

which transforms (4.17) into (4.19). By substituting into (4.27), the explicit form for u, i.e., (4.16), one sees that the transformations though involving different variables, really are identical.

More generally, with a bit of thought, it is not difficult to see that (4.20) with (4.21), is actually the same transformation as (4.25) just using different variables.

To summarize this section, we have seen, by example, that given a 3-dimensional Lorentzian manifold, and any complete integral (a *one-parameter* family of solutions) of its eikonal equation, there is a third order ODE with the *parameter* as independent variable, whose three-constant of integration solution is the same as the complete integral. Furthermore, all complete integrals can be obtained from any one, by a classical envelope construction; and the associated third order ODEs are related by contact transformations. In each of the cases, by direct calculation, the Wunschmann Invariant vanishes.

In the next section we will state the generalizations of these results.

4.3 Generalizations

What we have shown by example, can be greatly generalized. In the first part of this section we will give the main results of the generalizations for three-dimensional conformal Lorentzian geometries while in the second part, it is extended to four-dimensional conformal Lorentzian spaces. We state the results as local theorems. No proofs will be given here other than brief explanations — detailed proofs have been given elsewhere [4, 5]. The examples in the previous section are intended to give a heuristic argument for the truth of the theorems.

4.3.1 Three-Dimensional Conformal Lorentzian Geometries

Theorem 4.1. *Given any (local) three-dimensional Lorentzian metric on* $\mathfrak{M}_3$ *and any complete integral (a one-parameter, s, family of solutions) of the associated eikonal equation, there is a unique 3rd order ODE with vanishing Wunschmann Invariant (whose independent variable is the parameter s) that was obtained by eliminating the three local coordinates of* $\mathfrak{M}_3$*, whose solutions contain three constants of integration, which in turn can be taken as the local coordinates of* $\mathfrak{M}_3$ *and is identical to the complete integral. All other complete integrals can be obtained from any one complete integral by means of an envelope construction; the 3rd order ODEs associated with the complete integrals are transformed among each other by means of contact transformations. In other words, the equivalence classes of these 3rd order equations under the group of contact transformations, all have the same conformal Lorentzian geometry.*

Theorem 4.2. *(The converse of 4.1): Given any 3rd order ODE, with a vanishing Wunschmann Invariant, there is a local manifold* $\mathfrak{M}_3$ *given by the three constants of integration, on which there exists a unique conformal Lorentzian metric. This conformal metric is obtained from the demand that the solution to the ODE also be a solution of an eikonal equation with some unknown — to be determined — metric on* $\mathfrak{M}_3$*. Each member of the equivalence class of ODEs under the group of contact transformations yields, via this construction, the same conformal Lorentzian metric.*

Note that in either Theorem 4.1 or 4.2, if the contact transformation is restricted to a fiber preserving transformation, then the 3rd order ODE can be very much changed but the basic geometry of the associated null surfaces remains unchanged. If the vanishing of the Wunschmann Invariant is not satisfied there still remains a rich geometry [2–4].

Corollary 4.1. *Given a conformal Lorentzian 3-manifold, $\mathfrak{M}_3$, with a conformal metric $[g]$ having n conformal isometries and given a third order ODE, (any member of the equivalence class), whose solution yields a one parameter family of characteristic surfaces of the conformal metric $[g]$;*

Then, (a) the dimension of the symmetry group of the ODE (the contact transformations that preserve the ODE) is n.

(b) the solutions of the ODE, (as characteristic surfaces in $\mathfrak{M}_3$), transform in $\mathfrak{M}_3$ via the conformal isometries.

4.3.2 Four-Dimensional Conformal Lorentzian Geometries

In four dimensions the theorems are quite similar though they must be somewhat modified. The proofs however are very much more difficult and technically involved.

Theorem 4.3. *Given (locally) any four-dimensional manifold, $\mathfrak{M}_4$, with a (conformal) Lorentzian metric and any complete integral (a two-parameter family of solutions) of the associated eikonal equation, then there will be a unique pair of 2nd order PDEs with vanishing [generalized] Wunschmann Invariant (obtained by eliminating the four local coordinates of $\mathfrak{M}_4$ from the first and second derivatives with respect to the two-parameters). This pair of 2nd order equations form an overdetermined set so that the solution space is a four-dimensional manifold that we take as $\mathfrak{M}_4$. The solutions are identical to the complete integrals. All other complete integrals can be obtained from any one complete integral by means of an envelope construction; the pair of 2nd order PDEs associated with the complete integrals are transformed among each other by means of contact transformations. In other words, the equivalence classes of these 2nd order equations under the group of contact transformations, all have the same conformal Lorentzian geometry.*

Theorem 4.4. *(The converse of 4.3): Given any pair of 2nd order PDEs, (4.2) with vanishing [generalized] Wunschmann Invariant [4], satisfying a certain inequality, (4.3), then the solutions depend on four constants of integration, defining the local solution space, $\mathfrak{M}_4$. On $\mathfrak{M}_4$ there exists a unique conformal Lorentzian metric. This conformal metric is obtained from the demand that the solution to the pair of 2nd order PDEs also be a solution of an eikonal equation with some unknown — to be determined — metric on $\mathfrak{M}_4$. All members of the equivalence class of pairs of 2nd order PDEs under the group of contact transformations, yield, via this construction, the same conformal Lorentzian metric. The conformal Lorentzian metrics are invariant under the contact transformations applied to the PDEs.*

Corollary 4.2. *Corollary 4.1 generalizes, in a straightforward manner, from the three dimensional Lorentzian manifolds $\mathfrak{M}_3$ to $\mathfrak{M}_4$.*

Remark 4.3. Notice that Theorem 4.4 (and Theorem 4.2) allows one to adopt a radically different view towards differential geometry — with no implication that it is better or worse then the standard approach — just different:

Local coordinate patches of manifolds are *defined* as the solution spaces of differential equations in a completely coordinate independent manner. In a region where the solution breaks down (e.g., caustics or wave front singularities of the eikonal equation), one could construct overlap regions (without mentioning coordinates) where one chooses other differential equations from the same equivalence class. The change of one solution of the eikonal equation to another plays the role of standard coordinate transformations. In other words, an entire equivalence class of pairs of 2nd order PDEs, — satisfying the conditions of Theorem 4.4 — under the group of contact transformations, simply is the *definition* of a four-dimensional manifold.

Remark 4.4. Since, from Theorem 4.4, we can obtain by appropriate choice of the pairs of 2nd order PDEs *any* conformal Lorentzian 4-metric, one could then ask for further conditions on the PDEs so that the conformal Lorentzian metrics are conformally related to Einstein metrics.

In this manner, one could hope to study the Einstein equations via a *version* of the conformal Einstein equations. A first step (necessary, but not sufficient) in this direction would be to see how the PDEs could be restricted so that the solutions yield a conformal metric having a vanishing Bach tensor. Work on this problem — a very difficult one — has begun. In the case of asymptotic flatness, the vanishing of the Bach tensor is probably both necessary and sufficient for the metric to be conformally Einstein.

Note added in proof. There are further developments in the geometry of 3rd order ODEs so that one has, in a natural manner, the existence of a Cartan normal conformal connection on the solution space, arising from any 3rd order ODE with a vanishing Wunschmann invariant. This will be reported in a future paper by S.Frittelli, C.Kozameh, E.T. Newman and P. Nurowski.

Acknowledgments

This research was supported by NSERC grant number OGP0105490 and by the NSF under grants number PHY 97-22049 and PHY 98-03301.

References

1. K. Wünschmann, *Uber Berührungsbedingungen bei Integralkurven von Differentialgleichungen,* Inaug.Dissert., Leipzig, Teubner
2. E. Cartan, *La Geometria de las Ecuaciones Diferenciales de Tercer Orden,* Rev. Mat. Hispano-Amer. **4**,1-31,(1941).

3. Shiing-Shen Chern, *The Geometry of a Differential Equation,* $y''' = F(x, y, y', y'')$, Science Reports Tsing Hua Univ. **4**(1940), 97-111, reprinted in *Shiing-Shen Chern, Selected Papers* Springer-Verlag, 1978.
4. S. Frittelli, C. Kozameh and E. T. Newman, *Differential Geometry from Differential Equations*, submitted (2000, gr-qc/0012058.
5. S. Frittelli, N. Kamran and E. T. Newman, *Differential Equations and Conformal Geometry,*submitted (2001).
6. L.D. Landau and E.M. Lifschitz, *Mechanics,*Pergamon Press, Oxford, (1960).
7. P. J. Olver, *Equivalence, Invariants, and Symmetry,* Cambridge University Press, 1995.

5 Twistor Geometry of Conformal Infinity

Roger Penrose

[1] Mathematical Institute, Oxford
[2] Center for Gravitational Physics and Geometry, Penn State University

Abstract. A brief description is provided, of the present status of the twistor encoding of a general asymptotically flat Ricci-flat space-time M, in terms of the complexification $\mathbb{C}\mathscr{I}^+$ of its conformal infinity. The construction of the twistor space $\mathcal{T}$ is given explicitly, this generalizing the 1976 picture, which gave a "non-linear graviton construction" for anti-self-dual (complex) Ricci-flat space-times. The (suggested) converse determination of the space-time itself from $\mathcal{T}$ is indicated only.

5.1 Non-linear Gravitons

One of the major aims of twistor theory is to find a natural description of a general vacuum space-time, encoded into the structure of a deformed twistor space. This is taken as an essential step towards finding the appropriate union between Einstein's gravitational theory and quantum-theoretic principles, in accordance with the basic tenets of twistor theory. For about a quarter of a century a construction has been known which indeed translates the general (complex) *anti-self-dual* (ASD) vacuum space-time into a curved twistor-space structure. This has been known as "the non-linear graviton" construction [14], but since it encodes only the graviton of *left-handed* helicity (i.e. ASD), I refer to it here as the *leg-break* graviton construction ("leg-break" being a cricketing term, for a ball bowled with a left-handed spin). The full non-linear graviton would involve right-handed helicity also. This involves the solution of the problem of finding a corresponding twistor representation of the general *self-dual* (SD) vacuum space-time, which has been referred to as the *googly problem* (again a term from cricket, a "googly" being a right-spinning ball bowled with a leg-break-like action).

Twistor space has an inherently *chiral* character, and this underlies the qualitative difference between the coding of ASD and SD vacuums. By passing to the *dual* twistor space the chirality would be reversed so we could, if we wished, use the original 1976 construction to encode SD complex space-times. But this does not help us, because the *full* graviton, which includes both helicities, requires that we use the *same* twistor space to incorporate both. The googly construction is therefore needed, so as to allow for this combination of helicities. I should remark that the *ambitwistor* approach, based on the space of complex null geodesics and modeled on a canonical hypersurface in the product of twistor and dual twistor spaces [1, 9, 10], does not fit in with the general "graviton" philosophy of twistor theory, being (even if fully successful) more akin to a classical than a quantum description.

According to the philosophy that is being adhered to for the twistor constructions, the description of a "graviton" indeed requires the full non-linear (vacuum) Einstein equation to be solved, rather than just the *linearized* versions of this equation that conventional treatments use (see [14]). Further restrictions that we ought to demand (for a *free*graviton), over and above *Ricci-flatness* for the complex-Riemannian 4-manifold, would be appropriate conditions of asymptotic flatness and (though not discussed here) positive frequency. These would be the normal restrictions for the more conventional descriptions of gravitons in terms of solutions of the *linearized* Einstein equation.

The appropriate condition of asymptotic flatness seems to be the existence of a sufficiently extensive *conformal infinity* $\mathscr{I}^+$ [13], but where this must be taken as a *complex* null 3-space. In fact, the original "non-linear graviton construction" arose from considerations of ordinary *real* asymptotically flat vacuum space-times, via the $\mathcal{H}$-space concept of Newman [11]. Later work showed that the leg-break construction applied more generally, no assumption of asymptotic flatness for the ASD "complex space-time" being needed. Newman's construction, as applied to a real analytic asymptotically flat vacuum $\mathcal{M}$, takes the "ASD-part" $\mathcal{M}_L$ of $\mathcal{M}$, by defining the *points* of $\mathcal{M}_L$ to be the "good cuts" of the complexification $\mathbb{C}\mathscr{I}^+$ of $\mathcal{M}$'s $\mathscr{I}^+$. These "cuts" are complex cross-sections of $\mathbb{C}\mathscr{I}^+$, where $\mathbb{C}\mathscr{I}^+$ is taken as, in effect, a (mutilated) complex line bundle over a thickened-out anti-holomorphic diagonal $S(\cong S^2)$ of $\mathbb{CP}^1 \times \mathbb{CP}^1$. To be "good"[1], such a cut $\mathcal{C}$ must have vanishing asymptotic shear $\tilde{\sigma}$ (where $\tilde{\sigma}$ is the "freed-up" complex conjugate $\bar{\sigma}^0$ of the asymptotic complex shear of a complex null hypersurface meeting $\mathbb{C}\mathscr{I}^+$ in $\mathcal{C}$, cf. [8, 11, 12, 18]), and there is (in normal circumstances) a holomorphic 4-complex-dimensional family of such cuts, consistently with $\mathcal{M}_L$ being a complex 4-space. Newman's remarkable definition of the *metric* g_L of $\mathcal{M}_L$ can be expressed as (with $\mathcal{S}$ being the surface area 2-form)

$$8\pi^2\{g_L(V,V)\}^{-1} = \oint v^{-2}\,\mathcal{S} \tag{5.1}$$

Here V is a (non-null) vector in $\mathcal{M}_L$, at a point p of $\mathcal{M}_L$, where p is defined by a certain good cut $\mathcal{C}$ of $\mathbb{C}\mathscr{I}^+$ and where V specifies the displacement $v(\theta,\phi)$, on $\mathbb{C}\mathscr{I}^+$ (along generators of $\mathbb{C}\mathscr{I}^+$) from $\mathcal{C}$ to a "neighbouring" good cut $\mathcal{C}'$, "$\oint$" denoting integration over an appropriate 2-dimensional contour on $\mathcal{C}$. With this definition, it turns out that not only is g_L a (holomorphic) complex-Riemannian metric on $\mathcal{M}_L$, but it is automatically ASD and Ricci-flat.

[1] In Newman's original definition, a "good cut" would have vanishing σ, rather than $\tilde{\sigma}$, which would result in an SD complex vacuum space rather than an ASD one. I am reversing his descriptions here, so as to be consistent with the usual twistor conventions.

5.2 The Reasonableness of $\mathscr{I}^+$

Before relating this to twistor theory, it is appropriate to address the issue of the existence of $\mathscr{I}^+$, for an asymptotically flat space-time $\mathcal{M}$. Some people have expressed unease with the notion of $\mathscr{I}^+$, even for very well behaved asymptotically flat space-times. However, the mere *existence* of $\mathscr{I}^+$ for a space-time $\mathcal{M}$ that is asymptotically flat, even in a very weak sense, is not an issue of serious dispute, since the points of $\mathscr{I}^+$ can be defined in terms of TIPs (terminal indecomposable past-sets; cf. [7, 19]). What is at issue is the degree of *smoothness* that can be assigned to $\mathscr{I}^+$.

Now it is clear that there are many analytic space-times $\mathcal{M}$ that are asymptotically flat and vacuum in the asymptotic region, for which $\overline{\mathcal{M}}(= \mathcal{M} \cup \mathscr{I}^+)$ would have low differentiability at $\mathscr{I}^+$ (e.g. where the conformal metric is C^0 but not even C^1). Such could be the case if there were a tail of incoming radiation with logarithmic fall-off, or with some other rate of fall-off having an amplitude that is not adequately smooth (at $w = 0$) in a parameter w which is the reciprocal of advanced time. This is easily seen in the case of linearized general relativity. The issue is not whether regularity at $\mathscr{I}^+$ covers all situations that we would like to call "asymptotically flat"; it is whether this regularity condition allows all the freedom that we need in order to describe radiating isolated systems in general relativity.

We know from the work of Friedrich [3, 4] that vacuum space-times exist for which $\overline{\mathcal{M}}$ is even *analytic* at $\mathscr{I}^+$, and also at future time-like infinity i^+, for which there is the full functional freedom of outgoing radiation. This is what we shall need for the twistor construction referred to in Sect. 5.1 and which I shall outline in a little more detail in Sects 5.3 and 5.4. This analytic case does not include "time-bomb" type behaviour for the outgoing radiation, but we can guarantee that there are analytic solutions that are as close as we wish to such behavior. We are thus assured of the existence of large class of space-times, of the kind that we shall need, which cover what is physically required for the study of systems in which there is outgoing gravitational radiation.

A significant question does remain, however, concerning the appropriateness of the assumption of a smooth conformal infinity, namely the generality of an assumption of the *coexistence* of a sufficiently smooth (say analytic) behaviour at *both* $\mathscr{I}^+$ and $\mathscr{I}^-$. This is all tied up with the choice of appropriate conditions that should be imposed at space-like infinity i^0. There is much current interest in this question (see [5]), but the matter appears to be unresolved at the present time. Fortunately, the twistor construction that I wish to describe does not depend upon its resolution.

5.3 The Construction of Projective Twistor Space $\mathbb{P}\mathcal{T}$ from $\mathscr{I}^+$

Let us assume, in accordance with the discussion of Sect. 5.2, that our real space time $\mathcal{M}$ is vacuum and analytic, with analyticity holding throughout

$\overline{\mathcal{M}} = \mathcal{M} \cup \mathscr{I}^+ \cup i^+$ (with i^+ now included in the definition of $\overline{\mathcal{M}}$). We can thus form the complexification $\mathbb{C}\mathscr{I}^+$, which need only be some small extension ("thickening") of $\mathscr{I}^+$ into the complex. By the usual discussion [19, 20], we find that the complex 3-space $\mathbb{C}\mathscr{I}^+$ has a degenerate conformal metric, of rank 2, where $\mathbb{C}\mathscr{I}^+$ can be regarded as being composed of a family of α-planes (SD complex 2-planes), or alternatively of a family of β-planes (ASD complex 2-planes), where the intersection of an α-plane with a β-plane (except when vacuous–which can occur when they are "too far apart" for a particular thickening $\mathbb{C}\mathscr{I}^+$) will be a complex *generator* of $\mathbb{C}\mathscr{I}^+$, which is a null geodesic on $\mathbb{C}\mathscr{I}^+$ whose tangent directions are the null directions $\mathbf{n}$ of the singular metric of $\mathbb{C}\mathscr{I}^+$. There are also *other* complex null geodesics on $\mathbb{C}\mathscr{I}^+$. These lie either in β-planes, and are called *α-curves* (or *twistor lines*), or in α-planes, and are called *β-curves* (or *dual twistor lines*), but not both. Only the generators can be *real*, but generators can also be complex. The α-curves constitute a holomorphic 3-parameter family, and so do the β-curves. There is a holomorphic 2-parameter family of α-curves on each β-plane, the β-planes themselves forming a holomorphic 1-parameter family, and similarly for the β-curves on α-planes. The generators form a holomorphic 2-parameter family.

The "finite" part $\mathbb{P}\mathcal{T}_0$, of the *projective twistor space* $\mathbb{P}\mathcal{T}$, is the complex 3-manifold whose points represent the α-curves on $\mathbb{C}\mathscr{I}^+$. To obtain the full projective space $\mathbb{P}\mathcal{T}$, we have somehow to include the generators of $\mathbb{C}\mathscr{I}^+$ also, as limiting cases of α-curves; but taking them individually would give us what is referred to as the "blown-up" projective twistor space. $\mathbb{P}\mathcal{T}$ itself is obtained by *identifying* all the generators that belong to a single α-plane, so each α-plane produces one point of $\mathbb{P}\mathcal{T}$. The family of α-planes gives us a holomorphic curve $\mathbf{I}$ (a Riemann sphere) on $\mathbb{P}\mathcal{T}$, which we can think of as representing "infinity" on $\mathbb{P}\mathcal{T}$. It requires a little work to see that this results in a smooth complex manifold $\mathbb{P}\mathcal{T}$, but this is indeed the case (see [16]).

The projective twistor space $\mathbb{P}\mathcal{T}$ (in fact the finite part $\mathbb{P}\mathcal{T}_0$) is the essential part of the 1976 "leg-break" construction for ASD Ricci-flat 4-spaces. Newman's *good cuts* of $\mathbb{C}\mathscr{I}^+$ turn out to be complex 2-surfaces ruled by α-curves (or by β-curves, with Newman's original conventions). The entire family of such α-curves, for a particular cut, constitute a holomorphic 1-parameter family giving a Riemann sphere in $\mathbb{P}\mathcal{T}_0$. Thus, Newman's construction finds its twistor formulation in the finding of holomorphic curves ($\cong S^2$) belonging to the appropriate topological family. Each such holomorphic curve in $\mathbb{P}\mathcal{T}$ gives us a point of $\mathcal{M}_L$. This is the essence of the 1976 construction, although we need a little more information in order to get the scaling for the metric on $\mathcal{M}_L$. This extra information is often formulated in terms of a canonically defined *bundle* over $\mathbb{P}\mathcal{T}$. From the overall twistor perspective, this bundle structure is best presented as providing the *non*-projective twistor space, which is a complex 4-manifold $\mathcal{T}$. In the 1976 "leg-break" construction, $\mathcal{T}$ is taken to be a complex *line bundle* over $\mathbb{P}\mathcal{T}$ (with the zeros of the fibres omitted). The information contained in this $\mathcal{T}$ encodes only the ASD information in the radiation field (a remarkable enough fact, as it is), so we must do something more sophisticated if we are to encode (holomorphically[2]) the SD information also.

5.4 The Construction of the Full Twistor Space $\mathcal{T}$ from $\mathscr{I}^+$

How are we to encode the SD part of the radiation field? This is basically the "googly problem" referred to in Sect. 5.1. Let us return to the definition of $\mathbb{P}\mathcal{T}$, in Sect. 5.3, where each point of $\mathbb{P}\mathcal{T}_0$ is given by a null geodesic on $\mathbb{C}\mathscr{I}^+$. I shall use the 2-spinor abstract-index formalism [18], described with respect to some analytic choice of conformal factor Ω making the metric g_{ab} on $\overline{\mathcal{M}}$ well defined on $\mathbb{C}\mathscr{I}^+$ (patchwise, at least). Take the **n**-directions on $\mathbb{C}\mathscr{I}^+$ to be given by

$$n^a = o^A \tilde{o}^{A'} \tag{5.2}$$

(with $\tilde{o}^{A'}$ independent of o^A). The (null-geodesic) equation of an α-curve on $\mathbb{C}\mathscr{I}^+$ can be written

$$\mu^{B'} \nabla_{0B'} \mu_{A'} \propto \mu_{A'} \tag{5.3}$$

(where *suffixes* 0 and $0'$ denote components obtained by contraction with o^A and with $\tilde{o}^{A'}$, respectively), the complex null direction $o^A \mu^{A'}$ being *tangent* to the α-curve. This equation is conformally invariant, but it leaves the choice of *scaling* for the tangent vectors completely free.

In the leg-break construction, this is fixed by taking *zero* on the right-hand side of the above equation, so that $\mu_{A'}$ is parallel-propagated, and demanding that the (abstract-indexed co-spinor) $\mu_{A'}$ be a conformal *invariant* (rather than a conformal density). The different scalings for $\mu_{A'}$, for a given α-curve, then give the $(\mathbb{C} - 0)$-fibre of $\mathcal{T}$ that sits above the point of $\mathbb{P}\mathcal{T}$ defined by that α-curve. Here we need something more general, so that the SD radiation field can be incorporated.

This is achieved by taking the scalings to be defined by solutions of

$$\mu^{B'} \nabla_{0B'} \mu_{A'} = K \mu_{A'} \times \left(\mu^{0'}\right)^{-5} \text{þ}_c \tilde{\psi}_{0'0'0'0'} \tag{5.4}$$

(the *scaled twistor-line equation* STLE), along the various α-curves. Here þ_c is the conformally invariant "thorn" operator defined in [18, p. 395]), which is a modified version of the derivative operator $\nabla_{00'}$, and $\tilde{\psi}_{A'B'C'D'}$ is the (conformally invariant) helicity +2 massless field related to the SD Weyl spinor $\tilde{\Psi}_{A'B'C'D'}$ by

$$\tilde{\psi}_{A'B'C'D'} = \Omega^{-1} \tilde{\Psi}_{A'B'C'D'} \tag{5.5}$$

[2] Of course, for a real $\mathcal{M}$, the SD radiation information is simply the complex conjugate of the ASD information, but allowing the taking of complex conjugates at this stage would nullify the holomorphic nature of the construction, upon which the essential strength of the twistor formalism depends.

for any choice Ω of conformal factor. The quantity K is just a numerical constant whose most natural value is to be determined by later work.

The 1-parameter family of solutions of STLE, for a given α-curve, gives the (non-linear) fibre of $\mathcal{T}$ above that point of $\mathbb{P}\mathcal{T}$ which the α-curve defines. These fibres are called *Euler curves* on $\mathcal{T}$ (since they are integral curves of the "Euler homogeneity operator" of Sect. 5.5 which finds a natural interpretation in the present context). The whole construction is *conformally invariant* (independent of Ω) and not dependent upon the scalings of o^A and $\tilde{o}^{A'}$. The metric (as opposed merely to the conformal) structure of $\mathcal{M}$ *is* involved in the resulting structure of $\mathcal{T}$.

5.5 The Local Structure of Twistor Space $\mathbb{P}\mathcal{T}$

What is this structure? We find that there is a 1-form $\boldsymbol{\iota}$ and a 3-form $\boldsymbol{\theta}$, each defined *up to proportionality* on $\mathcal{T}$. There is, however, a restriction on how much rescaling can be applied to $\boldsymbol{\iota}$ and $\boldsymbol{\theta}$, the quantities

$$\boldsymbol{\Pi} = \mathrm{d}\boldsymbol{\theta} \otimes \boldsymbol{\iota} \qquad \text{and} \qquad \boldsymbol{\Sigma} = \mathrm{d}\boldsymbol{\theta} \otimes \mathrm{d}\boldsymbol{\theta} \otimes \boldsymbol{\theta} \tag{5.6}$$

being *invariant* structures on $\mathcal{T}$. Another way of putting this is to say that the only rescalings that are allowed have the form (locally)

$$\boldsymbol{\iota}' = k\boldsymbol{\iota}, \quad \boldsymbol{\theta}' = k^2\boldsymbol{\theta}, \quad \mathrm{d}\boldsymbol{\theta}' = k^{-1}\mathrm{d}\boldsymbol{\theta}, \tag{5.7}$$

where k is some (local) scalar function on $\mathcal{T}$. We also require the relations

$$\boldsymbol{\iota} \wedge \mathrm{d}\boldsymbol{\iota} = 0, \quad \boldsymbol{\iota} \wedge \boldsymbol{\theta} = 0, \quad \mathrm{d}\boldsymbol{\theta} \otimes \iota = -2\boldsymbol{\theta} \oslash \mathrm{d}\boldsymbol{\iota} \tag{5.8}$$

to hold between $\boldsymbol{\iota}$ and $\boldsymbol{\theta}$, where the bilinear operator "$\oslash$", acting between an n-form $\boldsymbol{\eta}$ and a 2-form, is defined by

$$\boldsymbol{\eta} \oslash (\mathrm{d}p \wedge \mathrm{d}q) = \boldsymbol{\eta} \wedge \mathrm{d}p \otimes \mathrm{d}q - \boldsymbol{\eta} \wedge \mathrm{d}q \otimes dp. \tag{5.9}$$

What is the geometrical significance of these various relations? First, the 3-form $\boldsymbol{\theta}$ determines the directions of the fibration of $\mathcal{T}$, above $\mathbb{P}\mathcal{T}$. These directions are those of the *Euler vector field* Υ ("Euler homogeneity operator"), defined by the formal relation

$$\Upsilon = 4\boldsymbol{\theta} \div \mathrm{d}\boldsymbol{\theta}. \tag{5.10}$$

(The precise interpretation of this quotient is: $\Upsilon(a)\,\mathrm{d}\boldsymbol{\theta} = 4\,\mathrm{d}a \wedge \boldsymbol{\theta}$, for all scalar fields a.) The fibres of $\mathcal{T}$ are the integral curves of Υ. We find that Υ is not absolutely determined, however, but scales according to

$$\Upsilon' = k^3\Upsilon. \tag{5.11}$$

Because of this, the notion of "homogeneity" is not an absolute thing on $\mathcal{T}$ (as it is in flat or "leg-break" twistor theory), a function "homogeneous of degree

m" being an eigenfunction of Υ for the eigenvalue m. Degree m homogeneity extends to non-scalar quantities $\boldsymbol{\Gamma}$ through the eigenfunction relation

$$\mathcal{L}_{\Upsilon}\boldsymbol{\Gamma} = m\boldsymbol{\Gamma}. \tag{5.12}$$

A significance of the relation $\mathrm{d}\boldsymbol{\theta} \otimes \iota = -2\boldsymbol{\theta} \oslash \mathrm{d}\iota$ is now found to be that this is equivalent to the assertion that the 1-form $\boldsymbol{\iota}$ is of homogeneity 2 (the fact that $\boldsymbol{\theta}$ has homogeneity 4 being automatic). The Frobenius relation $\boldsymbol{\iota} \wedge \mathrm{d}\boldsymbol{\iota} = 0$ informs us that $\boldsymbol{\iota}$ determines an integrable 3-foliation. The leaves of this simply describe the families of α-curves that lie in the same β-plane on $\mathbb{C}\mathscr{I}^+$. The relation $\boldsymbol{\iota} \wedge \boldsymbol{\theta} = 0$ tells us that the Euler curves lie in these leaves, a fact that is contained in the previous comment.

The restrictions on the rescalings are shown to imply that the (local) rescaling factors k can have only a very special and somewhat curious form, satisfying

$$\Upsilon(k) = 2k^{-2} - 2k. \tag{5.13}$$

In terms of a parameter z, defined along the Euler integral curves in a standard way

$$\Upsilon(z) = z \tag{5.14}$$

so that "Υ-homogeneity" of a scalar quantity is simply ordinary homogeneity in terms of z, we find

$$k^3 = 1 - Fz^{-6} \tag{5.15}$$

for some F, constant along each Euler integral curve. This amounts to saying that

$$k^3 = 1 - f_{-6}, \tag{5.16}$$

where f_{-6} is a function on twistor space, homogeneous of degree -6. The significance of the "(-6)-homogeneity" is that in ordinary flat-space twistor theory, right-handed "linear gravitons" (SD solutions of linearized gravity, i.e. of the spin-2 massless field equations) are described by means of holomorphic twistor functions homogeneous of degree -6. It had long been noted that the leg-break construction provides a non-linear version of the corresponding *left*-handed linear gravitons (ASD solutions of linearized gravity), for which the twistor functions have homogeneity $+2$. The *googly problem* of Sect. 5.1 is to find a corresponding non-linear role for (-6)-functions in generating the SD solutions of the non-linear Einstein equation. The present framework points to a plausible solution.

5.6 Present Status of the Role of $\mathcal{T}$ in Encoding Ricci-Flatness

The construction of $\mathcal{T}$ described in Sect. 5.4 and Sect. 5.5 indeed provides a complete coding (up to possible global issues) of a general analytic vacuum space-time $\mathcal{M}$, which is asymptotically flat in the strong sense of Sect. 5.3. (Both SD and ASD radiation fields are encoded, and $\mathscr{I}^+$ can be used as a final data hypersurface.) However, it is not yet completely clear that the geometric characterizations of $\mathcal{T}$ described in Sect. 5.5 (together with suitable global requirements) are sufficient. More seriously, we as yet have no explicit construction of the *points* of $\mathcal{M}$ (or of its complexification $\mathbb{C}\mathcal{M}$) in terms of geometrical structures within $\mathcal{T}$, analogous to the holomorphic curves in $\mathbb{P}\mathcal{T}$ that worked in the SD case. There are, however some "informed guesses" (which I shall not glorify by the term "conjecture"), that I have largely described elsewhere ([16]; cf. also [2, 15, 17] for some details), and which offer some hope of progress in the correct direction.

It is not appropriate that I attempt to spell these out here in any detail, but a few comments may give something of their flavour. In the first place, we appear to need an analogue $\boldsymbol{\xi}$ of the 1-form $\boldsymbol{\iota}$, for each point x of $\mathcal{M}$. (In fact, $\boldsymbol{\iota}$ may be thought of as representing the point i^+ at future-time-like infinity.) Each form $\boldsymbol{\xi}$ is to satisfy the same algebraic and differential relations (with $\boldsymbol{\theta}$) as does $\boldsymbol{\iota}$ (i.e. $\boldsymbol{\xi} \wedge \mathrm{d}\boldsymbol{\xi} = 0$, $\boldsymbol{\xi} \wedge \boldsymbol{\theta} = 0$, and $\mathrm{d}\boldsymbol{\theta} \otimes \boldsymbol{\xi} = -2\boldsymbol{\theta} \oslash \mathrm{d}\boldsymbol{\xi}$), and also scales in the same way (i.e. $\boldsymbol{\xi}' = k\boldsymbol{\xi}$). We also do something very similar for null infinity, providing a 2-form $\boldsymbol{\eta}$ for each point y of $\mathbb{C}\mathscr{I}^+$. Each form $\boldsymbol{\xi}$ is to specify a "surgery" of $\mathcal{T}$, which is characterized by certain global properties that relate to integrals of these various forms on $\mathcal{T}$. The details of these matters are still "work in progress", but it is perhaps worth mentioning that the Newman integral for the metric described in Sect. 5.1 (which at least works in both the SD and ASD cases) can be re-expressed in the twistorial form (cf. [6]):

$$16\pi^2 \mathbf{g}^{-1} = \oint \gamma^{-2} \boldsymbol{\Theta}, \quad \text{where} \quad \boldsymbol{\xi} \wedge \boldsymbol{\xi}^0 \wedge \boldsymbol{\iota} = \gamma \boldsymbol{\Theta}. \tag{5.17}$$

Here, $\mathbf{g}$ is the infinitesimal squared metric distance of the displacement from a point x of $\mathcal{M}$ to a neighbouring point x^0, these corresponding to the 1-forms $\boldsymbol{\xi}$ and $\boldsymbol{\xi}^0$, as indicated above.

Acknowledgments

I am grateful to many colleagues, especially E.T. Newman, K.P. Tod, F. Hadrovich, J. Frauendiener, and L.J. Mason. I should also like to thank NSF for support under contract PHY 93-96246 and the Leverhulme Foundation for an Emeritus Fellowship.

References

1. R.J. Baston and L.J. Mason (1987) *Conformal gravity, the Einstein equations and spaces of complex null geodesics.* Class. Quantum Grav. **4**, 815-26.
2. J. Frauendiener and R. Penrose (2001) *Twistors and general relativity.* In *Mathematics Unlimited—2001 and Beyond; Part 1*, eds. B. Enquist and W. Schmid (Springer-Verlag, Berlin), 479-505.
3. H. Friedrich (1986) *On the existence of n-geodesically complete or future complete solutions of Einstein's field equations with smooth asymptotic structure.* Commun. Math. Phys., **107**, 587-609.

4. H. Friedrich (1998) *Einstein's equation and conformal structure.* In *The Geometric Universe; Science, Geometry, and the Work of Roger Penrose*, eds. S.A. Huggett, L.J. Mason, K.P. Tod, S.T. Tsou, and N.M.J. Woodhouse (Oxford Univ. Press, Oxford).
5. H. Friedrich (2002), *Conformal Einstein evolution* This volume.
6. S. Frittelli, F. Hadrovich, L.J. Mason, and E.T. Newman (2000) *Twistors and Legendre transformations.* Twistor Newsletter **45**, 25-31.
7. R. Geroch, E.H. Kronheimer and R. Penrose (1972) *Ideal Points in Space-time*, Proc. Roy. Soc (Lond.) **A347**, 545.
8. R.O. Hansen., E.T. Newman, R. Penrose and K.P. Tod (1978) *The metric and curvature properties of H-space.* Proc. Roy. Soc. Lond.**A363**, 445-68.
9. C.R LeBrun (1985) *Ambi-twistors and Einstein's equations.* Class. Quant. Grav. **2**, 555-63.
10. C.R LeBrun (1990) *Twistors, ambitwistors, and conformal gravity*, in *Twistors in Mathematical Physics*, LMS Lec. note ser. 156, eds T.N.Bailey and R.J.Baston (Cambridge Univ. Press, Cambridge).
11. E.T. Newman (1976) *Heaven and its properties* Gen. Rel. Grav. **7**, 107-11.
12. E.T. Newman and R. Penrose (1962) *An approach to gravitational radiation by a method of spin coefficients*, J. Math. Phys **3**, 896-902 (Errata **4** (1963) 998).
13. R. Penrose (1965) *Zero rest-mass fields including gravitation: asymptotic behaviour*, Proc. Roy. Soc. London, **A284**, 159-203.
14. R. Penrose (1976) *Non-linear gravitons and curved twistor theory* Gen. Rel. Grav. **7**, 31-52.
15. R. Penrose (1999) *Twistor theory and the Einstein vacuum* Class. Quantum Grav.**16**A113-A130.
16. R. Penrose (2000) *On extracting the googly information.* Twistor Newsletter**45**, 1-24.
17. R. Penrose (2001) *Towards a twistor description of general space-times; introductory comments.* In *Further Advances in Twistor Theory, Vol.III: Curved Twistor Spaces* Chapman & Hall/CRC Research Notes in Mathematics 424, eds. L.J. Mason, L.P. Hughston, P.Z. Kobak, and K. Pulverer (Chapman& Hall/CRC, London), 239-255.
18. R. Penrose and W. Rindler (1984) *Spinors and Space-Time, Vol. 1: Two-Spinor Calculus and Relativistic Fields* (Cambridge University Press, Cambridge).
19. R. Penrose and W. Rindler (1986) *Spinors and Space-Time*, Vol. 2: *Spinor and Twistor Methods in Space-Time Geometry* (Cambridge University Press, Cambridge).
20. R. Penrose and R.S. Ward (1980). *Twistors for flat and curved space-time.* In *General Relativity and Gravitation, One Hundred Years after the Birth of Albert Einstein*, ed. A. Held (Plenum, New York).

6 Isotropic Cosmological Singularities

K. Paul Tod

Mathematical Institute, Oxford

Abstract. An isotropic cosmological singularity is a cosmological singularity which can be removed by conformally-rescaling the metric. In the rescaled metric, the singularity is required to occur on a space-like hypersurface. Thus in particular the conformal structure can be extended through the singularity. One can now study the Cauchy problem for the cosmology with data at the singularity for various matter models. I review previous work on this problem and consider various directions for future research.

6.1 Introduction

Most of the talks at this workshop are concerned with Roger Penrose's original idea of extending the conformal structure of an asymptotically-flat space-time M to the null infinity $\mathscr{I}$ of M. As we all know, this extension is achieved by conformally-rescaling the physical metric by a conformal factor which goes to zero at large distances and makes infinity finite. It was (and is) remarkable that the conditions for the existence of this extension are so closely related to the asymptotic conditions already given e.g. by Bondi and coworkers for solutions of the Einstein field equations (EFEs from now on) with an isolated source. As we have heard, it is still uncertain precisely which data on a space-like Cauchy surface give a solution of the EFEs whose conformal structure extends to (and across) $\mathscr{I}$, though there is a widespread belief that the existence of this extension of the conformal structure correctly geometrizes the idea of an isolated source.

There is another situation, related to another suggestion of Penrose, where one might wish to extend the conformal structure of a space-time beyond where the metric is defined, and this is at a space-time singularity. The situation at a singularity is different from the asymptotic situation in two ways: first, the conformal factor goes to infinity rather than zero and 'blows up' the shrinking metric to make it finite; second, there is no expectation that this is the general case i.e. that such a singularity is a general space-time singularity. Rather the idea is that a singularity through which the conformal structure can be extended is quite special, but that initial singularities should be special in just this way.

This second point is the burden of Penrose's Weyl Curvature Hypothesis [17] (see also [18]). In [17], Penrose noted that in the standard picture of the Big Bang, the matter content of the universe initially was in thermal equilibrium and had high entropy, so that the gravitational field initially must have low entropy, in a sense to be explained, in order for there to be a Second Law of Thermodynamics. Penrose goes on:

> I propose then that there should be a complete lack of chaos in the initial geometry... this restriction on the early geometry should be something like: the Weyl curvature C_{abcd} vanishes at any initial singularity.

Initial singularities are then contrasted with final singularities, formed in gravitational collapse or the Big Crunch, where the chaotic 'Mixmaster' picture is more likely to hold, and the Weyl tensor to be singular. (In [17], Penrose consistently writes the Weyl tensor as C_{abcd} which is not the form in which it is conformally-invariant, but he has told me not to see particular significance in this choice!)

Part of the Weyl Curvature Hypothesis is that singularities should be like this as a result of the correct quantum theory of gravity, which must therefore be time-asymmetric. However, we can leave these kinds of question aside and take the Weyl Curvature Hypothesis as a selection principle: Can we characterize singularities with finite or zero Weyl curvature? Can we find large classes of examples? Can we explain other features of the early universe by the assumption that the Big Bang had this character?

In this talk, I shall review what has been done in this programme. I begin in Sect. 6.2 with some general formalism, including two items of conformally-invariant calculus. One of these is the theory of tractors from [3], which has not so far found application in the study of the Einstein equations, and the other is a conformally-invariant propagation along curves due to Schmidt [23] which has. With the aid of these and something from Clarke [6,7] and Racz [20], I show how one might get from the assumption of finite Weyl tensor to an isotropic singularity according to the usual definition ([14]). This definition is essentially that the metric can be conformally-rescaled so that the singularity occurs on a smooth space-like hypersurface in the unphysical, rescaled space-time. Thus in particular the conformal structure can be extended through the singularity, and one can now contemplate a Cauchy problem with data actually at the singularity, in the sense of on this space-like hypersurface. In Sect. 6.3, we assume the usual definition and summarise the consequences for the EFEs in two cases: with source a perfect-fluid with a linear equation of state, or with source consisting of massless collisionless matter (the Einstein-Vlasov equations). We are led to a singular Cauchy problem with data on the singularity surface, but this can be solved with the use of an existence theorem due to Newman-Claudel. We go on to consider a wider class of matter models in Sect. 6.4, confining ourselves to looking for power series solutions to the Cauchy problem. Finally, in Sect. 6.5, there are some conclusions and suggestions for further work, including the possibility of numerical studies.

I am grateful to the organisers of the Workshop on Conformal Structure in Tübingen for the invitation to attend this excellent workshop and to present this material.

6.2 Formalism and Extensions

We want to define an isotropic cosmological singularity in the conventional way ([13-16, 24-26]) as one where the metric can be rescaled as in (6.1) below to become non-singular, with Ω vanishing on a space-like hypersurface Σ in the unphysical space-time. The conformal structure therefore extends to and through the singularity surface Σ. In this section, I want to suggest how one might get from an assumption on finiteness of the Weyl tensor at a singularity to a result on extendibility of the conformal structure through the singularity. This work is at a preliminary stage, but presenting it will enable me to review some parts of conformal geometry.

Let us begin by fixing conventions, which will generally follow Penrose and Rindler [19]. Given the metric g_{ab}, we write the metric covariant derivative as ∇_a and define the Riemann tensor by

$$(\nabla_a \nabla_b - \nabla_b \nabla_a)V^c = -R^c{}_{dab}V^d .$$

Conformal rescaling will be

$$\tilde{g}_{ab} = \Omega^2 g_{ab} . \tag{6.1}$$

When dealing with $\mathcal{I}$, g_{ab} is the physical metric and $\mathcal{I}$ is located where Ω vanishes, so that rescaling pulls infinity in to a finite place and the unphysical metric $\tilde{g}_{ab}$ extends the conformal structure through $\mathcal{I}$. When dealing with isotropic singularities, it is convenient to suppose that the singularity is where Ω vanishes but then $\tilde{g}_{ab}$ must be the physical metric and the unphysical metric g_{ab} extends the conformal structure through the singularity. The change in the Christoffel symbols under (6.1) is readily calculated and leads to the change in the Riemann tensor. This is most simply expressed by decomposing the Riemann tensor as

$$R_{abcd} = C_{abcd} - 2g_{c[a}P_{b]d} + 2g_{d[a}P_{b]c} \tag{6.2}$$

where the Rho-tensor P_{ab} may be given in terms of the Ricci tensor R_{ab} and Ricci scalar R by

$$P_{ab} = -(1/2)R_{ab} + (1/12)Rg_{ab} \tag{6.3}$$

(The Rho-tensor, called L_{ab} by Friedrich and Schmidt ([10-12]) among others, can be defined in any dimension; this is the correct definition in dimension 4). The virtue of the decomposition (6.2) is that under (6.1) the transformations simplify as

$$\tilde{C}^a{}_{bcd} = C^a{}_{bcd} , \tag{6.4}$$

$$\tilde{P}_{ab} = P_{ab} - \nabla_a \Upsilon_b + \Upsilon_a \Upsilon_b - (1/2)(g^{cd}\Upsilon_c \Upsilon_d)g_{ab} \tag{6.5}$$

where, as usual, $\Upsilon_a = \nabla_a(\log \Omega)$. In terms of the Rho-tensor, the (differential) Bianchi identity is

$$\nabla_d C^d{}_{cab} = 2\nabla_{[a}P_{b]c} .$$

Next I wish to introduce two conformally-invariant notions. The first of these are the *tractors* of [3]. A tractor is a section of a rank-6 vector bundle over space-time, and can be written as a column vector:

$$U^I = \begin{pmatrix} \sigma \\ \mu^a \\ \rho \end{pmatrix} \tag{6.6}$$

where the components transform under (6.1) as

$$\begin{aligned} \tilde{\sigma} &= \sigma, \\ \tilde{\mu}^a &= \mu^a + \sigma \Upsilon^a, \\ \tilde{\rho} &= \rho - \mu^a \Upsilon_a - (1/2)\sigma \Upsilon^a \Upsilon_a \ . \end{aligned} \tag{6.7}$$

There is a conformally-invariant tractor-metric g_{IJ} (we use capital letters I, J, K etc as tractor indices) defined by

$$U_J = g_{IJ} U^I = (\rho, \mu_a, \sigma) \tag{6.8}$$

so that

$$g_{IJ} U^I U^J = 2\rho\sigma + g_{ab}\mu^a \mu^b \ . \tag{6.9}$$

This metric is easily seen to have signature $(+ + - - - -)$. The connection on tractors is defined by

$$\nabla_a U^J = \begin{pmatrix} \nabla_a \sigma - \mu_a \\ \nabla_a \mu^b + \delta_a^{\ b} \rho + P_a^{\ b} \sigma \\ \nabla_a \rho - P_{ab} \mu^b \end{pmatrix} \tag{6.10}$$

and the commutator of derivatives leads to a curvature which can be written as

$$R_{abI}{}^J = \begin{pmatrix} 0 & 0 & 0 \\ 2\nabla_{[a} P_{b]}^{\ c} & -C^c{}_{dab} & 0 \\ 0 & -2\nabla_{[a} P_{b]d} & 0 \end{pmatrix} \tag{6.11}$$

(Tractors were given this name in [3] in honour of their inventor Tracey Thomas; they are sections of a vector-bundle associated to the principal bundle on which the normal conformal Cartan connection is defined. This is the connection defined above. It is closely related to local twistor transport, which is defined on an associated spinor bundle [10]).

Now a condition of finiteness of the Weyl tensor and its first derivative will lead, via (6.11), to an assumption of finiteness of the tractor curvature.

The second conformally-invariant notion was introduced by Schmidt [23] and has been used subsequently by Schmidt and Friedrich on a number of occasions (e.g. [10-12]). Given a parametrised curve γ with tangent vector v^a consider the following propagation equation for a 1-form b_a along γ:

$$v^b \nabla_b b_a = (b_c v^c) b_a - (1/2)(g^{cd} b_c b_d) g_{ab} v^b + P_{ab} v^b \ . \tag{6.12}$$

Under 6.1, a solution of this equation is easily seen to transform as

$$\tilde{b}_a = b_a - \Upsilon_a \, . \tag{6.13}$$

Given b_a along γ, we use it to define a Weyl connection, that is to say a torsion-free connection preserving the conformal metric, by

$$v^b \nabla_b e^a = -(b_c e^c) v^a - (b_c v^c) e^a + (g_{cd} e^c v^d) g^{ab} b_b \tag{6.14}$$

on vectors e^a and

$$v^b \nabla_b \theta_a = (b_c v^c) \theta_a + (\theta_c v^c) b_a - (g^{cd} \theta_c b_d) g_{ab} v^b \tag{6.15}$$

on 1-forms.

The interpretation of b_a follows from (6.13): given b_a along γ we may find Ω in a neighbourhood of γ such that the first and second derivatives of Ω satisfy, at γ, the equations

$$\nabla_a \Omega = \Omega b_a , \tag{6.16}$$
$$\nabla_a \Upsilon_b = b_a b_b - (1/2) g^{cd} b_c b_d g_{ab} + P_{ab} \, . \tag{6.17}$$

This is an application of the Whitney Extension Theorem. It is easy to check that with this Ω the quantities $\Omega^2 g_{ab} e^a e^b$ and $\Omega^{-2} g^{cd} \theta_c \theta_d$ are constant along γ by virtue of (6.14) and (6.15). If we rescale the metric with this Ω according to (6.1) then in the rescaled space-time, the Rho-tensor will vanish at γ and so will b_a. Thus the propagation (6.14) and (6.15) should be thought of as parallel propagation in a metric for which the Ricci tensor vanishes at γ.

Schmidt [23] introduced this formalism in order to define a conformal boundary or c-boundary. His definition, following his earlier definition [22] of b-boundary was as follows:

> A curve $x(\lambda)$, $0 \leq \lambda < 1$, which is inextendible in M terminates at a point of the c-boundary $\partial_c M$ if a metric $\tilde{g}_{ab}$ exists in the conformal class such that the Ricci tensor of $\tilde{g}_{ab}$ vanishes on $x(\lambda)$ for $\lambda_0 \leq \lambda < 1$, and the generalised affine parameter length of $x(\lambda), \lambda_0 \leq \lambda < 1$, calculated with the connection of $\tilde{g}_{ab}$ is finite.

Now suppose we have a singularity, say in the sense of a point of the b- boundary, at which the Weyl tensor and some set of its derivatives are finite. How are we to recognise this finiteness? It is necessary to look at components in a frame and the formalism of Schmidt above gives a supply of frames defined in a conformally-invariant way. Can we extend the conformal structure to this point? For theorems on extensions, we turn to the work of Clarke and in the abstract to [6] we find the statement:

> A space-time has a local extension through a point on its b- boundary if and only if an appropriate number of covariant derivatives of the Riemann tensor have limiting values on a curve ending at the boundary point, measured in a parallelly-propagated frame.

Now to be fair, this is not a precise statement and there is more precision in [7] (see also [20] where there is an extension theorem quite close to the quoted passage above). However, if we take this as essentially true for the moment then we can put this with the definition of c-boundary to make a conjecture: suppose we have a singular point p, so a point p in a space-time M and a curve γ of finite generalised-affine-parameter length terminating at p; suppose there is a $\tilde{g}_{ab}$ as in the definition of c-boundary so that p is in the c-boundary; finally suppose that the components of the Weyl tensor and an appropriate number of its derivatives have limiting values at p in a frame which is parallelly-propagated in the metric connection of $\tilde{g}_{ab}$; then there is a local extension of the space-time with metric $\tilde{g}_{ab}$ through p, so that there is a local extension of the given conformal structure through the boundary point.

Such a conjecture would follow from the quoted statement on extensions because in the metric $\tilde{g}_{ab}$ the Ricci tensor is zero along γ, so that if the Weyl tensor and its first derivatives are bounded then so are the Riemann tensor and its first derivatives. It would more-or-less say that if the tractor curvature was bounded in the frames of (6.14) and (6.15) then the conformal structure could be extended. Of course if one wanted to extend across a whole surface like Σ above one would need stronger assumptions, for example of finite Weyl tensor along a whole congruence.

6.3 Review of Polytropic Perfect Fluid Case

We turn now to reviewing what has been done on the EFEs with a polytropic perfect fluid source or massless, collisionless matter and an isotropic singularity. Thus we can rescale with a conformal factor Ω to obtain a nonsingular unphysical metric g_{ab} with Ω vanishing on a smooth space-like surface Σ. The first result for perfect fluid is that the twist of the congruence defined by the fluid flow-lines must vanish, [13]. Now the fluid flows orthogonally to a family of space-like hypersurfaces which will provide a 'cosmic time' coordinate.

Next we want the EFEs with the perfect-fluid stress tensor. If the fluid 4-velocity is $\tilde{u}_a$ as a 1-form in the physical space-time then it must be $u_a = \Omega^{-1}\tilde{u}_a$ in the unphysical space-time in order to preserve normalisation. The stress tensor can therefore be written

$$\tilde{T}_{ab} = \Omega^{-2}((\rho + p)u_a u_b - p g_{ab}) \,. \tag{6.18}$$

The Einstein field equations can be written as

$$\tilde{P}_{ab} = 4\pi G\left(\tilde{T}_{ab} - (1/3)\tilde{g}_{ab}\tilde{g}^{cd}\tilde{T}_{cd}\right) \tag{6.19}$$

and we can use (6.18) and (6.19) in (6.5) to obtain a set of equations wholly framed in the unphysical space-time:

$$\begin{aligned} &\Omega^2 P_{ab} - \Omega\nabla_a\nabla_b\Omega + 2\nabla_a\Omega\nabla_b\Omega - (1/2)g_{ab}g^{cd}\nabla_c\Omega\nabla_d\Omega \\ &= 4\pi G(\Omega^4(\rho + p)u_a u_b - (1/3)\Omega^4\rho g_{ab}) \,. \end{aligned} \tag{6.20}$$

Now let us try to follow the procedure used in the discussion of $\mathcal{I}$ and assume that Ω is smooth at Σ with non-zero derivative. Then, near to Σ we have

$$\nabla_a \Omega = \Theta N_a + O(\Omega) \ , \ \nabla_a \nabla_b \Omega = O(1) \tag{6.21}$$

for some positive function Θ, where N_a is the normal to Σ. If we substitute Eq. (6.21) into (6.20) and take Ω to zero, we see that to get a non-trivial limit we must have $N^a = u^a$ at Σ, and $4\pi G\Omega^4(\rho+p)$ and $4\pi G\Omega^4\rho$ tending respectively to $2\Theta^2$ and $(3/2)\Theta^2$ as Ω goes to zero. This is only possible if, in the limit as the density ρ becomes very large, the pressure p tends to $\rho/3$. Thus the assumptions made on Ω, that it be smooth at Σ with nonzero derivative, are only compatible with equations of state subject to this restriction. The condition $p = \rho/3$ implies that the stress-tensor be trace-free, which one expects to be a better-behaved situation conformally.

Suppose for the moment we impose the condition $p = \rho/3$, then in the limit the fluid flows orthogonally to the singularity, and we can change the choice of conformal factor and choose it always to be a function of the cosmic time introduced above. Let us call it t and use it as the time coordinate, and let us perform the usual '3+1-decomposition' of the EFEs with respect to the coordinate t and comoving coordinates $x^i, i = 1, 2, 3$. The EFEs are just (6.20) and the equations of motion of the fluid follow from the Bianchi identities. In the 3+1-splitting, the variables are the spatial metric which we can write h_{ij} and the second fundamental form say K_{ij}. If, for simplicity, we take the equation of state to be precisely the radiation equation of state, $p = \rho/3$, then p and ρ can be written explicitly in terms of Ω and a function V related to the lapse and defined by $V^2 = g^{ab}t_{,a}t_{,b}$. We may introduce extra variables to write the EFEs in a first-order form, but the presence of explicit factors of Ω in (6.20) means that the first-order system has singularities in the time. After some work, and introducing more variables to make the system symmetric and to include the propagation of constraints, Newman [16] was able to write the whole system in the form

$$A^0(u)\partial_t u = A^i(u)\partial_i(u) + B(u)u + \frac{1}{t}C(u)u \tag{6.22}$$

where u stands for the field variables written as a vector, A^0 is positive definite and symmetric, the A^i are symmetric, and A^0, A^i, B, C are polynomial in their arguments.

Given the EFEs as the system (6.22), one may seek solutions as a power series in t. If the initial data at $t = 0$ is $u(0) = u_0$ then necessarily $C(u_0)u_0$ must vanish. This is a constraint on the initial data which we can call for convenience *the Fuchsian constraint*. Being algebraic, it is easy to solve: the free data consists just of the initial 3-metric $h_{ij}(0)$; the initial second fundamental form K_{ij} must vanish and the initial V is a non-zero constant which, without loss of generality, may be taken to be one. This is to be contrasted with the situation when data is given at any non-zero value of t. Then there are more data but the data are subjected to the usual Hamiltonian and momentum constraints.

We shall see that this is typical of the Cauchy problem with data at the singularity surface. The singularities in the equations at $t = 0$ impose different and more stringent constraints on the data, but the constraints are typically easy to solve.

There was no existence and uniqueness theorem in the literature for an evolution equation like (6.22), so that Newman set about providing one and the successful outcome of this was [8]. Given the conditions of this theorem, solutions of (6.22) exist and are unique given the data described above, so that there are many cosmological models with the radiation equation of state.

The next step was to change the equation of state of the fluid, [24-26]. We saw above that the equation of state is fixed, at least asymptotically for large ρ, if we require that Ω be smooth with non-vanishing derivative at Σ. Consequently we must drop these conditions if we want to change the equation of state, and a consideration of the Friedman-Robertson-Walker (FRW) cosmologies, which certainly have isotropic cosmological singularities, indicates the correct route to take. Essentially, if the equation of state is linear with $p = (\gamma - 1)\rho$ then one requires

$$\Omega = t^{2/(3\gamma-2)} \tag{6.23}$$

where Σ is the surface $t = 0$ and t is a smooth function in the unphysical space-time. One now proceeds as before to write the EFEs in terms of variables in the unphysical space-time and, after some labour, one arrives at another system of the form of (6.22). Again the theorem of [8] can be applied and one has existence and uniqueness of solutions given just the unconstrained initial 3-metric - there are no further degrees of freedom for the matter [2].

It would be possible to go on and consider more complicated equations of state or, for example, multi-component fluids but it seems very likely that the general picture will not change. Now recall the Weyl curvature hypothesis: these solutions necessarily have vanishing magnetic part of the Weyl tensor at Σ, while the electric part is finite and equal to the trace-free Ricci tensor of Σ. If this is zero, then Σ is a space of constant curvature and uniqueness of solution forces the cosmology to be FRW. Summarising in two sentences: the metric determines the matter; and if the Weyl tensor is initially zero, then it is always zero. Next we shall consider massless, collisionless matter as the source and move away from both of these statements.

Massless, collisionless matter as a source of the EFEs is determined by a distribution function $f(x^a, p_b)$, a non-negative function supported on the light-cone in the cotangent bundle of space-time. The collisionless condition is that f satisfies the Liouville or Vlasov equation, that is that f be constant along the geodesic flow of the physical metric:

$$\mathcal{L}f = \{\tilde{g}^{ab}p_a p_b, f\} = \tilde{g}^{ab}p_a\frac{\partial f}{\partial x^b} - p_a p_b\frac{\partial \tilde{g}^{ab}}{\partial x^c}\frac{\partial f}{\partial p_c} = 0 \tag{6.24}$$

where { , } means the Poisson bracket, and the stress-tensor which goes into the EFEs is

$$\tilde{T}_{ab} = \int p_a p_b f \tilde{\omega}$$

where $\tilde{\omega}$ is the invariant volume form on the light-cone defined with the respect to the physical metric $\tilde{g}_{ab}$. Under conformal rescaling as in (6.1), f is unchanged and continues to satisfy (6.24) just because it is supported on the light-cone, while the stress tensor transforms as

$$T_{ab} = \Omega^{-2}\tilde{T}_{ab}$$

because this is how ω transforms.

We can use this in (6.5) and (6.19) to arrive, once again, at the EFEs written in terms of quantities in the unphysical space-time. The analysis now proceeds as before. After some labour, one arrives at a system like (6.22) (the integrals involved in defining T_{ab} mean that $C(u)$ contains some integral expressions; for details see [1,2]). The data are the initial values for the distribution function, say f_0, and the first and second fundamental forms of the singularity surface, say $h_{ij}(0), K_{ij}(0)$. These are related by the Fuchsian constraint, but this constraint can be solved to determine $h_{ij}(0)$ and $K_{ij}(0)$ from f_0. There remains a constraint on f_0 derived from the part of the Fuchsian constraint related to the familiar momentum constraint:

$$\int p_i f_0(x^j, p_k) d^3p = 0. \tag{6.25}$$

The equation determining the metric from f_0, which involves a gauge-choice, is worth giving. If a_{ij} is the initial metric and b^{ij} is its inverse then

$$a_{ij}(det(a))^{1/2} = \int \frac{p_i p_j f_0}{(b^{km}p_k p_m)^{1/2}} d^3p \tag{6.26}$$

which remarkably has a unique solution for a_{ij} given f_0 (the trace of (6.26) is fixed by gauge-choice).

Recall again the Weyl curvature hypothesis. This time, one can see from examples that there are choices of f_0 for which the Weyl tensor vanishes at the singularity surface but is subsequently nonzero. Summarising again in two sentences: this time the matter, in the form of f_0, determines the metric; and f_0 can be such that the Weyl tensor vanishes initially but is subsequently non-zero.

Summarising this section, we have seen how to use the Newman-Claudel theorem and its extension by Anguige to produce cosmological models with perfect fluid or massless collisionless matter as source. The Fuchsian condition constrains the allowed data differently in the different cases, but the allowed data may be freely given.

6.4 Further Matter Models

We want briefly to consider further matter models to see how robust the picture developed in Sect. 6.3 is. In the following cases, I have taken the calculations far enough to find the Fuchsian constraints but not far enough to reduce the EFEs to the form of (6.22). These results should therefore be regarded as tentative.

6.4.1 Massive Einstein-Vlasov

Suppose we have massive, collisionless matter with one particle species as source, so that the distribution function f is supported on the corresponding mass hyperboloid rather than the light-cone. Then the Liouville equation (6.24) in the unphysical space-time has an explicit appearance of Ω and its derivative, and the volume form $\tilde{\omega}$ has an explicit appearance of the mass m of the particle. The conformal transformation is more cumbersome but, as one might anticipate, the particle rest-mass is not significant at the singularity and the calculation proceeds as before.

6.4.2 Scalar Fields

Suppose the source is a single scalar field ϕ with the stress-tensor

$$\tilde{T}_{ab} = \frac{c^2}{8\pi G}(\phi_a\phi_b - (1/2)\tilde{g}_{ab}(\phi^c\phi_c + V(\phi))).$$

The situation is very much like a stiff perfect fluid unless the potential grows very rapidly with increasing ϕ (ϕ diverges like $(\sqrt{3}/2)\log t$ and, to be like a stiff fluid, we need V to diverge, if at all, slower than t^{-2}).

6.4.3 Einstein-Yang-Mills-Vlasov

This is a matter model considered by Choquet-Bruhat and coworkers [4,5]. The model consists of massless, collisionless particles with colour charge, moving in the Yang-Mills field generated by the charge. A limit would be massless, collisionless, electromagnetically charged particles. The problem is like the Einstein-Vlasov case with data f_0, plus an initial 3-potential $\mathcal{A}_i$ and a densitised electric field $\mathcal{E}^i$. There is a Gauss-law constraint as well as a modification of the momentum constraint (6.25), and the equations determining the initial first and second fundamental form are modified.

6.4.4 Einstein-Boltzmann

It was suggested by Rendall and Anguige [21] that the Einstein-Vlasov picture found above might become more like the perfect fluid picture if collisions were included. This would mean modifying the Liouville equation (6.24) by including a collision term:

$$\mathcal{L}f = C(f, f)$$

where $C(f, f)$ is a bilinear term with a kernel related to a collision cross-section σ (see e.g. [9]). Now the conformal rescaling behaviour of this depends on the behaviour of σ for large values of the energy. If σ grows slower than a certain rate with energy, then the collision term is negligible near the singularity and the situation is as for collisionless matter. However, if σ grows sufficiently rapidly then the existence of an isotropic singularity implies that the distribution function f must be isotropic (that is, the collision-term must vanish by reason of thermal equilibrium). Once f is isotropic, we are back to the perfect fluid case.

6.5 Conclusion and Future Possibilities

To conclude, we have seen that for a variety of matter models the EFEs for a cosmology with an isotropic singularity can be written as a singular Cauchy problem with data at the singularity. The data are subject to one or more Fuchsian conditions, which can often be solved in terms of free functions. In some cases, existence and uniqueness of solutions has been proved, so that there are lots of examples. Even without proofs of existence, it can be straightforward to develop power-series solutions.

What does one want to do next? Analytically, one could seek to extend the results obtained to more complicated matter models, but there is also scope for numerical calculations. One could consider more sophisticated matter models and look at the formation of structure. In particular, one could seek to relate observational constraints to the initial metric or distribution function. There are a range of problems connected with inflation. One could consider a mixture of scalar fields with radiation and look for inflationary solutions. One could ask: Does inflation occur for an open set of data? does it lead to isotropy and homogeneity? Equally one could see whether, given an isotropic singularity, inflation is actually necessary to obtain the observed universe. One reason that this formalism might be computationally simpler is that most current calculations of inflation typically give data at some time after the singularity, when the nonlinear constraints of GR have to be solved. With an isotropic singularity the data is often freely specifiable.

References

1. K. Anguige, *Isotropic cosmological singularities 3: The Cauchy problem for the inhomogeneous conformal Einstein-Vlasov equations*, Ann.Phys.(NY) **282** 395-419 (2000)
2. K. Anguige and K.P. Tod, *Isotropic cosmological singularities 1: Polytropic perfect fluid spacetimes; 2: the Einstein-Vlasov system* Ann.Phys.(NY) **276** 294-320 (1999)
3. T.N. Bailey, M.G. Eastwood and A.R. Gover, *Thomas's structure bundle for conformal, projective and related structures*, Rocky Mountain Jour. Math. **24** 1191-1217 (1994)
4. Y. Choquet-Bruhat, 'Yang-Mills plasmas' in *Global structure and evolution in general relativity* Springer Lecture Notes in Physics **460** ed. by S. Cotsakis and G.W. Gibbons (Springer, Berlin, Heidelberg 1996)
5. Y. Choquet-Bruhat and N. Noutchegueme, *Système de Yang-Mills-Vlasov en jauge temporelle*, Ann.Inst.H Poincaré **55** 759-787 (1991)
6. C.J.S. Clarke, *Local extensions in singular space-times*, Comm.Math.Phys. **32** 203-214 (1973)
7. C.J.S. Clarke, *The analysis of space-time singularities* Cambridge Lecture Notes in Physics **1** (CUP, Cambridge 1993)
8. C.M.Claudel and K.P.Newman, *The Cauchy problem for quasi-linear hyperbolic evolution problems with a singularity in the time*, Proc.Roy.Soc.Lond. **A 454** 1073-1107 (1998)

9. S.R. de Groot, W.A. van Leeuwen and Ch.G. van Weert: *Relativistic Kinetic Theory* (North-Holland Publishing Co, Amsterdam 1980)
10. H. Friedrich, *Twistor connection and normal conformal Cartan connection*, Gen.Rel.Grav. **8** 303-312 (1977)
11. H. Friedrich, *Einstein equations and conformal structure: existence of anti-de Sitter-type space-times*, J.Geom.Phys. **17** 125-184 (1995)
12. H. Friedrich and B.G. Schmidt, *Conformal geodesics in general relativity* Proc.Roy.Soc.Lond **A 414** 171-195 (1987)
13. S.W. Goode, *Vorticity and isotropic singularities*, Gen.Rel.Grav.**19** 1075-1082 (1987)
14. S.W. Goode and J. Wainwright, *Isotropic singularities in cosmological models*, Class.Quant.Grav. **2** 99-115 (1985)
15. R.P.A.C. Newman, *Conformal singularities and the Weyl curvature hypothesis*, Rend.Sem.Mat.Univ.Pol.Tor. **50** 61-67 (1992)
16. R.P.A.C. Newman, *On the structure of conformal singularities in classical general relativity: 1 and 2*, Proc.Roy.Soc.Lond.**A 443** 473-492 and 493-515 (1993)
17. R. Penrose, 'Singularities and time-asymmetry' in: *General Relativity and Gravitation: an Einstein centennial volume* ed. by W. Israel and S.W. Hawking (CUP, Cambridge 1979)
18. R. Penrose: *The Emperor's New Mind* (OUP, Oxford 1989)
19. R. Penrose and W. Rindler: *Spinors and space-time 2: spinor and twistor methods in space-time geometry* (CUP, Cambridge 1986)
20. I. Racz, *Space-time extensions I*, J. Math. Phys. **34** 2448-2464 (1993)
21. A. Rendall: private communication.
22. B.G. Schmidt, *A new definition of singular points in general relativity*, Gen.Rel.Grav. **1** 269-280 (1971)
23. B.G. Schmidt, *A new definition of conformal and projective infinity of space-times*, Comm.Math.Phys. **36** 73-90 (1974)
24. K.P. Tod, *Isotropic singularities and the $\gamma = 2$ equation of state*, Class.Quant.Grav.**7** L13-L16 (1990)
25. K.P. Tod, *Isotropic singularities and the polytropic equation of state*, Class.Quant.Grav.**8** L77-L82 (1991)
26. K.P. Tod, *Isotropic singularities*, Rend.Sem.Mat.Univ.Pol.Tor.**50** 69-92 (1992)

7 Polyhomogeneous Expansions Close to Null and Spatial Infinity

Juan Antonio Valiente Kroon

Max Planck Institute für Gravitationsphysik, Albert Einstein Institut,
Am Mühlenberg 1, 14476 Golm bei Potsdam, Germany

Abstract. A study of the linearised gravitational field (spin 2 zero-rest-mass field) on a Minkowski background close to spatial infinity is done. To this purpose, a certain representation of spatial infinity in which it is depicted as a cylinder is used. A first analysis shows that the solutions generically develop a particular type of logarithmic divergence at the sets where spatial infinity touches null infinity. A regularity condition on the initial data can be deduced from the analysis of some transport equations on the cylinder at spatial infinity. It is given in terms of the linearised version of the Cotton tensor and symmetrised higher order derivatives, and it ensures that the solutions of the transport equations extend analytically to the sets where spatial infinity touches null infinity. It is later shown that this regularity condition together with the requirement of some particular degree of tangential smoothness ensures logarithm-free expansions of the time development of the linearised gravitational field close to spatial and null infinities.

7.1 Introduction

In [7] an analysis of the behaviour of the gravitational field close to null infinity and spatial infinity has been given. To this end, a new representation of spatial infinity was introduced. In this representation spatial infinity is depicted as a cylinder, as opposed to the standard representation of it as a point (see e.g. [21]). This representation has, among other things, the following important feature: it allows to formulate an initial value problem in which the data and the equations are regular; furthermore, space-like infinity and null infinity have a finite representation with their structure and location known *a priori*. The aforementioned analysis shows that a certain type of logarithmic divergences arise at the sets where spatial infinity "touches" null infinity (which we will denote by I^+ and I^-) unless a certain regularity condition is satisfied by the initial data. The sources of these logarithmic divergences can be ultimately traced back to the fact that some of the evolution equations degenerate at the sets $I^{\pm}$ (the symbol of the system looses rank). This precludes the direct application of standard techniques of partial differential equations if one wishes to push the solutions of the field equations all the way up to null infinity. This kind of degeneracies is a peculiarity of not only the gravitational field, it is also shared by linear massless fields. In order to shed some light on the nature and consequences of these degeneracies, here we will look at a simpler situation. Namely, the linearised gravitational field (spin 2 zero-rest-mass field) propagating on a Minkowski background. The final product of our analysis will be isolation of a series of requirements one needs to impose in order to obtain logarithmic-free expansions close to null and spatial infinity. One expects, in principle, that it is possible to extend this discussion to the gravitational field.

7.2 Minkowski Space-Time Close to Null and Spatial Infinity

Our point of departure is the usual representation of Minkowski space-time $\widetilde{\mathbb{M}}$ in Cartesian coordinates:

$$\widetilde{g} = \widetilde{\eta}_{\mu\nu} dy^\mu dy^\nu,$$

where $\widetilde{\eta}_{\mu\nu} = \mathrm{diag}(1, -1, -1, -1)$. We are interested in analysing the geometry of this space-time close to both null and spatial infinities. Intuitively one expects this region of space-time to contain the domain $D = \{y_\mu y^\mu < 0\}$. Let us start by considering the inversion in D given by

$$x^\mu = -\frac{y^\mu}{y^\lambda y_\lambda}, \quad y^\mu = -\frac{x^\mu}{x^\lambda x_\lambda}.$$

This inversion clearly maps D onto itself. Observe how a point that is far from the origin of the y^μ coordinates seems to be close to the x^μ origin. In this sense, the x^μ coordinates convey the notion of "lying at infinity". A simple calculation shows that:

$$\widetilde{g} = \frac{1}{(x^\lambda x_\lambda)^2} \widetilde{\eta}_{\mu\nu} dx^\mu dx^\nu.$$

Let now ρ be the standard radial coordinate associated with the spatial coordinates x^a, $a = 1, 2, 3$. Then $x^\lambda x_\lambda = (x^0)^2 - \rho^2$. Finally, the introduction of a new coordinate τ ($|\tau| < 1$) via $x^0 = \rho\tau$ yields the Minkowski metric in the form:

$$\widetilde{g} = \frac{1}{\rho^4(1-\tau^2)^2} \left(d(\rho\tau)^2 - d\rho^2 - \rho^2 d\sigma^2 \right),$$

where $d\sigma^2$ is the standard line element of the unit 2-sphere in spherical coordinates. The latter line element suggests the introduction of two different conformal factors:

$$\begin{aligned} \Xi &= \rho^2(1-\tau^2), \\ \Theta &= \rho(1-\tau^2). \end{aligned}$$

By means of these two conformal factors one obtains two different extensions $\mathbb{M}_\Xi$ and $\mathbb{M}_\Theta$ of the original Minkowski space-time near spatial infinity (see Fig. 7.1). The choice of the first conformal factor Ξ yields the standard representation of spatial infinity as a point. Consider the conformally rescaled metric:

$$\begin{aligned} g_\Xi &= \Xi^2 \widetilde{g} = d(\rho\tau)^2 - d\rho^2 - \rho^2 d\sigma^2, \\ &= (dx^0)^2 - (dx^1)^2 - (dx^2)^2 - (dx^3)^2. \end{aligned} \tag{7.1}$$

Here and all throughout this chapter tilded quantities will always refer to quantities in the *physical* (i.e. not conformally rescaled space-time). It can be shown that all radial space-like geodesics in $\widetilde{\mathbb{M}}$ map to space-like curves (not necessarily geodesics) in $\mathbb{M}_{\Xi}$ ending at $\{\rho = 0\}$. From (7.1) we can see that the set $\{\rho = 0\}$ is in fact a point, which we denote by i^0. On the other hand, consider the rescaled metric:

$$g_{\Theta} = \Theta^2 \widetilde{g} = \frac{1}{\rho^2}\left(d(\rho\tau)^2 - d\rho^2 - \rho^2 d\sigma^2\right).$$

It can also be checked that all radial space-like geodesics in $\widetilde{\mathbb{M}}$ map to space-like curves with endpoints at $\{\rho = 0\}$. Now, the metric seems to be singular at $\rho = 0$, but introducing a new coordinate $r = -\log(\rho)$ one gets:

$$g_{\Theta} = d\tau^2 - 2\tau d\tau dr - (1-\tau)^2 dr^2 - d\sigma^2.$$

Observe that $g_{\Theta}|_{\rho=0} = d\tau^2 - d\sigma^2$, thus the set $\{\rho = 0\}$ has now the topology of a cylinder with S^2 as spatial sections. We define $I = \{|\tau| < 1, \rho = 0\}$, $I^+ = \{\tau = 1, \rho = 0\}$, $I^- = \{\tau = -1, \rho = 0\}$, $I^0 = \{\tau = 0, \rho = 0\}$.

Let u, v be the null coordinates given by

$$u = \rho(1+\tau), \quad v = \rho(1-\tau).$$

The curves given by $u = u'$, u' a constant, and fixed angular coordinates are null geodesics for which it can be checked that map to outgoing (future oriented) null geodesics in $\widetilde{\mathbb{M}}$. Similarly, the null geodesics $v = v'$, with v' constant, and fixed angular coordinates map to incoming (past oriented) null geodesics in $\widetilde{\mathbb{M}}$. Thus the set $\mathscr{I}^+ = \{\tau = 1, \rho \neq 0\}$ from which the geodesics with $u =$ constant start corresponds to future null infinity. Similarly, the set $\mathscr{I}^- = \{\tau = -1, \rho \neq 0\}$ from which the geodesics with $v =$ constant emanate corresponds to past null infinity. The sets $I^{\pm}$ where spatial infinity "touches" null infinity are in a particular sense (to be discussed later) special. Therefore they are considered neither belonging to the cylinder at spatial infinity I nor to $\mathscr{I}^{\pm}$. Notice that the conformal factor Θ vanishes on $\mathscr{I}^{\pm}$, I and $I^{\pm}$. Most of the discussion here presented will be concerned with the extension $\mathbb{M}_{\Theta}$ of Minkowski space-time produced by the conformal factor $\Theta = \rho(1-\tau^2)$. Using the null geodesics discussed in the previous paragraph one can construct an adapted null (NP) tetrad such that the real vector l is tangent along the incoming null geodesics, and the vector n points along the outgoing null geodesics [1]. The complex vectors m and $\overline{m}$ are then usually chosen so that they span the tangent space of the two dimensional spheres $S_{\rho',\tau'} = \{\rho = \rho', \tau = \tau'\}$, with ρ', τ' given. This construction has the problem that m should vanish at least in a point on $S_{\rho',\tau'}$. For technical reasons it will be therefore convenient to coordinatise the 2-spheres $S_{\rho,\tau}$ by means of unitary representations of $SU(2,\mathbb{C})$. This can be done as the unit sphere S^2 can be identified with $SU(2,\mathbb{C})/U(1)$ which is obtained from $S^3 \simeq SU(2,\mathbb{C})$ via the

[1] The use of the vectors l and n here is reverse with respect to what it is the standard notation [11, 12]. The vector l is generally chosen to lie along outgoing null geodesics. This is done to agree with the spinorial notation used in [7].

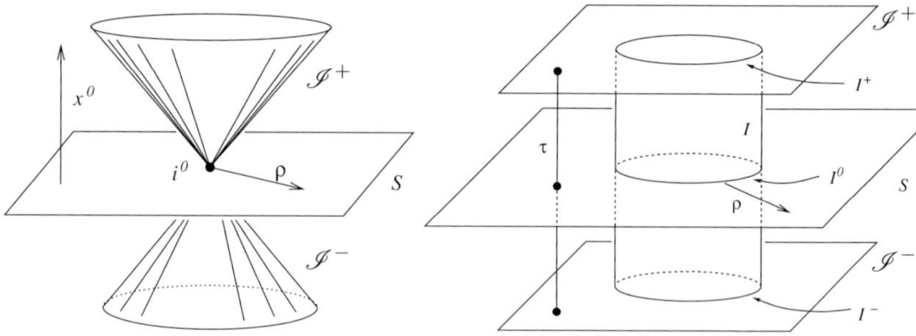

Fig. 7.1. Space-Time close to spatial and null infinities: to the left the standard representation of spatial infinity as a point i^0; to the right the representation where spatial infinity is envisaged as a cylinder

Hopf map. Any real analytic function f on $SU(2,\mathbb{C})$ admits an expansion of the form:

$$f = \sum_{m=0}^{\infty}\sum_{j=0}^{m}\sum_{k=0}^{m} f_{m,k,j} T_m{}^k{}_j,$$

with complex coefficients $f_{m,k,j}$. The functions $T_m{}^k{}_j$, $m = 0,1,2,\ldots$, $j,k = 0,\ldots,m$ can be related to the spherical harmonics $Y_{l,n}$ [8]. The function f will be said to have *spin-weight* s if its expansion in terms of the $T_m{}^k{}_j$ functions is of the form:

$$f = \sum_{q=|s|}^{\infty}\sum_{k=0}^{2q} f_{q,k} T_{2q}{}^k{}_{q-s}.$$

Important for later discussion will be the fact that any C^∞ function $\phi = \phi(\rho,t)$ on $\mathbb{R}_0^+ \times SU(2,\mathbb{C})$ such that for fixed ρ it is of spin-weight s, with s independent of ρ, has a normal expansion of the form:

$$\phi = \sum_{p\geq|s|}\sum_{q=|s|}^{p}\sum_{k=0}^{2q} \phi_{p;q,k} T_{2q}{}^k{}_{q-s}\rho^p,$$

with $\phi_{p;q,k}$ a complex number. This last result will be extensively used. For further details on the $T_m{}^k{}_j$ functions, the reader is referred to [5].

7.3 Linearised Gravity in the F-Gauge

We will describe the linearised gravitational field by means of a spin 2 zero-rest-mass field ϕ_{abcd} satisfying

$$\nabla^{aa'}\phi_{abcd} = 0.$$

Using the null tetrad described in the previous section, one can recover the $(D, \overline{\delta})$ and (Δ, δ) equations of the NP formalism (see e.g. [11, 12]) [2]:

$$(1+\tau)\partial_\tau \phi_n - \rho\partial_\rho \phi_n + X_+\phi_{n+1} = (n-2)\phi_n, \tag{7.2}$$

$$(1-\tau)\partial_\tau \phi_{n+1} + \rho\partial_\rho \phi_{n+1} + X_-\phi_n = (n-1)\phi_{n+1} \tag{7.3}$$

where $n = 0, 1, 2, 3$. The coefficients ϕ_n can be shown to have spin-weight $2-n$. The operators X_+ and X_- are complex linear combinations of left invariant vector fields on $SU(2, \mathbb{C})$, and can be related to the $\eth$ and $\overline{\eth}$ operators of Newman & Penrose [8]. The NP equations (7.2) and (7.3) imply a set of five propagation equations:

$$(1+\tau)\partial_\tau \phi_0 - \rho\partial_\rho \phi_0 + X_+\phi_1 = -2\phi_0, \tag{7.4}$$

$$\partial_\tau \phi_1 + \tfrac{1}{2}X_-\phi_0 + \tfrac{1}{2}X_+\phi_2 = -\phi_1, \tag{7.5}$$

$$\partial_\tau \phi_2 + \tfrac{1}{2}X_-\phi_1 + \tfrac{1}{2}X_+\phi_3 = 0, \tag{7.6}$$

$$\partial_\tau \phi_3 + \tfrac{1}{2}X_-\phi_2 + \tfrac{1}{2}X_+\phi_4 = \phi_3, \tag{7.7}$$

$$(1-\tau)\partial_\tau \phi_4 + \rho\partial_\rho \phi_4 + X_-\phi_3 = 2\phi_4, \tag{7.8}$$

and a set of three constraint equations:

$$\tau\partial_\tau \phi_1 - \rho\partial_\rho \phi_1 + \tfrac{1}{2}X_+\phi_2 - \tfrac{1}{2}X_-\phi_0 = 0, \tag{7.9}$$

$$\tau\partial_\tau \phi_2 - \rho\partial_\rho \phi_2 + \tfrac{1}{2}X_+\phi_3 - \tfrac{1}{2}X_-\phi_1 = 0, \tag{7.10}$$

$$\tau\partial_\tau \phi_3 - \rho\partial_\rho \phi_3 + \tfrac{1}{2}X_+\phi_4 - \tfrac{1}{2}X_-\phi_2 = 0. \tag{7.11}$$

This latter set of equations gives rise to equations on the initial hypersurface $S = \{\tau = 0\}$ which correspond to the constraints coming from the linearised Bianchi identities.

Using the equations (7.4)-(7.8) one obtains a symmetric hyperbolic system [2, 9] by simply multiplying (7.5)-(7.7) by a factor of 2. The resulting system is of the form:

$$A^0\partial_\tau \Phi + A^1\partial_\rho \Phi + A^+X_+\Phi + A^-X_-\Phi + B\Phi = 0, \tag{7.12}$$

where A^μ ($\mu = 0, 1, 2, 3$) and B are 5×5 matrices. It can be more concisely written in the form of $(A^\mu)_{\alpha\beta}\partial_\mu\phi_\beta + B_\alpha = 0$. Crucial for our discussion will be to realize that the matrix $A^0 = \mathrm{diag}(1+\tau, 2, 2, 2, 1-\tau)$ looses rank at $\tau = \pm 1$. This degeneracy will be the source of most of our problems as this fact precludes the direct use of the standard theory of symmetric hyperbolic systems.

Let Q be the 5×5 matrix with components given by

$$(Q)_{\alpha\beta} = (A^\mu)_{\alpha\beta}\partial_\mu\phi.$$

[2] Here again, in order to agree with reference [7] our notation is reversed with respect to the standard one of the NP formalism. Our ϕ_0, ϕ_1, ϕ_2, ϕ_3 and ϕ_4 correspond respectively to the ϕ_4, ϕ_3, ϕ_2, ϕ_1 and ϕ_0 of the standard NP notation.

Then the characteristics of the system (7.12) are the hypersurfaces $\phi = k$, with k a real constant, and such that the scalar field ϕ satisfies

$$\det Q = 0.$$

It can be readily checked that the hypersurfaces $u = k$ and $v = k'$, with u and v defined as in the preceeding section, are characteristics of the symmetric hyperbolic system (7.12). In particular, the hypersurfaces defined by $v = 0$ and $u = 0$ with $\rho \neq 0$ correspond to $\mathscr{I}^+$ and $\mathscr{I}^-$ respectively. Hence, $\mathscr{I}^+$ and $\mathscr{I}^-$ are both characteristic hypersurfaces of the system (7.12). The evaluation of the NP equations (7.3) on $\mathscr{I}^+$ gives rise to transport equations which enable us to calculate the value of the components $\phi_1, \ldots, \phi_4$ on $\mathscr{I}^+$ from a knowledge of the radiation field ϕ_0 on $\mathscr{I}^+$. Similarly, using equations (7.2) one obtains a corresponding set of transport equations on $\mathscr{I}^-$ by means of which it is possible to calculate the values of $\phi_0, \ldots, \phi_3$ from a knowledge of ϕ_4 on $\mathscr{I}^-$.

A further look to the system (7.12) reveals that at $\rho = 0$ the whole system reduces to transport equations which enable us to calculate the value of ϕ_n ($n = 0, \ldots, 4$) from their value at the initial hypersurface S. Hence, we say that I is a total characteristic of (7.12). As a consequence, no boundary data can be prescribed on I.

7.3.1 Initial Data for Linearised Gravity

That the spin 2 zero-rest-mass field ϕ_{abcd} can be used to describe the linearised gravitational field can be seen as follows. Consider a family $\widetilde{g}_{\mu\nu}(\lambda)$ of space-times depending in a C^1 fashion on the parameter λ, and such that at $\lambda = 0$ one obtains Minkowski space-time. Therefore,

$$\widetilde{g}_{\mu\nu} = \widetilde{\eta}_{\mu\nu} + \lambda \widetilde{h}_{\mu\nu} + \mathcal{O}(\lambda^2).$$

The symmetric rank 2 tensor $\widetilde{h}_{\mu\nu}$ describes the first order deviation from flatness of the metric $\widetilde{g}_{\mu\nu}$. The computation of the curvature to first order in λ yields [12]:

$$\widetilde{K}_{\mu\nu\lambda\rho} = 2\widetilde{\nabla}_{[\mu}\widetilde{\nabla}_{|[\nu}\widetilde{h}_{\lambda]|\rho]},$$

where $\widetilde{\nabla}_\mu$ is the covariant derivative of the flat space-time. Then the linearised field equations take the form:

$$\widetilde{K}_{\mu\nu\lambda}{}^{\rho} = 0. \tag{7.13}$$

The tensor $\widetilde{K}_{\mu\nu\lambda\rho}$ is trace-free and possesses the same symmetries of the Riemann tensor. It can be described in spinorial language by a totally symmetric spinor $\widetilde{\phi}_{abcd}$ satisfying the spin-2 zero-rest-mass field equation:

$$\widetilde{\nabla}^{aa'}\widetilde{\phi}_{abcd} = 0. \tag{7.14}$$

Equation (7.14) is a sufficient condition for the tensor $\widetilde{K}_{\mu\nu\lambda\rho}$ to be derivable (locally) from some symmetric tensor $\widetilde{h}_{\mu\nu}$ [13]. If the spinorial field $\widetilde{\phi}_{abcd}$ is set to transform as $\phi_{abcd} = \Omega^{-1}\widetilde{\phi}_{abcd}$ then the equation (7.14) is conformally covariant.

Let us assume that the family of metrics $\widetilde{g}_{\mu\nu}(\lambda)$ arise from the Einstein evolution of a corresponding 1-parameter family of asymptotically flat, time symmetric initial data given by the 3-dimensional metric

$$^{(3)}\widetilde{g}_{\alpha\beta} = \widetilde{\delta}_{\alpha\beta} + \lambda \ ^{(3)}\widetilde{h}_{\alpha\beta} + \mathcal{O}(\lambda^2).$$

It will be assumed that $^{(3)}\widetilde{g}_{\alpha\beta}$ satisfies the time symmetric vacuum constraint equations. Then, the symmetric tensor $^{(3)}\widetilde{h}_{\alpha\beta}$ is a solution of the corresponding linearised vacuum constraint equations. In the present discussion we will consider 1-parameter families of 3-metrics $^{(3)}\widetilde{g}_{\alpha\beta}$ such that their suitably conformally rescaled counterparts $^{(3)}g_{\alpha\beta}$ extend analytically near i, the infinity of the initial Cauchy hypersurface. For conceptual reasons it is convenient to distinguish i from i^0, the spatial infinity of the whole space-time. It can be proved [7] that under this assumption the components of the conformally rescaled Weyl spinor of the family of metrics $g_{\mu\nu}(\lambda)$ in the unphysical space-time evaluated on the initial hypersurface $S = \{\tau = 0\}$ are of the form:

$$\psi_n(\lambda)|_S = \sum_{p=|2-n|}^{\infty} \psi_n^p(\lambda)|_S \rho^n,$$

where

$$\psi_n^p(\lambda)|_S = \sum_{q=|2-n|}^{p} \sum_{k=0}^{2q} c_{n,p;q,k}(\lambda) T_{2q}{}^{k}{}_{q-2+n}$$

with complex coefficients $c_{n,p;q,k}$ $(n = 0, 1, 2, 3)$. Now, upon linearization one obtains a similar behaviour for the components of the spinorial field ϕ_{abcd}. Namely,

$$\phi_n|_S = \sum_{p=|2-n|}^{\infty} \phi_n^p|_{\tau=0}\rho^n, \tag{7.15}$$

where

$$\phi_n^p|_S = \sum_{q=|2-n|}^{p} \sum_{k=0}^{2q} d_{n,p;q,k} T_{2q}{}^{k}{}_{q-2+n}.$$

The components of the Weyl spinor ψ_n satisfy the constraints coming from the Bianchi identities. Hence, the components ϕ_n satisfy automatically their linearised version —equations (7.9)-(7.11) on $S = \{\tau = 0\}$.

Later considerations will make use of a particular spinorial object: the Cotton spinor (sometimes also called Bach spinor). The Cotton spinor B_{abcd} is the 3-dimensional analogue of the Weyl spinor [6]. It locally characterizes conformally flat 3-metrics in the sense that it vanishes identically if and only if the 3-metric is locally conformally flat. One can construct its corresponding linearised version b_{abcd}. Linearization around Minkowski yields the following expression:

$$b_{abcd} = 2D_{e(a}\Omega\phi_{bcd)}{}^{e} + \Omega D_{e(a}\phi_{bcd)}{}^{e},$$

where $\Omega = \rho^2$ is the conformal factor of the 3-metric of the Minkowski initial data, and D_{ab} its corresponding spinorial covariant derivative.

Inspired in a similar result by Friedrich we present now the following rather technical result. It will be of much use in later discussions.

Lemma 7.1. *The coefficients $d_{j,p;p,k}$ of the spin-2 zero-rest-mass (7.15) field satisfy the antisymmetry condition*

$$d_{0,p;p,k}|_{\tau=0} = -d_{4,p;p,k}|_{\tau=0}, \qquad p = 0, 1, \dots, \quad k = 0, \dots 2p. \tag{7.16}$$

Furthermore,

$$d_{j,p;p,k}|_{\tau=0} = 0, \qquad p = 0, \dots, s^*, \quad k = 0, \dots, 2p, \quad j = 0, \dots, 4,$$

if and only if

$$D_{(a_q b_q} \cdots D_{a_1 b_1} b_{abcd)}(i) = 0, \qquad q = 0, \dots, s^*.$$

The proof follows from direct linearization of theorem 4.1 in [7].

7.4 A Regularity Condition at Spatial Infinity

We now proceed to carry out an analysis of the transport equations one obtains upon evaluation of the field equations (7.4)-(7.8) and (7.9)-(7.11). Consider the equations (7.4)-(7.8) and (7.9)-(7.11). Differentiating them formally with respect to ρ and evaluating at the cylinder at $\rho = 0$ one gets:

$$(1+\tau)\partial_\tau \phi_0^p + X_+\phi_1^p - (p-2)\phi_0 = 0, \tag{7.17}$$
$$\partial_\tau \phi_1^p + \tfrac{1}{2}(X_-\phi_0^p + X_+\phi_2^p) + \phi_1 = 0, \tag{7.18}$$
$$\partial_\tau \phi_2^p + \tfrac{1}{2}(X_-\phi_1^p + X_+\phi_3^p) = 0, \tag{7.19}$$
$$\partial_\tau \phi_3^p + \tfrac{1}{2}(X_-\phi_2^p + X_+\phi_4^p) - \phi_3^p = 0, \tag{7.20}$$
$$(1-\tau)\partial_\tau \phi_4^p + X_-\phi_3^p + (p-2)\phi_4^p = 0, \tag{7.21}$$

and

$$\tau\partial_\tau \phi_1^p + \tfrac{1}{2}(X_+\phi_2^p - X_-\phi_0^p) - p\phi_1^p = 0, \tag{7.22}$$
$$\tau\partial_\tau \phi_2^p + \tfrac{1}{2}(X_+\phi_3^p - X_-\phi_1^p) - p\phi_2^p = 0, \tag{7.23}$$
$$\tau\partial_\tau \phi_3^p + \tfrac{1}{2}(X_+\phi_4^p - X_-\phi_2^p) - p\phi_3^p = 0, \tag{7.24}$$

where we have set $\phi_n^p = \partial_\rho^{(p)} \phi_n|_{\rho=0}$. One can expand these coefficients using the functions $T_k{}^l{}_m$ in the form:

$$\phi_n^p = \sum_{q=|2-n|}^{p} \sum_{k=0}^{2q} a_{n,p;q,k} T_{2q}{}^k{}_{q-2+n}. \tag{7.25}$$

The coefficients $a_{n,p;q,k}$ are, in principle, complex functions of τ. Using these expansions and the equations (7.18)-(7.20) and (7.22)-(7.24), one can calculate the coefficients $a_{1,p;q,k}$, $a_{2,p;q,k}$, $a_{3,p;q,k}$ from a knowledge of the $a_{0,p;q,k}$ and $a_{4,p;q,k}$ via a first order algebraic linear system. Some more algebra leads to

$$(1-\tau^2)\ddot{a}_{0,p;q,k} + (4+2(p-1)\tau)\dot{a}_{0,p;q,k} + (q+p)(q-p+1)a_{0,p;q,k} = 0, \quad (7.26)$$
$$(1-\tau^2)\ddot{a}_{4,p;q,k} + (-4+2(p-1)\tau)\dot{a}_{4,p;q,k} + (q+p)(q-p+1)a_{4,p;q,k} = 0, \quad (7.27)$$

for $p \geq 2$, $2 \leq q \leq p$, the overdot denoting differentiation with respect to τ. The equations (7.26) and (7.27) are examples of Jacobi equations. A canonical parametrisation for this class of ordinary differential equations is:

$$D_{(n,\alpha,\beta)}a \equiv (1-\tau^2)a'' + \{\beta-\alpha-(\alpha+\beta+2\tau)\}a' + n(n+\alpha+\beta+1)a = 0. \quad (7.28)$$

In our case the parameters are given by: $\alpha_0 = \beta_4 = -p-2$, $\beta_0 = \alpha_4 = -p+2$, $n_1 = p+q$, and $n_2 = p-q-1$. Regular solutions for equations (7.26) and (7.27) exist for $q \neq p$, and are given quite concisely in terms of Jacobi polynomials [14]. Namely,

$$a_{0,p;q,k} = C_1 P_{p-q-1}^{(-p-2,-p+2)}(\tau) + C_2 \left(\frac{1-\tau}{2}\right)^{p+2} P_{q-2}^{(p+2,-p+2)}(\tau), \quad (7.29)$$

$$a_{4,p;q,k} = D_1 P_{p-q-1}^{(-p+2,-p-2)}(\tau) + D_2 \left(\frac{1-\tau}{2}\right)^{p-2} P_{q+2}^{(p-2,-p-2)}(\tau), \quad (7.30)$$

where C_1, C_2, D_1 and D_2 are constants which can be determined from the initial data at $\tau = 0$. In the case $q = p$, the use of some identities of the Jacobi equation (7.28) leads to:

$$a_{0,p;p,k} = \left(\frac{1-\tau}{2}\right)^{p+2}\left(\frac{1+\tau}{2}\right)^{p-2}\left(E_0 + E_1\int_0^\tau \frac{ds}{(1+s)^{p-1}(1-s)^{p+3}}\right) \quad (7.31)$$

$$a_{4,p;p,k} = \left(\frac{1-\tau}{2}\right)^{p-2}\left(\frac{1+\tau}{2}\right)^{p+2}\left(F_0 + F_1\int_0^\tau \frac{ds}{(1+s)^{p+3}(1-s)^{p-1}}\right). \quad (7.32)$$

The use of Taylor expansions in the integrals shows that they give rise to logarithmic terms. More precisely, recalling that:

$$\int_0^\tau \frac{ds}{(1\pm s)^{p-1}(1\mp s)^{p+3}} = A_* \ln(1-\tau) + \frac{A_{p\pm 2}}{(1-\tau)^{p\pm 2}} + \cdots + \frac{A_1}{(1-\tau)} + A_0$$
$$+ B_* \ln(1+\tau) + \frac{B_{p\mp 2}}{(1+\tau)^{p\mp 2}} + \cdots + \frac{B_1}{(1+\tau)},$$

where the A's and B's are some constants. From the latter expansions one can conclude that the only non-regular coefficients in $a_{0,p;q,k}$ and $a_{4,p;q,k}$ in the $q = p$ case are of logarithmic nature. Analytic solutions arise if and only if $a_{0,p;p,m}(0) = a_{4,p;p,m}(0)$. That is only the case if the constants E_1 and F_1 in equations (7.31) and (7.32) are both zero. Now, using the antisymmetry condition (7.16) of lemma

7.1 one learns that in fact $a_{0,p;p,m}(0) = a_{4,p;p,m}(0) = 0$. It is not hard to see that if this is the case, then $a_{j,p;p,m}(0) = 0$ for $j = 0, \dots, 4$. Finally, using the second part of lemma 7.1 one can relate this behaviour of the initial data to the vanishing of the linearised Cotton tensor and its symmetrised derivatives on i. The results of the previous discussion are summarized in the following theorem.

Theorem 7.1. *The solutions of the transport equations on I corresponding to the system (7.4)-(7.8) and (7.9)-(7.11) extend analytically to $I^{\pm}$ if and only if the* regularity condition

$$D_{(a_s b_s} \cdots D_{a_1 b_1} b_{abcd)}(i) = 0, \quad s = 0, 1, \dots \;, \tag{7.33}$$

holds.

A peculiarity of the logarithms appearing in the solutions of the transport equations is that they only occur at the highest spherical harmonics sector at each order. This will be of importance later when discussing the asymptotic expansions of the field ϕ_{abcd} close to $\mathscr{I}^+$.

The afore discussed solutions of the transport equations allow us to obtain normal expansions,

$$\phi_n = \sum_{p \geq |2-n|} \phi_n^p \rho^p, \tag{7.34}$$

with the coefficients ϕ_n^p given by (7.25). Now, it is of interest to see how the form of these normal expansions and the regularity condition (7.33) reflect on the structure of the asymptotic expansions close to null infinity. In order to do this, one has essentially to reshuffle the normal expansions. So far, our discussion has applied to both future and past null infinities. The forthcoming analysis will without loss of generality be focused on $\mathscr{I}^+$. Nevertheless, a totally analogous treatment can be performed for $\mathscr{I}^-$.

Putting together the results from the analysis of the transport equations (7.17)-(7.21) and (7.22)-(7.24) one obtains normal expansions for the coefficients ϕ_0 and ϕ_4 which are of the form:

$$\phi_0 = \sum_{p=2} \left(S[0](1-\tau) + a_*^p \ln(1-\tau) P[p+2, 2p](1-\tau) \right) \rho^p,$$

$$\phi_4 = \sum_{p=2} \left(S[0](1-\tau) + b_*^p \ln(1-\tau) P[p-2, 2p](1-\tau) \right) \rho^p.$$

By $S[n](x)$ it will be understood a generic infinite series in x starting with x^n. Similarly, $P[n_1, n_2](x)$ denotes generic polynomials in x of order n_2 and whose lowest order term is x^{n_1}. The coefficients a_*^p, b_*^p and those in the series/polynomials are given in terms of the functions $T_m{}^k{}_j$. Matters of convergence of the normal expansions will be addressed in the next section. Now, a careful reshuffling in order to obtain expansions in $(1-\tau)$ reveals that,

$$\phi_0 = \sum_{p=0}^{3} S[2](\rho)(1-\tau)^p + \sum_{p\geq 4}\left(S[2](\rho) + \sum_{s=[t/2]}^{t-2} a_*^s \rho^s \ln(1-\tau)\right)(1-\tau)^p,$$

$$\phi_4 = \sum_{p\geq 0}\left(S[2](\rho) + \sum_{s=[t/2]}^{t+2} b_*^s \rho^s \ln(1-\tau)\right)(1-\tau)^p,$$

where $[t/2] = t/2$ if t is even, and $[t/2] = (t+1)/2$ if t is odd. Notice that the component ϕ_4 happens to be the most singular one of the field. If the regularity condition (7.33) is satisfied up to $s = s_*$, then it is not hard to see that,

$$\phi_4 = \sum_{k=0}^{s_*+1} c_k (1-\tau)^k + d_{s_*+1}(1-\tau)^{s_*+1}\ln(1-\tau) + \cdots, \tag{7.35}$$

where the coefficients c_k are C^∞ functions of ρ and the angular variables, and

$$d_{s_*+1} = \rho^{s_*+3} \sum_{m=0}^{2s_*+6} D_{s_*+1,m} T_{2s_*+6}{}^{m}{}_{s_*+5}, \tag{7.36}$$

with $D_{s_*+1,m}$ complex constants. Thus, the logarithmic term in the expansion has only dependence in the highest spherical harmonic sector possible at this order, i.e. $q = s_* + 3$. Finally, if the regularity condition holds also for $s = s_* + 1$ then $d_{s_*+1} = 0$.

7.5 Polyhomogeneous Expansions

7.5.1 A Substraction Argument

So far, nothing has been said about the convergence of the normal expansions (7.34) calculated in the previous section. We will now discuss how this can be done. As before, we write $\Phi = (\phi_0, \phi_1, \phi_2, \phi_3, \phi_4)$. Let

$$\Phi_N = \sum_{p=0}^{N} \frac{1}{p!}\Phi^p \rho^p$$

be the N order partial sum. The linear field equations (7.4)-(7.8) and (7.9)-(7.11) are such that Φ_N itself happens to be a solution of the field equations with truncated (to order N) initial data. The regularity of Φ_N will depend on whether the initial data satisfies the regularity condition (7.33) to a given order or not. For example, if the condition is satisfied up to $s = N$, then no $\ln(1-\tau)$ term will be present in the coefficients Φ^p, and thus Φ_N will be C^∞. How could one estimate the rest? Recall that the symmetric hyperbolic system derived from equations (7.4)-(7.8) breaks down precisely at $\tau = 1$. This is rather unfortunate, as we are mainly interested in observing the behaviour of the solutions on null infinity.

Bounds for the rest can be obtained by considering the conformal extension of Minkowski, $\mathbb{M}_{\Xi}$, in which spatial infinity is represented by a point i^0. The spin 2 zero-rest-mass field equations given in spherical coordinates are formally singular at $\rho = 0$ in this representation. Therefore, we resort to Cartesian coordinates (x^μ). In this way, the propagation equations read

$$
\begin{aligned}
&\partial_0\widehat{\phi}_0 - \partial_3\widehat{\phi}_0 + \partial_1\widehat{\phi}_1 + i\partial_2\widehat{\phi}_1 = 0,\\
&2\partial_0\widehat{\phi}_1 + \partial_1\widehat{\phi}_2 + \partial_1\widehat{\phi}_0 + i\partial_2\widehat{\phi}_2 - i\partial_2\widehat{\phi}_0 = 0,\\
&2\partial_0\widehat{\phi}_2 + \partial_1\widehat{\phi}_3 + \partial_1\widehat{\phi}_1 + i\partial_2\widehat{\phi}_3 - i\partial_2\widehat{\phi}_1 = 0,\\
&2\partial_0\widehat{\phi}_3 + \partial_1\widehat{\phi}_4 + \partial_1\widehat{\phi}_2 + i\partial_2\widehat{\phi}_4 - i\partial_2\widehat{\phi}_2 = 0,\\
&\partial_0\widehat{\phi}_4 + \partial_3\widehat{\phi}_4 + \partial_1\widehat{\phi}_3 - i\partial_2\widehat{\phi}_3 = 0,
\end{aligned}
$$

while the constraint equations are given by

$$
\begin{aligned}
&2\partial_3\widehat{\phi}_1 + \partial_1\widehat{\phi}_0 - \partial_1\widehat{\phi}_2 - i\partial_2\widehat{\phi}_0 - i\partial_2\widehat{\phi}_2 = 0,\\
&2\partial_3\widehat{\phi}_2 + \partial_1\widehat{\phi}_1 - \partial_1\widehat{\phi}_3 - i\partial_2\widehat{\phi}_1 - i\partial_2\widehat{\phi}_3 = 0,\\
&2\partial_3\widehat{\phi}_3 + \partial_1\widehat{\phi}_2 - \partial_1\widehat{\phi}_4 - i\partial_2\widehat{\phi}_2 - i\partial_2\widehat{\phi}_4 = 0.
\end{aligned}
$$

The components of the linearised Weyl tensor rescale as follows: $\widehat{\phi}_n = \rho^{-1-n}\phi_n$, $n = 0, \ldots, 4$, where $\rho^2 = (x^1)^2 + (x^2)^2 + (x^3)^2$. Thus, the smooth data in $\mathbb{M}_{\Theta}$, discussed in §3.1, are such that $\phi_0 = \mathcal{O}(\rho^2)$, $\phi_1 = \mathcal{O}(\rho)$, $\phi_2 = \mathcal{O}(1)$, $\phi_3 = \mathcal{O}(\rho)$ and $\phi_4 = \mathcal{O}(\rho^2)$, becomes singular in the $\mathbb{M}_{\Xi}$ picture at i (i.e. $\rho = 0$). However, if one provides initial data with sufficiently fast decay at i^0 then one can obtain energy estimates of the form:

$$
\|\widehat{\Phi}\|^2_{L^2(\mho)} \leq C\|\widehat{\Phi}\|^2_{L^2(\Sigma_1)}, \tag{7.37}
$$

where $\|\ \|^2_{L^2(\mho)}$ and $\|\ \|^2_{L^2(\Sigma_1)}$ denote the L^2 norms over the region $\mho$ [3] and the hypersurface Σ_1 respectively. Higher order estimates for the derivatives can be similarly obtained. The domain $\mho$ is as shown in figure (7.2). Notice how in this way one can obtain estimates that go up to $\mathscr{I}^+$ —and through it. In this spirit consider

$$
\widehat{X} = \widehat{\Phi} - \widehat{\Phi}_N.
$$

A close look to the rescaling rules will convince us that

$$
\widehat{X}|_{\tau=0} = \mathcal{O}(\rho^{N-5}).
$$

It can be readily checked that such a decay implies $\widehat{X}|_{\tau=0} \in H^{N-4}(\Sigma_1)$, where $H^k(\Sigma_1)$ denotes the k-th Sobolev space over the domain Σ_1. Standard theory of linear, symmetric hyperbolic systems (see for example [15]) then guarantees the existence of $\widehat{X}$ such that $\widehat{X} \in C^{N-7}(\mho)$. Hence, we have estimated the rest of

[3] The symbol $\mho$ is to be read "mho".

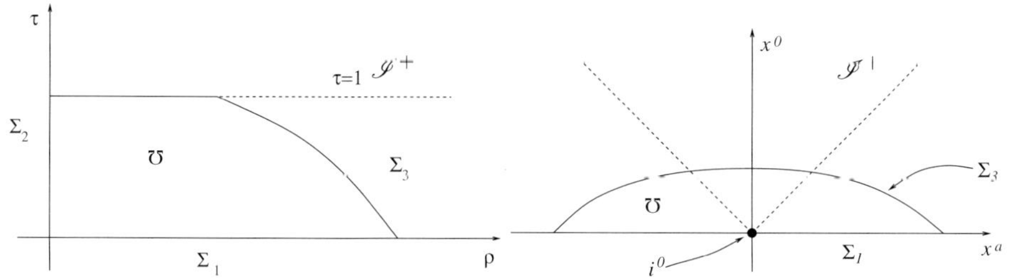

Fig. 7.2. The region of integration used to obtain the energy estimate (7.37). The first picture shows the domain of integration in the representation where spatial infinity is envisaged as a cylinder, and the second in the in the standard representation of spatial infinity as a point. Note that Σ_2 which seems to be an extended 2 dimensional surface is in fact a point: i^0. The surface Σ_3 is space-like and almost flat, and may reach and go through null infinity.

the partial sum Φ_N. If one wants the solution to be of class $C^k(\mho)$, $k \in \mathbb{N}$, then the previous discussion shows that one needs the regularity condition (7.33) to hold up to order $s = k + 7$.

The disadvantage of this approach is that it is hard to extend for the case of quasi-linear equations. Therefore, it will be of little use when we eventually try to address the full non-linear gravitational field. We need to devise something else.

7.5.2 An Investigation of Expansions Close to Null Infinity

The analysis of the solutions of the transport equations on the spatial infinity cylinder I show that logarithmic divergences will generically arise at $I^{\pm}$ for smooth initial data of the sort considered in §3.1 unless the regularity condition (7.33) is satisfied. Accordingly, if the regularity condition is satisfied, then the solutions to the transport equations extend smoothly through I^+. This is a hint that the evolution of generic smooth initial data will be polyhomogeneous, i.e. its expansions will contain $\ln(1-\tau)$ terms. An analysis and a discussion of the properties of such space-times close to null infinity, including the existence of an intriguing set of conserved quantities analogous to the Newman-Penrose constants has been given in [1, 16–18, 20]. The regularity condition given in theorem 7.1 ensures that the solutions of the transport equations (7.17)-(7.21) and (7.22)-(7.24) extend analytically to $I^{\pm}$. Now, with the discussion on I^+ more or less settled, one would like to study the effects the degeneracy of the field equations on I^+ has one null infinity. In particular, one would like to know what do the logarithmic terms on I^+ imply on $\mathscr{I}^+$. How do they propagate? In order to investigate this point, we resort to some ideas coming from the asymptotic characteristic initial value problem [3, 4, 10]. Consider a future oriented null hypersurface $\mathcal{N}_0$ intersecting $\mathscr{I}^+$ at $\mathcal{Z}_0$. One can provide characteristic initial data in the following way: ϕ_4 is prescribed on $\mathcal{N}_0$, ϕ_0 on the portion of $\mathscr{I}^+$ comprised between $\mathcal{Z}_0$ and I^+, and ϕ_2 and ϕ_1 on $\mathcal{Z}_0$. Theorems providing existence in a

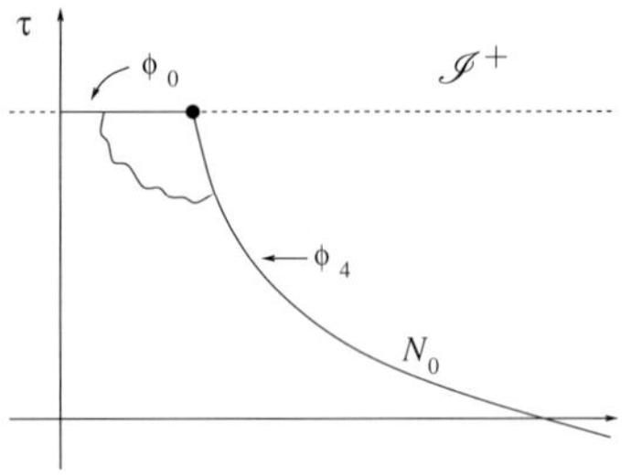

Fig. 7.3. The asymptotic characteristic initial value problem close to spatial infinity.

neighbourhood $\mathcal{U} \subset J^-(\mathcal{Z}_0)$ of $\mathcal{Z}_0$ for the conformal Einstein equations have been given in [3, 4, 10]. An existence theorem for polyhomogeneous Maxwell fields has been given in [19]. Problems with the characteristic approach have mainly to do with the ample freedom one has in choosing the initial data on $\mathcal{N}_0$. As an added factor, the theorems by Friedrich and Kánnár give only local existence. This is somehow a problem, as one would like to take the field ϕ_{abcd} all along $\mathscr{I}^+$ and evaluate at the set I^+, and then compare with the results obtained from the transport equations on I.

Inspired by the aforementioned characteristic approach, we will make use of the transport equations obtained on $\mathscr{I}^+$ arising from the NP equations (7.3). In this way once the radiation field ϕ_0 is provided, the remaining field components can be obtained essentially by a mere integration along the generators of null infinity. Expansions containing $\ln(1-\tau)$ terms appeared in the normal expansions (7.34). However, it is of interest to consider a more general type of situation in which terms of the form $\ln^k(1-\tau)$ can arise. So, we will assume that the components of the spin 2 zero-rest-mass field ϕ_{abcd} have asymptotic expansions near to $\mathscr{I}^+$ of the form:

$$\phi_n = \sum_{i\geq 0}\sum_{j=0}^{N_i} \phi_n^{(i,j)}(1-\tau)^i \ln^j(1-\tau), \tag{7.38}$$

$n = 0, \ldots, 4$, where the coefficients $\phi_n^{(i,j)}$ contain only ρ and angular dependence. They are assumed to be smooth at least in $\{\tau = 1, 0 < \rho \leq \rho_0\}$ for ρ_0 a given non-negative real number. Some calculus rules will be needed in order to manipulate the polyhomogeneous expansions (7.38).

Assumption. *For the expansions (7.38), it will be assumed that*

$$\partial_\tau\phi_n = -\sum_{i\geq 1}\sum_{j=0}^{N_i}\left(i\ln^j(1-\tau) + j\ln^{j-1}(1-\tau)\right)\phi_n^{(i,j)}(1-\tau)^{i-1}, \tag{7.39}$$

$$\partial_\rho\phi_n = \sum_{i\geq 0}\sum_{j=0}^{N_i}\phi_n^{(i,j)\prime}(1-\tau)^i\ln^j(1-\tau), \tag{7.40}$$

$$X_\pm\phi_n = \sum_{i\geq 0}\sum_{j=0}^{N_i} X_\pm\phi_n^{(i,j)}(1-\tau)^i\ln^j(1-\tau) \tag{7.41}$$

hold, where $'$ denotes differentiation with respect to ρ.

The above calculus rules can be used together with the transport equations on $\mathscr{I}^+$ to obtain equations for the coefficients $\phi_n^{(i,j)}$ in the Ansatz (7.38). At each order $(1-\tau)^p$ the solutions to these equations will be calculated. The conditions under which the coefficients $\phi_n^{(i,j)}$ can be pushed down to will then be investigated. The idea is then to compare the expansions obtained by means of this procedure with those arising from solving the transport equations on I. If an identification of the two different expansions is possible, then one can analyse the effect the regularity condition (7.33) has on the asymptotic expansions close to spatial infinity $\mathscr{I}^+$. Now, if at each order $(1-\tau)^p$ it is assumed that no $\log^k(1-\tau)$ terms are present in the lower terms, then it is possible to implement an inductive argument. The results of our forthcoming argument are summarized in the following theorem.

Theorem 7.2. *Let*

i) the components of the spin 2 zero-rest-mass field ϕ_{abcd} have an evolution expansion of the form:

$$\phi_n = \sum_{i\geq 0}\sum_{j=0}^{N_i} \phi_n^{(i,j)}(1-\tau)^i \ln^j(1-\tau),$$

$n = 0,\ldots,4$ with the coefficients $\phi_n^{(i,j)}$ at least C^1 in $\{\tau = 1, 0 < \rho \leq \rho_0\}$ for ρ_0 a non-negative constant;

ii) the calculus rules (7.39)-(7.41) hold;

iii) $\phi_0^{(0,0)}$ (the radiation field) be C^∞ in $\{\tau = 1, 0 \leq \rho \leq \rho_0\} \subset \mathscr{I}^+ \cup I^+$;

iv) $\phi_1^{(0,0)}$ and $\phi_3^{(0,0)}$ be bounded in $\{\tau = 1, 0 \leq \rho \leq \rho_0\}$;

v) the initial data $\phi_n|_{\tau=0}$, $n = 0,\ldots,4$, be obtained by linearization around flat space of asymptotically flat time symmetric initial data satisfying the Einstein vacuum constraints, and analytic in a neighbourhood of i;

vi) furthermore, the initial data satisfy the regularity condition

$$D_{(a_s b_s}\cdots D_{a_1 b_1} b_{abcd)}(i) = 0 \qquad s = 0,1,\ldots;$$

vii) $\partial_\rho^{(k)}\phi_4^{(i,0)}$ be continuous in $\{\tau = 1, 0 \leq \rho \leq \rho_0\}$, for $k = 0,\ldots,i+2$ (tangential smoothness).

Then the expansions of ϕ_n are in fact logarithm-free, that is:

$$\phi_n = \sum_{i\geq 0} \phi_n^{(k,0)}(1-\tau)^k,$$

where the coefficients $\phi_n^{(i,0)}$ are C^∞ in $\{\tau = 1, 0 \leq \rho \leq \rho_0\}$.

A proof of the latter shall be provided by resorting to an inductive argument. To this we turn now our attention.

The base step $(1-\tau)^0$. In agreement with the Ansatz (7.38) let the leading terms of the spin 2 zero-rest-mass field be of the form:

$$\phi_n = \sum_{j=0}^{N_0} \phi_n^{(0,j)} \ln^j(1-\tau) + \sum_{i\geq 1}\sum_{j=0}^{N_i} \phi_n^{(i,j)}(1-\tau)^i \ln^j(1-\tau),$$

with $n=0,\ldots,4$. Direct substitution into equations (7.2) show that

$$\phi_0^{(0,j)} = \phi_1^{(0,j)} = \phi_2^{(0,j)} = \phi_3^{(0,j)} = 0,$$

for $j\geq 1$. The coefficients $\phi_0^{(0,0)}$, $\phi_1^{(0,0)}$, $\phi_2^{(0,0)}$ can be determined from the equations (7.3) $n=0,1,2$. The coefficients satisfy

$$\rho\phi_1^{(0,0)\prime} + X_-\phi_0^{(0,0)} + \phi_1^{(0,0)} = 0, \tag{7.42}$$

$$\rho\phi_2^{(0,0)\prime} + X_-\phi_1^{(0,0)} = 0, \tag{7.43}$$

$$\rho\phi_3^{(0,0)\prime} + X_-\phi_2^{(0,0)} - \phi_3^{(0,0)} = 0, \tag{7.44}$$

which can in principle be solved if the coefficient $\phi_0^{(0,0)} = \phi_0|_{\mathscr{I}^+}$ (the radiation field) is given. It will be assumed that this is the case, and moreover, that it is C^∞. Being $\phi_0^{(0,0)}$ smooth, one has that

$$\phi_0^{(0,0)} = \sum_{k\geq 2} f_{0,k}^{(0)}\rho^k = \sum_{k\geq 2}\left(\sum_{q=2}^{k}\sum_{m=0}^{2q} F_{0,k;q,m}^{(0)} T_{2q}{}^{m}{}_{q-2}\right)\rho^k, \tag{7.45}$$

with $F_{0,k;q,m}^{(0)}$ complex constants. Using this expansion and equations (7.42)-(7.44) plus the further requirement of ϕ_1 and ϕ_3 to be smooth at I^+ one finds that

$$\phi_1 = \mathcal{O}(\rho^2), \quad \phi_2 = \mathcal{O}(1), \quad \phi_3 = \mathcal{O}(\rho). \tag{7.46}$$

The substitution of the Ansatz for the time development in equation (7.3), $n=3$, yields as a result the following hierarchy:

$$\begin{aligned} &-2\phi_4^{(0,N_0)} + \rho\phi_4^{(0,N_0)\prime} = 0,\\ &-2\phi_4^{(0,N_0-1)} - \phi_4^{(0,N_0)} + \rho\phi_4^{(0,N_0-1)\prime} = 0,\\ &\quad\vdots\\ &-2\phi_4^{(0,0)} - \phi_4^{(0,1)} + \rho\phi_4^{(0,0)\prime} + X_-\phi_3^{(0,0)} = 0. \end{aligned}$$

It can be solved from top to bottom, yielding:

$$\begin{aligned} \phi_4^{(0,N_0)} &= \widehat{\phi}_4^{(0,N_0)}\rho^2,\\ \phi_4^{(0,N_0-1)} &= \widehat{\phi}_4^{(0,N_0-1)}\rho^2 + \widehat{\phi}_4^{(0,N_0)}\rho^2\ln\rho,\\ &\vdots\\ \phi_4^{(0,0)} &= \left[\frac{1}{N_0!}\widehat{\phi}_4^{(0,N_0)}\rho^2\ln^{N_0}\rho + \cdots + \widehat{\phi}_4^{(0,0)}\rho^2\right] - \rho^2\int X_-\phi_3^{(0,0)}\rho^{-3}d\rho \end{aligned} \tag{7.47}$$

where the hatted quantities denote functions which appear during the integration and thus contain only angular dependence. The coefficients $\phi_4^{(0,j)}$ can be then "pushed down" to I^+, but not in general in such a way that they are smooth at I^+. The presence of terms of the form $\ln^k(1-\tau)$ in the Ansatz for the evolution yields as a result the presence of $\ln^{N_0-k}\rho$ terms in the corresponding expansions. This is an indication that if $\ln\rho$ terms are present in the initial data then these will propagate in the evolution via some particular $\ln^k(1-\tau)$ terms. Logarithmic terms of the form $\ln^j\rho$ do not appear in the normal expansions (7.34). Thus, in order to be able of performing an identification of the expansions on $\mathscr{I}^+$ and I some degree of tangential smoothness at I^+ will be assumed. A close look to formulae (7.47) shows that one needs $\phi_4^{(0,0)}$ to have two continuous ρ-derivatives at I^+. We will denote this by $\phi_4^{(0,0)} \in C^2_\rho(I^+)$. This requirement now implies that $\widehat{\phi}_4^{(0,j)} = 0$ for $j \geq 2$, and thus also $\phi_4^{(0,j)} = 0$ for $j \geq 2$. Note that the "integration constant" $\widehat{\phi}_4^{(0,1)}$ does not need to vanish as the $\rho^2\ln\rho$ term in $\phi_4^{(0,0)}$ may be canceled with a similar one coming from $\rho^2\int X_-\phi_3^{(0,0)}\rho^{-3}d\rho$. Indeed, the integral $\rho^2\int X_-\phi_3^{(0,0)}\rho^{-3}d\rho$ produces a term of the form $\rho^2\ln\rho$ from the ρ^2 term in $\phi_3^{(0,0)}$. As discussed previously,

$$\phi_3^{(0,0)} = \sum_{k\geq 1} f_{3,k}^{(0)}\rho^k, \tag{7.48}$$

where the coefficients $f_{3,k}^{(0)}$ are calculated from the $f_{0,k}^{(0)}$ of equation (7.45) by means of equations (7.42)-(7.44). Setting $\widehat{\phi}_4^{(0,1)} = X_- f_{3,2}^{(0)}$ in accordance with our previous discussion one finds that

$$\phi_4^{(0,1)} = X_- f_{3,2}^{(0)}\rho^2, \tag{7.49}$$

$$\phi_4^{(0,0)} = \widehat{\phi}_4^{(0,0)}\rho^2 - \sum_{\substack{k\geq 1\\ k\neq 2}} \frac{X_- f_{3,k}^{(0)}}{k-2}\rho^k. \tag{7.50}$$

One expects to be able to deduce the required degree of tangential smoothness form the smoothness properties of the initial data.

Finally, one is now in the position of performing an identification of the expansions obtained from the analysis of the transport equations on the cylinder, and those obtained from the transport equations at $\mathscr{I}^+$. As discussed, both expansions contain $\ln(1-\tau)$ terms. However, the $\ln(1-\tau)$ terms in the normal expansions appear only at the highest spherical sector at each order. So do those in our asymptotic expansions: from equation (7.48) one has

$$f_{3,2}^{(0)} = \sum_{q=1}^{2}\sum_{m=0}^{2q} F_{3,2;q,m}^{(0)} T_{2q}{}^{m}{}_{q+1}.$$

And hence

$$X_- f_{3,2}^{(0)} = -\sum_{q=1}^{2}\sum_{m=0}^{2q} F_{3,2;q,m}^{(0)}\beta_{2q,q+2} T_{2q}{}^{m}{}_{q+2},$$

with $\beta_{2q,q+2} = \sqrt{(q+2)(q-1)}$. Thus, X_- cancels the sector given by $q=1$, and as a consequence the only non-zero sector in (7.49) is $q=2$. It is noticed by passing that $X_-\phi_4^{(0,1)} = 0$ whence it follows that (tangential smoothness assumed) $X_-\phi_4$ is bounded at I^+. Hence, the leading term of the evolution expansion of ϕ_4 has the same form of the expansion given in (7.35) and (7.36). Consequently, if $b_{abcd}(i) = 0$ (regularity condition at order $s=0$) holds, we have shown that

$$\phi_n = \phi_n^{(0,0)} + \sum_{i\geq 1}\sum_{j=0}^{N_i} \phi_n^{(i,j)}(1-\tau)^i \ln^j(1-\tau), \tag{7.51}$$

$n = 0,\ldots,4$. That is, there are no logarithms in the leading term of the expansions.

The step $(1-\tau)^1$. The base induction step is somehow exceptional. In order to understand better the situation for the general case it is convenient to take a look to what happens at the order $(1-\tau)^1$. To begin with, let us assume that the analysis of base step $(1-\tau)^0$ has been carried out, and no $\ln(1-\tau)$ are present at that order. Consequently, the expansion (7.51) holds.

Using equations (7.3), $n = 0,1,2$, and a similar approach to the one used in the base step, one readily finds that:

$$\phi_0^{(1,j)} = \phi_1^{(1,j)} = \phi_2^{(1,j)} = \phi_3^{(1,j)} = 0,$$

for $j \geq 1$. This is a direct consequence of the fact that there are no logarithms at order $(1-\tau)^0$. Furthermore,

$$\phi_0^{(1,0)} = \frac{1}{2}\left[2\phi_0^{(0,0)} - \rho\phi_0^{(0,0)\prime} + X_+\phi_1^{(0,0)}\right], \tag{7.52}$$

$$\phi_1^{(1,0)} = \frac{1}{2}\left[\phi_1^{(0,0)} - \rho\phi_1^{(0,0)\prime} + X_+\phi_2^{(0,0)}\right], \tag{7.53}$$

$$\phi_2^{(1,0)} = \frac{1}{2}\left[-\rho\phi_2^{(0,0)\prime} + X_+\phi_3^{(0,0)}\right], \tag{7.54}$$

$$\phi_3^{(1,0)} = \frac{1}{2}\left[-\phi_3^{(0,0)} - \rho\phi_3^{(0,0)\prime} + X_+\phi_4^{(0,0)}\right]. \tag{7.55}$$

Thus, one can calculate the coefficients $\phi_0^{(1,0)}$, $\phi_1^{(1,0)}$, $\phi_2^{(1,0)}$ and $\phi_3^{(1,0)}$ directly from a knowledge of the coefficients at order $p=0$. Using (7.46) in equations (7.52)-(7.55) one concludes that

$$\phi_n^{(1,0)} = \sum_{k\geq|2-n|} f_{n,k}^{(1)}\rho^k.$$

The coefficients $f_{n,k}^{(1)}$ given in terms of the coefficients $f_{n,k}^{(0)}$ of the previous order. As in the base step, the logarithmic dependence comes from equation (7.3) with

$n = 3$. From it one obtains

$$\begin{aligned}
&-3\phi_4^{(1,N_1)} + \rho\phi_4^{(1,N_1)\prime} = 0,\\
&-3\phi_4^{(1,N_1-1)} - \phi_4^{(1,N_1)} + \rho\phi_4^{(1,N_1-1)\prime} = 0,\\
&\vdots\\
&-3\phi_4^{(1,0)} - \phi_4^{(1,1)} + \rho\phi_4^{(1,0)\prime} + X_-\phi_3^{(1,0)} = 0,
\end{aligned}$$

which again is solved from top to bottom so that

$$\begin{aligned}
\phi_4^{(1,N_1)} &= \widehat{\phi}_4^{(1,N_1)}\rho^3,\\
\phi_4^{(1,N_1-1)} &= \widehat{\phi}_4^{(1,N_1-1)}\rho^3 + \widehat{\phi}_4^{(1,N_1)}\rho^3\ln\rho,\\
&\vdots\\
\phi_4^{(1,0)} &= \left[\frac{1}{N_1!}\widehat{\phi}_4^{(1,N_1)}\rho^3\ln^{N_1}\rho + \cdots + \widehat{\phi}_4^{(1,0)}\rho^3\right] - \rho^3\int X_-\phi_3^{(1,0)}\rho^{-4}d\rho.
\end{aligned}$$

Whence one observes one more time the appearance of $\ln^k\rho$ dependence associated with the $\ln^{N_1-k}(1-\tau)$ in the Ansatz for the evolution. The expression $\rho^3\int X_-\phi_3^{(1,0)}\rho^{-4}d\rho$ can only account for the cancellation of the $\rho^3\ln\rho$ term in $\phi_4^{(1,0)}$ via the ρ^3 term in $\phi_3^{(1,0)}$. The coefficient $\phi_3^{(1,0)}$ itself contains no logarithmic dependence, as a consequence of equation (7.55) and the smoothness of the coefficients calculated at the base step $p = 0$. Following the spirit of previous discussions, we require $\phi_4^{(1,0)} \in C^3_\rho(I^+)$. Therefore $\widehat{\phi}_4^{(1,j)} = 0$ for $j = 2, \ldots, N_1$, and furthermore $\phi_4^{(1,j)} = 0$ for $j = 2, \ldots, N_1$. So, one is left with only two non-zero coefficients. Namely,

$$\begin{aligned}
\phi_4^{(1,1)} &= \widehat{\phi}_4^{(1,1)}\rho^3,\\
\phi_4^{(1,0)} &= \widehat{\phi}_4^{(1,1)}\rho^3\ln\rho + \widehat{\phi}_4^{(1,0)}\rho^3 - \rho^3\int X_-\phi_3^{(1,0)}\rho^{-4}d\rho
\end{aligned}$$

Now, writing

$$\phi_3^{(1,0)} = \sum_{k\geq 1} f_{3,k}^{(1)}\rho^k, \tag{7.56}$$

one has to set

$$\widehat{\phi}_4^{(1,1)} = X_- f_{3,3}^{(1)} \tag{7.57}$$

in order to eliminate the remaining logarithmic term in $\phi_4^{(1,0)}$.

In a similar way to what happened in the base step, one can further show that the logarithmic dependence is only found at the highest spherical harmonic sector. However, in this case and also in the general case, the analysis is a bit more elaborate. From equations (7.49) and (7.55) one finds that:

$$\phi_3^{(1,0)} = \frac{1}{2}\left[\sum_{k\geq 1}(-1-k)f_{3,k}^{(0)}\rho^k + X_+\widehat{\phi}_1^{(0,0)}\rho^2 - \sum_{k\geq 3}\frac{X_+X_-f_{3,k}^{(0)}}{k-2}\rho^k\right].$$

The spin weight of ϕ_3 is -1, and accordingly,

$$f^{(0)}_{3,k} = \sum_{q=1}^{k}\sum_{m=0}^{2q} F^{(0)}_{3,k;q,m} T_{2q}{}^{m}{}_{q+1}.$$

Thus, at the end of the day one has the following rather complicate-looking expression:

$$\begin{aligned}\phi_3^{(1,0)} &= \sum_{k\geq 2} f^{(1)}_{3,k}\rho^k \\ &= \tfrac{1}{2}(X_+\widehat{\phi}_4^{(0,0)} - f^{(0)}_{3,2})\rho^2 \\ &+\tfrac{1}{2}\sum_{\substack{k\geq 1\\ k\neq 2}}\sum_{q=1}^{k}\sum_{m=0}^{2q}\left(\tfrac{(q+2)(q-1)}{k-2} - (k+1)\right) F^{(0)}_{3,k;q,m} T_{2q}{}^{m}{}_{q+1}\rho^k.\end{aligned}$$

The term $(q+2)(q-1)-(k+1)(k-2)$ vanishes whenever $q=k-1$. Hence, if one writes

$$f^{(1)}_{3,k} = \sum_{q=1}^{k}\sum_{m=0}^{2q} F^{(1)}_{3,k;q,m} T_{2q}{}^{m}{}_{q+1},$$

one has $F^{(1)}_{3,k;k-1,m} = 0$ for $k = 1, 2, \ldots$. Therefore, the ρ^3 term in $X_-\phi_3^{(1,0)}$ contains only $q=3$ and $q=1$ spherical harmonic dependence, i.e.

$$f^{(1)}_{3,3} = \sum_{m=0}^{2} F^{(1)}_{3,3;1,m} T_{2}{}^{m}{}_{2} + \sum_{m=0}^{6} F^{(1)}_{3,3;3,m} T_{6}{}^{m}{}_{4}.$$

Therefore, using (7.57) one finally arrives at

$$\phi_4^{(1,1)} = -4\rho^3 \sum_{m=0}^{6} F^{(1)}_{3,3;3,m} T_{6}{}^{m}{}_{5}. \tag{7.58}$$

Again, the expansions have the same form of (7.35) and (7.36) with $s_* = 0$. If the regularity condition is satisfied up to $s=1$, i.e. if $b_{abcd}(i) = 0$ and $D_{(a_1b_1}b_{abcd)}(i) = 0$ then

$$\phi_n = \phi_n^{(0,0)} + \phi_n^{(1,0)}(1-\tau) + \sum_{i\geq 2}\sum_{j=0}^{N_i}\phi_n^{(i,j)}(1-\tau)^i \ln^j(1-\tau).$$

The general step $(1-\tau)^{p+1}$. The procedures of the general step are lengthier but nevertheless of a similar nature to those of the analysis of the $(1-\tau)^1$ terms. We begin by assuming that a similar analysis to that one carried for $(1-\tau)^1$ has been carried up to $(1-\tau)^p$ inclusive. Accordingly, the components of the field are assumed to have an expansion of the form:

$$\phi_n = \sum_{k=0}^{p}\phi_n^{(k,0)}(1-\tau)^k + \sum_{k\geq p+1}\sum_{j=0}^{N_k}\phi_n^{(k,j)}(1-\tau)^k \ln^j(1-\tau). \tag{7.59}$$

The coefficients $\phi_n^{(k,0)}$ with $k \leq p$ are C^∞, for they have been constructed out of the radiation field $\phi_0^{(0,0)}$ in a way that preserves the smoothness. Hence,

$$\phi_n^{(p,k)} = \sum_{k\geq|2-n|} f_{n,k}\rho^k,$$

where the coefficient $f_{n,k}$ contains all the angular dependence. In particular, for the component ϕ_3 they are such that

$$f_{3,k}^{(p)} = \sum_{q=1}^{k}\sum_{m=0}^{2q} F_{3,k;q,m}^{(p)} T_{2q}{}^{m}{}_{q+1},$$

with coefficients $F_{3,k;q,m}^{(p)}$ such that

$$F_{3,k;k-1,m}^{(p)} = F_{3,k;k-2,m}^{(p)} = \cdots = F_{3,k;k-p,m}^{(p)} = 0, \tag{7.60}$$

as suggested by the analysis of the $(1-\tau)^0$ and $(1-\tau)^1$ cases. Substitution of the Ansatz (7.59) in equations (7.2) with $n = 0, 1, 2, 3$ yields a hierarchy of $N_{p+1}+1$ equations. A similar analysis to that performed at order $(1-\tau)$ readily shows that

$$\phi_0^{(p+1,j)} = \phi_1^{(p+1,j)} = \phi_2^{(p+1,j)} = \phi_3^{(p+1,j)} = 0,$$

for $1 \leq j \leq N_{p+1}$. One also obtains a set of 4 recurrence relations which allows to calculate the coefficients $\phi_n^{(p+1,0)}$, $n = 0, 1, 2, 3$, from (in principle known) lower order terms. In future discussions only the expression for $\phi_3^{(p+1,0)}$ will be needed,

$$\phi_3^{(p+1,0)} = \frac{1}{2(p+1)}\left[(p-1)\phi_3^{(p,0)} - \rho\phi_3^{(p,0)\prime} + X_+\phi_4^{(p,0)}\right]. \tag{7.61}$$

The coefficients $\phi_4^{(p+1,j)}$ are on the other hand calculated using equation (7.3) with $n = 3$. Similar computations to those described in the $(1-\tau)^0$ and $(1-\tau)^1$ steps lead to:

$$\begin{aligned}
\phi_4^{(p+1,N_{p+1})} &= \widehat{\phi}_4^{(p+1,N_{p+1})}\rho^{p+3}, \\
\phi_4^{(p+1,N_{p+1}-1)} &= \widehat{\phi}_4^{(p+1,N_{p+1}-1)}\rho^{p+3} + \widehat{\phi}_4^{(p+1,N_{p+1})}\rho^{p+3}\ln\rho, \\
&\vdots \\
\phi_4^{(p+1,0)} &= \frac{1}{N_p!}\widehat{\phi}_4^{(p+1,N_{p+1})}\rho^{p+3}\ln^{N_{p+1}}\rho + \cdots + \widehat{\phi}_4^{(p+1,0)}\rho^{p+3} \\
&\quad - \rho^{p+3}\int X_-\phi_3^{(p+1,0)}\rho^{-4-p}d\rho.
\end{aligned} \tag{7.62}$$

Again, the term $\rho^{p+3}\int X_-\phi_3^{(p+1,0)}\rho^{-4-p}d\rho$ in the hierarchy (7.62) can only be used to cancel out the $\widehat{\phi}_4^{(p+1,1)}\rho^{p+3}\ln\rho$ term in the expression for the coefficient $\phi_4^{(p+1,0)}$. Thus, in order to perform an identification at I^+, some tangential smoothness will be required. More precisely, we will demand $\phi_4^{(p+1,0)} \in$

$C^{p+3}_\rho(I^+)$. This assumption implies that $\widehat{\phi}_4^{(p+1,N_{p+1})} = \ldots = \widehat{\phi}_4^{(p+1,2)} = 0$, with the further consequence of yielding $\phi_4^{(p+1,N_{p+1})} = \ldots = \phi_4^{(p+1,2)} = 0$. Hence as before, we are left with only two non-zero coefficients:

$$\phi_4^{(p+1,1)} = \widehat{\phi}_4^{(p+1,1)}\rho^{p+3}, \tag{7.63}$$

$$\begin{aligned}\phi_4^{(p+1,0)} = \widehat{\phi}_4^{(p+1,1)}\rho^{p+3}\ln\rho + \widehat{\phi}_4^{(p+1,0)}\rho^{p+3} \\ - \rho^{p+3}\int X_-\phi_3^{(p+1,0)}\rho^{-4-p}d\rho.\end{aligned} \tag{7.64}$$

The coefficient $\phi_3^{(p+1,0)}$ is smooth, as it is constructed from the $\phi_n^{(p,0)}$'s using equation (7.61). Consequently,

$$\phi_3^{(p+1,0)} = \sum_{k\geq 1} f_{3,k}^{(p+1)}\rho^k, \tag{7.65}$$

with the coefficients $f_{3,k}^{(p+1)}$ containing only angular dependence. Hence, one sees that

$$\rho^{p+3}\int X_-\phi_3^{(p+1,0)}\rho^{-4-p}d\rho = X_- f_{3,p+3}^{(p+1)}\rho^{p+3}\ln\rho + \sum_{\substack{k\geq 1\\ k\neq p+3}} \frac{X_- f_{3,k}^{(p+1)}}{k-p-3}\rho^k.$$

In accordance with our previous discussion one sets

$$\widehat{\phi}_4^{(p+1,1)} = X_- f_{3,p+3}^{(p+1)}.$$

In this way we have accounted for the remaining logarithm in $\phi_4^{(p+1,0)}$ —see equation (7.64). Therefore, at the end of the day one has that

$$\begin{aligned}\phi_4^{(p+1,1)} &= X_- f_{3,p+3}^{(p+1)}\rho^{p+3}, \\ \phi_4^{(p+1,0)} &= \widehat{\phi}_4^{(p+1,0)}\rho^{p+3} - \sum_{\substack{k\geq 2\\ k\neq p+3}} \frac{X_- f_{3,k}^{(p+1)}}{k-p-3}\rho^k.\end{aligned}$$

The component ϕ_4 is the most irregular one of the field, and thus, the $\ln(1-\tau)$ terms will appear firstly there. Carrying out the previous discussion to the following orders, one finds that the first $\ln(1-\tau)$ terms in ϕ_3, ϕ_2, ϕ_1 and ϕ_0 appear at orders $(1-\tau)^{p+2}$, $(1-\tau)^{p+3}$, $(1-\tau)^{p+4}$ and $(1-\tau)^{p+5}$ respectively.

Finally, in order to make use of the regularity condition (7.33) it is again necessary to identify the $\ln(1-\tau)$'s appearing in our expansions with those coming from the analysis of the transport equations on the cylinder I. Using equations (7.61) and (7.65) one finds:

$$\phi_3^{(p+1,0)} = \frac{1}{p+1}\left[G_{p+2}\rho^{p+2} + \sum_{k\geq 1}(p-1-k)f_{3,k}^{(p)}\rho^k - \sum_{\substack{k\geq 1\\ k\neq p+2}} \frac{X_+X_- f_{3,k}^{(p)}}{k-p-2}\rho^k \right].$$

One can make explicit the spherical harmonics dependence:

$$\phi_3^{(p+1,0)} = \frac{1}{p+1}\Bigg[G_{p+2}\rho^{p+2} + \sum_{\substack{k\geq 1\\ k\neq p+2}} \sum_{q=1}^{k}\sum_{m=0}^{2q} \left(\frac{(q+2)(q-1)}{k-p-2} + (p-k-1) \right) F^{(p)}_{3,k;q,m} T_{2q}{}^{m}{}_{q+1}\rho^{k} \Bigg],$$

where the coefficients $F^{(p)}_{3,k;q,m}$ satisfy (7.60). The expression $(k-p+1)(k-p-2)-(q+2)(q-1)$ is zero whenever $q=p-k$ and/or $q=k-p-1$. Hence, writing

$$f^{(p+1)}_{3,k} = \sum_{q=1}^{k}\sum_{m=0}^{2q} F^{(p+1)}_{3,k;q,m} T_{2q}{}^{m}{}_{q+1},$$

then one finds that

$$F^{(p+1)}_{3,k;k-p-1,m} = 0.$$

Furthermore, using the induction hypothesis (7.60) in equation (7.66) one finds that

$$F^{(p+1)}_{3,k;k-1,m} = F^{(p+1)}_{3,k;k-2,m} = \cdots = F^{(p+1)}_{3,k;k-p,m} = F^{(p+1)}_{k;k-p-1,m} = 0.$$

In particular if $k=p+3$ then $X_- f^{(p+1)}_{3,p+3}\rho^{p+3}\ln\rho$ will only contain spherical harmonic dependence at the $q=p+3$ sector. The other possibly remaining sector, the $q=1$ one, is annihilated by the X_- derivative. Accordingly, it has been proved that

$$\phi_4^{(p+1,1)} = -\beta_{2p+6,p+5}\rho^{p+3}\sum_{m=0}^{2p+6} F^{(p+1)}_{3,p+3;p+3,m} T_{2p+6}{}^{m}{}_{p+5}.$$

Hence, the expansions we have obtained so far are similar to those given in (7.35) and (7.36). Finally, if the regularity condition (7.33) holds up to order $p+1$ then one has:

$$\phi_n = \sum_{k=0}^{p+1}\phi_n^{(k,0)}(1-\tau)^k + \sum_{k\geq p+2}\sum_{j=0}^{N_k}\phi_n^{(k,j)}(1-\tau)^k\ln^j(1-\tau).$$

7.5.3 Concluding Remarks

Some final remarks come now into place. One would expect to be able to deduce the smoothness of ϕ_0 on null infinity, the boundedness of ϕ_1 and ϕ_3 and the tangential smoothness of ϕ_4 from the smoothness of the initial data. This would require, in principle, the implementation of some energy estimates which would allow us to reach null infinity (see the discussion in §3). This is at the time of writing still an open problem. The present analysis is nevertheless valuable in the sense that it focuses our attention on the facts/properties one should be able to prove, and their interconnections. It should be, in principle, possible to undertake an analogue of the above discussion for the non-linear gravitational field. The analysis of the linearised gravitational field has shown us the way to lead. These matters will be the subject of future work.

Acknowledgments

I would like to thank H. Friedrich for suggesting me this research topic, for numerous discussions and criticisms which lead to the substantial improvement of this work.

References

1. P. T. Chruściel, M. A. H. MacCallum, & D. B. Singleton, *Gravitational waves in general relativity XIV. Bondi expansions and the "polyhomogeneity" of $\mathscr{I}$*, Phil. Trans. Roy. Soc. Lond. A **350**, 113 (1995).
2. R. Courant & D. Hilbert, *Methods of Mathematical Physics*, volume II, John Wiley & Sons, 1962.
3. H. Friedrich, *The asymptotic characteristic initial value problem for Einstein's vacuum field equations as an initial value problem for a first-order quasilinear symmetric hyperbolic system*, Proc. Roy. Soc. Lond. A **378**, 401 (1981).
4. H. Friedrich, *On the regular and the asymptotic characteristic initial value problem for Einstein's vacuum field equations*, Proc. Roy. Soc. Lond. A **375**, 169 (1981).
5. H. Friedrich, *On purely radiative space-times*, Comm. Math. Phys. **103**, 35 (1986).
6. H. Friedrich, *On static and radiative space-times*, Comm. Math. Phys. **119**, 51 (1988).
7. H. Friedrich, *Gravitational fields near space-like and null infinity*, J. Geom. Phys. **24**, 83 (1998).
8. H. Friedrich & J. Kánnár, *Bondi-type systems near space-like infinity and the calculation of the NP-constants*, J. Math. Phys. **41**, 2195 (2000).
9. F. John, *Partial differential equations*, Springer, 1991.
10. J. Kánnár, *On the existence of C^∞ solutions to the asymptotic characteristic initial value problem in general relativity*, Proc. Roy. Soc. Lond. A **452**, 945 (1996).
11. E. T. Newman & R. Penrose, *An approach to gravitational radiation by a method of spin coefficients*, J. Math. Phys. **3**, 566 (1962).
12. R. Penrose & W. Rindler, *Spinors and space-time. Volume 1. Two-spinor calculus and relativistic fields*, Cambridge University Press, 1984.
13. R. K. Sachs & P. G. Bergmann, *Structure of particles in linearized gravitational theory*, Phys. Rev. **112** (1958).
14. G. Szegö, *Orthogonal polynomials*, volume 23 of *AMS Colloq. Pub.*, AMS, 1978.
15. M. E. Taylor, *Partial differential equations I*, Springer, 1996.
16. J. A. Valiente Kroon, *Conserved Quantities for polyhomogeneous space-times*, Class. Quantum Grav. **15**, 2479 (1998).
17. J. A. Valiente Kroon, *A Comment on the Outgoing Radiation Condition and the Peeling Theorem*, Gen. Rel. Grav. **31**, 1219 (1999).
18. J. A. Valiente Kroon, *Logarithmic Newman-Penrose Constants for arbitrary polyhomogeneous space-times*, Class. Quantum Grav. **16**, 1653 (1999).

19. J. A. Valiente Kroon, *On the existence and convergence of polyhomogeneous expansions of zero-rest-mass fields*, Class. Quantum Grav. **17**, 4365 (2000).
20. J. A. Valiente Kroon, *Polyhomogeneity and zero-rest-mass fields with applications to Newman-Penrose constants*, Class. Quantum Grav. **17**, 605 (2000).
21. R. M. Wald, *General Relativity*, The University of Chicago Press, 1984.

8 Asymptotically Flat and Regular Cauchy Data

Sergio Dain

Max-Planck-Institut für Gravitationsphysik, Am Mühlenberg 1, 14476 Golm, Germany

Abstract. I describe the construction of a large class of asymptotically flat initial data with non-vanishing mass and angular momentum for which the metric and the extrinsic curvature have asymptotic expansions at space-like infinity in terms of powers of a radial coordinate. I emphasize the motivations and the main ideas behind the proofs.

8.1 Introduction

Suppose we want to describe an isolated self-gravitating system. For example a star, a binary system, a black hole or colliding black holes. Typically these astrophysical systems are located far away from the Earth, so that we can receive from them only electromagnetic and gravitational radiation. How is this radiation? For example one can ask how much energy is radiated, or which are the typical frequencies for some systems. This is the general problem we want to study. These systems are expected to be described by the Einstein field equations. It is in principle possible to measure this radiation and compare the results with the predictions of the equations.

There are no explicit solution to the Einstein equations that can describe such systems. Since the equations are very complicated it is hard to believe that such explicit solution can ever be found. Instead of trying to solve the complete equations at once, we use the so called 3+1 decomposition. We split the equations into " constraint" and "evolution". First we give appropriate initial data: that is, a solution of the constraint equations. Once we have chosen the initial data, the problem is completely fixed. Then, we use the evolution equations to calculate the whole space-time. From the evolution we can compute physical relevant quantities like the wave form of the emitted gravitational wave. This method of solving the equations is consistent with the idea that in physics we want to make predictions, that is: knowing the system at a given time we want to predict its behavior in the future. It is of course in general very hard to compute the evolution of the data, one has to use numerical techniques. The question we want to study here is: what are the appropriate initial data for an isolated system?

We can think of initial data for Einstein's equation as given a picture of the space time at a given time. It consists of a three dimensional manifold $\tilde{S}$ with some fields on it. The fields must satisfy the constraint equations. If the data describe an isolated system, the manifold $\tilde{S}$ is naturally divided in two regions. One compact region Ω which "contains the sources", and its complement $\tilde{S} \setminus \Omega$

which is unbounded. We will call the latter one the asymptotic region. Of course, Ω can be as large as we want, the only requirement is that it is bounded. The fact that $\tilde{S} \setminus \Omega$ is unbounded means that we can go as far a we want from the source region Ω, this capture the idea that the sources in Ω are isolated. That "the sources are in Ω" suggest that the field decays in $\tilde{S} \setminus \Omega$. Then, the initial data approach flat initial data near infinity. We call them *asymptotically flat initial data.* An opposite situation is when we want to study the universe as a whole. In this case one usually consider initial data where the manifold $\tilde{S}$ is compact without boundary.

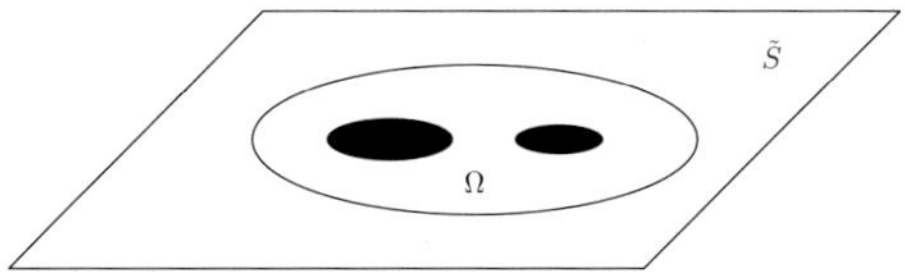

Fig. 8.1. Initial data for two stars. The region Ω contains the matter sources, which are represented by black regions. The manifold $\tilde{S}$ is $\mathbb{R}^3$

The simplest and most important example of asymptotically flat initial data is the case when $\tilde{S}$ is $\mathbb{R}^3$ and Ω some ball. Such data can describe, for example, ordinary stars. Consider an initial data set for a binary system, as is shown in Fig. 8.1. The stars are inside Ω. Ordinary stars emit light, then in the asymptotic region we will have electromagnetic field beside the gravitational one. We can also have some dilute gas in this region. It is important to recall that we do not require vacuum in the asymptotic region. We only require that the field decays properly outside Ω. However, in many situations it is a good approximation to assume vacuum in that region. When they evolve, the stars will follow some trajectory. Presumably a spiral orbit. The radius of the orbit will decrease with time, since the system loses energy in the form of gravitational waves. At late times, the system will settle down to a final stationary regime. This final state can be a rotating, stationary star. Or, if the initial data have enough mass concentrated in a small Ω, a spinning black hole. We show the conformal diagrams of this two cases in Fig. 8.2 and Fig. 8.3. The gravitational radiation is measured at null infinity, where the observer is placed.

Since the sources are in Ω, the fields in this region can be very complicated. They depend on the kind of matter that forms the stars. Remarkable, even the topology of Ω can be complicated. The example in Fig. 8.1 has trivial topology, however non trivial topologies are relevant for black hole initial data. Consider the initial data for the Schwarzschild and Kerr black hole. In this case $\tilde{S} = S^2 \times \mathbb{R}$ and $\Omega = S^2 \times [-a, a]$ for some constant a, as it is shown in Fig. 8.4. The asymptotic region of these data has two disconnected components $\tilde{S} \setminus \Omega = \tilde{S}_1 + \tilde{S}_2$. Each component is diffeomorphic to $\mathbb{R}^3$ minus a compact ball. We say that the data have two asymptotic ends. That is, there exist two disconnected "infinities". Observations can be made in one of them but not in both. It is not clear the physical relevance of this extra asymptotic end. A normal collapse, as is shown in

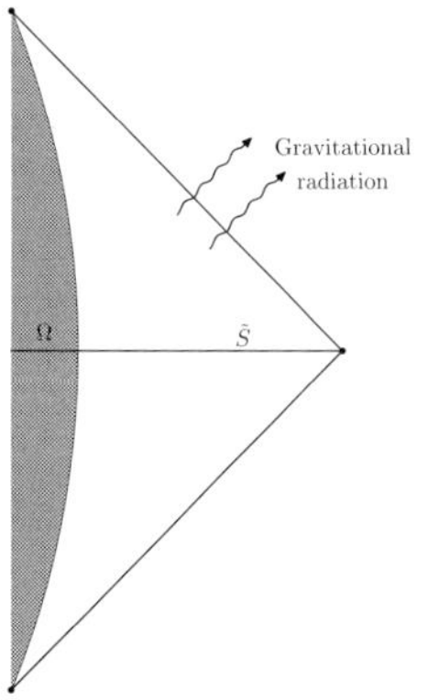

Fig. 8.2. Conformal diagram of the space time generated by the initial data given in Fig. 8.1, in the case in which the final state is a stationary star

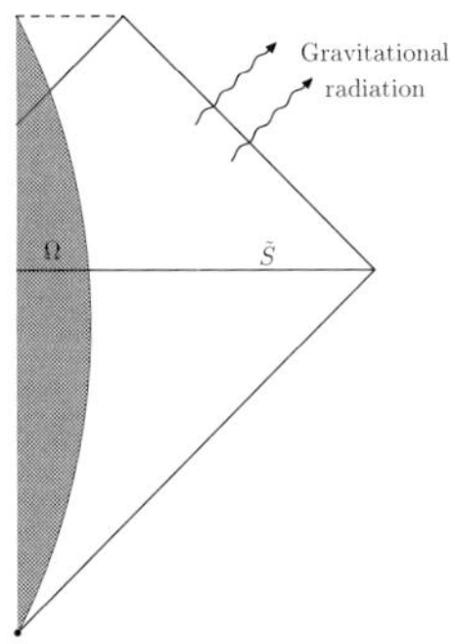

Fig. 8.3. Conformal diagram of the space time generated by the initial data given in Fig. 8.1, in the case in which the final state is a black hole

Fig. 8.3, will not have it. Therefore, one can think that only one asymptotic end is astrophysically relevant, being the other one just a mathematical peculiarity. The topology is not fixed by the Einstein equations. In this sense it remains quite arbitrary, one can thing that it is a kind of boundary conditions that has to be extra imposed to the equations. However, non trivial topologies appear naturally in the study of vacuum stationary black holes. A space time is stationary if it admits a time-like Killing vector field. The Schwarzschild and Kerr metrics are stationary. One can prove that every vacuum stationary asymptotically flat space time with trivial topology must be Minkowski. That is the non trivial topology is the "source" of the gravitational field in the vacuum stationary space times. In Fig. 8.5 we show another example with different topology, in this case we have only one asymptotic end. Other topologies with many asymptotic ends are of interest because their evolution in time may describe the collision of several black holes. In Fig. 8.9 and Fig. 8.11 we show some examples.

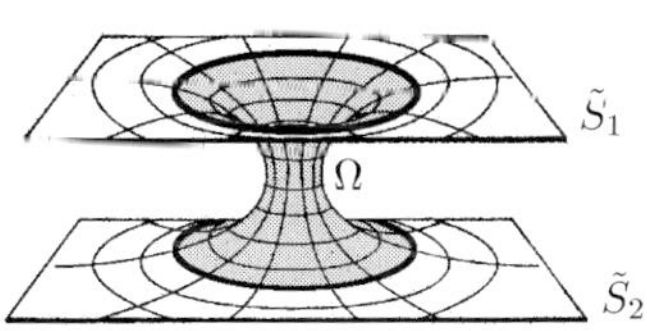

Fig. 8.4. One black hole initial data. The asymptotic region has two disjoint components $\tilde{S}_1$ and $\tilde{S}_2$. The compact set Ω is represented by the shadowed region

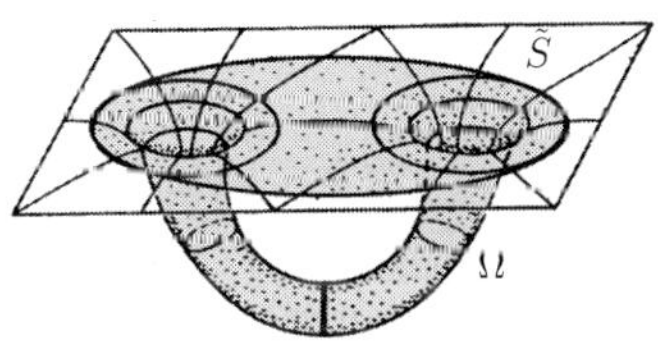

Fig. 8.5. Initial data for a Misner wormhole [31]. The asymptotic region has only one component. The compact set Ω is represented by the shadowed region

One usually has the idea that the matter sources generate gravitational field. To some extend this is true for many cases of astrophysical interest. However, as we have seen, a pure vacuum initial data with non trivial topologies can have mass. Moreover, one can also have a pure gravitational radiation initial data, that is a vacuum initial data with trivial topology with non zero mass. These kind of data are important in order to understand properties of the radiation itself which do not depend on the specific matter models. It is even possible to form a black hole with these type of data (see [5]).

In contrast to Ω, the asymptotic region is very simple. The fields on it are approximately flat and its topology is $\mathbb{R}^3$ minus a compact ball. We want to analyze the fields in this region. It is of course true that the fields are determined by the fields in Ω. But some important properties of them do not depend very much on the detailed structure of the sources in Ω. The most important of these properties is that the initial data will have a positive mass if the sources satisfies some energy condition. In this article we want to study some other properties of the fall of the fields in the asymptotic region which will be also independent on the sources in Ω. It is important to note that we can only measure precisely this kind of properties, because we can not prepare the initial conditions of an astrophysical system in the laboratory and then it is impossible to find the exact initial data for a real system. We can only analyze those effects which do not depend very much on fine details of the sources.

In order to simplify the hypothesis of the theorems, we will assume in this article that there are no matter sources in $\tilde{S}$. All the results will remain valid under suitable assumptions on the matter sources.

We summarize the discussion above in the following definition. An *initial data set* for the Einstein vacuum equations is given by a triple $(\tilde{S}, \tilde{h}_{ab}, \tilde{\Psi}_{ab})$ where $\tilde{S}$ is a connected 3-dimensional manifold, $\tilde{h}_{ab}$ a (positive definite) Riemannian metric, and $\tilde{\Psi}_{ab}$ a symmetric tensor field on $\tilde{S}$. They satisfy the vacuum constraint equations

$$\tilde{D}^b\tilde{\Psi}_{ab} - \tilde{D}_a\tilde{\Psi} = 0, \tag{8.1}$$

$$\tilde{R} + \tilde{\Psi}^2 - \tilde{\Psi}_{ab}\tilde{\Psi}^{ab} = 0, \tag{8.2}$$

on $\tilde{S}$, where $\tilde{D}_a$ is the covariant derivative with respect to $\tilde{h}_{ab}$, $\tilde{R}$ is the trace of the corresponding Ricci tensor, and $\tilde{\Psi} = \tilde{h}^{ab}\tilde{\Psi}_{ab}$. The data will be called *asymptotically flat* with N *asymptotic ends* if for some compact set Ω we have that $\tilde{S}\setminus\Omega = \sum_{k=1}^{N}\tilde{S}_k$, where $\tilde{S}_k$ are open sets such that each $\tilde{S}_k$ can be mapped by a coordinate system $\tilde{x}^j$ diffeomorphically onto the complement of a closed ball in $\mathbb{R}^3$ such that we have in these coordinates:

$$\tilde{h}_{ij} = (1 + \frac{2m_k}{\tilde{r}})\delta_{ij} + O(\tilde{r}^{-2}), \tag{8.3}$$

$$\tilde{\Psi}_{ij} = O(\tilde{r}^{-2}), \tag{8.4}$$

as $\tilde{r} = (\sum_{j=1}^{3}(\tilde{x}^j)^2)^{1/2} \to \infty$ in each set $\tilde{S}_k$. Here the constant m_k denotes the mass of the data at each end, $a, b, c...$ denote abstract indices, $i, j, k...$, which take values $1, 2, 3$, denote coordinate indices while δ_{ij} denotes the flat metric with respect to the given coordinate system $\tilde{x}^j$. Tensor indices of quantities with tilde will be moved with the metric $\tilde{h}_{ab}$ and its inverse $\tilde{h}^{ab}$. We set $x_i = x^i$ and $\partial^i = \partial_i$. These conditions guarantee that the mass, the momentum, and the angular momentum of the initial data set are well defined in every end.

We want to analyze the higher order terms in (8.3) and (8.4). For example, the terms $O(\tilde{r}^{-2})$ could have the form $\tilde{r}^{-2}\sin(\tilde{r}^{10})$. This function is certainly $O(\tilde{r}^{-2})$, but any derivative of it will blow up at infinity. Should such complicated terms be admissible in a description of a realistic isolated system? In Newton's theory and Electromagnetism one can give a definitive answer to this question: the fields have a fall off behavior like powers in a radial coordinate. Take for example a matter density ρ with compact support. The Newton's gravitational potential u satisfies the Poisson equation

$$\Delta u = 4\pi\rho, \tag{8.5}$$

where Δ is the Laplacian. Outside the support of ρ the potential u satisfies $\Delta u = 0$. It is well known that every harmonic function that goes to zero at infinity has an expansion of the form:

$$u = \sum_{k\geq 1} \frac{u^{k-1}}{\tilde{r}^k}, \tag{8.6}$$

where u^k are the spherical harmonics of order k. In this case the field equations force the potential to have the fall off behavior (8.6).

The situation for the Einstein equation is more complicate. In analogy with (8.6), one can ask the question whether there exist a class of initial data such that the metric and the extrinsic curvature admit near space-like infinity asymptotic expansions of the form:

$$\tilde{h}_{ij} \sim (1+\frac{2m}{\tilde{r}})\delta_{ij} + \sum_{k\geq 2} \frac{\tilde{h}^k_{ij}}{\tilde{r}^k}, \tag{8.7}$$

$$\tilde{\Psi}_{ij} \sim \sum_{k\geq 2} \frac{\tilde{\Psi}^k_{ij}}{\tilde{r}^k}, \tag{8.8}$$

where $\tilde{h}^k_{ij}$ and $\tilde{\Psi}^k_{ij}$ are smooth function on the unit 2-sphere (thought as being pulled back to the spheres $\tilde{r} = const.$ under the map $\tilde{x}^j \to \tilde{x}^j/\tilde{r}$). In this article I want to give an answer to this question. It is not only for convenience or aesthetic reasons that we want to avoid terms like $\tilde{r}^{-2}\sin(\tilde{r}^{10})$ in the expansions. They are also very difficult to handle numerically.

In order to see how the gravitational field behaves near infinity, it is natural to study first some examples. Consider asymptotically flat static space-times. We say that a space-time is static if it admits a hypersurface orthogonal time like Killing vector. One can take one of this hypersurfaces and analyze the fields $\tilde{h}_{ab}$ and $\tilde{\Psi}_{ab}$ on it. The simplest static space time is the Schwarzschild metric. In this case we have $\tilde{h}_{ij} = (1+m/(2\tilde{r}))^4\delta_{ij}$, $\tilde{\Psi}_{ij} = 0$, which is certainly of the form

(8.7), (8.8). For general static, asymptotically flat space times it can be proved that the initial data have also asymptotic expansions of the form (8.7) and (8.8). Moreover, in this case the fields are analytic functions of the coordinates. Then the expansions (8.7) and (8.8) are in fact convergent power series, in analogy with (8.6). This result is far from being obvious. It was proved by Beig and Simon in [8] based on an early work by Geroch [24].

One can go a step further and consider asymptotically flat stationary solutions. A stationary space time admits a time-like Killing field which is in general non hypersurface orthogonal. These space times describe rotating stars in equilibrium. The twist of the Killing vector is related to the angular velocity of the star. In this case there are no preferred hypersurfaces. The most important example of stationary solutions is the Kerr metric. It can be seen that for $t = const.$ slice of the Kerr metric in the standard Boyer-Lindquist coordinates the fields $\tilde{h}_{ab}$ and $\tilde{\Psi}_{ab}$ satisfy (8.7) and (8.8). For general stationary asymptotically flat solutions there also exist slices such that (8.7) and (8.8) holds. The essential part of this result was also proved by Beig and Simon in [9] (see also [28]). However, in these works the expansions are made in the abstract manifold of the orbits of the Killing vector field. In contrast with the static case this manifold does not correspond to any hypersurface of the space time. One has to translate this result in terms of the metric and the extrinsic curvature of some slices. This last step was made in [17]. As in the static case, the fields are analytic functions of the coordinates.

We have shown that there exist non trivial examples of initial data which satisfy (8.7) and (8.8). But how general are these expansions? For example, is it possible to have data with gravitational radiation that satisfy (8.7) and (8.8)? I want to show that in fact there exists a large class of asymptotically flat initial data which have the asymptotic behavior (8.7) and (8.8). These data will not be, in general, stationary.

The interest in such data is twofold. First, the evolution of asymptotically flat initial data near space-like and null infinity has been studied in considerable detail in [22]. In that article has been derived in particular a certain "regularity condition" on the data near space-like infinity, which is expected to provide a criterion for the existence of a smooth asymptotic structure at null infinity. To simplify the lengthy calculations, the data considered in [22] have been assumed to be time-symmetric and to admit a smooth conformal compactification. With these assumptions the regularity condition is given by a surprisingly succinct expression. With the present work we want to provide data which will allow us to perform the analysis of [22] without the assumption of time symmetry but which are still "simple" enough to simplify the work of generalizing the regularity condition to the case of non-trivial second fundamental form. Second, the "regular finite initial value problem near space-like infinity", formulated and analyzed in [22], suggests how to calculate numerically entire asymptotically flat solutions to Einstein's vacuum field equations on finite grids. In the present article I provide data for such numerical calculations which should allow us to study interesting situations while keeping a certain simplicity in the handling of the initial data.

The difficulty of constructing data with the asymptotic behavior (8.7), (8.8) arises from the fact that the fields need to satisfy the constraint equations (8.1) and (8.2). Part of the data, the " free data", can be given such that they are compatible with (8.7), (8.8). However, the remaining data are governed by elliptic equations (the constraint equations will reduce to elliptic equations as we will see below) and we have to show that (8.7), (8.8) are in fact a consequence of the equations and the way the free data have been prescribed.

To employ the standard techniques to provide solutions to the constraints, we assume

$$\tilde{\Psi} = 0, \tag{8.9}$$

such that the data correspond to a hypersurface which is maximal in the solution space-time.

I give an outline of the results which are available so far (see [18] and [15] for more details and proofs). Because of the applications indicated above, we wish to control in detail the conformal structure of the data near space-like infinity. Therefore we shall analyze the data in terms of the conformal compactification (S, h_{ab}, Ψ_{ab}) of the "physical" asymptotically flat data. Here S denotes a smooth, connected, orientable, compact 3-manifold. Take an arbitrary point i in S and define $\tilde{S} = S\setminus\{i\}$. The point i will represent, in a sense described in detail below, space-like infinity for the physical initial data. The physical manifold $\tilde{S}$ is essentially the stereographic projection of the compact manifold S (see Fig. 8.6). For example if we chose S to be S^3 then $\tilde{S}$ will be $\mathbb{R}^3$. Working with S and

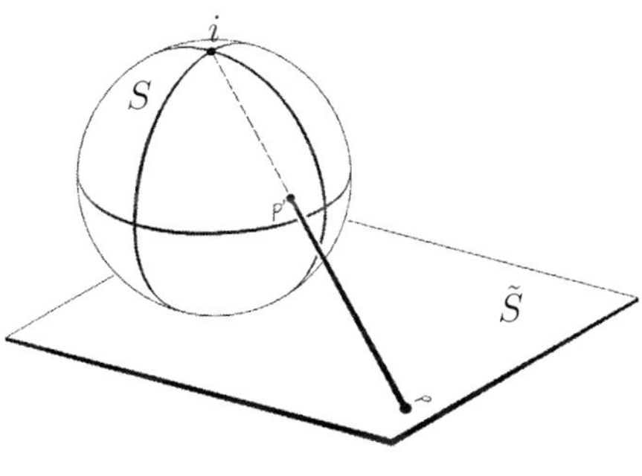

Fig. 8.6. The stereographic projection of S^3 in $\mathbb{R}^3$. A neighborhood B_a of the point i is mapped into the asymptotic region of the physical manifold $\tilde{S}$

not with $\tilde{S}$ has several technical advantage. It is simpler to prove existence of solutions for an elliptic equation on a compact manifold than in a non compact one. The price that we have to pay is that the equations will be singular at i. However this singularity is mild. It is also simpler to analyze the fields in terms of local differentiability in a neighborhood of i than in terms of fall off expansions at infinity in $\tilde{S}$. Moreover, this technique is also useful to construct initial data with non-trivial topology. By singling out more points in S and by treating the fields near these points in the same way as near i we could construct data with several asymptotically flat ends. In Fig. 8.7 – 8.13 we show some examples. All the following arguments equally apply to such situations, however, for convenience we restrict ourselves to the case of a single asymptotically flat end.

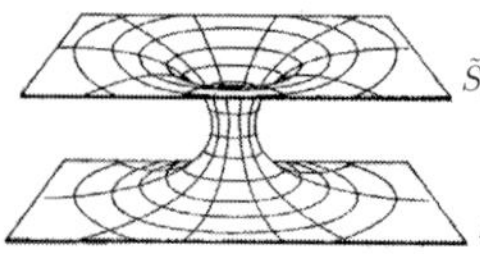

Fig. 8.7. Initial data with two asymptotic ends

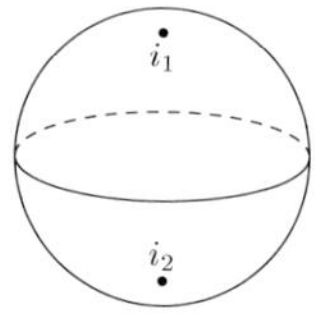

Fig. 8.8. A compactification of these data in the sphere

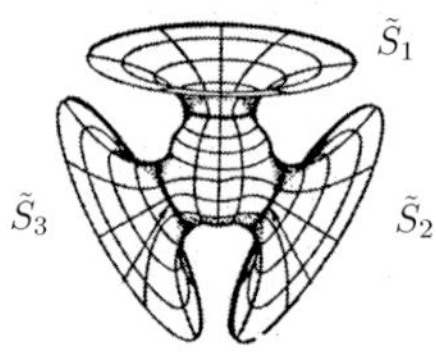

Fig. 8.9. The Brill-Lindquist initial data for two black holes [12]. The data have three asymptotic ends

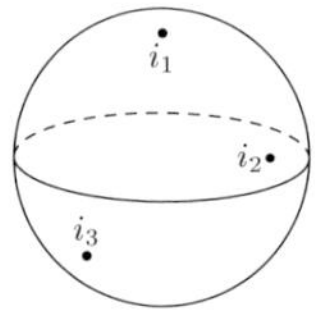

Fig. 8.10. A compactification of these data in the sphere

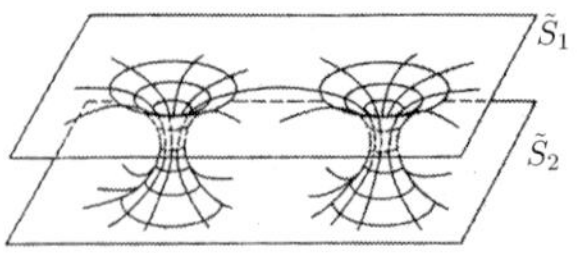

Fig. 8.11. The Misner initial data for two black holes [32]. The data have three asymptotic ends

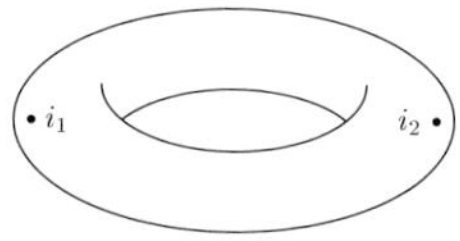

Fig. 8.12. A compactification of these data in the torus

We assume that h_{ab} is a positive definite metric on S with covariant derivative D_a and Ψ_{ab} is a symmetric tensor field which is smooth on $\tilde{S}$. In agreement with (8.9) we shall assume that Ψ_{ab} is trace-free:

$$h^{ab}\,\Psi_{ab} = 0.$$

The fields above are related to the physical fields by rescaling

$$\tilde{h}_{ab} = \theta^4\, h_{ab}, \quad \tilde{\Psi}_{ab} = \theta^{-2}\,\Psi_{ab}, \tag{8.10}$$

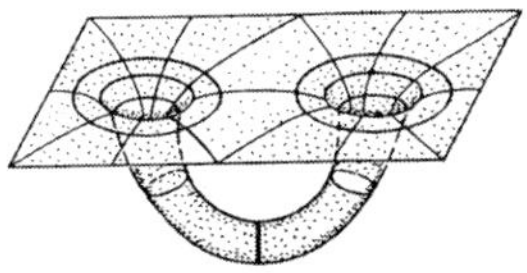

Fig. 8.13. The Misner wormhole initial data. The data have only one asymptotic end

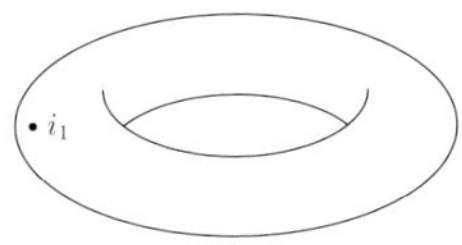

Fig. 8.14. A compactification of these data in the torus

with a conformal factor θ which is positive on $\tilde{S}$. For the physical fields to satisfy the vacuum constraints we need to assume that

$$D^a \Psi_{ab} = 0 \quad \text{on} \quad \tilde{S}, \tag{8.11}$$

$$(D_b D^b - \frac{1}{8} R)\theta = -\frac{1}{8} \Psi_{ab} \Psi^{ab} \theta^{-7} \quad \text{on} \quad \tilde{S}. \tag{8.12}$$

Equation (8.12) for the conformal factor θ is the Lichnerowicz equation, transferred to our context. Tensor indices of quantities without tilde will be moved with the conformal metric h_{ab} and its inverse h^{ab}.

Let x^j be h-normal coordinates centered at i such that $h_{kl} = \delta_{kl}$ at i and set $r = (\sum_{i=1}^{3} (x^j)^2)^{1/2}$. To ensure asymptotical flatness of the data (8.10) we require

$$\Psi_{ab} = O(r^{-4}) \quad \text{as} \quad r \to 0, \tag{8.13}$$

$$\lim_{r \to 0} r\theta = 1. \tag{8.14}$$

In the coordinates $\tilde{x}^j = x^j / r^2$ the fields (8.10) will then satisfy (8.3), (8.4) (cf. [21], [22] for this procedure).

Not all data as given by (8.10), which are derived from data h_{ab}, Ψ_{ab} as described above, will satisfy conditions (8.7), (8.8). We will have to impose extra conditions and we want to keep these conditions as simple as possible.

Since we assume the metric h_{ab} to be smooth on S, it will only depend on the behavior of θ near i whether condition (8.7) will be satisfied. Via equation (8.12) this behavior depends on Ψ_{ab}. What kind of condition do we have to impose on Ψ_{ab} in order to achieve (8.7) ?

The following space of functions will play an important role in our discussion. Denote by B_a the open ball with center i and radius $r = a > 0$, where a is chosen small enough such that B_a is a convex normal neighborhood of i. A function $f \in C^\infty(\tilde{S})$ is said to be in $E^\infty(B_a)$ if on B_a we can write $f = f_1 + r f_2$ with $f_1, f_2 \in C^\infty(B_a)$ (cf. definition 8.1). An answer to our question is given by following theorem:

Theorem 8.1. *Let h_{ab} be a smooth metric on S with positive Ricci scalar R. Assume that Ψ_{ab} is smooth in $\tilde{S}$ and satisfies on B_a:*

$$r^8 \Psi_{ab} \Psi^{ab} \in E^\infty(B_a). \tag{8.15}$$

Then there exists on $\tilde{S}$ a unique solution θ of equation (8.12), which is positive, satisfies (8.14), and has in B_a the form:

$$\theta = \frac{\hat{\theta}}{r}, \quad \hat{\theta} \in E^\infty(B_a), \quad \hat{\theta}(i) = 1. \tag{8.16}$$

In fact, we will get slightly more detailed information. We find that $\hat{\theta} = u_1 + r\, u_2$ on B_a with $u_2 \in E^\infty(B_a)$ and a function $u_1 \in C^\infty(B_a)$ which satisfies $u_1 = 1 + O(r^2)$ and

$$(D_b D^b - \frac{1}{8} R) \frac{u_1}{r} = \theta_R,$$

in $B_a \setminus \{i\}$, where θ_R is in $C^\infty(B_a)$ and vanishes at any order at i.

If θ has the form (8.16) then (8.7) will be satisfied due to our assumptions on h_{ab}.

Note the simplicity of condition (8.15). If the metric is analytic on B_a it can be arranged that $\theta_R = 0$ and u_1 is analytic on B_a (and unique with this property, see [23]). The requirement $R > 0$, which ensures the solvability of the Lichnerowicz equation, could be reformulated in terms of a condition on the Yamabe number (cf. [29]).

Theorem 8.1 has two parts. One is the existence and uniqueness of the solution. This part depends on global properties of the fields on S. It can be proved under much weaker assumptions on the differentiability of h_{ab} and Ψ_{ab}. The second part concerns the regularly in B_a, and depends only on local properties of the fields in B_a. This part can also be proved under weaker hypothesis. These generalizations have physical relevance, I will come back to this point in the final section.

It remains to be shown that condition (8.15) can be satisfied by tensor fields Ψ_{ab} which satisfy (8.11), (8.13). A special class of such solutions, namely those which extend smoothly to all of S, can easily be obtained by known techniques (cf. [14]). However, in that case the initial data will have vanishing momentum and angular momentum. To obtain data without this restriction, we have to consider fields $\Psi_{ab} \in C^\infty(\tilde{S})$ which are singular at i in the sense that they admit, in accordance with (8.4), (8.10), (8.14), at $i = \{r = 0\}$ asymptotic expansions of the form:

$$\Psi_{ij} \sim \sum_{k \geq -4} \Psi^k_{ij} r^k \quad \text{with} \quad \Psi^k_{ij} \in C^\infty(S^2). \tag{8.17}$$

It turns out that condition (8.15) excludes data with non-vanishing linear momentum, which requires a non-vanishing leading order term of the form $O(r^{-4})$. In section 8.2 we will show that such terms imply terms of the form $\log r$ in θ and thus do not admit expansion of the form (8.7). However, this does not necessarily indicate that condition (8.15) is overly restrictive. In the case where the metric h_{ab} is smooth it will be shown in Sect. 8.2 that a non-vanishing linear momentum always comes with logarithmic terms, irrespective of whether condition (8.15) is imposed or not.

There remains the question whether there exist fields Ψ_{ab} which satisfy (8.15) and have non-trivial angular momentum. The latter requires a term of the form

$O(r^{-3})$ in (8.17). It turns out that condition (8.15) fixes this term to be of the form:

$$\Psi^{ab}_{AJ} = \frac{A}{r^3}\,(3n^a n^b - \delta^{ab}) + \frac{3}{r^3}(n^a \epsilon^{bcd} J_c n_d + n^b \epsilon^{acd} J_c n_d), \tag{8.18}$$

where $n^i = x^i/r$ is the radial unit normal vector field near i and J^k, A are constants, the three constants J^k specifying the angular momentum of the data. The spherically symmetric tensor which appears here with the factor A agrees with the extrinsic curvature for a maximal (non-time symmetric) slice in the Schwarzschild solution (see for example [7]). Note that the tensor Ψ^{ab}_{AJ} satisfies condition (8.15) and the equation $\partial_a \Psi^{ab}_{AJ} = 0$ on $\tilde{S}$ for the flat metric, hence it is a non-trivial example. We want to study more general situations.

In the existence proof for equations (8.11) and (8.13), the possible existence of conformal Killing vectors ξ^a of the metric h_{ab} will play an important role. A conformal Killing vector is a solution on S of $\mathcal{L}_h \xi = 0$, where we have defined the conformal Killing operator

$$(\mathcal{L}_h v)^{ab} \equiv D^a v^b + D^b v^a - \frac{2}{3}\, h^{ab}\, D_c v^c. \tag{8.19}$$

Given ξ^a we define the following constants

$$S^a = \epsilon^a{}_{bc} D^b \xi^c(i), \quad a = \frac{1}{3} D_a \xi^a(i). \tag{8.20}$$

The free-data in the solution Ψ^{ab} consist in two pieces: a singular and a regular one, which we will denote by Φ^{ab}_{sing} and Φ^{ab}_{reg} respectively. We define Φ^{ab}_{sing} in $B_a \setminus \{i\}$ by

$$\Phi^{ab}_{sing} = \chi\,(\Psi^{ab}_{AJ} - \frac{1}{3}\, h_{cd}\, \Psi^{cd}_{AJ}\, h^{ab}), \tag{8.21}$$

and vanishes elsewhere. Here χ denotes a smooth function of compact support in B_a equal to 1 on $B_{a/2}$. We assume that Φ^{ab}_{reg} can be written near i in the form

$$\Phi^{ab}_{reg} = r^{-3}\, \Phi^{ab}_{1reg} + \Phi^{ab}_{2reg}, \tag{8.22}$$

where Φ^{ab}_{1reg}, Φ^{ab}_{2reg} are smooth symmetric trace-free tensors in S and such that $\Phi^{ij}_{1reg} = O(r^2)$, and $x_i\, x_j\, \Phi^{ij}_{1reg} - r^2\, \Phi$ with some $\Phi \in C^\infty(B_a)$.

Theorem 8.2. *Let h_{ab} a smooth metric in S. Let Φ^{ab}_{sing} and Φ^{ab}_{reg} given by (8.21) and (8.22).*

i) If the metric h admits no conformal Killing fields on S, then there exists a unique vector field v^a such that the tensor field

$$\Psi^{ab} = \Phi^{ab}_{sing} + \Phi^{ab}_{reg} + (\mathcal{L}_h v)^{ab} \tag{8.23}$$

satisfies the equation $D_a \Psi^{ab} = 0$ in $S \setminus \{i\}$. The solution satisfies $\Psi_{ab} = O(r^{-3})$ at i and condition (8.15).

ii) If the metric h admits conformal Killing fields ξ^a on S, a vector field v^a as specified above exists if and only if the constants J^a, A (partly) characterizing the tensor field Φ^{ab}_{sing} satisfy the equation

$$J^a\, S_a + A\, a = 0, \tag{8.24}$$

for any conformal Killing field ξ^a of h, where the constants S_a and a are given by (8.20).

In both cases the angular momentum of Ψ^{ab} is given by J^a. This quantity can thus be prescribed freely in case (i).

Theorem 8.2, as Theorem 8.1, contains two parts. One is the existence and uniqueness of the solution v^a. The restriction (8.24) appears in this part. See [6], [18] and [19] for a generalization and interpretation of this condition. The other part concerns the regularity of the solution in B_a, namely that it satisfies condition (8.15). This part depends only on local assumptions of the fields in B_a. We prove a more detailed version of this theorem in [18].

8.2 Solution of the Hamiltonian Constraint with Logarithmic Terms

Assume, for simplicity, that the metric h_{ab} is conformally flat in B_a. Then the operator that appears in the left hand side of equation (8.12) reduce to the Laplacian Δ. This simplification is of course a restriction on the allowed initial data, but it already contains all the main problems and the essential features of the more general case. It is also important to remark that even under this assumption it is possible to describe a rich family of initial data, since we are making restrictions only on B_a. Consider the Poisson equation (8.5). The example

$$\Delta(\log r h_m) = r^{-2} h_m (2m+1), \tag{8.25}$$

where h_m is an harmonic polynomial of order m (i.e. $\Delta h_m = 0$) shows that logarithmic terms can occur in solutions to the Poisson equation even if the source has only terms of the form $r^s p$ where p is some polynomial. That is, even if our free data have expansion in powers of r logarithmic terms will appear in the conformal factor θ. We shall use this to show that logarithmic terms can occur in the solution to the Lichnerowicz equation if the condition $r^8 \Psi_{ab}\Psi^{ab} \in E^\infty(B_a)$ is not satisfied. Our example will be concerned with initial data with non-vanishing linear momentum.

We assume that in a small ball B_a centered at i the tensor Ψ^{ab} has the form:

$$\Psi^{ab} = \Psi^{ab}_P + \Psi^{ab}_R, \tag{8.26}$$

where Ψ^{ab}_P is given in normal coordinates by

$$\Psi^{ab}_P = \frac{3}{2r^4}\left(-P^a n^b - P^b n^a - (\delta^{ab} - 5n^a n^b)\, P^c n_c\right), \tag{8.27}$$

and $\Psi_R^{ab} = O(r^{-3})$ is a tensor field such that Ψ^{ab} satisfies equation (8.11). We will assume also that Ψ_R^{ab} satisfies some mild smoothness condition (cf. [18]).

Since Ψ_P^{ab} is trace-free and divergence-free with respect to the flat metric, we could, of course, choose h_{ab} to be the flat metric and $\Psi_R^{ab} = 0$. This would provide one of the conformally flat initial data sets discussed in [10]. We are interested in a more general situation.

Lemma 8.1. *Let h_{ab} be a smooth metric, and let Ψ^{ab} be given by (8.26). Then, there exists a unique, positive, solution to the Hamiltonian constraint (8.12). In B_a it has the form:*

$$\theta = \frac{w_1}{r} + \frac{1}{2}m + \frac{1}{32}\,r\,\left(9\,(P^i n_i)^2 - 33\,P^2\right) + \frac{7}{16}\,m\,r^2\left(\frac{5}{4}\,P^2 + \frac{3}{5}\,(3\,(P^i n_i)^2 - P^2)\,\log r\right) + u_R, \quad (8.28)$$

where $P^iP_i = P^2$, the constant m is the total mass of the initial data, w_1 is a smooth function with $w_1 = 1 + O(r^2)$, and $u_R \in C^{2,\alpha}(B_a)$ with $u_R(0) = 0$.

Since w_1 is smooth and u_R is in $C^{2,\alpha}(B_a)$, there cannot occur a cancellation of logarithmic terms. For non-trivial data, for which $m \neq 0$, the logarithmic term will always appear. In the case where h_{ab} is flat (in this case we have $w_1 = 1$) and $\Psi_R^{ab} = 0$ an expansion similar to (8.28) has been calculated in [26].

8.3 Explicit Solutions of the Momentum Constraint

Instead of given the proof of Theorem 8.2, in this section I want to present some explicit solutions of equation (8.11) for conformally flat and also for axially symmetric metrics. In the first case we will construct all the solutions, in the second one only some of them. We will show how to achieve condition (8.15) in terms of the free data. In both cases the solution is constructed in terms of derivatives of some free functions. That makes them suitable for explicit computations. In particular, if we assume that these free functions have compact support, then Ψ^{ab} will also have compact support. We note incidentally that, as an application, one can easily construct regular hyperboloidal initial data with non-trivial extrinsic curvature.

8.3.1 The Momentum Constraint on Euclidean Space

In the following we shall give an explicit construction of the smooth solutions to the equation $\partial_a \Psi^{ab} = 0$ on the 3-dimensional Euclidean space $\mathbb{E}^3$ (in suitable coordinates $\mathbb{R}^3$ endowed with the flat standard metric) or open subsets of it. Another method to obtain such solutions has been described in [4], multipole expansions of such tensors have been studied in [6].

Let i be a point of $\mathbb{E}^3$ and x^k a Cartesian coordinate system with origin i such that in these coordinates the metric of $\mathbb{E}^3$, denoted by δ_{ab}, is given by the standard form δ_{kl}. We denote by n^a the vector field on $\mathbb{E}^3 \setminus \{i\}$ which is given in these coordinates by $x^k/|x|$.

Denote by m_a and its complex conjugate $\bar{m}_a$ complex vector fields, defined on $\mathbb{E}^3$ outside a lower dimensional subset and independent of $r = |x|$, such that

$$m_a m^a = \bar{m}_a \bar{m}^a = n_a m^a = n_a \bar{m}^a = 0, \quad m_a \bar{m}^a = 1. \tag{8.29}$$

There remains the freedom to perform rotations $m_a \to e^{if} m_a$ with functions f which are independent of r.

The metric has the expansion

$$\delta_{ab} = n_a n_b + \bar{m}_a m_b + m_a \bar{m}_b,$$

while an arbitrary symmetric, trace-free tensor Ψ_{ab} can be expended in the form:

$$r^3 \Psi_{ab} = \xi(3n_a n_b - \delta_{ab}) + \sqrt{2}\eta_1 n_{(a}\bar{m}_{b)} + \sqrt{2}\bar{\eta}_1 n_{(a} m_{b)} + \bar{\mu}_2 m_a m_b + \mu_2 \bar{m}_a \bar{m}_b, \tag{8.30}$$

with

$$\xi = \frac{1}{2} r^3 \Psi_{ab} n^a n^b, \quad \eta_1 = \sqrt{2} r^3 \Psi_{ab} n^a m^b, \quad \mu_2 = r^3 \Psi_{ab} m^a m^b.$$

Since Ψ_{ab} is real, the function ξ is real while η_1, μ_2 are complex functions of spin weight 1 and 2 respectively.

Using in the equation

$$\partial_a \Psi^{ab} = 0 \tag{8.31}$$

the expansion (8.30) and contracting suitably with n^a and m^a, we obtain the following representation of (8.31):

$$4r\partial_r \xi + \bar{\eth}\eta_1 + \eth\bar{\eta}_1 = 0, \tag{8.32}$$

$$r\partial_r \eta_1 + \bar{\eth}\mu_2 - \eth\xi = 0. \tag{8.33}$$

Here ∂_r denotes the radial derivative and $\eth$ the edth operator of the unit two-sphere (cf. [33] for definition and properties). By our assumptions the differential operator $\eth$ commutes with ∂_r.

Let ${}_sY_{lm}$ denote the spin weighted spherical harmonics, which coincide with the standard spherical harmonics Y_{lm} for $s = 0$. The ${}_sY_{lm}$ are eigenfunctions of the operator $\bar{\eth}\eth$ for each spin weight s. More generally, we have

$$\bar{\eth}^p \eth^p ({}_sY_{lm}) = (-1)^p \frac{(l-s)!}{(l-s-p)!} \frac{(l+s+p)!}{(l+s)!} {}_sY_{lm}. \tag{8.34}$$

If μ_s denotes a smooth function on the two-sphere of integral spin weight s, there exists a function μ of spin weight zero such that $\eta_s = \eth^s \eta$. We set $\eta^R = \mathrm{Re}(\eta)$ and $\eta^I = i\,\mathrm{Im}(\eta)$, such that $\eta = \eta^R + \eta^I$, and define

$$\eta_s^R = \eth^s \eta^R, \quad \eta_s^I = \eth^s \eta^I,$$

such that $\eta_s = \eta_s^R + \eta_s^I$. We have

$$\overline{\eth^s \eta^R} = \bar{\eth}^s \eta^R, \quad \overline{\eth^s \eta^I} = -\bar{\eth}^s \eta^I.$$

Using these decompositions now for η_1 and μ_2, we obtain equation (8.32) in the form:

$$2r\partial_r \xi = -\bar{\eth}\eth\eta^R. \tag{8.35}$$

Applying $\bar{\eth}$ to both sides of equation (8.33) and decomposing into real and imaginary part yields

$$r\partial_r \bar{\eth}\eth\eta^I = -\bar{\eth}^2\eth^2\mu^I, \tag{8.36}$$

$$2r\partial_r(r\partial_r\xi) + \bar{\eth}\eth\xi = \bar{\eth}^2\eth^2\mu^R. \tag{8.37}$$

Since the right hand side of (8.35) has an expansion in spherical harmonics with $l \geq 1$ and the right hand sides of (8.36), (8.37) have expansions with $l \geq 2$, we can determine the expansion coefficients of the unknowns for $l = 0, 1$. They can be given in the form:

$$\xi = A + rQ + \frac{1}{r}P, \quad \eta^I = iJ + const., \quad \eta^R = rQ - \frac{1}{r}P + const.,$$

with

$$P = \frac{3}{2}P^a n_a, \quad Q = \frac{3}{2}Q^a n_a, \quad J = 3J^a n_a, \tag{8.38}$$

where A, P^a, Q^a, J^a are arbitrary constants. Using (8.30), we obtain the corresponding tensors in the form (cf. [10]):

$$\Psi_P^{ab} = \frac{3}{2r^4}\left(-P^a n^b - P^b n^a - (\delta^{ab} - 5n^a n^b)\, P^c n_c\right), \tag{8.39}$$

$$\Psi_J^{ab} = \frac{3}{r^3}(n^a \epsilon^{bcd} J_c n_d + n^b \epsilon^{acd} J_c n_d), \tag{8.40}$$

$$\Psi_A^{ab} = \frac{A}{r^3}\,(3n^a n^b - \delta^{ab}), \tag{8.41}$$

$$\Psi_Q^{ab} = \frac{3}{2r^2}\left(Q^a n^b + Q^b n^a - (\delta^{ab} - n^a n^b)\, Q^c n_c\right). \tag{8.42}$$

We assume now that ξ and η^I have expansions in terms of spherical harmonics with $l \geq 2$. Then there exists a smooth function λ_2 of spin weight 2 such that

$$\xi = \bar{\eth}^2 \lambda_2^R, \quad \eta_1^I = \bar{\eth}\lambda_2^I.$$

Using these expressions in equations (8.35) – (8.37) and observing that for smooth spin weighted functions μ_s with $s > 0$ we can have $\bar{\eth}\mu_s = 0$ only if $\mu_s = 0$, we obtain:

$$\eth\eta^R = -2r\partial_r\bar{\eth}\lambda_2^R, \quad \eth^2\mu^I = -r\partial_r\lambda_2^I,$$

$$\eth^2\mu^R = 2r\partial_r(r\partial_r\lambda_2^R) - 2\lambda_2^R + \eth\bar{\eth}\lambda_2^R.$$

We are thus in a position to describe the general form of the coefficients in the expression (8.30):

$$\xi = \bar{\eth}^2 \lambda_2^R + A + r\,Q + \frac{1}{r}\,P, \tag{8.43}$$

$$\eta_1 = -2\,r\,\partial_r\,\bar{\eth}\lambda_2^R + \bar{\eth}\lambda_2^I + r\,\bar{\eth}Q - \frac{1}{r}\,\bar{\eth}P + i\,\bar{\eth}J, \tag{8.44}$$

$$\mu_2 = 2\,r\,\partial_r(r\,\partial_r\,\lambda_2^R) - 2\,\lambda_2^R + \eth\bar{\eth}\lambda_2^R - r\,\partial_r\,\lambda_2^I. \tag{8.45}$$

Theorem 8.3. *Let λ be an arbitrary complex C^∞ function in $B_a \setminus \{i\} \subset \mathbb{E}^3$ with $0 < a < \infty$, and set $\lambda_2 = \eth^2\lambda$. Then the tensor field*

$$\Psi^{ab} = \Psi_P^{ab} + \Psi_J^{ab} + \Psi_A^{ab} + \Psi_Q^{ab} + \Psi_\lambda^{ab}, \tag{8.46}$$

where the first four terms on the right hand side are given by (8.39) – (8.41) while Ψ_λ^{ab} is obtained by using in (8.30) only the part of the coefficients (8.43) – (8.45) which depends on λ_2, satisfies the equation $D^a\Psi_{ab} = 0$ in $B_a \setminus \{i\}$. Conversely, any smooth solution in $B_a \setminus \{i\}$ of this equation is of the form (8.46).

Obviously, the smoothness requirement on λ can be relaxed since $\Psi_\lambda^{ab} \in C^1(B_a \setminus \{i\})$ if $\lambda \in C^5(B_a \setminus \{i\})$. Notice, that no fall-off behavior has been imposed on λ at i and that it can show all kinds of bad behavior as $r \to 0$.

Since we are free to choose the radius a, we also obtain an expression for the general smooth solution on $\mathbb{E}^3 \setminus \{i\}$. By suitable choices of λ we can construct solutions Ψ_λ^{ab} which are smooth on $\mathbb{E}^3$ or which are smooth with compact support. Finally we provide tensor fields which satisfy condition (8.15) and thus prove a special case of theorem 8.2, see [18] for the proof.

Theorem 8.4. *Denote by Ψ^{ab} a tensor field of the type (8.46). If $r\lambda \in E^\infty(B_a)$ and $P^a = 0$, then $r^8\Psi_{ab}\Psi^{ab} \in E^\infty(B_a)$.*

We wish to point out a further application of the results above. Given a subset S of $\mathbb{R}^3$ which is compact with boundary, we can use the representation (8.46) to construct hyperboloidal initial data ([20]) on S with a metric h which is Euclidean on all of S or on a subset U of S. In the latter case we would require Ψ_λ^{ab} to vanish on $S\backslash U$. In the case where the trace-free part of the second fundamental form implied by h on ∂S vanishes and the support of Ψ^{ab} has empty intersection with ∂S the smoothness of the corresponding hyperboloidal initial data near the boundary follows from the discussion in ([2]). Appropriate requirements on h and Ψ^{ab} near ∂S which ensure the smoothness of the hyperboloidal data under more general assumptions can be found in ([1]).

8.3.2 Axially Symmetric Initial Data

The momentum constraint with axial symmetry has been studied in [3], [11] and [16]. Assume that the metric h_{ab} has a Killing vector η^a, which is hypersurface orthogonal. We define η by $\eta = \eta^a\eta^b h_{ab}$. Following [27], consider Ψ^{ab} defined by

$$\Psi^{ab} = \frac{2\Psi^{(a}\eta^{b)}}{\eta}, \tag{8.47}$$

where Ψ^a satisfies

$$\pounds_\eta \Psi^a = 0, \quad \Psi^a \eta_a = 0, \quad D_a \Psi^a = 0, \tag{8.48}$$

with the Lie derivative $\pounds_\eta$ with respect η^a.

We use the Killing equation $D_{(a}\eta_{b)} = 0$, the fact that η^a is hypersurface orthogonal, (i.e. it satisfies $D_a\eta_b = -\eta_{[a}D_{b]} \ln \eta$) and equations (8.48) to conclude that Ψ^{ab} is trace-free and divergence-free with respect to the metric h_{ab}.

The solution of equations (8.48) can be written in terms of a scalar potential ω:

$$\Psi^a = \frac{1}{\eta}\epsilon^{abc}\eta_b D_c\omega, \quad \pounds_\eta \omega = 0. \tag{8.49}$$

Using this equation we find that

$$\Psi^{ab}\Psi_{ab} = 2\frac{D_c\omega D^c\omega}{\eta^2}. \tag{8.50}$$

We want to find now which conditions we have to impose in ω in order to achieve (8.15). The metric h_{ab} has the form:

$$h_{ab} = e_{ab} + \frac{\eta_a\eta_b}{\eta}, \tag{8.51}$$

where e_{ab} is the two dimensional metric induced on the hypersurfaces orthogonal to η^a, its satisfies $\pounds_\eta e_{ab} = 0$. All two dimensional metrics are locally conformally flat, we will assume here that e_{ab} is globally conformally flat. Then, we can perform a conformal rescaling of h_{ab} such that in the rescaled metric the corresponding intrinsic metric e_{ab} is flat. We will denote this rescaled metric again by h_{ab}. Assume that η^a is a rotation. Take spherical coordinates (r, ϑ, ϕ) such that $\eta^a = (\partial/\partial\phi)^a$. In this coordinates the metric has the form:

$$h = dr^2 + r^2 d\vartheta^2 + \eta d\phi^2. \tag{8.52}$$

The norm η can be written as

$$\eta = r^2 \sin^2\vartheta(1 + f), \tag{8.53}$$

where f satisfies $1 + f > 0$. Note that r is the geodesic distance with respect to the origin.

In these coordinates equation (8.50) has the form:

$$\Psi^{ab}\Psi_{ab} = \frac{2}{(1+f)^2}\frac{(r^2(\partial_r\omega)^2 + (\partial_\vartheta\omega)^2)}{r^6\sin^4\vartheta}. \tag{8.54}$$

In the flat case (i.e. $f = 0$) this solution reduces to the one given by Theorem 8.3 with

$$P = Q = A = 0, \quad \lambda_2^R = 0, \tag{8.55}$$

and

$$i\omega = \lambda_2^I \sin^2\vartheta + iJ^z(-3\cos\vartheta + \cos^3\vartheta), \tag{8.56}$$

where J^z is the only non-vanishing component of the vector J^a, and $\pounds_\eta \lambda^I_2 = 0$.

Motivated by this expression we write ω in the form

$$\omega = \eth^2 \lambda \sin^2 \vartheta + J^z(-3\cos\vartheta + \cos^3\vartheta), \tag{8.57}$$

where λ is an arbitrary zero spin real function which depends on r and ϑ. The constant J^z will give the angular momentum of the data. This can be seen from the following expression:

$$J^z = \frac{1}{8\pi}\int_\Sigma \Psi_a n^a \, dS, \tag{8.58}$$

where Σ is any closed two-surface in the asymptotic region and n^a its normal.

Using equation (8.57) and lemma 9 of [18] we prove the following lemma.

Theorem 8.5. *Assume that $f \in E^\infty(B_a)$. Let Ψ^{ab} be given by (8.47), (8.49) and (8.57), where J^z is an arbitrary constant. If $r\lambda \in E^\infty(B_a)$ then $r^8 \Psi_{ab}\Psi^{ab} \in E^\infty(B_a)$.*

Note that in Theorem 8.5 we have assumed only $f \in E^\infty(B_a)$, that is, f and hence the conformal metric h_{ab} is not required to be smooth. I will come back to this point in the final section.

8.4 Main Ideas in the Proof of Theorem 8.1

I want to describe in this section the main idea in the proof of theorem 8.1. I will concentrate on the regularity part of this theorem and not on the existence part, since the later is more or less standard. I will give an almost self contained proof of a simplified version of theorem 8.1, which contains all the essential elements of the general proof.

Consider the semi-linear equation

$$\Delta u = f(u, x), \tag{8.59}$$

on B_a, where the function f is given by

$$f(x, u) = -\frac{\Psi^2(x)}{(1 + ru)^7}, \quad \Psi(x) \in E^\infty. \tag{8.60}$$

This equation is similar to equation (8.12) when the metric h_{ab} is flat in B_a. Assume that we have a positive solution $u \in C^{2,\alpha}(B_a)$. There exist several methods to prove existence of solutions for semi-linear equations, see for example [30] for an elementary introduction to the subject. We want to prove the following theorem.

Theorem 8.6. *If $u \in C^{2,\alpha}(B_a)$ and $u \geq 0$ is a solution of equation (8.59), then $u \in E^\infty(B_a)$.*

The important feature of equation (8.59) is that the radial coordinate r appears explicitly in f. As a function of the Cartesian coordinates x^i, r is only in C^α. Thus, we can not use standard elliptic estimates in order to improve the regularity of our solution. Instead of this we use the following spaces.

Definition 8.1. For $m \in \mathbb{N}_0$ and $0 < \alpha < 1$, we define the space $E^{m,\alpha}(B_a)$ as the set $E^{m,\alpha}(B_a) = \{f = f_1 + rf_2 \, : \, f_1, f_2 \in C^{m,\alpha}(B_a)\}$. Furthermore we set $E^\infty(B_a) = \{f = f_1 + rf_2 \, : \, f_1, f_2 \in C^\infty(B_a)\}$.

The space $E^{m,\alpha}$ has two important properties. The first one is given by the following lemma which is an easy consequence of definition 8.1.

Lemma 8.2. *For $f, g \in E^{m,\alpha}(B_a)$ we have*
(i) $f + g \in E^{m,\alpha}(B_a)$.
(ii) $fg \in E^{m,\alpha}(B_a)$.
(iii) If $f \neq 0$ in B_a, then $1/f \in E^{m,\alpha}(B_a)$.
Analogous results hold for functions in $E^\infty(B_a)$.

The second important property is related to elliptic operators. Let u be a solution of the Poisson equation

$$\Delta u = f. \tag{8.61}$$

Then we have the following lemma.

Lemma 8.3. $f \in E^{m,\alpha}(B_a) \Rightarrow u \in E^{m+2,\alpha}(B_a)$.

Proof. Since $f_2 \in C^{m,\alpha}(B_a)$ we can define the corresponding Taylor polynomial T_m of order m. Define f_2^R by $f_2 = T_m + f_2^R$. It can be seen that $f_2^R = O(r^{m+\alpha})$. By explicit calculation we can prove that there exist a polynomial p_m of order m such that $\Delta(r^3 p_m) = rT_m$. Set $u = r^3 p_m + u_R$. Then u_R satisfies the equation

$$\Delta u_R = f_1 + rf_2^R. \tag{8.62}$$

One can prove that $rf_2^R \in C^{m,\alpha}(B_a)$. This is not trivial because r is only in $C^\alpha(B_a)$. Here we use that $f_2^R = O(r^{m+\alpha})$ (see lemma 3.6 in [18]). Then the right-hand side of equation (8.62) is in $C^{m,\alpha}(B_a)$. By the standard Schauder elliptic regularity (see [25]) we conclude that $u_R \in C^{m+2,\alpha}(B_a)$. Thus $u \in E^{m+2,\alpha}(B_a)$. □

As an application of lemmas 8.2 and 8.3 we can prove theorem 8.6. Using lemma 8.2, we have that the function f satisfies the following property:

$$u \in E^{m,\alpha}(B_a) \Rightarrow f(u(x), x) \in E^{m,\alpha}(B_a), \tag{8.63}$$

for every positive u. Assume that we have a solution $u \in C^{2,\alpha}$ of equation (8.59). Using (8.63), lemma 8.3 and induction in m we conclude that $u \in E^\infty(B_a)$. Here we have used that if $f \in E^{m,\alpha}(B_a)$ for all $m \in \mathbb{N}_0$, then $f \in E^\infty(B_a)$, see lemma 3.8 in [18].

Note that the only property of f that we have used in order to prove theorem 8.6 is (8.63). Lemma 8.3 can be generalized for second order elliptic operators with smooth coefficients (cf. [18]).

8.5 Final Comments

In this section I want to make some remarks concerning the differentiability of the initial data. It is physically reasonable to assume that the fields are smooth in the asymptotic region $\tilde{S} \setminus \Omega$. However in the source region Ω smoothness is a restriction. For example, the matter density of a star is discontinuous at the boundary. Generalizations of the existence theorems to include such situations have been made in [13] and [19].

It is also a restriction to assume that the conformal metric is smooth in a neighborhood B_a of the point i. The point i is the infinity of the data, hence there is no reason a priori to assume that the fields there will have the same smoothness as in the interior. Moreover it has been proved in [17] that the stationary space times do not have a smooth conformal metric. In this case the metric h_{ab} in B_a has the following form:

$$h_{ij} = h^1_{ij} + r^3 h^2_{ij}, \tag{8.64}$$

where h^1_{ij} and h^2_{ij} are analytic functions of the coordinates x^i. Theorems 8.1 and 8.2 are generalized in [15] for metrics that satisfy (8.64). An example of this generalization is Theorem 8.5, in which we have assumed that $f \in E^\infty$, this assumption allows metrics of the form (8.64).

References

1. L. Andersson and P. Chruściel. *On hyperboloidal Cauchy data for vacuum Einstein equations and obstructions to the smoothness of scri.* Commun. Math. Phys., **161**, 533–568, 1994.
2. L. Andersson, P. Chruściel, and H. Friedrich. *On the regularity of solutions to the Yamabe equation and the existence of smooth hyperboloidal initial data for Einstein's field equations.* Commun. Math. Phys., **149**, 587–612, 1992.
3. J. Baker and R. Puzio. *A new method for solving the initial value problem with application to multiple black holes.* Phys. Rev. D, **59**, 044030, 1999.
4. R. Beig. *TT-tensors and conformally flat structures on 3-manifolds.* In P. Chruściel, ed., *Mathematics of Gravitation, Part 1*, vol. 41 (Banach Center Publications, Polish Academy of Sciences, Institute of Mathematics, Warszawa, 1997). Gr-qc/9606055.
5. R. Beig and N. O. Murchadha. *Trapped surfaces due to concentration of gravitational radiation.* Phys. Rev. Lett., **66**, 2421–2424, 1991.
6. R. Beig and N. O. Murchadha. *The momentum constraints of general relativity and spatial conformal isometries.* Commun. Math. Phys., **176**, 723–738, 1996.
7. R. Beig and N. O. Murchadha. *Late time behavior of the maximal slicing of the Schwarzschild black hole.* Phys. Rev. D, **57**, 4728–4737, 1998.
8. R. Beig and W. Simon. *Proof of a multipole conjecture due to Geroch.* Commun. Math. Phys., **78**, 75–82, 1980.
9. R. Beig and W. Simon. *On the multipole expansion for stationary space-times.* Proc. R. Lond. A, **376**, 333–341, 1981.
10. J. M. Bowen and J. W. York, Jr. *Time-asymmetric initial data for black holes and black-hole collisions.* Phys. Rev. D, **21**, 2047–2055, 1980.

11. S. Brandt and E. Seidel. *The evolution of distorted rotating black holes III: Initial data.* Phys. Rev. D, **54**, 1403–1416, 1996.
12. D. Brill and R. W. Lindquist. *Interaction energy in geometrostatics.* Phys.Rev., **131**, 471–476, 1963.
13. Y. Choquet-Bruhat, J. Isenberg, and J. W. York, Jr. *Einstein constraint on asymptotically euclidean manifolds.* Phys. Rev. D, **61**, 084034, 1999. Gr-qc/9906095.
14. Y. Choquet-Bruhat and J. W. York, Jr. *The Cauchy problem.* In A. Held, ed., *General Relativity and Gravitation*, vol. 1, 99–172 (Plenum, New York, 1980).
15. S. Dain. *Asymptotically flat initial data with prescribed regularity II*, 2001. In preparation.
16. S. Dain. *Initial data for a head on collision of two kerr-like black holes with close limit.* Phys. Rev. D, **64**, 124002, 2001. Gr-qc/0103030.
17. S. Dain. *Initial data for stationary space-time near space-like infinity.* Class. Quantum Grav., **18**, 4329–4338, 2001. Gr-qc/0107018.
18. S. Dain and H. Friedrich. *Asymptotically flat initial data with prescribed regularity.* Comm. Math. Phys., **222**, 569–609, 2001. Gr-qc/0102047.
19. S. Dain and G. Nagy. *Initial data for fluid bodies in general relativity.* Phys. Rev. D, **65**, 084020, 2002. Gr-qc/0201091.
20. H. Friedrich. *Cauchy problems for the conformal vacuum field equations in general relativity.* Commun. Math. Phys., **91**, 445–472, 1983.
21. H. Friedrich. *On static and radiative space-time.* Commun. Math. Phys., **119**, 51–73, 1988.
22. H. Friedrich. *Gravitational fields near space-like and null infinity.* J. Geom. Phys., **24**, 83–163, 1998.
23. P. R. Garabedian. *Partial Differential Equations* (John Wiley, New York, 1964).
24. R. Geroch. *Multipole moments. II. curved space.* J. Math. Phys., **11**, 2580–2588, 1970.
25. D. Gilbarg and N. S. Trudinger. *Elliptic Partial Differential Equations of Second Order* (Springer-Verlag, Berlin, 1983).
26. R. J. Gleiser, G. Khanna, and J. Pullin. *Evolving the Bowen-York initial data for boosted black holes.* gr-qc/9905067, 1999.
27. S. W. Hawking. *The event horizon.* In C. DeWitt and B. S. DeWitt, eds., *Black Holes*, 1–56 (Gordon and Breach Science Publishers, New York, 1973).
28. P. Kundu. *On the analyticity of stationary gravitational fields at spatial infinity.* J. Math. Phys., **22**, 2006–2011, 1981.
29. J. M. Lee and T. H. Parker. *The Yamabe problem.* Bull. Amer. Math. Soc., **17**, 37–91, 1987.
30. R. C. McOwen. *Partial Differential Equation* (Prentice Hall, New Jersey, 1996).
31. C. W. Misner. *Wormhole initial conditions.* Phys.Rev., **118**, 1110–1111, 1960.
32. C. W. Misner. *The method of images in geometrostatics.* Ann. Phys., **24**, 102–117, 1963.
33. E. T. Newman and R. Penrose. *Note on the Bondi-Metzner-Sachs group.* J. Math. Phys., **7**, 863–870, 1966.

9 Construction of Hyperboloidal Initial Data

Lars Andersson

Department of Mathematics, University of Miami, Coral Gables, FL 33124, USA

Abstract. In this note I discuss the conformal rescaling of spacetimes, with focus on the constraint equations. I give conditions on the Cauchy data at the conformal boundary which are sufficient for regularity, and show how to construct solutions to the conformal constraint equations. Finally, I discuss gauge conditions in the conformal setting and indicate the first steps in analyzing the conformally rescaled version of the Einstein vacuum evolution equations.

9.1 Introduction

Let $(\mathbf{M}, \mathbf{g}, \Omega)$ be the conformal rescaling of a 3+1 dimensional vacuum space-time $(\tilde{\mathbf{M}}, \tilde{\mathbf{g}})$, with $\mathbf{g} = \Omega^2 \tilde{\mathbf{g}}$, where Ω is a smooth function on $\mathbf{M}$ vanishing on the boundary $\partial\mathbf{M}$. If $\mathbf{M}$ has smooth null boundary $\mathscr{I}$, with connected components of topology $S^2 \times \mathbb{R}$, it follows that the Weyl tensor of $(\mathbf{M}, \mathbf{g})$ vanishes on $\mathscr{I}$. From this follows peeling properties for $(\tilde{\mathbf{M}}, \tilde{\mathbf{g}})$. This picture of isolated systems in general relativity, developed by Penrose, see [9, 16], is useful for studying the mass and angular momentum of space-times, as well as gravitational radiation.

Friedrich has found a first order symmetric–hyperbolic version of the Einstein evolution equations, called the conformally regular field equations, which may be extended through $\mathscr{I}$ [11]. The Cauchy data for this system on a future Cauchy surface M, asymptotic to $\mathscr{I}$ so that $\bar{M} \cap \partial\mathbf{M} = \partial M$, include g_{ab}, K_{ab}, Ω as well as components of the rescaled Weyl tensor $\Omega^{-1}\mathbf{C}^{\alpha}{}_{\beta\gamma\delta}$. In order to get a regular evolution at $\mathscr{I}$, these data must be regular up to ∂M. It was shown by the author, Chrusciel and Friedrich [3–6] that under certain conditions on the boundary geometry of M, the Cauchy data for the conformally regular field equations are smooth up to ∂M.

Using the conformally regular field equations, Friedrich has shown that the maximal vacuum development of data (called hyperboloidal data) on a future Cauchy surface intersecting $\mathscr{I}$, regular up to ∂M is a space-time which has a "smooth piece of $\mathscr{I}$". Further, for small data the maximal vacuum development has a null boundary with future complete null generators and a regular time-like infinity.

A programme has been initiated to numerically evolve the Einstein equations using the conformally regular field equations [12–15]. This approach may have advantages over the "traditional" approach which treats the asymptotic flatness condition by introducing boundary conditions far away from the isolated system under study, and attempts to observe for example gravitational wave signatures on this boundary.

In this note we will discuss the conformal procedure for constructing solutions (g_{ab}, K_{ab}, Ω) to the constraint equations and the geometric conditions for regularity at $\mathscr{I}$. I will also briefly discuss the Cauchy problem for the Einstein evolution equations at $\mathscr{I}$, directly from the point of view of the Einstein equations in $(\mathbf{M}, \mathbf{g})$.

9.2 Preliminaries

The following index conventions will be used. Greek indices $\alpha, \beta, \ldots$ run over $0, \ldots, 3$, lower case Latin indices $a, b, c, \ldots$ run over $1, 2, 3$, and upper case Latin indices $A, B, C, \ldots$ run over $2, 3$.

Let $(\mathbf{M}, \mathbf{g})$ be a 3+1 dimensional globally hyperbolic space–time with covariant derivative $\mathbf{D}$. Introduce coordinates $(x^\alpha) = (t, x^a)$ on $\mathbf{M}$, and consider the foliation M_t consisting of level sets of t. We will often drop the subscript t on M_t and associated fields. M will be a compact manifold with boundary ∂M, and closure $\bar{M} = M \cup \partial M$. Let T be the normal of M, assumed to be time-like, and let g be the induced positive definite metric on M, with covariant derivative ∇. We will use an index T to denote contraction with T, for example $t_T = T^\alpha t_\alpha$. Let $\Sigma = \mathbf{D}_T \Omega$ and assume $|\Sigma| > 0$. The lapse function N and shift vectorfield X are defined by $\partial_t = NT + X$. The second fundamental form K is defined by $K_{ab} = \langle \mathbf{D}_a e_b, T \rangle$. Then we have $K_{ab} = -\frac{1}{2}\mathcal{L}_T \mathbf{g}_{ab} = -\frac{1}{2}N^{-1}(\partial_t - X) g_{ab}$ and $\mathbf{D}_a e_b = \nabla_a e_b - T K_{ab}$.

Assume that $\mathbf{M}$ has null boundary $\partial\mathbf{M}$ and let $\bar{\mathbf{M}} = \mathbf{M} \cup \partial\mathbf{M}$. Let Ω be a positive function on $\mathbf{M}$ with $\Omega = 0$, $d\Omega \neq 0$ on $\partial\mathbf{M}$ and denote by $\tilde{\mathbf{g}}$ the conformally related metric $\tilde{\mathbf{g}} = \Omega^{-2}\mathbf{g}$. We will call $\mathbf{g}$ the unphysical metric and $\tilde{\mathbf{g}}$ the physical metric. We will use $\langle \cdot, \cdot \rangle$ for the inner product induced on $T\mathbf{M}$ by $\mathbf{g}$. Geometric quantities associated with $\tilde{\mathbf{g}}$ will be decorated with a tilde $\sim$, for example the covariant derivative $\tilde{\mathbf{D}}$ and the Ricci tensor $\tilde{\mathbf{R}}_{\alpha\beta}$. We will consider only the case when $(\tilde{\mathbf{M}}, \tilde{\mathbf{g}})$ satisfies the vacuum Einstein equations $\tilde{\mathbf{R}}_{\alpha\beta} = 0$. Let $\omega = \Omega\big|_M$. Assume that $\partial M = \bar{M} \cap \partial\mathbf{M}$, has positive definite induced metric and that ω is a defining function for ∂M, i.e. $d\omega\big|_{\partial M} \neq 0$. We will refer to $\partial\mathbf{M}$ as $\mathscr{I}$ and the surface $\partial M \subset \partial\mathbf{M}$ as a cross section of $\mathscr{I}$.

There is a gauge ambiguity in the conformal rescaling. Let $\bar{\Omega} = \Omega\Theta^{-1}$, $\bar{\mathbf{g}} = \Theta^{-2}\mathbf{g}$ for some positive function Θ which is bounded and bounded away from zero on $\mathbf{M}$. Then $\tilde{\mathbf{g}} = \bar{\Omega}^{-2}\bar{\mathbf{g}}$ is another conformal rescaling of $\mathbf{g}$. A conformal gauge change can be used to control the mean curvature of ∂M, as well as the mean curvature of M. In particular, the conformal gauge freedom can be used so that the level sets of ω are the leaves of the gauss foliation w.r.t. ∂M, i.e. there are coordinates $(x^a) = (x, y^A)$, so that in a neighborhood of ∂M, we have $\omega = x$, and the metric takes the form:

$$g_{ab} dx^a dx^b = dx^2 + h_{AB} dy^A dy^B. \tag{9.1}$$

This possible without loss of generality, cf. [6, Lemma 2.1]. By construction we have $x(p) = d(p, \partial M)$ and the unit normal ∂M is $\eta = \partial_x$. We use the index 1 for contraction with η, for example $t_{1b} = \nabla^a x t_{ab} = \eta^a t_{ab}$. Let $\not{\nabla}_A$ and $\not{R}_{AB}$ denote the covariant derivative with respect to h_{AB} and the Ricci tensor of h_{AB}. The second fundamental form of ∂M with respect to η is defined by $\lambda_{AB} = \langle \nabla_A e_B, \eta \rangle$. Then we have $\nabla_A e_B = \not{\nabla}_A e_B + \eta \lambda_{AB}$ and $\lambda_{AB} = -\frac{1}{2}\mathcal{L}_\eta h_{AB}$.

The quantities associated to (g_{ij}, K_{ij}, N, T) in the physical space–time are

$$\tilde{g}_{ij} = \omega^{-2} g_{ij}, \qquad \tilde{K}_{ij} = \omega^{-1} K_{ij} + \Sigma \tilde{g}_{ij}, \tag{9.2a}$$

$$\tilde{N} = \omega^{-1} N, \qquad \tilde{T} = \omega T. \tag{9.2b}$$

The shift vectorfield X does not scale, so $\tilde{X} = X$. The physical Weyl tensor satisfies $\tilde{\mathbf{C}}_{\alpha\beta\gamma\delta} = \tilde{\mathbf{R}}_{\alpha\beta\gamma\delta}$ since $\tilde{\mathbf{R}}_{\alpha\beta} = 0$. Further, $\tilde{\mathbf{C}}^\alpha{}_{\beta\gamma\delta} = \mathbf{C}^\alpha{}_{\beta\gamma\delta}$.

A symmetric tensor field A_{ab} on M is said to satisfy the shear free condition if

$$(A_{AB} - \frac{1}{2} h^{CD} A_{CD} h_{AB})\big|_{\partial M} = \frac{\Sigma}{|\Sigma|} (\lambda_{AB} - \frac{1}{2} h^{CD} \lambda_{CD} h_{AB})\big|_{\partial M}. \tag{9.3}$$

In the following we will consider only hypersurfaces intersecting $\mathscr{I}_+$ which in view of our conventions force $\Sigma < 0$, so that $\Sigma/|\Sigma| = -1$. The space-time $(\mathbf{M}, \mathbf{g})$ is said to be shear free at ∂M if K_{ab} satisfies the shear free condition. In order for $\mathbf{g}$ to be in $C^2(\bar{\mathbf{M}})$, it is necessary that K_{ab} satisfy the shear free condition, cf. [4, Prop. 3.1]. In the following we only consider initial data which are shear free. Note that a change of conformal gauge can be used to get $h^{AB}\lambda_{AB}\big|_{\partial M} = 0$.

9.3 Conformal Rescalings of Minkowski Space

In order to understand the Einstein equations near $\mathscr{I}$ it is important to choose an appropriate conformal compactification and a suitable foliation near $\mathscr{I}$. In this section we will display a few examples in the simplest case, Minkowski space.

Minkowski space is $\mathbb{R}^4$ with the flat line element

$$d\tilde{s}^2 = -dt^2 + dr^2 + r^2 d\sigma^2,$$

in radial coordinates (t, r, θ, ϕ), where $d\sigma^2 = d\theta^2 + \sin^2(\theta) d\phi^2$ is the line element on S^2. We will consider two conformal rescalings of Minkowski space. First, define new coordinates τ, ρ by

$$t + r = \tan\left(\frac{\tau + \rho}{2}\right), \qquad t - r = \tan\left(\frac{\tau - \rho}{2}\right),$$

on

$$U = \{\tau, \rho : -\pi < \tau < \pi, \quad 0 \le \rho < \pi, \quad |\tau + \rho| < \pi, \quad |\tau - \rho| < \pi\}. \tag{9.4}$$

Rescaling with the conformal factor

$$\Omega(\tau, \rho) = 2 \cos\left(\frac{\tau + \rho}{2}\right) \cos\left(\frac{\tau - \rho}{2}\right),$$

which is positive on U, gives [11, p. 450-451]

$$ds^2 = \mathbf{g}_{\alpha\beta} dx^\alpha dx^\beta = -d\tau^2 + d\rho^2 + \sin^2(\rho) d\sigma^2,$$

which is the line element of the Einstein universe $\mathbb{R} \times S^3$. We have $N \equiv 1$, $X \equiv 0$.

The null part of the boundary of U has two components with $\tau > 0$ or $\tau < 0$, called future and past null infinity, denoted $\mathscr{I}_\pm$. The point $(\tau, \rho) = (0, \pi)$ is called spatial infinity and denoted i_0. The points $i_\pm$ with $(\tau, \rho) = (\pm\pi, 0)$ are called future and past time-like infinity. All inextendible future directed null curves start at $\mathscr{I}_-$ and end at $\mathscr{I}_+$.

The hyperboloids $t^2 - r^2 = u^2$ in Minkowski space correspond to hypersurfaces in U intersecting ∂U at $\tau = \pi/2$ at different angles depending on u, cf. Fig. 9.1. In particular, the hyperboloid with sectional curvature -1 corresponds to $\tau = \pi/2$. On the other hand, the time translated hyperboloids given

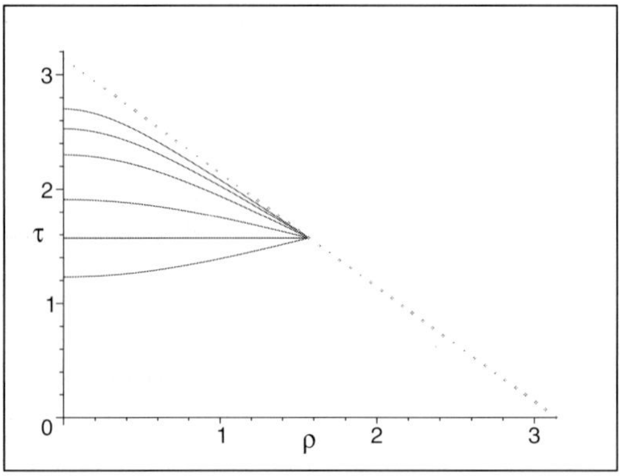

Fig. 9.1. Hyperboloids $t^2 - r^2 = u^2$ in conformally compactified Minkowski space

by $t = \sqrt{1 + r^2} + u$ intersect $\mathscr{I}$ at $\tau = \pi/2 + \arctan(u)$, see Fig. 9.2. In this coordinate system, the future translated hyperboloids approach time-like infinity i_+ as u increases. The Lapse N is given at $\mathscr{I}$ by

$$\lim_{r\to\infty} N = \frac{1}{\sqrt{1+u^2}} = \cos(\tau - \pi/2)$$

which tends to zero as τ increases to π.

For the second conformal rescaling, introduce new coordinates ψ, τ by

$$\begin{aligned} t &= \coth(\psi) + \tau, \\ r &= 1/\sinh(\psi). \end{aligned}$$

This is a singular coordinate transformation, but this does not cause a problem as we are only interested in the behavior near infinity. In terms of the new coordinates, the Minkowski line element takes the form:

$$d\tilde{s}^2 = -d\tau^2 + 2\sinh^{-2}(\psi) d\psi d\tau + \sinh^{-2}(\psi)(d\psi^2 + d\sigma^2).$$

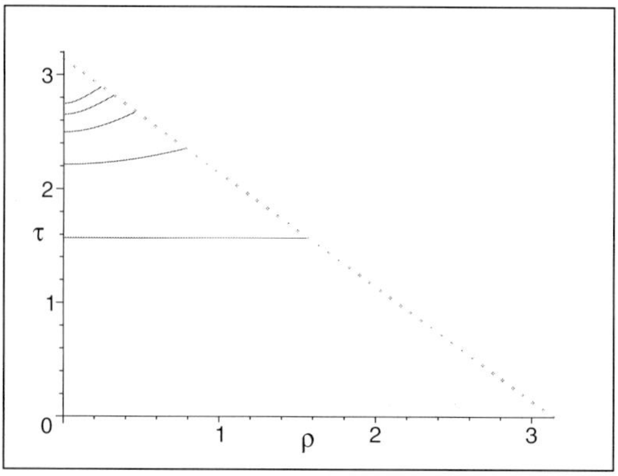

Fig. 9.2. Time translated hyperboloids $t = \sqrt{1+r^2} + u$ in conformally compactified Minkowski space

After a conformal rescaling with

$$\Omega = \sinh(\psi),$$

we get the line element $ds^2 = \Omega^2 d\tilde{s}^2$ given by

$$ds^2 = -\sinh^2(\psi)d\tau^2 + 2d\psi d\tau + (d\psi^2 + d\sigma^2). \tag{9.5}$$

The hypersurfaces τ =constant correspond to the time translated hyperboloids $t = \sqrt{1+r^2} + \tau$ in Minkowski space. The conformally rescaled space-time $\mathbf{M}$ has null boundary $\mathscr{I}_+ = \{\psi = 0\}$, and so the boundary of this conformal rescaling does not include $\mathscr{I}_-$ or $i_0, i_\pm$. Further, the closure of $\mathbf{M}$ is not compact.

The foliation M_τ of level sets of τ has Lapse $N = \cosh(\psi)$, and Shift $X = \partial_\psi$ so that the time like normal T is $T = \cosh^{-1}(\psi)(\partial_\tau - \partial_\psi)$. The foliation M_τ is static so that $K_{ab} = 0$. The induced geometry on M_τ is the cylinder $d\psi^2 + d\sigma^2$, with totally geodesic boundary $\{\psi = 0\}$. The future sheets of the hyperboloids $t^2 - r^2 = u^2$ of mean curvature $-3/u$ intersect $\mathscr{I}$ at $\tau = 0$, see Fig. 9.3.

9.4 Conformal Constraint Equations

The Einstein vacuum equations $\tilde{\mathbf{R}}_{\alpha\beta} = 0$ imply the constraint equations

$$\tilde{R} + (\tilde{K}^a{}_a)^2 - \tilde{K}_{ab}\tilde{K}^{ab} = 0, \tag{9.6a}$$

$$\tilde{\nabla}^a \tilde{K}_{ab} - \tilde{\nabla}_b \tilde{K}^c{}_c = 0. \tag{9.6b}$$

We will consider the constraint equations, in the special case of constant mean curvature $\nabla_a \mathrm{tr}\tilde{K} = 0$, as a system of equations for the unphysical data and rewrite them in a form which exhibits the freely specifiable data.

In the form of the constraint equations that we will present, the freely specifiable data are (g_{ij}, A_{ij}), where g_{ij} is a metric on M, and A_{ij} is a trace free,

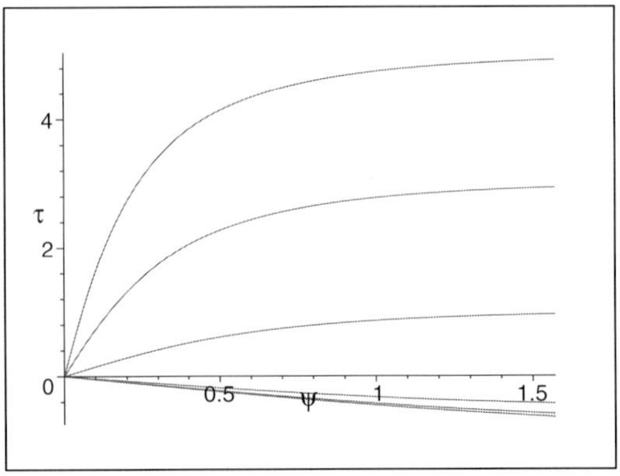

Fig. 9.3. Hyperboloids $t^2 - r^2 = u^2$ in the conformal rescaling (9.5)

symmetric tensor field on M satisfying the shear free condition. We will assume for simplicity $A_{1a}\big|_{\partial M} = 0$, which does not restrict the degrees of freedom.

Let ρ be a defining function for ∂M, $\rho = x$ near ∂M. Assume $\Sigma\big|_M$ is a nonzero constant. Set $\hat{g}_{ab} = \rho^{-2} g_{ab}$, define $\hat{\nabla}, \hat{R}$ etc. w.r.t. $\hat{g}$ and raise and lower indices on hatted fields with $\hat{g}$. Given data g_{ij}, A_{ij}, a solution (g_{ij}, K_{ij}, ω) to the constraint equations is constructed by solving the system (9.7) below. We will refer to the system (9.7) as the conformal constraint equations. Existence, uniqueness and regularity of solutions to the conformal constraint equations has been analyzed in the papers [4, 6]. For a vector field Y, let

$$L(Y)_{ij} = \nabla_i Y_j + \nabla_j Y_i - \frac{2}{3} \nabla_k Y^k g_{ij}. \tag{9.7a}$$

For a solution Y to the system

$$\hat{\nabla}^i \hat{L}(Y)_{ij} = \hat{\nabla}^i (\rho^{-1} A_{ij}), \tag{9.7b}$$

define σ_{ij} by

$$\sigma_{ij} = A_{ij} - \rho^{-1} L(Y)_{ij}. \tag{9.7c}$$

Then $\hat{\sigma}_{ij} = \rho^{-1}\sigma_{ij}$ satisfies $\hat{\nabla}^a \hat{\sigma}_{ab} = 0$. Let u be the solution to the semi-linear elliptic equation

$$-8\hat{\Delta} u + \hat{R} u - \hat{\sigma}_{ij}\hat{\sigma}^{ij} u^{-7} + 6\Sigma^2 u^5 = 0. \tag{9.7d}$$

Finally, define (g_{ij}, K_{ij}, ω) by $\omega = u^{-2}\rho$, $K_{ij} = u^{-4}\sigma_{ij} + \Sigma \tilde{g}_{ij}$. The fields $(\tilde{g}_{ij}, \tilde{K}_{ij})$ corresponding to a solution g_{ij}, K_{ij}, ω of the conformal constraint equations (9.7) solve the physical constraint equations (9.6) with mean curvature $\tilde{g}^{ij}\tilde{K}_{ij} = 3\Sigma$, under the constant mean curvature condition $\nabla_a \Sigma = 0$.

9.4.1 Constant Mean Curvature Hypersurfaces

The existence of complete constant mean curvature hypersurfaces in asymptotically flat space-times was studied in [7]. There it was proved that in an asymptotically Schwarzschild space-time satisfying a natural causal condition (domain of dependence of compact sets is compact), given a cross section of $\mathscr{I}_+$, and a number $\Sigma < 0$, there is a unique CMC hypersurface intersecting $\mathscr{I}_+$ at the given cross section. The proof uses a barrier construction, together with the causality condition, to prove that the solutions to a sequence of boundary value problems converges to a complete CMC hypersurface.

This result indicates that a wide class of space-times with regular conformal compactification are foliated by CMC hypersurfaces. The regularity of the height functions of these CMC hypersurfaces at $\mathscr{I}$ has not been studied in detail, nor has the existence proof been carried out in the conformally compactified setting. This is a natural problem which should be studied further.

The conditions for regularity of CMC solutions of the conformal constraint equations constructed using the procedure discussed above is well understood, see the discussion in Sect. 9.4.3 below. However, it is not at all well understood how these results relate to the regularity of CMC foliations in the physical or unphysical space-time, evolved from these data, nor what spatial gauge conditions are suitable in this situation, see Sect. 9.5 below.

9.4.2 Degenerate Elliptic Equations

The system (9.7) of conformal constraint equations is a degenerate elliptic system for (u, Y^i). We will briefly describe the regularity properties of solutions to systems of this type. As an example consider the Lichnerowicz equation (9.7d). The principal part of the right hand side of (9.7d) is essentially of the form:

$$L = L_x + B,$$

where L_x is the ordinary differential operator $L_x = x^2\partial_x^2 + ax\partial_x + b$, and B is of lower order in ∂_x. In case $(1-a)^2 > 4b$, the equation for $L_x x^\alpha = 0$ has real distinct characteristic roots $\alpha_\pm$ and the equation $L_x u = f$ has a solution $u = o(x^{\alpha_-})$ for sufficiently regular $f = o(x^{\alpha_-})$.

The operator appearing in the Lichnerowicz equation has critical exponents $\alpha_- = -1$, $\alpha_+ = 3$. Let L be this operator and consider the equation $Lu = f$, for $f \in C^\infty(\overline{M})$. Suppose f is of the form $f = f_1 x + f_2 x^2 + f_3 x^3$. Then a solution $u = o(x^{-1})$ is of the form

$$u = 1 + u_1 x + u_2 x^2 + u_{3,1} x^3 \ln x + u_3 x^3 + \text{higher order terms},$$

where $u_{3,1} = c f_3$ for some explicit constant c. This example reflects the fact, cf. [4–6], that solutions to degenerate elliptic systems are in general non-smooth at ∂M, instead the general form of the solution has a polylogarithmic expansion of the form $u = \sum u_{i,j} x^i \ln^j x$. In case the RHS and the coefficients are smooth, the logarithm terms in u appear first at the critical exponent α_+.

The terms in the expansion of u up to and including the first logarithm term are computable in terms of the data, which implies that explicit geometric conditions for the regularity of solutions of the conformal constraint equations (9.7), and hence of the physical space-time constructed from these data, can be found.

9.4.3 Regularity of Solutions to the Conformal Constraint Equations

We will now discuss the detailed conditions for regularity of the conformal data (g_{ab}, K_{ab}, Ω) at ∂M. We will assume that free data $(g_{ij}, A_{ij}) \in C^\infty(\bar{M})$ are given and consider the conditions needed for the solution (g_{ij}, K_{ij}, ω) of the conformal constraint equations, produced by solving the system (9.7), to be in $C^\infty(\bar{M})$.

In the case when M is a moment of time symmetry in the unphysical space–time, i.e. $A_{ij} = K_{ij} = 0$, there is a simple criterion for regularity of the solution to the constraint equations. Let

$$\lambda^*_{AB} = \lambda_{AB} - \frac{1}{2} h^{CD} \lambda_{CD} \lambda_{AB}$$

denote the trace free part of λ_{AB}. Then the solution to the conformal constraint equation is in $C^\infty(\bar{M})$ if and only if the conformal density $\mathcal{C}$ on ∂M, defined by

$$\mathcal{C} = \nabla\!\!\!\!/\,^A \nabla\!\!\!\!/\,^B \lambda_{AB} + \lambda^{*AB} R_{AB} - \frac{1}{2} h^{CD} \lambda_{CD} \lambda^{AB} \lambda_{AB}$$

vanishes. In particular, this holds if ∂M is totally umbilic, i.e. if

$$(\lambda_{AB} - \frac{1}{2} h^{CD} \lambda_{CD} \lambda_{AB})\big|_{\partial M} = 0.$$

In general if this condition does not hold, then the solution has an expansion in powers of x and $\ln x$ near ∂M and is thus of finite regularity at ∂M.

In the case where $K_{ij} \neq 0$ and consequently $A_{ij} \neq 0$, we make the following simplifying assumptions. We consider shear free data (g_{ij}, A_{ij}), such that

$$h^{AB} \lambda_{AB}\big|_{\partial M} = 0, \tag{9.8a}$$

$$A_{1j}\big|_{\partial M} = 0. \tag{9.8b}$$

Equation (9.8a) should be viewed as a conformal gauge condition, which can always be satisfied if the data (g_{ij}, A_{ij}) is smooth up to boundary. Similarly, equation (9.8b) may be viewed as a gauge type condition as it is possible to achieve this by subtracting a term $\rho^{-1} L_{ij}$ from A_{ij}. If (9.8) holds, then the fields ω, Y^i satisfy

$$\omega = |\Sigma| x + O(x^3), \tag{9.9a}$$

$$Y^i = O(x^3), \tag{9.9b}$$

and K_{ab} is determined up to first order by A_{ab}. If (9.8) holds, then the fields $\omega, K_{ij}, \omega^{-1} \mathbf{C}_{\alpha\beta\gamma\delta}$ are in $C^\infty(\bar{M})$ if and only if the equations

$$\partial_x K_{AB} - \frac{1}{2} h^{CD} \partial_x K_{CD} h_{AB} = 0, \tag{9.10a}$$

$$\nabla\!\!\!\!/\,^A \nabla\!\!\!\!/\,^B \lambda_{AB} + R_{AB} \lambda^{AB} = 0, \tag{9.10b}$$

$$\nabla\!\!\!\!/\,^B (\partial_x K_{AB} + \frac{3}{2} \lambda_{CD} \lambda^{CD} h_{AB}) = 0, \tag{9.10c}$$

cf. [4, Eq. (4.21)–(4.23), (4.33)], hold on ∂M. Note that our sign conventions for ϵ and K_{ab} are opposite those of [4]. The particular form we present here holds assuming (9.8). Equation (9.10a), which states that $\partial_x K_{AB}$ is proportional to h_{AB} on ∂M, is a consequence of the vanishing of the Weyl tensor on ∂M, while equations (9.10b) and (9.10c) follow from smoothness of K_{ab} up to ∂M. In particular, it can be shown that if $\partial M \cong S^2$, then

$$\left(\partial_x K_{AB} + \frac{3}{2}\lambda^{CD}\lambda_{CD}h_{AB}\right)\big|_{\partial M} = 0. \tag{9.11}$$

9.5 The Initial Value Problem

In this section we will consider the Einstein evolution equations directly from the point of view of the unphysical Cauchy data. Since we know from the work of Friedrich [11] that the maximal vacuum extension of conformally regular data on M has regular conformal boundary near ∂M, it follows that solutions to a well–posed formulation of the Einstein evolution equations in the unphysical space–time will have this property. Working directly in terms of the unphysical Cauchy data allows one to relate this problem to the extensive literature on the numerical solution of the Einstein evolution equations. In particular, it is interesting to consider gauge choices and hyperbolic reformulations.

Consider a foliation M_t of $(\mathbf{M}, \mathbf{g})$ with $\partial_t = NT + X$. The structure equations of the foliation imply the Einstein evolution equations

$$\partial_t g_{ij} = -2NK_{ij} + \mathcal{L}_X g_{ij}, \tag{9.12a}$$
$$\partial_t K_{ij} = -\nabla_i\nabla_j N + N(R_{ij} + \mathrm{tr}K K_{ij} - 2K_{im}K^m_{\ j} - \mathbf{R}_{ij}) + \mathcal{L}_X K_{ij}, \tag{9.12b}$$

and the constraint equations

$$R - |K|^2 + (\mathrm{tr}K)^2 = 2\mathbf{R}_{TT} + \mathbf{R}, \tag{9.13a}$$
$$\nabla_i \mathrm{tr}K - \nabla^j K_{ij} = \mathbf{R}_{Ti}. \tag{9.13b}$$

The unphysical Ricci tensor $\mathbf{R}_{\alpha\beta}$ is given by

$$\mathbf{R}_{\alpha\beta} = \tilde{\mathbf{R}}_{\alpha\beta} - \Omega^{-1}[2\mathbf{D}_\alpha\mathbf{D}_\beta\Omega + \mathbf{D}_\gamma\mathbf{D}^\gamma\Omega\mathbf{g}_{\alpha\beta}] + 3\Omega^{-2}\mathbf{D}_\gamma\Omega\mathbf{D}^\gamma\Omega\mathbf{g}_{\alpha\beta},$$

the unphysical Ricci scalar is $\mathbf{R} = \mathbf{g}^{\alpha\beta}\mathbf{R}_{\alpha\beta}$, and $|K|^2 = K_{ij}K^{ij}$. The condition $\mathbf{D}_T\Omega = \Sigma$ may be viewed as an evolution equation for Ω,

$$\partial_t\Omega = N\Sigma + X\Omega. \tag{9.13c}$$

9.5.1 Gauge Condition at ∂M

It is natural to require $\partial_t\Omega\big|_{\partial M} = 0$. From (9.13c) and $\omega = |\Sigma|x + h.o.$, $\partial_t\Omega\big|_{\partial M} = N\Sigma + |\Sigma|\big|_{\partial M}$, so $\partial_t\Omega\big|_{\partial M} = 0$ implies

$$N\big|_{\partial M} = \langle X, \eta\rangle\big|_{\partial M}.$$

In view of this, a natural boundary condition for N and X is

$$N\big|_{\partial M} = 1, \qquad X\big|_{\partial M} = \eta, \tag{9.14}$$

which corresponds to the asymptotic behavior of Lapse and Shift in the second conformal compactification discussed in Sect. 9.3. In a neighborhood of ∂M, we may decompose X as $X = \alpha\eta + \beta$. Then the boundary condition for X is $\alpha = 1, \beta = 0$ on ∂M.

9.5.2 Evolution at ∂M

As discussed in [4, §5], the shear free condition is necessary in order for the development of the data on M to have a regular conformal boundary. It is convenient to introduce the notation $\lambda = h^{CD}\lambda_{CD}$, $\kappa = h^{CD}K_{CD}$. We relax the condition $\lambda\big|_{\partial M} = 0$ used in Sect. 9.4.3.

Imposing the boundary gauge condition (9.14), a calculation shows that

$$\partial_t h_{AB}\big|_{\partial M} = -(\kappa + \lambda)h_{AB}\big|_{\partial M}.$$

In particular, $\partial_t h_{AB}\big|_{\partial M}$ is pure trace. For a symmetric tensor t_{AB}, let $t^*_{AB} = t_{AB} - \frac{1}{2}h^{CD}t_{CD}h_{AB}$ denote the trace free part. Define the shear tensor σ_{AB} by

$$\sigma_{AB} = \lambda^*_{AB} + K^*_{AB},$$

so that the shear free condition can be formulated as

$$\sigma_{AB}\big|_{\partial M} = 0.$$

From the fact that $\partial_t h_{AB}\big|_{\partial M}$ is pure trace together with the fact that M is 3–dimensional, it follows that

$$\partial_t \sigma_{AB}\big|_{\partial M} = [\partial_t(K_{AB} + \lambda_{AB})]^*.$$

Thus, if $\partial_t(K_{AB} + \lambda_{AB})\big|_{\partial M}$ is pure trace, then the shear free condition is conserved by the evolution. The evolution equations and the boundary conditions imply

$$\partial_t \lambda_{AB}\big|_{\partial M} = \partial_x(\lambda_{AB} + K_{AB}) + N_1 K_{AB} - \frac{1}{2}\mathcal{L}_{X_1} g_{AB}\big|_{\partial M},$$

where $N_1 = \partial_x N\big|_{\partial M}$, $X_1 = [\partial_x, X]\big|_{\partial M}$. After a lengthy calculation one finds:

$$\partial_t \sigma_{AB}\big|_{\partial M} = N_1 K^*_{AB} - \frac{1}{2}[\mathcal{L}_{X_1} g_{AB}]^*.$$

The leading order term N_1 is determined by the regularity of $\mathbf{R}_{\alpha\beta}$ up to $\partial\mathbf{M}$. Therefore the condition for preserving the shear free condition under evolution, $\partial_t \sigma_{AB}\big|_{\partial M} = 0$ shows up as a gauge condition for X. Due to the shear free condition, the equation

$$N_1 K^*_{AB} - \frac{1}{2}[\mathcal{L}_{X_1} g_{AB}]^*\big|_{\partial M} = 0,$$

can be solved by putting $X_1\big|_{\partial M} = N_1\eta$, in other words by imposing the condition that X is parallel to η at ∂M to first order, and that ∂_t is null to first order.

9.6 Discussion

In this note, I have shown how to construct solutions of the conformal constraint equations, and indicated the first steps in analyzing the evolution problem for the Einstein vacuum evolution equations, directly in the unphysical setting. It seems worthwhile to explore gauge choices, hyperbolic reformulations, boundary conditions etc. for the evolution equations directly in this setting. This makes it possible to tie in to the extensive literature on the standard form of the Einstein evolution equations. As an example, it is interesting to look at reformulations of the Einstein equations, derived by adding multiples of the constraints in the evolution equations, which have improved stability of the constraints, see for example [10]. It is natural to ask here if there are such reformulations which are also better behaved at $\mathscr{I}$.

The gauge conditions that have been considered so far in the literature on the numerical solution of the Einstein equations are mainly hyperbolic in nature, and therefore suffer from well known problems such as gauge singularities [1, 2]. It is known that CMC time gauge gives a well posed form of the Einstein evolution equations. Therefore it seems natural to consider this time gauge also for the hyperboloidal IVP.

Choosing a time gauge condition such as the CMC condition, leaves open the choice of spatial gauge. It seems likely that it is advantageous to work with a "fully gauge fixed formulation" for numerical work. The combination of CMC time gauge and harmonic spatial gauge gives a well posed elliptic–hyperbolic system in case of compact Cauchy surface [8]. Does this gauge condition give a well posed system also for the hyperboloidal IVP, with boundary conditions given by a foliation of $\mathscr{I}$? The defining equation for the Shift vector field given in [8] is derived from the evolution equation for the metric and therefore does not depend on the matter content of the space-time. This makes it possible to apply this formulation both in the physical and unphysical space-time. It is an interesting problem to explore the consequences of this gauge choice for the hyperboloidal IVP, both from the point of view of the physical and unphysical formulation of the IVP and gauge conditions.

Acknowledgments

This paper is based on a talk given at the Workshop on the Conformal Structure of Space-Times held at Tübingen, April 2–4, 2001. I am grateful to the organizers, Jörg Frauendiener and Helmut Friedrich, for their hospitality and support. This work is supported in part by the Swedish Natural Sciences Research Council (SNSRC), contract no. R-RA 4873-307 and NSF, contract no. DMS 0104402.

References

1. M. Alcubierre. *Appearance of coordinate shocks in hyperbolic formalisms of general relativity.* Phys. Rev. D (3), **55**, 5981–5991, 1997.

2. M. Alcubierre and J. Massó. *Pathologies of hyperbolic gauges in general relativity and other field theories.* Phys. Rev. D (3), **57**, R4511–R4515, 1998.
3. L. Andersson and P. T. Chruściel. *Hyperboloidal Cauchy data for vacuum Einstein equations and obstructions to smoothness of null infinity.* Phys. Rev. Lett., **70**, 2829–2832, 1993.
4. L. Andersson and P. T. Chruściel. *On "hyperboloidal" Cauchy data for vacuum Einstein equations and obstructions to smoothness of scri.* Comm. Math. Phys., **161**, 533–568, 1994.
5. L. Andersson and P. T. Chruściel. *Solutions of the constraint equations in general relativity satisfying "hyperboloidal boundary conditions".* Dissertationes Math. (Rozprawy Mat.), **355**, 100, 1996.
6. L. Andersson, P. T. Chruściel, and H. Friedrich. *On the regularity of solutions to the Yamabe equation and the existence of smooth hyperboloidal initial data for Einstein's field equations.* Comm. Math. Phys., **149**, 587–612, 1992.
7. L. Andersson and M. S. Iriondo. *Existence of constant mean curvature hypersurfaces in asymptotically flat spacetimes.* Ann. Global Anal. Geom., **17**, 503–538, 1999.
8. L. Andersson and V. Moncrief. *Elliptic–hyperbolic systems and the Einstein equations.* gr-qc/0110111, 2001.
9. J. Frauendiener. *Conformal infinity.* Living Rev. Relativ., **3**, 2000–4, 92 pp. (electronic), 2000.
10. S. Frittelli and O. Reula. *Well-posed forms of the* 3 + 1 *conformally-decomposed Einstein equations.* J. Math. Phys., **40**, 5143–5156, 1999.
11. H. Friedrich. *Cauchy problems for the conformal vacuum field equations in general relativity.* Comm. Math. Phys., **91**, 445–472, 1983.
12. P. Hübner. *How to avoid artificial boundaries in the numerical calculation of black hole spacetimes.* Class. Quant. Grav., **16**, 2145–2164, 1999.
13. P. Hübner. *A scheme to numerically evolve data for the conformal Einstein equation.* Class. Quant. Grav., **16**, 2823–2843, 1999.
14. P. Hübner. *From now to timelike infinity on a finite grid.* Class. Quant. Grav., **18**, 1871–1884, 2001.
15. P. Hübner. *Numerical calculation of conformally smooth hyperboloidal data.* Class. Quant. Grav., **18**, 1421–1440, 2001.
16. R. Penrose and W. Rindler. *Spinors and space-time. Vol. 2* (Cambridge University Press, Cambridge, 1988), second edn. Spinor and twistor methods in space-time geometry.

10 Exploring the Conformal Constraint Equations

Adrian Butscher

Albert-Einstein-Institut, Max-Planck-Institut für Gravitationsphysik
Am Mühlenberg 1, 14476 Golm, Germany

Abstract. One method of studying the asymptotic structure of spacetime is to apply Penrose's conformal rescaling technique. In this setting, the rescaled Einstein equations for the metric and the conformal factor in the unphysical spacetime degenerate where the conformal factor vanishes, namely at the boundary representing null infinity. This problem can be avoided by means of a technique of H. Friedrich, which replaces the Einstein equations in the unphysical spacetime by an equivalent system of equations which is regular at the boundary. The initial value problem for these equations produces a system of constraint equations known as the conformal constraint equations. This work describes some of the properties of the conformal constraint equations and develops a perturbative method of generating solutions near Euclidean space under certain simplifying assumptions.

10.1 Introduction

A model for the asymptotic structure of space-time was suggested by Roger Penrose in [22] (see also [15] for a review of the development of these ideas) using the technique of conformal rescaling. Since the reader is by now familiar with the details of the conformal rescaling construction, only enough will be said here to fix the notation to be used in the remainder of this article. The object under study will consist of a *physical space-time* – a smooth, time- and space-orientable Lorentz manifold $(\tilde{M}, \tilde{g})$ satisfying the vacuum Einstein equation $Ric(\tilde{g}) = 0$ and that is *asymptotically simple*. In other words, it is, conformally diffeomorphic to a Lorentz manifold (M, g), called the *unphysical space-time*, in such a way that $\tilde{g} = \Omega^{-2} g$, where the *conformal factor* is a smooth function $\Omega : M \to \mathbb{R}_+$. In addition, the boundary of M is non-empty and is associated to points at null infinity by requiring that Ω be a *defining function* for ∂M; that is, $\Omega\big|_{\partial M} = 0$ while $\mathrm{d}\Omega$ never vanishes identically along ∂M.

The purpose of the conformal boundary is to study asymptotic properties of the physical space-time in null directions. To this end, one could use the fact that, due to the conformal equivalence with the physical space-time, the quantities Ω and g must satisfy the conformally rescaled version of Einstein's equation, namely that $Ric(\Omega^{-2} g) = 0$. However, this equation has the drawback that it is degenerate near the boundary of M because there $\Omega \to 0$, and is thus not ideally suited for analytic investigations of the nature of the space-time at null infinity. One possible means of avoiding this difficulty is to use a technique developed by Friedrich [16], which aims to describe the geometry of the unphysical space-time by means of a new, yet fully equivalent system of equations derived from the equation $Ric(\Omega^{-2}g) = 0$ that is formally regular at the boundary of the unphysical space-time. These equations involve g, Ω and several additional quantities and are known as the *conformal Einstein equations.*

As with Einstein's equations in the physical space-time, it is possible to attempt to solve the conformal Einstein equations in the unphysical space-time by means of an initial value formulation, where appropriate initial data are defined on a space-like hypersurface $\mathcal{Z}$ in M and then evolved in time. Again as in the physical space-time, the conformal equations induce certain constraint equations on the initial data; these equations are known as the *conformal constraint equations* and consist of a complicated system of coupled nonlinear differential equations for the induced metric h and second fundamental form χ of $\mathcal{Z}$, the conformal factor restricted to $\mathcal{Z}$, and several additional quantities. A particular case of interest is when $\mathcal{Z}$ is asymptotically *hyperboloidal,* i. e. $\mathcal{Z}$ intersects ∂M transversely. In this case, the evolution of the boundary of $\mathcal{Z}$ forward in time produces the conformal boundary of the unphysical space-time, and global questions concerning the existence of classes of space-times satisfying the definition of asymptotic simplicity can be addressed. See [15] or [17, 18] for a review of these ideas.

The purpose of this article is twofold. First, it is to introduce the conformal constraint equations and to investigate some of their properties, which will be done in Sect. 10.2. It will be found that, in a certain sense, they describe in a coupled way two mathematical problems – namely, the elliptic boundary value problem for the conformal factor Ω and the constraint problem arising from the Gauss-Codazzi equations of $\mathcal{Z}$. Furthermore, a simple geometric assumption will be shown to lead to a special case of the equations in which the first problem does not appear and the second is in the forefront. In this special case, the full system of conformal constraint equations reduces to a much simpler and smaller system of equations that will be called the *extended constraint equations* because they will turn out to be equivalent to the usual vacuum Einstein constraint equations satisfied by the metric and second fundamental form of $\mathcal{Z}$. (Tackling the boundary value problem is at present beyond the scope of this article but will be considered in the future.)

The second goal of this article is to set up a perturbative approach for generating solutions of the extended constraint equations in the neighbourhood of a known solution, but only in the case of time-symmetric data – the more general case will be handled in another future paper [7]. This task will be accomplished in Sect. 10.3 and the main theorem proved in this section appears on page 10.3.1. Because the extended constraint equations are equivalent to the usual constraint equations, the Main Theorem can be interpreted as a new way of finding solutions of these equations, and furthermore, it will turn out to be a way that is completely different from the 'classical' (i. e. Lichnerowicz-York) method of handling them. This issue will be discussed further in Sect. 10.3.

10.2 The Conformal Constraint Equations

10.2.1 Deriving the Equations

Suppose (M, g, Ω) is an unphysical space-time satisfying the assumptions of asymptotic simplicity and thus that the metric and conformal factor satisfy the rescaled version of Einstein's equation

$$Ric(\Omega^{-2}g) = 0\,. \tag{10.1}$$

This section sketches briefly how equation (10.1) for g and Ω leads first to the conformal Einstein equations for g, Ω and additional quantities, and then to the conformal constraint equations. Begin by expanding (10.1) to obtain

$$R_{\mu\nu} = -\frac{\Box\Omega}{\Omega}\,g_{\mu\nu} - \frac{2}{\Omega}\nabla_\mu\nabla_\nu\Omega + \frac{3\,\nabla^\lambda\Omega\nabla_\lambda\Omega}{\Omega^2}\,g_{\mu\nu}\,, \tag{10.2}$$

where $R_{\mu\nu}$ are the components of the Ricci tensor in the unphysical space-time, ∇_μ is the covariant derivative of the four-metric and $\Box$ is its D'Alembertian operator. Notice that, as it is written, equation (10.2) contains terms with negative powers of Ω which tend to infinity near the boundary of the unphysical space-time. Alternatively, if the equation is multiplied through by Ω^2, then the principal parts of the differential operators acting on g and Ω, would tend to zero at the boundary. Either way, equation (10.2) degenerates near the boundary of the unphysical space-time, and as mentioned in the Introduction, this makes it an unwieldy choice for studying the geometry of the space-time near null infinity.

Helmut Friedrich's procedure for obtaining a system of equations equivalent to the rescaled Einstein equations (10.2) and formally regular at the boundary of the unphysical space-time can be found in several papers, see for example [16]. His derivation proceeds in the following way. Let $C_{\mu\nu\lambda\rho}$ be the Weyl tensor of the metric g and define the quantities

$$\begin{aligned} L_{\mu\nu} &= \frac{1}{2}R_{\mu\nu} - \frac{1}{12}Rg_{\mu\nu}, \\ S_{\mu\nu\lambda\rho} &= \Omega^{-1}C_{\mu\nu\lambda\rho}, \\ \psi &= \frac{1}{4}\Box\Omega + \frac{1}{24}R\Omega\,. \end{aligned} \tag{10.3}$$

The tensor $S_{\mu\nu\lambda\rho}$ is smooth on ∂M because under the assumptions of asymptotic simplicity, Penrose has shown that $C_{\mu\nu\lambda\rho}$ vanishes at the boundary of M [23] (a further condition on the topology of ∂M – that ∂M admits spherical sections – is also needed, and will be assumed to hold). Then, by rephrasing (10.2) in terms of the quantities (10.3), adjoining the Bianchi identity for the curvature tensors in the physical and unphsyical space-times, and adjoining the well-known decomposition of the curvature tensor

$$R_{\mu\nu\lambda\rho} = C_{\mu\nu\lambda\rho} + g_{\mu\lambda}L_{\nu\rho} - g_{\mu\rho}L_{\nu\lambda} + g_{\nu\rho}L_{\mu\lambda} - g_{\nu\lambda}L_{\mu\rho}\,,$$

one obtains the system of equations

$$
\begin{aligned}
\nabla_\mu \nabla_\nu \Omega &= -\Omega L_{\mu\nu} + \psi g_{\mu\nu}, \\
\nabla_\mu \psi &= -L_{\mu\nu} \nabla^\nu \Omega, \\
\nabla_\lambda L_{\mu\nu} - \nabla_\mu L_{\lambda\nu} &= \nabla^\rho \Omega \, S_{\mu\lambda\nu\rho}, \\
\nabla^\rho S_{\mu\lambda\nu\rho} &= 0, \\
2\Omega\psi - \nabla_\mu \Omega \nabla^\mu \Omega &= 0, \\
R_{\mu\nu\lambda\rho} &= \Omega S_{\mu\nu\lambda\rho} + g_{\mu[\lambda} L_{\nu]\rho} - L_{\mu[\lambda} g_{\nu]\rho}\,,
\end{aligned}
\tag{10.4}
$$

by means of lengthy, though straightforward algebraic manipulations. The equations above are known as the *conformal Einstein equations.*

The equivalence of (10.4) to (10.2) is confirmed as follows. Suppose the quantities L, S and ψ as well as g and Ω satisfy (10.4). Then by algebra, it can be shown that the pair (g, Ω) satisfies (10.2) and that L, S and ψ relate to Ω and the curvature quantities in the manner indicated in (10.3). (The algebra is fairly straightforward: for instance, the last equation in (10.4) identifies L and S as components of the curvature tensor; then it is a matter of computation to recover equation (10.2) from the remaining five.)

It is immediately clear that the equations in (10.4) are regular when $\Omega = 0$. Furthermore, not only do the conformal Einstein equations contain the rescaled vacuum Einstein equations, but they also contain the Bianchi identity for the curvature tensor, though expressed in the new unknowns. Thus one can consider (10.4) to contain *integrability conditions* since the Bianchi identity is in some sense a integrability condition for the curvature tensor – meaning that the Bianchi identity is a result of requiring second covariant derivatives to commute properly (this can best be seen explicitly by rewriting the curvature tensor in terms of the vector-valued connection 1-forms as in [6], whereby the Bianchi identity becomes an incarnation of the identity $\mathrm{d}^2 = 0$ satisfied by the exterior differential operator).

Suppose now that $\mathcal{Z}$ is a space-like hypersurface in M. The fact that the conformal Einstein equations constrain certain initial data on $\mathcal{Z}$ can be seen by performing a $3+1$ splitting of the space-time near $\mathcal{Z}$. Choose a frame E_a, $a = 1, 2, 3$, for the tangent space of $\mathcal{Z}$ and complete this to a frame for the unphysical space-time by adjoining the forward-pointing unit normal vector field n of $\mathcal{Z}$. Use this frame to decompose the equations (10.4) into components parallel and perpendicular to $\mathcal{Z}$. The constraint equations induced by the conformal Einstein equations are those equations arising in this process in which no second normal derivatives of g or Ω, and no first normal derivatives of L, S or ψ appear. These equations are:

$$\begin{aligned}
\nabla_a \nabla_b \Omega &= \Sigma \chi_{ab} - \Omega L_{ab} + \psi g_{ab}, \\
\nabla_a \Sigma &= \chi_a^c \nabla_c \Omega - \Omega L_a, \\
\nabla_a \psi &= -\nabla^b \Omega L_{ba} - \Sigma L_a, \\
\nabla_a L_{bc} - \nabla_b L_{ac} &= \nabla^e \Omega S_{ecab} - \Sigma S_{cab} - (\chi_{ac} L_b - \chi_{bc} L_a), \\
\nabla_a L_b - \nabla_b L_a &= \nabla^e \Omega\, S_{eab} + \chi_a^c L_{bc} - \chi_b^c L_{ac}, \\
\nabla^a \bar{S}_{abc} &= \chi_b^a S_{ac} - \chi_c^a S_{ab}, \\
\nabla^a S_{ab} &= -\chi^{ac} \bar{S}_{abc}, \\
0 &= 2\Omega\psi + \Sigma^2 - \|\nabla\Omega\|^2, \\
\nabla_c \chi_{ba} - \nabla_b \chi_{ca} &= \Omega \bar{S}_{abc} + g_{ab} L_c - g_{ac} L_b, \\
R_{ab} &= \Omega S_{ab} + L_{ab} + \frac{1}{4} L_c^c g_{ab} - \chi_c^c \chi_{ab} + \chi_{ca} \chi_b^c,
\end{aligned} \tag{10.5}$$

where ∇ now denotes the covariant derivative operator on $\mathcal{Z}$ corresponding to its induced metric g and R_{ab} is its Ricci curvature. The unknown quantities appearing in these equations are the initial data. They are:

- the induced metric of $\mathcal{Z}$, which is still called g (no confusion will arise because the 4-dimensional setting will not be considered further in the remainder of this article),
- the second fundamental form χ of $\mathcal{Z}$,
- the function Ω restricted to $\mathcal{Z}$,
- the normal derivative $n(\Omega)\big|_{\mathcal{Z}}$, denoted Σ,
- the tensors $L_{ab} = E_a^\mu E_b^\nu L_{\mu\nu}$ and $L_a = n^\mu E_a^\nu L_{\mu\nu}$,
- the tensors $\bar{S}_{abc} = n^\mu E_a^\nu E_b^\lambda E_c^\rho S_{\nu\mu\lambda\rho}$ and $S_{ab} = n^\mu n^\nu E_a^\lambda E_b^\rho S_{\lambda\mu\rho\nu}$,
- and the function ψ restricted to $\mathcal{Z}$.

The equations (10.5) are known as the *conformal constraint equations.* The derivation of these equations will not be reproduced here – the reader is asked to consult [16] for this material. However, it is fairly easy to recognize the origin of the various terms appearing there. For example, the first two equations arise as the tangential and tangential-normal components of the first equation of (10.4). Furthermore, and more importantly for the sequel, the last two equations arise as the Gauss and Codazzi equations applied to the decomposition of the curvature tensor given by the last equation of (10.4).

NOTE: The various tensor quantities that appear in (10.5) possess certain symmetries as a result of their origin as components of the curvature tensor: L_{ab} is symmetric; S_{ab} is symmetric and trace-free; and $\bar{S}_{abc}$ is antisymmetric on its last two indices, satisfies the Jacobi symmetry $\bar{S}_{abc} + \bar{S}_{cab} + \bar{S}_{bca} = 0$ and is trace free on all its indices. (Tensors with these symmetries will appear often in the sequel. Tensors of rank three that are antisymmetric on their last two indices and satisfy the Jacobi symmetry will be called Jacobitensors for short while those which are in addition trace-free will be called traceless Jacobitensors.) Note that even though the tensor $S_{abcd} = E_a^\mu E_b^\nu E_c^\lambda E_d^\rho S_{\mu\nu\lambda\rho}$ appears in the constraint equations, it is not a truly independent initial datum because, thanks to the symmetries of $S_{\mu\nu\lambda\rho}$, it can be written as $S_{abcd} = g_{a[c}S_{d]b} - S_{a[c}g_{d]b}$.

The system (10.5) is clearly exceedingly complicated because it is quasi-linear and highly coupled. However, the advantage provided by (10.5) is once again that it is formally regular at the boundary of $\mathcal{Z}$. For the sake of comparison, recall that the interior of $\mathcal{Z}$ can be viewed as a space-like hypersurface of the physical space-time, and as such, satisfies the usual Einstein constraint equations there. In other words, if its induced metric is denoted by $\tilde{g}$ and its second fundamental form by $\tilde{\chi}$, then

$$\begin{aligned} \tilde{\nabla}^a \tilde{\chi}_{ab} - \tilde{\nabla}_b \tilde{\chi}_a^{\ a} &= 0, \\ \tilde{R} + (\tilde{\chi}_a^{\ a})^2 - \tilde{\chi}^{ab}\tilde{\chi}_{ab} &= 0\,, \end{aligned} \tag{10.6}$$

where $\tilde{\nabla}$ is the covariant derivative operator of the metric $\tilde{g}$ and $\tilde{R}$ is its scalar curvature. These equations can be rephrased in terms of g, χ and Ω in the unphysical space-time by conformal transformation. The necessary transformation rules are that $\tilde{g} = \Omega^{-2} g$ and also that $\tilde{\chi} = \Omega^{-1}\chi + \Sigma\Omega^{-2} g$ (which can be found by conformally transforming the definition of the second fundamental form as the normal component of the covariant derivative restricted to $\tilde{\mathcal{Z}}$). The resulting equations are:

$$\begin{aligned} \Omega^2\big(R + (\chi_a^{\ a})^2 - \chi^{ab}\chi_{ab}\big) + 4\Omega\Delta_g\Omega - 6\|\nabla\Omega\|^2 + 4\Omega\Sigma\chi_a^{\ a} + 6\Sigma^2 &= 0, \\ \Omega\big(\nabla_a\chi_b^{\ a} - \nabla_b\chi_a^{\ a}\big) - 2\nabla_b\Sigma - 2\chi_b^{\ a}\nabla_a\Omega &= 0\,, \end{aligned} \tag{10.7}$$

where $\Sigma = n(\Omega)\big|_{\mathcal{Z}}$ and Δ_g is the Laplacian of the metric g. Once again, the principal parts of these equations contain factors of Ω and thus degenerate as $\Omega \to 0$ near the boundary of $\mathcal{Z}$. This behaviour does not arise in the conformal constraint equations.

The conformal constraint equations listed in (10.5) are equivalent to the usual constraint equations (10.7) because if $(g, \chi, \Omega, \Sigma)$ solves (10.7) and the additional quantities S, $\bar{S}$, L and ψ are defined as indicated in (10.5) (e. g. the last equation defines ψ; then the first equation defines the 2-tensor L_{ab}, etc.), then straightforward computation shows that the conformal constraint equations are satisfied; furthermore, if $(g, \chi, \Omega, \Sigma, S, \bar{S}, L, \psi)$ satisfies (10.5), then it can be shown that $(g, \chi, \Omega, \Sigma)$ satisfies (10.7), and consequently, $\tilde{g}$ and $\tilde{\chi}$, given by the transformation rules above, satisfy the usual constraint equations (10.6). These considerations thus suggest one method for constructing solutions of the conformal constraint equations: construct any solution $(\tilde{g}, \tilde{\chi})$ of the usual constraint equations using standard techniques, choose a conformal factor, perform the transformations to the unphysical space-time and use the conformal constraint equations to define the subsidiary quantities in terms of $(\tilde{g}, \tilde{\chi})$. Then these new quantities satisfy the conformal constraint equations.

Consequently, it is possible to *assume* the existence of initial data satisfying (10.5) with well-defined asymptotic properties (essentially given by the transformation rules above) and study only the time evolution of the data according to the conformal Einstein equations (10.4). This is the idea behind the work of

Friedrich in [16] (extended in [19]), where the time evolution of suitably small initial data on an asymptotically hyperboloidal hypersurface was studied and a complete future development was found. The nature of the asymptotic structure of this class of solutions near null infinity, and in particular the relationship between the asymptotic structure of the solution and the asymptotic structure of the initial data, was then analyzed extensively by Andersson, Chruściel and Friedrich in [4] (extended by Andersson and Chruściel in [2, 3]), and was based on the rescaled Einstein equations (10.4) and their constraints (10.7). However, the problem of the vanishing of the conformal factor near the boundary of the unphysical space-time and the resultant degeneration of these equations remains a part of the ACF methods. Thus they are not ideally suited for certain applications, in particular for implementing numerical studies of asymptotically hyperboloidal data near null infinity where the presence of negative powers of Ω can cause computational codes to crash (see [15] for details). It is for this reason that new methods for solving (10.5) directly, rather than through the usual constraint equations, must be developed. This question will begin to be tackled in the remainder of this article.

10.2.2 Reduction to the Extended Constraint Equations

The complexity of the conformal constraint equations makes it a daunting task to attempt to develop any methods for obtaining solutions of the equations in their full generality. However, a great deal of structure is contained within these equations, and the hope is that this structure can be exploited in the search for solutions. For instance, it is possible to disentangle in some sense the equations relating to the conformal factor and its associated boundary value problem from the equations related to the Gauss-Codazzi equations of $\mathcal{Z}$ by restricting to a special case of the equations.

The special case that will be considered in the rest of this article is to assume that the conformal diffeomorphism between $\tilde{M}$ and M is the identity, and consequently that the conformal factor is trivial (i. e. $\Omega = 1$) in the unphysical space-time. This is somewhat of a strange simplification, because it requires that the space-time M have empty boundary (since $\Omega^{-1}(0) = \partial M$)! One would thus not find oneself in this special case in practice since the whole point of the conformal constraint equations is to study hyperboloidal initial data in a conformally rescaled space-time that has a boundary at null infinity. Nevertheless, the simplification afforded by the assumption $\Omega = 1$ is worthwhile to consider from a mathematical point of view because it accomplishes the disentanglement described above and allow the Gauss-Codazzi-type equations within the conformal constraint equations to be studied in isolation.

To see this explicitly, one must substitute $\Omega = 1$ and $\Sigma = 0$ (which is consistent with the assumption that $\Omega = 1$ in space-time since $\Sigma = n(\Omega)\big|_{\mathcal{Z}} = 0$ where n is the forward-pointing unit normal of $\mathcal{Z}$) into the equations (10.5). One first sees that L_{ab}, L_a and ψ are forced to vanish under this assumption, and then that the conformal constraint equations reduce to the following system of four coupled equations:

$$\begin{aligned} R_{ab} &= S_{ab} - \chi_c^c \chi_{ab} + \chi_a^c \chi_{cb}, \\ \nabla_c \chi_{ab} - \nabla_b \chi_{ac} &= \bar{S}_{abc}, \\ \nabla^a \bar{S}_{abc} &= \chi_b^a S_{ac} - \chi_c^a S_{ab}, \\ \nabla^a S_{ab} &= -\chi^{ac} \bar{S}_{abc} \,. \end{aligned}$$

Here, covariant derivatives are taken with respect to the induced metric g_{ab} of $\mathcal{Z}$ and χ_{ab} is the second fundamental form of $\mathcal{Z}$. As before, the tensor S_{ab} is symmetric and trace-free with respect to g_{ab} whereas the tensor $\bar{S}_{abc}$ is a traceless Jacobi tensor. These four quantities are the unknowns for which these equations must be solved. For reasons that will become apparent later on, it will be helpful to work instead with the equivalent system obtained by replacing S_{ab} and S_{ac} in the third equation by R_{ab} and R_{ac} from the first equation. The system one obtains is actually just:

$$\begin{aligned} R_{ab} &= S_{ab} - \chi_c^c \chi_{ab} + \chi_a^c \chi_{cb}, \\ \nabla_c \chi_{ab} - \nabla_b \chi_{ac} &= \bar{S}_{abc}, \\ \nabla^a \bar{S}_{abc} &= \chi_b^a R_{ac} - \chi_c^a R_{ab}, \\ \nabla^a S_{ab} &= -\chi^{ac} \bar{S}_{abc} \,, \end{aligned} \tag{10.8}$$

because the terms cubic in χ vanish.

Notice that because of the symmetries of S and $\bar{S}$, if the traces of the first two equations of (10.8) are taken, then the usual constraint equations (10.6) result. Furthermore, if g_{ab} and χ_{ab} satisfy the usual constraint equations and one *defines* $\bar{S}_{abc}$ and S_{ab} by the first two equations of (10.8) respectively, then the remaining two equations follow by straightforward algebra and the Bianchi identity. Thus equations (10.8) are equivalent to the usual vacuum Einstein constraint equations and for this reason are called the *extended constraint equations.*

10.2.3 Properties of the Extended Constraint Equations

The extended constraint equations (10.8) are clearly formally much simpler than the full system of conformal constraint equations. However, several essential features of the full equations remain. These features refer to the ellipticity properties of the various differential operators appearing in (10.8) as well as to the integrability conditions built into these equations.

ELLIPTICITY PROPERTIES

One must consider the principal symbols of the operators that appear on the left hand sides of the extended constraint equations in order to understand their ellipticity properties. Begin with a definition of the symbol. Recall that if $P : C^\infty(\mathbb{R}^n, \mathbb{R}^N) \to C^\infty(\mathbb{R}^n, \mathbb{R}^M)$ is a linear differential operator of order m with constant coefficients, then it can be expressed as

$$P(u) = \sum_{\alpha_1 + \cdots + \alpha_n = m} \left(\sum_{i=1}^N b_i^{\alpha_1 \cdots \alpha_n} \frac{\partial^m u^i}{(\partial x^1)^{\alpha_1} \cdots (\partial x^n)^{\alpha_n}} \right) + P_0(u) \,,$$

where P_0 is a differential operator of order less than or equal to $m-1$ and the $b_i^{\alpha_1\cdots\alpha_n}$ are elements of $\mathbb{R}^M$. The *principal symbol* of P is the family of linear maps $\sigma_\xi : \mathbb{R}^N \to \mathbb{R}^M$ given by

$$\sigma_\xi(v) = \sum_{\alpha_1+\cdots+\alpha_n=m} \left(\sum_{i=1}^N b_i^{\alpha_1\cdots\alpha_n} \xi_1^{\alpha_1} \cdots \xi_n^{\alpha_n} v^i \right)$$

for any non-zero $(\xi_1,\ldots,\xi_n) \in \mathbb{R}^n$ and $v \in \mathbb{R}^N$. The operator P is called *underdetermined elliptic* if the symbol is surjective for each non-zero ξ, *overdetermined elliptic* if the symbol is injective for each non-zero ξ and simply *elliptic* if the symbol is bijective for each non-zero ξ. An operator with non-constant coefficients has a symbol at each point of the domain, while for a nonlinear operator, it is the linearization which has a symbol at each given $u \in C^\infty(\mathbb{R}^n, \mathbb{R}^N)$. Such operators are overdetermined, underdetermined or elliptic if their symbols possess these properties uniformly.

To understand the ellipticity properties of the conformal constraint equations, begin with the equation for the metric g_{ab}. It is quasi-linear in g, with highest-order terms given by

$$g_{ab} \mapsto g^{cd} \left(\frac{\partial^2 g_{ad}}{\partial x^b \partial x^c} + \frac{\partial^2 g_{bd}}{\partial x^a \partial x^c} - \frac{1}{2} \frac{\partial^2 g_{ab}}{\partial x^c \partial x^d} - \frac{1}{2} \frac{\partial^2 g_{cd}}{\partial x^a \partial x^b} \right) .$$

The linearization of this expression at a given metric is neither over- nor underdetermined elliptic, nor is it elliptic. However, it is well known that the Ricci curvature is degenerate as an operator on metrics because it is invariant under changes of coordinates of the metric, and that the Ricci curvature operator can be made formally elliptic by making an appropriate choice of coordinate gauge. The standard choice is to require that the metric be expressed in *harmonic coordinates*, which are defined by the requirement that the coordinate functions x^a are harmonic functions, i. e. that $\Delta_h x^a = 0$ for each a. (Since the metric itself depends on the coordinate functions, the requirement that the coordinates be harmonic is in fact a nonlinear condition. Nevertheless, the existence of such coordinates, defined outside sufficiently large balls in $\mathbb{R}^3$ for any asymptotically flat metric, has been guaranteed by Bartnik in [5].)

To show that the Ricci operator is elliptic in harmonic coordinates, first note that a straightforward calculation implies that the harmonic coordinate condition $\Delta_g x^a = 0$ for all a is equivalent to the condition $g^{bc}\Gamma^a_{bc} = 0$ for all a on the Christoffel symbols of g. Now set $\Gamma^a = g^{bc}\Gamma^a_{bc}$ (and also $\Gamma_a = g_{as}\Gamma^s$), and then recall that the components of the Ricci tensor satisfy

$$R_{ab} = R^H_{ab} + \frac{1}{2}(\Gamma_{a;b} + \Gamma_{b;a}), \tag{10.9}$$

where R^H_{ab} are the components of the *reduced* Ricci operator defined by

$$R^H_{ab} = -\frac{1}{2} g^{rs} g_{ab,rs} + q(\Gamma) . \tag{10.10}$$

In the expressions above, a comma denotes ordinary differentiation with respect to the coordinates, a semicolon denotes covariant differentiation (since Γ^a is not a tensor, this is to be taken formally; i. e. $\Gamma_{a;b} = \Gamma_{a,b} - \Gamma_s \Gamma^s_{ab}$), and $q(\Gamma)$ denotes a term that is quadratic in the components Γ^a. The reduced Ricci operator is clearly elliptic in g. Since $\Gamma^a = 0$ for all a in harmonic coordinates, $R_{ab}(g) = R^H_{ab}(g)$ in these coordinates, and thus the Ricci operator is elliptic in g when g satisfies the harmonic coordinate condition.

The second equation in the extended constraint equations is linear in χ_{ab} and its left hand side defines a differential operator $\chi_{ab} \mapsto \nabla_c \chi_{ab} - \nabla_b \chi_{ac}$ from the space of symmetric tensors to the space of Jacobi tensors. (It can be easily verified that the left hand side of the first equation in (10.8) satisfies the relevant symmetries. However, it can also be verified that the left hand side is not *a priori* traceless on all its indices – this is only a requirement on the eventual solution since the left hand side is equated with a traceless Jacobi tensor.) The principal symbol of this operator is

$$\sigma_\xi : \chi_{ab} \mapsto \xi_c \chi_{ab} - \xi_b \chi_{ac} \,.$$

By the following simple argument, one can show that σ_ξ has a one-dimensional kernel and is not surjective.

Suppose first that $\sigma_\xi(\chi_{ab}) = 0$ for some non-zero ξ. Since $\xi_a \xi^a \neq 0$, one can write uniquely $\chi_{ab} = \chi^0_{ab} + c\xi_a \xi_b$ for some c, where χ^0_{ab} is trace-free. Substituting this expression for χ_{ab} yields

$$\xi_b \chi^0_{ac} - \xi_c \chi^0_{ab} = 0 \,. \tag{10.11}$$

Taking the trace over a and b implies that $\xi^c \chi^0_{ac} = 0$. Then, contracting with ξ^c gives $\xi^c \xi_c \chi^0_{ab} = 0$, or $\chi^0_{ab} = 0$. Consequently, the kernel of the symbol σ_ξ is one-dimensional, and consists of tensors of the form $c\xi_a \xi_b$. Next, since the space of symmetric 2-tensors is six-dimensional, the image of the symbol is five-dimensional. Now, the target space of Jacobitensors is eight-dimensional because any Jacobitensor can be decomposed as $T_{abc} = \varepsilon^e{}_{bc} F_{ae} + A_b g_{ac} - A_c g_{ab}$ where F_{ae} is a trace-free and symmetric tensor (accounting for five dimensions), A_b is a 1-form (accounting for the remaining three), and ε_{abc} is the fully antisymmetric permutation symbol. The symbol can thus not be surjective. Note, however, that when it is restricted to trace-free tensors, the principal symbol *is* at least injective. Consequently, the first equation of (10.8) is overdetermined elliptic when restricted to the space of trace-free symmetric 2-tensors.

The third and fourth equations in (10.8) are linear in $\bar{S}_{abc}$ and S_{ab} respectively. It can be shown that the operators $\bar{S}_{abc} \mapsto \nabla^a \bar{S}_{abc}$ and $S_{ab} \mapsto \nabla^a S_{ab}$ are underdetermined elliptic by demonstrating that their principal symbols $\bar{S}_{abc} \mapsto \xi^a \bar{S}_{abc}$ and $S_{ab} \mapsto \xi^a S_{ab}$ are surjective maps from the space of symmetric, trace-free tensors onto the space of 1-forms and from the space of traceless Jacobi tensors onto the space of antisymmetric 2-tensors, respectively. These are fairly straightforward calculations and left to the reader.

INTEGRABILITY CONDITIONS

As mentioned in Sect. 10.2.1, the conformal Einstein equations (10.4) satisfied by the unphysical space-time contain the Bianchi identity which was interpreted as being a integrability condition. Integrability conditions are also to be found in the conformal constraint equations – at least in the special case $\Omega \equiv 1$. (It is also true that many integrability conditions are contained within the full equations, but these will not be explicitly demonstrated here). To exhibit these integrability conditions, begin by considering the first and fourth equations in (10.8). The Bianchi identity for the Ricci curvature is

$$\nabla^a R_{ab} - \frac{1}{2}\nabla_b R = 0\,,$$

whereby the first equation of (10.8) implies:

$$\begin{aligned}
0 &= \nabla^a\big(S_{ab} - \chi_c^c\chi_{ab} + \chi_a^c\chi_{cb}\big) - \frac{1}{2}\nabla_b\big(-(\chi_c^c)^2 + \chi^{ac}\chi_{ac}\big)\\
&= \nabla^a S_{ab} - \big(\chi_c^c\delta_b^a - \chi_b^a\big)\big(\nabla^u\chi_{au} - \nabla_a\chi_u^u\big) - \chi^{ca}\big(\nabla_b\chi_{ac} - \nabla_c\chi_{ab}\big)\\
&= \nabla^a S_{ab} - \big(\chi_c^c\delta_b^a - \chi_b^a\big)h^{uv}\bar{S}_{uav} + \chi^{ac}\bar{S}_{abc}
\end{aligned} \tag{10.12}$$

using the second equation in (10.8) and its trace. By the symmetries of $\bar{S}_{abc}$, the middle term in (10.12) vanishes, leaving

$$0 = \nabla^a S_{ab} + \chi^{ac}\bar{S}_{abc}\,,$$

which is exactly the fourth equation of (10.8). Thus the fourth equation of (10.8) encodes the Bianchi identity, the integrability condition for the curvature.

The second and third equations of (10.8) also consist of a constraint equation and its integrability condition as well, but in a different sense. Recall that what a integrability condition should reflect is that second covariant derivatives commute properly. Consider then the result of commuting the second covariant derivatives of the second equation of (10.8). Begin with

$$\bar{S}_{abc} = \nabla_c\chi_{ab} - \nabla_b\chi_{ac}$$

and compute:

$$\begin{aligned}
\varepsilon^{ebc}\nabla_e\bar{S}_{abc} &= 2\varepsilon^{ebc}\nabla_e\nabla_c\chi_{ab}\\
&= \varepsilon^{ebc}\big(\nabla_e\nabla_c\chi_{ab} - \nabla_c\nabla_e\chi_{ab}\big)\\
&= \varepsilon^{ebc}\big(R_{eca}{}^s\chi_{sb} + R_{ecb}{}^s\chi_{as}\big)\\
&= \varepsilon^{ebc}R_{eca}{}^s\chi_{sb},
\end{aligned} \tag{10.13}$$

since the symmetries of R_{abcd} imply that $\varepsilon^{abc}R_{abcd} = 0$. Now substitute in (10.13) the well-known decomposition of the curvature tensor in three dimensions, namely that

$$R_{eca}{}^s = g_{ea}R_c^s - \delta_e^s R_{ca} + \delta_c^s R_{ea} - g_{ca}R_e^s - \frac{1}{2}R\big(g_{ea}\delta_c^s - \delta_e^s g_{ca}\big)\,,$$

to obtain

$$\varepsilon^{ebc}\nabla_e \bar{S}_{abc} = 2\varepsilon^{bc}{}_a \mathcal{X}^s_b R_{cs} \,. \tag{10.14}$$

CLAIM: equation (10.14) is exactly the third equation of (10.8). To see this, recall that a traceless Jacobi tensor can be decomposed as $\bar{S}_{abc} = \varepsilon^e{}_{bc} F_{ae}$ where F_{ae} is trace-free and symmetric. Consequently,

$$\begin{aligned}
\varepsilon^{ebc}\nabla_e \bar{S}_{abc} &= \varepsilon^{ebc}\nabla_e \varepsilon^u{}_{bc} F_{au} \\
&= 2\nabla^e F_{ae} \\
&= 2\nabla^e F_{ea} && \text{(by symmetry)} \\
&= \varepsilon_a{}^{bc}\nabla^e \varepsilon^u{}_{bc} F_{eu} \\
&= \varepsilon_a{}^{bc}\nabla^e \bar{S}_{ebc} \,. && (10.15)
\end{aligned}$$

Thus (10.15) together with (10.14) implies that

$$\varepsilon_a{}^{bc}\nabla^e \bar{S}_{ebc} = 2\varepsilon^{bc}{}_a \mathcal{X}^s_b R_{cs} \,,$$

which is the third equation of (10.8) (or at least its dual, but this is equivalent).

10.3 Asymptotically Flat Solutions of the Extended Constraint Equations in the Time Symmetric Case

10.3.1 Statement of the Main Theorem

Because the conformal boundary of the space-time $\tilde{M}$ is absent under the triviality assumptions that have been made on the conformal diffeomorphism, a natural setting in which to investigate the extended constraint equations (10.8) is the case in which $\tilde{M}$ is asymptotically Minkowski space and that $\mathcal{Z}$ is asymptotically flat. In fact, one solution of the extended constraint equations satisfying these conditions is when $\mathcal{Z} = \mathbb{R}^3$ and the initial data is the Euclidean metric $g = \delta$ with vanishing tensors $\mathcal{X}$, S and $\bar{S}$. Neighbouring asymptotically flat solutions are those whose metric g is a small perturbation of δ that decays suitably to δ near infinity, and $\mathcal{X}$, $\bar{S}$ and S are also small and decay suitably. These solutions are in addition called *time symmetric* if their second fundamental form $\mathcal{X}$ actually vanishes identically.

The theorem that will be proved in the remainder of this article is a characterization of the space of asymptotically flat *and* time-symmetric solutions of the extended constraint equations in the neighbourhood of the trivial solution given above. The case of non-time-symmetric solutions is as yet beyond the scope of this article, though a future paper by the Author will clear this up [7].

Under the assumption of time-symmetry, the requirement that $\mathcal{X} = 0$ implies that $\bar{S} = 0$ as well, and so the extended constraint equations further reduce to the following system of equations:

$$\begin{aligned}
\nabla^a S_{ab} &= 0 \\
R_{ab}(g) &= S_{ab}
\end{aligned}$$

for the unknown metric g and unknown trace-free and symmetric tensor S. Since these equations will be solved for metrics near the Euclidean metric, it will be preferable to write metrics as small perturbations of the Euclidean metric of the form $\delta + h$ where h is a symmetric tensor suitably near 0. Thus the above system should be replaced with the system:

$$\begin{aligned} \nabla^a S_{ab} &= 0, \\ R_{ab}(\delta + h) &= S_{ab} . \end{aligned} \tag{10.16}$$

The covariant derivative here corresponds to the metric $\delta + h$. The theorem that will be proved is the following.

Main Theorem: *There exists a Banach space B of free data along with a neighbourhood U of zero in B, Banach spaces Y and Y' of symmetric 2-tensors, and smooth functions $\psi : U \to Y$ and $\psi' : U \to Y'$ with $\psi(0) = \psi'(0) = 0$ so that for every $b \in U$, the following hold:*

1. *$\psi(b)$ and $\psi'(b)$ tend asymptotically towards zero;*
2. *$g \equiv \delta + \psi(b)$ defines an asymptotically flat Riemannian metric on $\mathbb{R}^3$;*
3. *$S \equiv \psi'(b)$ defines a symmetric tensor that is trace-free with respect to g;*
4. *g and S satisfy the equations* (10.16).

The proof of this theorem will be presented in the remaining sections of this article, and consists of essentially two steps. Since (10.16) is not an elliptic system (as outlined in the previous section), the first step of the proof consists of exploiting the elliptic properties of these equations to define a closely related system of equations, called the *associated system*, which is elliptic. In it, the tensor S is decomposed into a sum of two components of the form $T + \mathcal{L}^g(X)$, where T is a symmetric and trace-free tensor, X is a 1-form and $\mathcal{L}^g$ is the *conformal Killing operator* which is also the adjoint of the divergence operator $S_{ab} \mapsto \nabla^a S_{ab}$. The system (10.16), written in terms of this decomposition, yields equations for g, X, and T whose linearization in the g and X directions will be seen to be bijective (or near enough to being bijective – the details will be seen in due course). Thus the Implicit Function Theorem can be invoked to find solutions where the quantities g and X are expressed as functions of T. The second step is then to show that all solutions of the associated system are also solutions of the original system (10.16). This will turn out to be true when the metric $g = \delta + h$ is sufficiently close to δ, and relies on a Poincaré inequality and the integrability condition. The Author wishes to thank H. Friedrich for suggesting this approach for solving (10.16)

The method outlined above for solving the extended constraints in the time symmetric case is in fact a method for solving the usual vacuum constraint equations in the time-symmetric case (namely the equation $R(g) = 0$, which follows from (10.16) by taking a trace) because of the equivalence of the extended constraints and the usual constraints described earlier. The differences between this method and the 'classical' Lichnerowicz-York method for solving the constraint equations are now readily apparent. In the classical method, one freely prescribes a metric g_0 on $\mathbb{R}^3$ and considers the conformally rescaled metric $g = u^4 g_0$, where $u : \mathbb{R}^3 \to \mathbb{R}$ is an unknown function. One then reads the equation $R(u^4 g_0) = 0$ as a semi-linear elliptic equation for u. In contrast, the present method treats the metric g and the one-form X as the unknowns and leads to a quasi-linear elliptic system for these quantities in terms of the freely prescribable quantity T, which is a component of the curvature of the solution.

REMARK: The Main Theorem does *not* fall into the domain of *prescribed Ricci curvature* as, for example, do the results of De Turck and his collaborators [9, 12–14]. In these papers, the authors suppose a fixed symmetric tensor S is given on a set $\mathcal{O}$ and attempt to find conditions under which a metric g exists on $\mathcal{O}$ so that $Ric(g) = S$. In the Main Theorem, by contrast, the tensor S is itself an unknown quantity and only a component is prescribed ahead of time by the free data. Furthermore, De Turck's results are local in nature since $\mathcal{O}$ is usually an open set in $\mathbb{R}^n$, while the Main Theorem of this article gives a global (though perturbative) result.

10.3.2 Formulating an Elliptic Problem

The first task in the proof of the Main Theorem is to construct the associated elliptic system that is to be solved by the Implicit Function Theorem. To this end, recall that the Ricci curvature operator in (10.16) is not elliptic, but that imposing the coordinate gauge choice defined by the harmonic coordinate condition makes the Ricci operator elliptic. As indicated in Sect. 10.2.3, assuming *a priori* that the harmonic coordinate condition is satisfied by the metric $\delta + h$ is equivalent to replacing the Ricci operator by the reduced Ricci operator $Ric^H(\delta + h)$. (Of course, this assumption must be justified later on; i. e. it must be shown that $\delta + h$ does indeed satisfy the harmonic coordinate condition, and this is precisely what the second step of the proof of the Main Theorem will accomplish). The remaining operator in (10.16) is underdetermined elliptic, and an elliptic operator can be constructed from this by using a standard technique known as the *York decomposition* (see [26] but also [8, 11] for a thorough analysis of this method). Write a symmetric, trace-free tensor S in terms of a 1-form X and a freely prescribed symmetric tensor T as

$$S(h, X, T) = T^* + \mathcal{L}^{\delta+h}(X)\,, \tag{10.17}$$

where $T^* = T - \frac{1}{3}\mathrm{Tr}_{\delta+h}(T)(\delta+h)$ is the trace-free part of T and $\mathcal{L}^{\delta+h}(X)$ is the conformal Killing operator with respect to the metric $\delta + h$ acting on X. This is defined for a general metric g by

$$\mathcal{L}^g_{ab}(X) = \nabla_a X_b + \nabla_b X_a - \frac{2}{3}\nabla^c X_c g_{ab}\,,$$

where ∇ is the covariant derivative of the metric g. The reason for making this choice is that the composition of the divergence operator in (10.16) and the conformal Killing operator, that is the composite operator $\mathrm{div}_g \circ \mathcal{L}^g$ given componentwise by

$$\begin{aligned}[\mathrm{div}_g \circ \mathcal{L}^g(X)]_b &= \nabla^a \left(\nabla_a X_b + \nabla_b X_a - \frac{2}{3}\nabla^c X_c g_{ab}\right) \\ &= \nabla^a \nabla_a X_b + \frac{1}{3}\nabla_b \nabla^a X_a + R^a_b(g) X_a \,,\end{aligned}$$

is elliptic, as can easily be seen by computing its symbol. However, the observation that the conformal Killing operator is the formal adjoint of the divergence operator $S_{ab} \mapsto \nabla^a S_{ab}$ taking symmetric, trace-free tensors to 1-forms obviates these calculations because it is well-known that the compositon PP^* of an underdetermined elliptic operator P with its adjoint is elliptic.

These considerations lead to the following definition of the associated system, given here in index-free notation for ease of presentation:

$$\begin{aligned}Ric^H(\delta + h) &= S(h, X, T), \\ \mathrm{div}_{\delta+h} \circ S(h, X, T) &= 0,\end{aligned} \tag{10.18}$$

where $S(h, X, T)$ is as in (10.17) and is called the *York operator*. As will be shown in due course, the map defined by

$$\Phi(h, X, T) \equiv \begin{pmatrix} Ric^H(\delta + h) - S(h, X, T) \\ \mathrm{div}_{\delta+h} \circ S(h, X, T) \end{pmatrix} \tag{10.19}$$

on appropriate Banach spaces has a bounded, elliptic linearization in the h and X directions and as a result, the Implicit Function Theorem yields solutions $h(T)$ and $X(T)$ as smooth functions of sufficiently small tensors T.

10.3.3 Choosing the Banach Spaces

Before proceeding with the solution of the equations (10.18), it is necessary to specify in what Banach spaces of tensors the equations are to be solved. The notion of asymptotic flatness in $\mathbb{R}^3$ should be encoded rigorously into the function spaces by requiring that the relevant objects belong to a space of tensors with built-in control at infinity. Furthermore, the spaces should be chosen to exploit the Fredholm properties of the operators appearing in the map Φ. Both these ends will be served by weighted Sobolev spaces, and an appropriate choice of these spaces for use in the Main Theorem will be made below. Begin, however, with a short introduction to these spaces.

Let T be any tensor on $\mathbb{R}^3$. (This tensor may be of any order — the norm $\|\cdot\|$ appearing in the following definition is then simply the norm on such tensors that is induced from the metric of $\mathbb{R}^3$. Also, Sobolev spaces for tensors on $\mathbb{R}^n$ can be defined equally well, but since solving the conformal constraint equations is explicitly a three-dimensional problem, all definitions and theorems concerning these spaces will be stated for $\mathbb{R}^3$.) The $H^{k,\beta}$ Sobolev norm of T is the quantity

$$\|T\|_{H^{k,\beta}} = \left(\sum_{l=0}^{k} \int_{\mathbb{R}^3} \|\nabla^l T\|^2 \sigma^{-2(\beta-l)-3}\right)^{1/2} ,$$

where $\sigma(x) = (1+r^2)^{1/2}$ is the *weight function* and $r^2 = (x^1)^2 + (x^2)^2 + (x^3)^2$ is the squared distance to the origin. Note that Bartnik's convention for describing the weighted spaces is being used (the reason for this is psychological: if $f \in H^{k,\beta}$ and f is smooth enough to invoke the Sobolev Embedding Theorem (see below), then $f(x) = o(r^\beta)$ as $r \to \infty$, which is easy to remember — see [5] for details).

The space of $H^{k,\beta}$ functions of $\mathbb{R}^3$ will be denoted by $H^{k,\beta}(\mathbb{R}^3)$ and the space of $H^{k,\beta}$ sections of a tensor bundle B over $\mathbb{R}^3$ will be denoted by $H^{k,\beta}(B)$. As an abbreviation, or where the context makes the bundle clear, such a space may be indicated simply by $H^{k,\beta}$. Note also that the following convention for integration will be used in the rest of this paper. An integral of the form $\int_{\mathbb{R}^3} f$, as in the definition above, denotes an integral of f with respect to the standard Euclidean volume form. Integrals of quantities with respect to the volume form of a different metric will be indicated explicitly, as, for example, $\int_{\mathbb{R}^3} f \, \mathrm{dVol}_g$.

The spaces of $H^{k,\beta}$ tensors satisfy several important analytic properties and the reader is asked to consult Bartnik's paper, or others on the same topic [5, 8, 10, 11], for details. The three most important properties that will be used in the sequel are the Sobolev Embedding Theorem, the Poincaré Inequality and Rellich's Lemma; these will be restated here for easy reference:

1. The Sobolev Embedding Theorem states that if $k > \frac{3}{2}$ and T is a tensor in $H^{k,\beta}$, then T is C^0. Furthermore, if the weighted C^k_β norm of a tensor T is given by
$$\|T\|_{C^k_\beta} = \sum_{l=0}^{k} \|\nabla^l f \sigma^{-\beta+l}\|_0 ,$$
where $\|T\|_0 = \sup\{\|T(x)\| : x \in \mathbb{R}^3\}$ (using the Euclidean metric of $\mathbb{R}^3$ to define and measure the pointwise norm of T and its derivatives), then in fact, $T \in C^0_\beta$ and $\|T\|_{C^0_\beta} \leq C\|T\|_{H^{k,\beta}}$.
2. The Poincare Inequality states that if $\beta < 0$, then
$$\|f\|_{H^{0,\beta}} \leq C\|\nabla f\|_{H^{0,\beta-1}} ,$$
whenever f is a function in $H^{1,\beta}(\mathbb{R}^3)$.
3. The Rellich Lemma states that the inclusion $H^{k,\beta}(B) \subseteq H^{k',\beta'}(B)$, for any tensor bundle B, is compact when $k' < k$ and $\beta' > \beta$. In other words, if T_i is a uniformly bounded sequence of tensors in $H^{k,\beta}$, then there is a subsequence $T_{i'}$ converging to a tensor T in $H^{k',\beta'}$.

REMARK: The constant C appearing in the estimates above is meant as a general numerical constant, independent of the tensors or functions measured in the estimate. In the remainder of this article, any such constant will be denoted by a generic C, unless it is important to emphasize otherwise.

In addition to the three properties above, two important results that are valid in weighted Sobolev spaces will be needed in the sequel. The first concerns integration.

Duality Lemma: *If $u \in H^{l,\gamma}(\mathbb{R}^3)$ and $v \in H^{l-2,-\gamma-3}(\mathbb{R}^3)$, then the integral $\int_{\mathbb{R}^3} u \cdot v$ is well defined. Furthermore, the functional analytic dual space of $H^{0,\gamma}(\mathbb{R}^3)$ is isomorphic to $H^{0,-\gamma-3}(\mathbb{R}^3)$ under the pairing $v \mapsto \phi_v$ where $\phi_v(u) = \int_{\mathbb{R}^3} u \cdot v$.*

Proof: Choose u and v as in the statement of the lemma. Then by Hölder's inequality,

$$\begin{aligned}\int_{\mathbb{R}^3} |u \cdot v| &\leq \int_{\mathbb{R}^3} |u|\sigma^{-\gamma-3/2} \cdot |v|\sigma^{-(-\gamma-3)-3/2} \\ &\leq \left(\int_{\mathbb{R}^3} u^2\sigma^{-2\gamma-3}\right)^{1/2} \left(\int_{\mathbb{R}^3} v^2\sigma^{-2(-\gamma-3)-3}\right)^{1/2} \\ &< \infty .\end{aligned}$$

The product $u \cdot v$ is thus in L^1 and so its integral is well defined. The statement about duality follows from the Riesz Representation Theorem for L^2 and the inequality above. See [20, 25] for details. □

The second result concerns the Fredholm properties of certain linear, elliptic partial differential operators on weighted Sobolev spaces.

Invertibility Theorem: *Suppose B is any tensor bundle over $\mathbb{R}^3$ and let $Q : H^{k,\beta}(B) \to H^{k-2,\beta-2}(B)$ be any linear, second order, elliptic, homogeneous, partial differential operator with constant coefficients mapping between weighted Sobolev spaces of sections of B, and $k \geq 2$. Then Q is surjective if $\beta \notin \mathbb{Z}$ and $\beta > -1$ and injective if $\beta \notin \mathbb{Z}$ and $\beta < 0$. It is thus bijective when $\beta \in (-1, 0)$. The operator Q is not Fredholm if $\beta \in \mathbb{Z}$.*

Proof: The proof of this result can be found in [10], but see also [21] for an excellent discussion of the intuitive foundation underlying the theory of elliptic operators on weighted spaces. □

Choice of Banach spaces

Denote by $S^2(\mathbb{R}^3)$ the symmetric tensors over $\mathbb{R}^3$ and by $\Lambda^1(\mathbb{R}^3)$ the 1-forms of $\mathbb{R}^3$. Let β be any number in $(-1, 0)$ and let k be any integer strictly larger than $\frac{7}{2}$. Solutions of the associated system will be found in the following Banach spaces:

- metrics $\delta + h$ will be found so that $h \in H^{k,\beta}\big(S^2(\mathbb{R}^3)\big)$;
- 1-forms X will be found in $H^{k-1,\beta-1}\big(\Lambda^1(\mathbb{R}^3)\big)$;
- tensors T will be found in $H^{k-2,\beta-2}\big(S^2(\mathbb{R}^3)\big)$.

The preceding choice of Banach spaces will be justified in the next section by showing that solutions of the associated system exist in these spaces. However, an argument can be made right now that suggests that the spaces above are indeed the correct ones in which to expect to find solutions. First, in order to ensure that the metric $\delta + h$ is asymptotically flat, h must decay as $r \to \infty$, and this holds by the Sobolev Embedding Theorem when $\beta < 0$. Next, a non-trivial, asymptotically flat metric satisfying the constraint equations must satisfy the Positive Mass Theorem [24] and consequently must have non-zero ADM mass. Thus the r^{-1} term in the asymptotic expansion of h must be allowed to be non-zero, which by the Sobolev Embedding Theorem imposes the further requirement that $\beta > -1$. Furthermore, $k \geq 4$ implies that the Sobolev Embedding Theorem can be applied to the second derivatives of the metric, and thus the curvature of the metric decays pointwise as $r \to \infty$. Finally, the h, X and T quantities are chosen in different Sobolev spaces because of the differing numbers of derivatives taken on these quantities in the associated system. For instance, the reduced Ricci curvature operator is homogeneous and of degree two and thus sends a metric in $H^{k,\beta}$ to a tensor in $H^{k-2,\beta-2}$. The operator $S(h,X,T)$ is homogeneous but is only of degree one in X and of degree zero in T; it thus maps to $H^{k-2,\beta-2}$ only when the weightings on X and T match together properly and match the weighting on the metric h as in the choice above.

10.3.4 First Attempt to Solve the Associated System

The Implicit Function Theorem, the tool which will be used to solve the associated system, is restated here for ease of reference.

Implicit Function Theorem: *Let $\Phi : A \times B \to C$ be a smooth map between Banach spaces and suppose that $\Phi(0,0) = 0$. If the restricted linearized operator $D\Phi(0,0)\big|_{A\times\{0\}} : A \to C$ is an isomorphism, then there exists an open set $\mathcal{U} \subset B$ which contains 0 and a smooth function $\phi : \mathcal{U} \to A$ with $\phi(0) = 0$ so that $\Phi\big(\phi(b), b\big) = 0$.*

For an excellent discussion and proof of this theorem, see [1]. In order to use this theorem, let

$$\begin{aligned} A &= \big\{(h,X) \in H^{k,\beta}\big(S^2(\mathbb{R}^3)\big) \times H^{k-1,\beta-1}\big(\Lambda^1(\mathbb{R}^3)\big)\big\}, \\ B &= \big\{T \in H^{k-2,\beta-2}\big(S^2(\mathbb{R}^3)\big)\big\}, \\ C &= H^{k-2,\beta-2}\big(S^2(\mathbb{R}^3)\big) \times H^{k-3,\beta-3}\big(\Lambda^1(\mathbb{R}^3)\big)\,; \end{aligned}$$

then the linearization of the operator Φ in the A direction at the origin must be calculated and its mapping properties understood.

The linearization of Φ is actually quite simple when evaluated at the origin because the only nonlinearities in Φ occur in the second order terms of the reduced Ricci operator and in terms that are quadratic in the derivatives of the metric (such as in products of Christoffel symbols or in the connection terms). Since the covariant derivative of the Euclidean metric is trivial, it is thus easy to see that the linearization of a covariant derivative operator at the Euclidean metric is just the Euclidean derivative operator, and it is now a straightforward matter to deduce from the definition of the associated system in (10.18) that the linearization of Φ in the $A \times \{0\}$ direction is

$$\mathrm{D}\Phi(0,0,0)\,(h,X,0) = \begin{pmatrix} -\frac{1}{2}\Delta h - \mathcal{L}(X) \\ \operatorname{div} \circ \mathcal{L}(X) \end{pmatrix}, \tag{10.20}$$

where Δ is the Euclidean Laplacian and $\mathcal{L}$ is the Euclidean conformal Killing operator.

Denote by P_δ the operator $D\Phi(0,0,0)(\cdot,\cdot,0)$. It is a bounded linear operator between the appropriate weighted Sobolev spaces because of the way in which the weights were chosen in Sect. 10.3.3. To determine whether P_δ is an isomorphism, one appeals to the Invertibility Theorem. Recall that the weight β in the domain spaces of P_δ has been chosen between -1 and 0.

INJECTIVITY OF P_δ

Suppose (h, X) belong to the kernel of $P_\delta(h, X)$. In other words, (h, X) solves the equation $P_\delta(h, X) = (0, 0)$, or

$$\begin{aligned} -\frac{1}{2}\Delta h - \mathcal{L}(X) &= 0, \\ \operatorname{div} \circ \mathcal{L}(X) &= 0\,. \end{aligned}$$

Since the operator $\operatorname{div}\circ\mathcal{L} : H^{k-1,\beta-1}\big(\Lambda^1(\mathbb{R}^3)\big) \to H^{k-3,\beta-3}\big(\Lambda^1(\mathbb{R}^3)\big)$ is a linear, elliptic, homogeneous, constant coefficient operator of second order, the Invertibility Theorem applies, and since $\beta - 1 \in (-2,-1)$ when $\beta \in (-1,0)$, it is thus injective. Hence $X = 0$. The remaining equation now reads $\Delta h = 0$ and again, since $\Delta : H^{k,\beta}\big(S^2(\mathbb{R}^3)\big) \to H^{k-2,\beta-2}\big(S^2(\mathbb{R}^3)\big)$ and $\beta \in (-1,0)$, Δ is an isomorphism and thus $h = 0$. Hence P_δ is injective.

SURJECTIVITY OF P_δ

Although the operator P_δ is injective, it is *not* surjective. First note that the Invertibility Theorem does not *guarantee* surjectivity in the same way that it guaranteed injectivity. To see this, attempt to solve the equations $P_\delta(h, X) = (f, g)$ for any $f \in H^{k-2,\beta-2}\big(S^2(\mathbb{R}^3)\big)$ and $g \in H^{k-3,\beta-3}\big(\Lambda^1(\mathbb{R}^3)\big)$. In other words, consider the system of equations:

$$\begin{aligned} -\frac{1}{2}\Delta h - \mathcal{L}(X) &= f, \\ \operatorname{div} \circ \mathcal{L}(X) &= g\,. \end{aligned}$$

Because $\beta - 1 \in (-2,-1)$, the operator $\operatorname{div}\circ\mathcal{L}$ is not necessarily surjective according to the Invertibility Theorem. The full equations $P_\delta(h, X) = (f, g)$ can thus not necessarily be solved.

To show that P_δ actually does fail to be surjective, it is necessary to show that the dimension of its cokernel in $H^{k,\beta}\big(S^2(\mathbb{R}^3)\big) \times H^{k-1,\beta-3}\big(\Lambda^1(\mathbb{R}^3)\big)$ is strictly greater than zero. First, note that if X_g satisfies $\operatorname{div}\circ\mathcal{L}(X_g) = g$, then the remaining equation $-\frac{1}{2}\Delta h = \mathcal{L}(X_g) + f$ *can* be solved by the Invertibility Theorem since the weight β is chosen such that Δ is an isomorphism. Thus the dimension of the cokernel of P_δ is equal to the dimension of the cokernel of $\operatorname{div}\circ\mathcal{L}$ as an operator between $H^{k-1,\beta-1}\big(\Lambda^1(\mathbb{R}^3)\big)$ and $H^{k-3,\beta-3}\big(\Lambda^1(\mathbb{R}^3)\big)$.

To characterize the cokernel of $\operatorname{div}\circ\mathcal{L}$, one appeals to general, function-theoretic properties of linear, second order, homogeneous, elliptic operators on weighted Sobolev spaces. The following lemma and its proof show how this is done.

Cokernel Lemma: *Suppose B is any tensor bundle over $\mathbb{R}^3$ and let*

$$Q : H^{k,\gamma}(B) \to H^{k-2,\gamma-2}(B)$$

be a linear, second order, homogeneous, elliptic operator on weighted Sobolev spaces of sections of B where $k \geq 2$ and $\gamma \notin \mathbb{Z}$, $\gamma < -1$. The image of the operator Q is the space:

$$\operatorname{Im}(Q) = \left\{ w \in H^{k-2,\gamma-2}(B) \; : \; \int_{\mathbb{R}^3} \langle w, z \rangle = 0 \quad \forall \quad z \in \operatorname{Ker}(Q^*; -1-\gamma) \right\}, \tag{10.21}$$

where the inner product $\langle \cdot, \cdot \rangle$ is induced on B from the Euclidean metric of $\mathbb{R}^3$, the operator Q^* is the formal adjoint of Q, and $\operatorname{Ker}(Q^*; -1-\gamma)$ *is its kernel as an operator from $H^{k,-1-\gamma}(B)$ to $H^{k-2,-3-\gamma}(B)$.*

Proof: Denote the space on the right hand side of equation (10.21) by W. Suppose that $k = 2$ and consider first the containment $\operatorname{Im}(Q) \subseteq W$. Choose $Q(y) \in \operatorname{Im}(Q)$ and $z \in Ker(Q^*; -1-\gamma)$. Since $Q(y) \in H^{2,\gamma-2}(B)$, the integral $\int_{\mathbb{R}^3} \langle Q(y), z \rangle$ is well defined by the Duality Lemma. □

CLAIM: This integral equals $\int_{\mathbb{R}^3} \langle y, Q^*(z) \rangle$.

The equality of the integrals on smooth, compactly supported sections of B is true by definition of the adjoint. The equality of the integrals for $H^{k,\gamma}$ sections follows because C_c^∞ sections of B are dense in $H^{k,\gamma}$ sections of B [5].

The integral $\int_{\mathbb{R}^3} \langle Q(y), z \rangle$ is thus zero and so $Q(y) \in W$.

The reverse containment $W \subseteq \operatorname{Im}(Q)$ is proved as follows. Suppose w_0 belongs to W; thus, $w_0 \in H^{0,\gamma-2}(B)$ and satisfies $\int_{R^3} \langle w_0, z \rangle = 0$ for all $z \in Ker(Q^*; -1-\gamma)$. Suppose also that $w_0 \notin \operatorname{Im}(Q)$. Since Q is elliptic, $\operatorname{Im}(Q)$ is closed; thus by the Hahn-Banach theorem, there exists a linear functional ϕ on $H^{0,\gamma-2}(B)$ so that $\phi(w_0) \neq 0$ but $\phi\big|_{\operatorname{Im}(Q)} = 0$. Again by the Duality Lemma, there is a unique $z_0 \in H^{0,-1-\gamma}(B)$ so that $\phi(w) = \int_{\mathbb{R}^3} \langle w, z_0 \rangle$ for all $w \in H^{0,\gamma-2}(B)$. Therefore, $\phi\big|_{\operatorname{Im}(Q)} = 0$ implies that

$$\begin{aligned} 0 &= \phi(Q(y)) \\ &= \int_{\mathbb{R}^3} \langle z_0, Q(y) \rangle \\ &= \int_{\mathbb{R}^3} \langle Q^*(z_0), y \rangle \end{aligned}$$

for all $y \in H^{2,\gamma}(B)$. Thus $Q^*(z_0) = 0$ or $z_0 \in \operatorname{Ker}(Q^*; -1-\gamma)$. But now, the assumptions $\phi(w_0) \neq 0$ and $\int_{\mathbb{R}^3} \langle w_0, z \rangle = 0$ for all $z \in \operatorname{Ker}(Q^*; -1-\gamma)$ are mutually contradictory. Thus it must be that $w_0 \in \operatorname{Im}(Q)$. Finally, the extension to $k > 2$ follows in a similar manner by standard functional analysis. □

Apply this theorem to the operator $Q = \operatorname{div} \circ \mathcal{L}$ with $\gamma = \beta - 1$. Now, $Q^* = Q$, so in order to solve the equation $\operatorname{div} \circ \mathcal{L}(X) = g$, the tensors g must satisfy the constraints

$$\int_{\mathbb{R}^3} g_a Y^a = 0\,,$$

where Y is any tensor in the kernel of $\operatorname{div} \circ \mathcal{L}$ in the space $H^{k-1,-1-\gamma}\big(\Lambda^1(\mathbb{R}^3)\big)$.

The kernel of $\operatorname{div} \circ \mathcal{L}$ is well known and consists of 1-forms dual to the conformal Killing fields of $\mathbb{R}^3$. There are precisely ten linearly independent families of such vector fields: the translation vector fields, the rotation vector fields, the dilation field and three so-called *special* conformal Killing fields (these correspond to transformations of the form $i \circ T \circ i$, where i is the inversion with respect to the unit circle and T is a translation). The asymptotic behaviour of these vector fields can thus be computed exactly: the translations have constant norm, the rotations and dilations have norm growing linearly in the distance from the origin, and the special vector fields have quadratic growth in the distance from the origin. Since $-1 - \gamma \in (0, 1)$ when $\beta \in (-1, 0)$, the only 1-forms dual to the conformal Killing fields in $H^{k-1,-1-\gamma}\big(\Lambda^1(\mathbb{R}^3)\big)$ are thus those spanned by the translation 1-forms $\mathrm{d}x^1$, $\mathrm{d}x^2$ and $\mathrm{d}x^3$. Consequently, the image of $Q = \operatorname{div} \circ \mathcal{L}$ in the space $H^{k-3,\gamma-2}\big(\Lambda^1(\mathbb{R}^3)\big)$ can be characterized as follows:

$$\operatorname{Im}(\operatorname{div} \circ \mathcal{L}) = \left\{ g \in H^{k-3,\beta-3}\big(\Lambda^1(\mathbb{R}^3)\big) \;::\; \int_{\mathbb{R}^3} g_a = 0\,,\ a = 1, 2, 3 \right\}\,,$$

where g_a are the components of g in the standard coordinates of $\mathbb{R}^3$.

The conclusion that can be drawn from the analysis of this section is that the equation $\Phi(h, X, T) = (0, 0)$ is *not* solvable near $(0, 0, 0)$ using the Implicit Function Theorem. The non-surjectivity of the linearized operator at $(0, 0, 0)$ is the essential obstruction. The best that can be achieved using the Implicit Function Theorem is thus that the equation $\Phi(h, X, T) = (0, 0)$ can be solved *up to* a term that is transverse to the space $\operatorname{Im}(\operatorname{div} \circ \mathcal{L})$. It will turn out that this is nevertheless sufficient for solving the full equations as a result of the integrability conditions built into the equations. But in order to show this, the associated system defined in the previous section must be modified somewhat.

10.3.5 Reestablishing Surjectivity and Solving the Associated System

In order to modify the associated system appropriately, first note that the space $H^{k-3,\beta-3}\big(\Lambda^1(\mathbb{R}^3)\big)$ can be written as $\operatorname{Im}(\operatorname{div} \circ \mathcal{L}) \oplus W$ in many different ways; but in each case, W is a three dimensional subspace of $H^{k-3,\beta-3}\big(\Lambda^1(\mathbb{R}^3)\big)$ whose members do not integrate to zero upon taking the Euclidean inner product with the translation 1-forms. One such choice is

$$W = \operatorname{span}\left\{\phi\, \mathrm{d}x^a\right\}_{a=1,2,3}\,,$$

where ϕ is any smooth, positive function of compact support whose integral over $\mathbb{R}^3$ is equal to 1.

Again, denote the domain space of the operator Φ by A. The previous paragraph suggests that one should attempt to construct a new associated operator Φ' that extends Φ in such a way that $\Phi' : A \times \mathbb{R}^3 \to \operatorname{Im}(P_\delta) \oplus W$, where the additional $\mathbb{R}^3$ factor in the domain should map under the linearization $\mathrm{D}\Phi'$ at the solution $(0,0,0;0) \in A \times \mathbb{R}^3$ onto the W factor in the image. If such a construction is possible, then the equation $\Phi'(h,X,T;\lambda) = (0,0)$ can be solved using the Implicit Function Theorem.

Construct the operator $\Phi' : A \times \mathbb{R}^3 \to H^{k-3,\beta-3}\big(\Lambda^1(\mathbb{R}^3)\big)$ according to the prescription

$$\Phi'(h,X,T;\lambda) = \begin{pmatrix} Ric^H(\delta + h) - S(h,X,T) \\ \operatorname{div}_{\delta+h} \circ S(h,X,T) - \sum_{a=1}^3 \lambda_a \phi \, \mathrm{d}x^a \end{pmatrix}, \tag{10.22}$$

where, as before, Ric^H is the reduced Ricci operator and $S(\cdot,\cdot,\cdot)$ is the York operator. The linearization of Φ' at $(0,0,0;0)$ in the directions transverse to the T direction is easily seen to be

$$\mathrm{D}\Phi'(\delta,0,0;0)(h,X,0;\lambda) = \begin{pmatrix} -\frac{1}{2}\Delta h - \mathcal{L}(X) \\ \operatorname{div} \circ \mathcal{L}(X) - \sum_{a=1}^3 \lambda_a \phi \, \mathrm{d}x^a \end{pmatrix}. \tag{10.23}$$

Denote this new operator by P'_δ. It is still bounded because ϕ has compact support, and it is now also bijective by the following arguments.

Injectivity of P'_δ

Suppose $P'_\delta(h,X;\lambda) = (0,0)$. Integrate the components of the second equation; by the divergence theorem for the Euclidean metric (valid because constant functions can be integrated against $H^{k-3,\beta-3}$ functions when $\beta \in (-1,0)$ according to the Duality Lemma), the divergence terms integrate to zero, yielding $\lambda_a = 0$ for all a. The argument that both X and h are then equal to zero follows as in Sect. 10.3.4.

Surjectivity of P'_δ

Suppose that $P'_\delta(h,X;\lambda) = (f,g)$. First choose the components λ_a so that

$$\int_{\mathbb{R}^3} (g_a + \lambda_a \phi) = 0$$

for each a. The equation $\operatorname{div} \circ \mathcal{L}(X) = g - \sum_{a=1}^3 \lambda_a \phi \, \mathrm{d}x^a$ can then be solved for X_g according to the characterization of the image of the operator $\operatorname{div} \circ \mathcal{L}$ from the previous section. The remaining equation $-\frac{1}{2}\Delta h = \mathcal{L}(X_g) + f$ can then be solved because $\beta \in (-1,0)$ makes Δ an isomorphism.

The Implicit Function Theorem can now be invoked to solve the equation $\Phi'(h,X,T;\lambda) = (0,0)$ near $(0,0,0;0)$. To be precise, there is a neighbourhood $\mathcal{U} \subset H^{k-2,\beta-2}\big(S^2(\mathbb{R}^3)\big)$ with the following property. If $T \in \mathcal{U}$, then there is a metric $\delta + h(T)$ with $h(T) \in H^{k,\beta}\big(S^2(\mathbb{R}^3)\big)$, a covector field $X(T) \in H^{k-1,\beta-1}\big(\Lambda^1(\mathbb{R}^3)\big)$, and three real numbers $\lambda_a(T)$ so that

$$\Phi'\big(h(T), X(T), T; \lambda(T)\big) = (0,0).$$

Furthermore, the various functions $T \mapsto h(T)$, etc., are smooth in the appropriate Banach space norms. In particular, there exists a constant C so that

$$\begin{aligned} \|h\|_{H^{k,\beta}} &\leq C\|T\|_{H^{k-2,\beta-2}}, \\ \|X\|_{H^{k-1,\beta-1}} &\leq C\|T\|_{H^{k-2,\beta-2}}, \\ \|\lambda\|_{\mathbb{R}^3} &\leq C\|T\|_{H^{k-2,\beta-2}}, \end{aligned} \tag{10.24}$$

where $\|\cdot\|_{\mathbb{R}^3}$ denotes the standard Euclidean norm of $\mathbb{R}^3$, as long as $T \in \mathcal{U}$.

10.3.6 Satisfying the Harmonic Coordinate Condition

Sect. 10.3.5 shows how the associated system (10.18) can be modified in such a way that it can be solved using the Implicit Function Theorem. This procedure results in a family of solutions of the equations

$$\begin{aligned} Ric^H(\delta + h) &= S(h, X, T), \\ \mathrm{div}_{\delta+h} \circ S(h, X, T) &= \lambda\phi, \end{aligned} \tag{10.25}$$

where $\lambda = \sum_{a=1}^3 \lambda_a \, dx^a$. It remains to show whether the original equations (10.16) are satisfied by the solution $\delta + h$ and $S(h, X, T)$. This will be done by showing that the integrability conditions built into the extended constraint equations (i.e. the Bianchi identity only, since the time-symmetric assumption has eliminated the other integrability condition) actually ensure that if $(h, X, T; \lambda)$ solves (10.25), then $\lambda = 0$ and $h + \delta$ satisfies the harmonic coordinate condition. Therefore $Ric^h(\delta + h) = Ric(\delta + h)$ and solutions of (10.25) are indeed solutions of the full equations.

To prove this claim, assume instead that λ and the quantities Γ^a are nonzero. Argue towards a contradiction as follows. First, write $g = \delta + h$ for short. The Bianchi identity $\mathrm{div}_g\big(Ric(g) - \frac{1}{2}R(g)g\big) = 0$, applied to equation (10.9) defining the reduced Ricci operator yields the identity

$$0 = \Big(R^H_{ab} - \frac{1}{2}R^H g_{ab}\Big)_{;}{}^{a} = \big(\Gamma_{a;b} + \Gamma_{b;a} - \Gamma^c{}_{;c} h_{ab}\big)_{;}{}^{a}$$

which is equivalent to

$$\Gamma_{b;a}{}^{a} + R^a_b \Gamma_a = 2\phi\lambda_b , \tag{10.26}$$

after using the modified associated system and commuting covariant derivatives appropriately. If Q_h denotes the operator $u_a \mapsto \Delta_{\delta+h} u_a + [Ric(\delta + h)]^b_a u_b$, then (10.26) asserts that $2\phi\lambda_a$ is in the image of $H^{k-1,\beta-1}(\Lambda^1(\mathbb{R}^3))$ under Q_h, because $h \in H^{k,\beta}\big(S^2(\mathbb{R}^3)\big)$ and the Γ^a are obtained from $\delta + h$ by differentiation. This, however, can be shown to violate the following basic result about elliptic operators.

Stability Lemma: *Let B be a tensor bundle over $\mathbb{R}^3$ and let $Q_\varepsilon : H^{l,\gamma}(B) \to H^{l-2,\gamma-2}(B)$, $\varepsilon \in [0, 1]$, be a continuous family of linear, homogeneous, second order, elliptic operators, for all $\gamma \notin \mathbb{Z}$, $\gamma < -1$. Furthermore, suppose Q_ε is* *uniformly injective for any ε whenever $\gamma < -1$; i.e. for each $\gamma \notin \mathbb{Z}$, $\gamma < -1$, there is a constant C independent of ε so that $\|Q_\varepsilon(y)\|_{H^{l-2,\gamma-2}} \geq C\|y\|_{H^{k,\gamma}}$. If $z \notin \mathrm{Im}(Q_0)$, then there exists $\varepsilon_0 > 0$ so that $z \notin \mathrm{Im}(Q_\varepsilon)$ for all $\varepsilon < \varepsilon_0$.*

Proof: Suppose the contrary; then for some $\gamma < -1$, there exists a sequence $\varepsilon_i \to 0$ and a sequence $y_i \in H^{l,\gamma}(B)$ so that $z = Q_{\varepsilon_i}(y_i)$. By the uniform injectivity of Q_ε, $\|y_i\|_{H^{l,\gamma}} \leq C\|z\|_{H^{l-2,\gamma-2}}$ and is thus uniformly bounded. By Rellich's Lemma, there exists a subsequence $y_{i'}$ which converges to an element y in $H^{l-1,\gamma+\rho}$, where ρ is small enough so that $\gamma + \rho < -1$. Again, by uniform injectivity,

$$\begin{aligned}\|y_{i'} - y_{j'}\|_{H^{l,\gamma+\rho}} &\leq C\|Q_{\varepsilon_{i'}}(y_{i'} - y_{j'})\|_{H^{l-2,\gamma+\rho-2}} \\ &\leq C\|(Q_{i'} - Q_{j'})(y_{j'})\|_{H^{l-2,\gamma+\rho-2}} \\ &\leq C\|Q_{i'} - Q_{j'}\|_{op} \cdot \|y_{j'}\|_{H^{l,\gamma+\rho}} \\ &\leq C\|Q_{i'} - Q_{j'}\|_{op} \cdot \|y_{j'}\|_{H^{l,\gamma}} \\ &\longrightarrow 0\,,\end{aligned}$$

by the continuity of Q_ε and the uniform boundedness of y_i. Here, $\|\cdot\|_{op}$ denotes the relevant operator norm. The subsequence $y_{i'}$ is thus Cauchy in the $H^{l,\gamma+\rho}$ norm and so $y_{i'} \to y$ in this norm. But now,

$$z = \lim_{i' \to \infty} Q_{\varepsilon_{i'}}(y_{i'}) = Q_0(y)\,,$$

contradicting the fact that $z \notin \mathrm{Im}(Q_0)$. □

In order to derive a contradiction from (10.26) using this lemma, the uniform injectivity of Q_h must be established and it must be shown that $\phi\lambda_a$ does not belong to the image of Q_0.

UNIFORM INJECTIVITY OF Q_h

Suppose that $Q_h(u) = 0$ for $u \in H^{k-1,\gamma}\big(\Lambda^1(\mathbb{R}^3)\big)$ where $\gamma < -1$. In other words, ${\Gamma_{b;a}}^a + R^a_b \Gamma_a = 0$. From this, one easily deduces:

$$-\Delta_g \|u\|^2 = 2\big(R_{ab}u^a u^b - \|\nabla u\|^2\big)\,. \tag{10.27}$$

Before continuing, recall the following facts about Green's identity in weighted Sobolev spaces. If functions u and v are chosen such that $v \in H^{k,\gamma}(\mathbb{R}^3)$ and $u \in H^{k,-1-\gamma}(\mathbb{R}^3)$ for some γ, then the integrals appearing in Green's identity for a general metric g on a large ball B_r, that is

$$\int_{B_r} u\Delta_g v \,\mathrm{dVol}_g + \int_{B_r} \nabla u \cdot \nabla v \,\mathrm{dVol}_g = \int_{\partial B_r} u \frac{\partial v}{\partial n} \,\mathrm{dA}_g\,, \tag{10.28}$$

where dA_g is the area form of the metric g, are all well defined as $r \to \infty$. Thus by applying a density argument as in the proof of the Cokernel Lemma, one can conclude that

$$\int_{\mathbb{R}^3} u\Delta_g v \,\mathrm{dVol}_g + \int_{\mathbb{R}^3} \nabla u \cdot \nabla v \,\mathrm{dVol}_g = 0\,,$$

in the limit of (10.28) as $r \to \infty$.

With this in mind, integrate both sides of equation (10.27) against the volume form of the metric $g = \delta + h$ to obtain

$$-\frac{1}{2}\int_{\mathbb{R}^3} \Delta_g \|u\|^2 \,\mathrm{dVol}_g = \int_{\mathbb{R}^3} R_{ab} u^a u^b \,\mathrm{dVol}_g - \int_{\mathbb{R}^3} \|\nabla u\|^2 \,\mathrm{dVol}_g \,. \tag{10.29}$$

Since $u \in H^{k-1,\gamma}$ and $1 \in H^{k-1,-\gamma-1}$ (true since $\gamma < -1$), Green's Identity applied to the left hand side of (10.29) gives $\int_{\mathbb{R}^3} \Delta_g \|u\|^2 = 0$. Consequently,

$$\begin{aligned} 0 &\le \int_{\mathbb{R}^3} \|Ric(g)\| \, \|u\|^2 \,\mathrm{dVol}_g - \int_{\mathbb{R}^3} \|\nabla u\|^2 \,\mathrm{dVol}_g \\ &\le \int_{\mathbb{R}^3} \|Ric(g)\| \, \|u\|^2 \,\mathrm{dVol}_g - C \int_{\mathbb{R}^3} \|\nabla \|u\|\,\|^2 \,\mathrm{dVol}_g \end{aligned} \tag{10.30}$$

for some constant C, by the Cauchy-Schwarz inequality and straightforward algebra. Next, assume that h is small in a pointwise sense (this assumption follows from the Sobolev Embedding Theorem if h is sufficiently small in the $H^{k,\beta}$ norm and $k\frac{3}{2}$). In fact, assume that h is sufficiently close to 0 so that all norms, derivatives and volume forms of the metric g can be replaced by their Euclidean counterparts (at the expense of changing C of course). Finally, since $\|u\|$ is a scalar function, the derivative operator in (10.30) can be replaced by the Euclidean derivative operator without introducing lower order terms. Thus, there exists a new constant C so that the estimate

$$0 \le \int_{\mathbb{R}^3} \|Ric(g)\| \, \|u\|^2 - C \int_{\mathbb{R}^3} \|\nabla \|u\|\,\|^2 \tag{10.31}$$

holds, where the norms and derivatives appearing here are those of the Euclidean metric. Next, $Ric(g) \in H^{k-2,\beta-2}$ because $g - \delta \in H^{k,\beta}$. But since $k > \frac{7}{2}$, the Sobolev Embedding Theorem gives $Ric(g) \in C^0_{-\beta+2}$. That is,

$$\sup_{\mathbb{R}^3} \left\| Ric(g) \cdot \sigma^{-\beta+2} \right\| \le C < \infty \,,$$

which implies that

$$\sup_{\mathbb{R}^3} \left\| Ric(g) \cdot \sigma^{2} \right\| \le C < \infty \,,$$

since $\beta < 0$. Finally, apply the Poincaré inequality for weighted Sobolev norms to the function $\|u\|$ to deduce

$$\begin{aligned} \int_{\mathbb{R}^3} \|Ric(g)\| \, \|u\|^2 &\le \|Ric(g) \cdot \sigma^2\|_0 \int_{\mathbb{R}^3} \|u\|^2 \sigma^{-2} \\ &\le C \|Ric(g)\|_{C^0_{-2}} \int_{\mathbb{R}^3} \|\nabla \|u\|\,\|^2 \\ &\le C \|g - \delta\|_{C^2_0} \int_{\mathbb{R}^3} \|\nabla \|u\|\,\|^2 \\ &\le C \|h\|_{H^{k,\beta}} \int_{\mathbb{R}^3} \|\nabla \|u\|\,\|^2 \end{aligned} \tag{10.32}$$

again by the Sobolev Embedding Theorem and the fact that $\beta < 0$. Using (10.32) in inequality (10.31) leads to the contradiction because the preceding estimates imply

$$0 \leq (C\|h\|_{H^{k,\beta}} - 1) \int_{\mathbb{R}^3} \| \nabla \|u\| \|^2 ,$$

while if $\|h\|_{H^{k,\beta}}$ is sufficiently small, the right hand side above is clearly negative. Avoiding this contradiction requires $\nabla \|u\| = 0$. But since the Sobolev Embedding Theorem applied to $u \in H^{k-1,\gamma}$ shows that $\|u\|$ decays at infinity when $\gamma < -1$, it must be true that $u = 0$.

The operator Q_h acting on $H^{k-1,\gamma}$ 1-forms is injective for all $\gamma < -1$ whenever h is sufficiently close to zero in the $H^{k,\beta}$ norm. The uniform injectivity follows in the standard way from the injectivity of each Q_h and the fact that the constant in the elliptic estimate for these operators is independent of h, again provided h is sufficiently near to 0.

IMAGE OF Q_0

The $\phi\lambda$ term in (10.25) was specifically chosen in Sect. 10.3.5 to satisfy the integral condition $\int_{\mathbb{R}^3} \langle \lambda\phi, \mathrm{d}x^b \rangle \neq 0$ (since $\lambda_a = 0$ for all a). This condition ensures that indeed $2\phi\lambda_a$ is not in the image of the operator $Q_0 = \Delta_\delta$ acting on the space of $H^{l,\gamma}$ 1-forms of $\mathbb{R}^3$ because the image of Δ_δ in $H^{l,\gamma}$ for $\gamma < -1$ is perpendicular to the harmonic polynomials of degree less than the nearest integer less than γ, and this always includes the constants.

The Stability Lemma thus applies to equation (10.26) and implies that $\phi\lambda$ can not be in the image of Q_h when h is sufficiently small in the $H^{k,\beta}$ norm, unless of course $\lambda = 0$. Now, by the injectivity of the operator Q_h, this in turn implies that $\|\Gamma\| = 0$, or that $\Gamma^a = 0$ for each a. Consequently, the harmonic coordinate condition for the metric $\delta + h$ is satisfied, and as indicated earlier, this implies that the metric $\delta + h$ and the tensor $S(h, X, T)$ satisfy the time-symmetric extended constraint equations (10.16). This completes the proof of the Main Theorem. □

Acknowledgements

I would like to thank Helmut Friedrich for suggesting the topic of the conformal constraint equations to me as well as for providing me with guidance and insight. I would also like to thank Piotr Chruściel, Jörg Frauendiener and Niall O'Murchadha for their valuable comments and suggestions.

References

1. R. Abraham, J. E. Marsden, and T. Ratiu, *Manifolds, Tensor Analysis, and Applications*, second ed., Springer-Verlag, New York, 1988.
2. L. Andersson and P. Chruściel, *On "hyperboloidal" Cauchy data for vacuum Einstein equations and obstructions to smoothness of Scri*, Commun. Math. Phys. **355** (1994), 1–102.
3. L. Andersson and P. Chruściel, *Solutions of the constraint equations in general relativity satisfying "hyperboloidal boundary conditions"*, Dissertationes Mathematicae **161** (1996), 533–568.
4. L. Andersson, P. Chruściel, and H. Friedrich, *On the regularity of solutions to the Yamabe equation and the existence of smooth hyperboloidal initial data for Einstein's field equations*, Commun. Math. Phys. **149** (1992), 587–612.
5. R. Bartnik, *The mass of an asymptotically flat manifold*, Comm. Pure Appl. Math. **39** (1986), 661–693.
6. W. M. Boothby, *An Introduction to Differentiable Manifolds and Riemannian Geometry*, 2nd ed., Academic Press, Inc, 1986.
7. A. Butscher, *Perturbative solutions of the extended constraint equations in General Relativity*, In preparation.
8. M. Cantor, *Elliptic operators and the decomposition of tensor fields*, Bull. Amer. Math. Soc. **5** (1981), no. 3, 235–262.
9. J. Cao and D. DeTurck, *Prescribing Ricci curvature on open surfaces*, Hokkaido Math. J. **20** (1991), no. 2, 265–278.
10. Y. Choquet-Bruhat and D. Christodoulou, *Elliptic systems in $H_{s,\delta}$ spaces on manifolds which are Euclidean at infinity*, Acta Math. **146** (1981), 129–150.
11. D. Christodoulou and N. O'Murchadha, *The boost problem in general relativity*, Commun. Math. Phys. **80** (1981), 271–300.
12. D. DeTurck and H. Goldschmidt, *Aspects of prescribing Ricci curvature*, CR-Geometry and Overdetermined Systems (Osaka, 1994), Math. Soc. Japan, Tokyo, 1997, pp. 60–84.
13. D. DeTurck and H. Goldschmidt, *Metrics with prescribed Ricci curvature of constant rank. I. The integrable case*, Adv. Math. **145** (1999), no. 1, 1–97.
14. D. M. DeTurck, *Metrics with prescribed Ricci curvature*, Seminar on Differential Geometry, Princeton Univ. Press, Princeton, N.J., 1982, pp. 525–537.
15. J. Frauendiener, *Conformal infinity*, Living Rev. Relativity **3** (2000), no. 4, Online Article: http://www.livingreviews.org/Articles/Volume3/2000-4frauendiener/.
16. H. Friedrich, *Cauchy problems for the conformal vacuum field equations in general relativity*, Commun. Math. Phys. **91** (1983), 445–472.
17. H. Friedrich, *Asymptotic structure of space time*, Recent Advances in General Relativity (A. Janis and J. Porter, eds.), Einstein Studies, vol. 4, Birkäuser, 1992.
18. H. Friedrich, *Einstein's equations and conformal structure*, The Geometric Universe. Science, Geometry and the work of Roger Penrose (S. Huggett, L. Mason, K. P. Tod, S. Tsou, and N. M. J. Woodhouse, eds.), Oxford University Press, 1998.
19. H. Friedrich, *On the existence of n-geodesically complete or future complete solutions of Einstein's field equations with smooth asymptotic structure*, J. Geom. Phys. **24** (1998), 83–163.
20. E. Hebey, *Sobolev Spaces on Riemannian Manifolds*, Springer-Verlag, Berlin, 1996.
21. R. C. McOwen, *The behavior of the Laplacian on weighted Sobolev spaces*, Comm. Pure Appl. Math. **32** (1979), 783–795.

22. R. Penrose, *Asymptotic properties of fields and spacetimes*, Phys. Rev. Lett. **66** (1963), no. 10.
23. R. Penrose, *Zero rest-mass fields including gravitation: asymptotic behaviour*, Proc. R. Soc. Lond. **A284** (1965), 159–203.
24. R. Schoen and S.-T. Yau, *On the proof of the positive mass conjecture in general relativity*, Commun. Math. Phys. **65** (1979), 45–76.
25. G. Schwarz, *Hodge Decomposition—A Method for Solving Boundary Value Problems*, Springer-Verlag, Berlin, 1995.
26. J. York, *Gravitational degrees of freedom and the initial-value problem*, Phys. Rev. Lett. **26** (1971), 1656–1658.

11 Criteria for (In)finite Extent of Static Perfect Fluids

Walter Simon

Institut für theoretische Physik der Universität Wien,
Boltzmanngasse 5, A-1090 Wien

Abstract. In Newton's and in Einstein's theory we give criteria on the equation of state of a barotropic perfect fluid which guarantee that the corresponding one-parameter family of static, spherically symmetric solutions has finite extent. These criteria are closely related to ones which are known to ensure finite or infinite extent of the fluid region if the assumption of spherical symmetry is replaced by certain asymptotic falloff conditions on the solutions. We improve this result by relaxing the asymptotic assumptions. Our conditions on the equation of state are also related to (but less restrictive than) ones under which it has been shown in Relativity that static, asymptotically flat fluid solutions are spherically symmetric. We present all these results in a unified way.

11.1 Introduction

In this work we consider static, self-gravitating perfect fluids with a barotropic equation of state (EOS) $\rho(p)$ relating density and pressure, in Newton's theory (NT) and in General relativity (GR). We are thus dealing with time-independent, globally regular solutions to the Euler-Poisson equations and the Euler-Einstein equations. Some problems which have been studied in this context are the following:

A. Do there exist (spherically symmetric, topologically $R^3 \times R$, asymptotically flat) solutions with a given equation of state ?
B. Are all solutions (in particular all asymptotically flat ones) necessarily topologically $R^3 \times R$ and spherically symmetric ?
C. Do fluid solutions (with parameters in some given range) have finite (or infinite) extent and a finite (or infinite) mass ?
D. Can solutions (if they exist) be (uniquely) determined by parameters like surface potential, or mass (or central pressure or -density in the spherically symmetric case)?
E. Are fluid solutions (with parameters in some given range) stable or unstable (against a certain classes of perturbations)?

Some of the questions listed above are clearly physically relevant when one develops models for stars or a model for the distribution of stars in a galaxy, while others are at least of mathematical interest. A recent review dealing with some of these issues can be found in [2]. We will (after making a few general remarks), focus here on the question of finiteness (point C.), summarizing previous work as well as presenting new results.

For smooth equations of state, existence of spherically symmetric solutions (point A) has been proven satisfactorily in NT and in GR [18], while for discontinuous equations of state an existence theory has been developed in NT only [19]. Regarding topology, partial results (in relativity) are given in [14]. As to spherical symmetry, a complete proof is available in NT [10], but the problem is not really settled in GR (see [12], and below). Conjectures on points A and B under rather general conditions will be formulated in the next section. In any case, it is useful to pose questions C, D and E in a general context, in particular without assuming spherical symmetry, since the latter is not required in some theorems. We also note that the available results on spherical symmetry in GR not only have to be formulated somewhat differently for infinite and finite configurations, but the proof of the finite case actually goes the detour via a uniqueness theorem for a given surface potential (point D). Thus, in GR points A and B have so far been studied jointly with (parts of) C and D.

Turning now to the details of the finiteness problem, an interesting aspect is the relevance of the behaviour of $\rho(p)$ for small p. In fact, by modifying the EOS suitably in just an arbitrarily small neighbourhood of $p = 0$, all corresponding static solutions become necessarily finite or infinite. To see this, recall that fluid balls with $\rho(0) \neq 0$ are always finite [18], so we can simply set $\rho = \rho(\epsilon) = const.$ for $p \in [0, \epsilon]$ to ensure finiteness. On the other hand, (as we shall see below) we can always produce infinite solutions e.g. by attaching a piece with a linear EOS $p = C\rho$ for $p \in [0, \epsilon]$. We also recall that there are nontrivial criteria sufficient for finiteness of static fluid configurations, for example (assuming spherical symmetry) the existence of the integral $\int_0^p dp'/\rho(p')^2$ for some $p > 0$ (due to Rendall and Schmidt [18]) or $(\rho/p)(dp/d\rho) = \gamma + O(\rho^{\gamma-1})$ for a constant γ with $4/3 < \gamma < 2$ (due to Makino [13]), which only involve the behaviour of the EOS at small pressures.

These facts notwithstanding, it is important to realize that for the finiteness question the behaviour of $\rho(p)$ *for all* p is relevant in general. The striking examples in this context are the polytropes of index 5 in NT and the "Buchdahl" EOS in GR [6], viz.

$$p = \frac{1}{6}\rho_-^{-\frac{1}{5}}\rho^{\frac{6}{5}}, \qquad\qquad p = \frac{1}{6}(\rho_-^{\frac{1}{5}} - \rho^{\frac{1}{5}})^{-1}\rho^{\frac{6}{5}} \tag{11.1}$$

where ρ_- is a positive constant, and $\rho < \rho_-$ in GR. All static solutions with these EOS have infinite extent but finite mass [4]. Criteria for (in)finiteness [22] (discussed and generalized below) imply, in particular, that these solutions are "on the verge" of having (in)finite extent and (in)finite mass in the following sense: Suitable, but arbitrarily small modifications of the EOS performed in an arbitrarily small neighbourhood of an arbitrary value p_0 of p, necessarily force *all* solutions with $p_0 \in$ *supp* ρ to be finite, while other small modifications near any p_0 force *all* solutions with $p_0 \in$ *supp* ρ to have infinite mass.

In this paper we discuss a theorem consisting of several parts. Under suitable conditions on the equation of state in each case, the theorem shows (roughly speaking) spherical symmetry from asymptotic flatness, finite and infinite extent of asymptotically flat fluid solutions, and finite extent in the spherically symmetric case. While the first part of the theorem is, in essence, just reproduced from [3, 11] and the second part is a technically improved version of [12, 22], the final part is the main new input of the present paper. The purpose of combining these three parts here is to exhibit the close relationship between the respective conditions on the equation of state (and in particular the role of (11.1) as a limiting case). As a complement to the present work, Heinzle has recently obtained results on finiteness and infiniteness of static fluids [8], assuming both asymptotic flatness and spherical symmetry, but less restrictive conditions on the equation of state than we do.

This report is organized as follows. In the next section we will give some preliminary material and quote the theorem discussed above. Crucial tools for the proof of the results on (in)finiteness are Pohozaev(-type) identities leading to virial(-type) theorems, which we discuss in Sect. 11.3. While in NT the virial theorem is an equality (without further assumptions), in GR one obtains in the same manner only a "virial inequality", (but equalities can still be derived when spherical symmetry is assumed [8]). Correspondingly, the "general" Newtonian virial theorem yields criteria both for finiteness and infiniteness of the fluid configuration, whereas in GR one only obtains a finiteness criterion in the same fashion, (but again an infiniteness criterion in the spherically symmetric case [8]). However, even without this latter assumption a criterion for infiniteness also exists in GR and is proven using the positive mass theorem with respect to a suitably conformally rescaled metric (a procedure which is, on the other hand, meaningless in NT). The proofs of the new parts of the main theorem are given in Sect. 11.4. In the final section we expose our results in the light of the "quasi-polytropic" family of EOS.

11.2 The Main Theorem

We will treat NT and GR in a "parallel" fashion as far as possible. This means that, if equations in NT and GR are direct analogues, we give them in left and right columns, respectively, and use the same symbols and the same numbers for corresponding quantities to facilitate comparison. In either situation, we consider a Riemannian manifold $(\mathcal{M}, g)$ with a scalar function V. In the Conjecture and the Theorem below, we will take V and g to be in some Sobolev space L^q_{k+2} (using this symbol we follow Lee and Parker [9]), with exponent $q \geq 4$ and with k, the degree of (weak) differentiability, equal to either $0, 1$ or 2. Furthermore, we consider on $(\mathcal{M}, g)$ the non-negative functions $p \in L^q_{k+1}$, and $\rho \in L^q_k$, related by an EOS $\rho(p)$ which is C^{k-1} (if $k \geq 1$) and piecewise C^k. This means that for $k = 0$ we allow the density to "jump", while $k \geq 1$ forbids such jumps. We also require that (g, V) satisfy on $\mathcal{M}$ (at least in a weak sense) the Euler-Poisson equations resp. the Euler-Einstein equations which we write as follows:

$$\Delta V^{-1} = -4\pi\rho, \qquad \Delta V = 4\pi(\rho + 3p)V, \tag{11.2}$$

$$R_{ij} = 0, \qquad R_{ij} = V^{-1}D_iD_jV + 4\pi(\rho - p)g_{ij}, \tag{11.3}$$

$$D_ip = \rho D_iV^{-1}, \qquad D_ip = -V^{-1}(\rho + p)D_iV. \tag{11.4}$$

The reason why we use here V^{-1} on the l.h. side will become clear shortly. In the GR case, $\mathcal{M}$ can be understood as hypersurface orthogonal to a time-like Killing vector ξ, and V denotes the norm of ξ. The covariant derivative D_i, the Laplacian $\Delta = D_iD^i$ and the Ricci tensor R_{ij} refer to g. As to the Euler equation, there is a difference between NT and GR: While in the former case we require that $p \in L_1^q$ and postulate (11.4), the corresponding (weak) version in GR can be derived as follows.

Lemma 11.1. *(C.f. Remark 1.8. of [14].) Let $(\mathcal{M}, g, V)$ be a L_2^q solution which satisfies the GR-side of the field equations (11.2) and (11.3) weakly. Then $p \in L_1^q$ and Euler's equation (11.4) also hold weakly.*

Proof. Since C_0^∞ functions lie dense in L_k^q we can approximate g by a sequence of such functions. For any test function $\phi \in C_0^\infty$ we have then the following weak form of the Bianchi identity:

$$\int_{\mathcal{M}} D^i\phi^j(R_{ij} - \frac{1}{2}Rg_{ij})d\mathcal{V} = 0, \tag{11.5}$$

where $d\mathcal{V}$ is the volume element w.r. to g. Approximating now also V, p and ρ by C_0^∞-functions we easily obtain the required result. □

While we allow ρ to have zeros in general (in particular at $p = 0$) we always assume that the integrals

$$\Gamma = \int_0^p \frac{dp'}{\rho(p')}, \qquad \Gamma = \int_0^p \frac{dp'}{\rho(p') + p'} \tag{11.6}$$

exist for some interval $p \in [0, p_{max}]$. (The integral does not exist e.g. for linear equations of state $\rho = Cp$). Obviously Γ is a continuous function of p and also on $\mathcal{M}$. In NT, from Euler's equation, we have $D_i(V^{-1} - \Gamma) = 0$ and so the expression in parenthesis is a constant V_s^{-1} equal to V^{-1} at $p = 0$, whence V_s is called the surface potential. In GR, the same conclusions can be drawn from $D_i(Ve^\Gamma) = 0$. Integrating (11.4) we obtain

$$\Gamma = V^{-1} - V_s^{-1}, \qquad \Gamma = \ln V_s - \ln V. \tag{11.7}$$

V is related to the standard Newtonian potential U by $U = 1 - V^{-1}$. To formulate our theorems we now introduce some quantities defined from the EOS, and proceed with a lemma on their mutual relationship. If $\rho(p)$ is C^0 and piecewise C^1 we define

$$\kappa = \frac{d\rho}{dp}, \qquad \kappa = \frac{\rho + p}{\rho + 3p}\frac{d\rho}{dp} \tag{11.8}$$

and if $\rho(p)$ is C^{k-1} (for $k=1,2$) and piecewise C^k (for $k=0,1,2$):

$$I_0 = \rho\Gamma - 6p, \qquad\qquad I_0 = \rho(e^{\Gamma}-1) - 6p, \tag{11.9}$$

$$I_1 = 6p\kappa - 5\rho, \qquad\qquad I_1 = 6\frac{\rho+3p}{\rho}p\kappa - 5\rho, \tag{11.10}$$

$$I_2 = 5\rho\frac{d\kappa}{dp} + \kappa^2, \qquad\qquad I_2 = 5(\rho+p)\frac{d\kappa}{dp} + \kappa^2 + 10\kappa. \tag{11.11}$$

The static spherically symmetric solutions corresponding to $I_0 \equiv 0$ and $I_1 \equiv 0$ are known explicitly and read (c.f. [5] for NT and [4,6] for GR),

$$1-V^{-1} = -\frac{M}{\sqrt{\frac{4\pi}{3}\rho_- M^4 + r^2}}, \qquad\qquad 1-V = \frac{M}{\sqrt{\frac{4\pi}{3}\rho_- M^4 + r^2} + \frac{M}{2}}, \tag{11.12}$$

where M is the mass which can take any positive value in NT. In GR, M is bounded from *below* by $3M^{-2} < 16\pi\rho_-$, and the metric on $\mathcal{M}$ is

$$g = [(2/(1+V)]^4\delta$$

where δ is the Kronecker symbol.

We continue with a result on the relation between the quantities I_k.

Proposition 11.1. *Assume that $\rho > 0$ for $p > 0$, and that $\lim_{p\to 0}\rho^{-1}p$ exists. Let $\{I_k \geq 0\}$, $\{I_k \equiv 0\}$ and $\{I_k \leq 0\}$ denote the sets of equations of state which are C^{k-1} (for $k=1,2$), piecewise C^k for $k=0,1,2$ and which satisfy, for all $p \in [0, p_{max}]$, $I_k \geq 0$, $I_k \equiv 0$, or $I_k \leq 0$, respectively. Then*

$$\{I_1 \geq 0\} \subseteq \{I_0 \geq 0\}, \tag{11.13}$$

$$\{I_2 \equiv 0\} \supseteq \{I_1 \equiv 0\} \equiv \{I_0 \equiv 0\}, \tag{11.14}$$

$$\{I_2 \leq 0\} \subseteq \{I_1 \leq 0\} \subseteq \{I_0 \leq 0\}. \tag{11.15}$$

Proof. These results are due to the fact that the I_j are in a certain sense the integrals of I_{j+1}. In particular, from (11.4), (11.7), (11.9) and (11.10) we obtain

$$\rho\frac{d}{dp}[\rho^{-1}I_0] = \rho^{-1}I_1, \qquad\qquad (\rho+p)e^{\Gamma}\frac{d}{dp}[\rho^{-1}e^{-\Gamma}I_0] = \rho^{-1}I_1. \tag{11.16}$$

To show (11.13) and the second inclusion in (11.15) we note that $\rho^{-1}I_0$ vanishes at $p=0$, which is obvious when $\rho_s \neq 0$. As to the case $\rho_s = 0$, we first note that the existence of Γ and of $\lim_{p\to 0}\rho^{-1}p$ implies $\lim_{p\to 0}\rho^{-1}p = 0$ (c.f. [8], Appendix A) and so the assertion follows from (11.9). Therefore, when $I_1 \geq 0$, $I_1 = 0$ or $I_1 \leq 0$, the same holds for I_0.

Each of the conditions $I_1 \equiv 0$ and $I_0 \equiv 0$ (together with the differentiability assumptions) characterizes the 1-parameter families of EOS (11.1) and so the sets $\{I_1 \equiv 0\} \equiv \{I_0 \equiv 0\}$ of corresponding solutions are given by (11.12). The condition $I_2 = 0$ admits the case $\rho = const.$ and a two-parameter family of EOS which reads $p = 1/6(\rho_-^{-1/5}\rho^{6/5} - \rho_+)$ in NT (where ρ_- and ρ_+ are constants with $\rho_- > \rho_+$), while in GR it is given in [21] (together with the corresponding static solutions).

Finally, the first inclusion in (11.15) is proven in [3] (in a similar way as above). □

We now impose falloff conditions on g and V in terms of weighted Sobolev spaces $L^q_{k,\beta}$ (defined as in Def. (2.1) of Bartnik [1], and Sect. 9 of Lee and Parker, [9] following again the notation of the latter paper).

Definition 11.1. A L^q_2 solution $(\mathcal{M}, g, V)$ of (11.2), (11.3) is called asymptotically flat (AF) if

1. The "end" $\mathcal{M}^\infty = \mathcal{M} \setminus \{\text{a compact set}\}$ is diffeomorphic to $R^3 \setminus \mathcal{B}$ where $\mathcal{B}$ is a closed ball.
2. On $\mathcal{M}^\infty$, $1 - V = o(1)$, and the metric g satisfies, for some $q \geq 4$ and $\alpha > 0$,

$$g - \delta \in L^q_{2,\alpha}. \tag{11.17}$$

Thus, in this definition of AF we have supplemented Bartnik's falloff condition [1] for g by the requirement that $V \to$ *const.* at infinity (with the convention that *const.* = 1).

Definition 11.2. An AF solution $(\mathcal{M}, g, V)$ of (11.2), (11.3) is called asymptotically flat with mass decay conditions (AFMD) if on $\mathcal{M}^\infty$ the fluid variables satisfy, for some $q \geq 4$ and $\alpha > 0$,

$$\rho \in L^q_{0,-3-\alpha}, \tag{11.18}$$

$$p \in L^q_{0,-3-\alpha}. \tag{11.19}$$

Here the name "mass decay conditions" for these falloff conditions is inspired by similar requirements on the Ricci scalar in eqns.(4.4) of [1].

We remark that, to obtain our results in NT, it would be sufficient to impose (11.18) and (11.19) with "3" replaced by "5/2", which would not a priori restrict us to finite mass. In order not to spoil the "parallel" presentation of NT and GR, we just restrict ourselves to remarks on this option (after the formulation of the theorem).

Together with the field equations (11.2), (11.3), the AFMD conditions lead to the following stronger falloff properties. (We only give the formulation in GR explicitly; the Newtonian case arises via obvious simplifications.)

Lemma 11.2. *Let $(\mathcal{M}, g, V)$ be an AFMD GR solution. Then there exist a harmonic coordinate chart $(\mathcal{M}^\infty, g)$ and a constant M ("the mass") such that*

$$1 - V - \frac{M}{r} \in L^q_{2,-1-\alpha}, \tag{11.20}$$

$$g - (1 + 2M/r)\delta \in L^q_{2,-1-\alpha}, \tag{11.21}$$

$$p \in L^q_{1,-4-\alpha}. \tag{11.22}$$

Proof. Inserting $1 - V = o(1)$, (11.18) and (11.19) in (11.2) we have $\Delta V \in L^q_{0,-3-\alpha}$. Inverting the Laplacian with Proposition 2.2 of [1] we obtain (11.20). The next step is to insert (11.18), (11.19) and (11.20) in (11.3) which yields $R_{ij} \in L^q_{0,-3-\alpha}$. By Proposition 3.3 of [1] we can now pass to the harmonic chart on $\mathcal{M}^\infty$ and get (11.21). Finally, (11.22) is obvious after inserting (11.20) and (11.21) in (11.4). □

If we replace "3" by "5/2" in the definition of AFMD, we get Lemma 11.2 without the mass term, and with falloff $L^q_{2,-1/2-\alpha}$ for V, g and $L^q_{1,-3-\alpha}$ for p.

We now formulate a conjecture on points A,B and C raised in the introduction, and compare it with our theorem.

Conjecture 11.1. Let $\rho(p)$ be a piecewise C^0 equation of state with $\rho \geq 0$, $p \geq 0$. Then

A: There exists a 1-parameter family of spherically symmetric, AF L^q_2 solutions $(\mathcal{M}, g, V)$ of (11.2,11.3,11.4).

B: Every L^q_2-solution $(\mathcal{M}, g, V)$ is necessarily spherically symmetric.

C: **I.** If $I_0 \leq 0$ holds for $p \in [0, p_{max}]$, then the solution either has finite extent or $\rho(p)$ satisfies $I_0 \equiv 0$.

II. If $I_0 \geq 0$ holds for $p \in [0, p_{max}]$, then either $\rho(p)$ satisfies $I_0 \equiv 0$ or the solution is not AF.

As remarked in Sect. 11.1, part A was shown under slightly stronger assumptions in NT [19] and for smooth EOS in GR [18]. We refer to these papers for details. As to spherical symmetry (B), the following theorem requires AFMD as well as a condition on the EOS, while the results on finiteness require either AFMD or spherical symmetry, together with stronger assumptions on the EOS than conjectured above.

Theorem 11.1. *Let $\rho(p)$ be given with $p \geq 0$ and $\rho \geq 0$ and assume that there is a solution $(\mathcal{M}, g, V)$ of (11.2), (11.3), (11.4) specified below.*

B: *Assume that $\rho(p)$ is C^1 with $d\rho/dp > 0$ and piecewise C^2 with $I_2 \leq 0$ for all $p \in [0, p_{max}]$, and assume that the solution is L^q_4 ($q \geq 4$) and AFMD. Then the solution is spherically symmetric.*

C: **I.** *If one of the following conditions 1. or 2. holds:*

1. *$\rho(p)$ is piecewise C^0, and $I_0 \leq 0$ holds for $p \in [0, p_{max}]$. The solution is AFMD.*

2. *$\rho(p)$ is C^0, piecewise C^1 with $d\rho/dp > 0$, and $I_1 \leq 0$ holds for $p \in [0, p_{max}]$. The solution is L^q_3 ($q \geq 4$) and spherically symmetric.*

Then the solution either has finite extent, or $\rho(p)$ satisfies $I_0 \equiv 0$.

II. *Assume that $\rho(p)$ is piecewise C^0, and $I_0 \geq 0$ holds for all $p \in [0, p_{max}]$. Then either $I_0 \equiv 0$, or the solution is not AFMD.*

Some comments on this theorem are in order.

Whenever $I_0 \equiv 0$, then the EOS reads (11.1), and the solutions are given by (11.12).

Part B of this theorem has been proven in [12] in the smooth case, but extends straightforwardly to the present setting.

Part C.I.1 has in essence been shown in [22], while C.II is proven in [22] and (assuming $I_1 \geq 0$) in [11]. Compared to these papers, we set out here from weaker assumptions on differentiability and on the falloff (as to NT, c.f. the remarks after Definition 11.2 and after Lemma 11.2).

The main new result of the present paper is part C.I.2. Its proof goes, in essence, the detour via C.I.1, i.e. we first show that assumptions C.I.2 (together with the field equations) imply falloff conditions similar to (11.20), (11.21) and (11.22) which can be used alternatively to the AFMD ones in C.I.1. in the spherically symmetric case. Since we also know from Proposition 1 that $I_1 \leq 0$ implies $I_0 \leq 0$, finite extent of the solution follows.

We note that C.I.1 and C.I.2 are also related in the sense that they require similar auxiliary results (in particular Lemma 4), and so we found it useful to repeat (from [22]) the whole proof of C.I.1 here.

A natural extension of part C of the above theorem has been obtained by Heinzle [8]. From the EOS he defines quantities J_{-1} in NT and J_0 in GR whose signs again determine the (in)finiteness properties of solutions which are now assumed to be both spherically symmetric and asymptotically flat. Moreover, as an extension of Proposition 11.1, it is shown in [8] that if I_0 has a sign (and if $\lim_{p\to 0} \rho^{-1} p$ exists), then J_{-1} and J_0 have the same sign as I_0. Hence these quantities extend our "I_k series" in a natural manner, and Heinzle's (in)finiteness results require rather weak conditions on the EOS.

11.3 The Virial Theorem

This section contains two easy technical lemmas followed by the "virial theorem" (as a proposition). The latter is a tool for the proof of the Theorem in Sect. 11.4.

Lemma 11.3 (The modified Pohozaev identity). ([22]; compare [16]) *On $(\mathcal{M}, g)$ with $\mathcal{M} \sim R^3$ and g flat, let ξ_i be a dilation, i.e. $D_{(i}\xi_{j)} = \delta_{ij}$ (in Cartesian coordinates x^i, $\xi_i = x^i$). Assume that $\sigma = \sigma(X)$ with $X \in L^q_2$, $\sigma \in L^q_0$ $(q \geq 4)$ satisfies $\Delta X = 4\pi\sigma$, and define functions $\tau(X) = \int^X \sigma(X')dX'$ and $Z = D_j X D^j X$. (Here τ is defined only up to an additive constant). Then we have*

$$D_i[(\xi^j D_j X + \frac{1}{2}X)D^i X - \frac{1}{2}Z\xi^i + 4\pi\tau\xi^i] = 2\pi(\sigma X + 6\tau). \tag{11.23}$$

This Lemma, which is proven by straightforward computation, will be applied below in NT, with $X = 1 - V^{-1}$, $\sigma = \rho$ and $\tau = p$.

In spaces in which there exists a (general) conformal Killing vector ξ, (i.e. $\mathcal{C}[D_i\xi_j] = 0$ where $\mathcal{C}$ denotes the symmetric, trace-free part) there is an obvious generalization of (11.23). A more subtle generalization is the following.

Lemma 11.4. *Let $(\mathcal{M}, g')$ be a Riemannian 3-manifold with $g' \in L^q_2$ and Ricci-tensor and -scalar R'_{ij}, $R' = g'^{ij}R'_{ij}$ and $B'_{ij} = \mathcal{C}'[R'_{ij}]$ (where $\mathcal{C}'$ denotes the* *symmetric,trace-free part w.r. to g'). Assume further that there exist a vector field $\eta'^i \in L^q_1$ and a function $\alpha' \in L^q_0$ such that*

$$\mathcal{C}'[D'_i\eta'_j] = \alpha' B'_{ij}. \tag{11.24}$$

Then

$$D'^i[(R'_{ij} - \frac{1}{2}g'_{ij}R')\eta'^j] = \alpha' B'_{ij}B^{ij\prime} - \frac{1}{6}R'D'_i\eta'^i \tag{11.25}$$

and if $g' \in L^q_3$,

$$D'^i[B'_{ij}\eta'^j] = \alpha' B'_{ij}B^{ij\prime} + \frac{1}{6}\eta'^i D'_i R'. \tag{11.26}$$

The derivatives on the l.h. sides on (11.25) and (11.26) are to be understood in the weak sense.

Proof. (11.25) and (11.26) follow from (11.24) by approximating η' with a sequence of C^∞_0 functions, and from the weak Bianchi identity (11.5) with respect to g'. □

If there exists a conformal Killing vector ξ then we could in principle take $\eta = \xi$ in Lemma 11.4. Instead, we will make later use of this lemma with the more sophisticated choice:

$$g' = (1 - V^{-1})^4 g, \qquad g' = \frac{1}{16}(1+V)^4 g, \tag{11.27}$$

$$\eta'_i = D_i(1 - V^{-1})^2, \qquad \eta'_i = \frac{1+V}{(1-V)^3}D_iV, \tag{11.28}$$

$$\alpha' = -(1 - V^{-1})^2, \qquad \alpha' = \frac{V(1+V)^2}{(1-V)^4}. \tag{11.29}$$

As to the GR case of (11.25) we note that, using (11.27)-(11.29), the l.h. side can be rewritten (after multiplying with $(det\ g'/det\ g)^{1/2}$) as

$$D_i\left[\frac{1}{(1-V^2)^2}\left(V^{-1}D^iW + 8\frac{WD^iV}{1-V^2} - 8\pi\left(\rho + \frac{1-5V}{1-V}p\right)D^iV\right)\right], \tag{11.30}$$

where $W = D_iVD^iV$. The quantities on the r.h. side of (11.25) read explicitly:

$$R' = 256\pi(1+V)^{-5}[\rho(1-V) - 6pV], \tag{11.31}$$

$$D'_i\eta'^i = 32(1-V^2)^{-4}[2\pi V(1-V^2)(\rho+3p) + 3W], \tag{11.32}$$

and so $D'_i\eta'^i \geq 0$.

We can now show the following

Proposition 11.2 (The "Virial theorem"). *Let $(\mathcal{M}, g, V)$ be AFMD. Then there is a $B \geq 0$ such that*

$$\int_\mathcal{M}[\rho(1-V^{-1}) + 6p]d\mathcal{V} = 0, \qquad \int_\mathcal{M} B[\rho(1-V^{-1}) + 6p]d\mathcal{V} \leq 0. \tag{11.33}$$

Proof. In NT we integrate the Pohozaev identity (11.23), apply the divergence theorem and note that the resulting surface terms vanish by virtue of the asymptotic properties derived in Lemma 11.2 (see also the remark after this lemma). In GR we apply the same reasoning to (11.25), using (11.31), $B = (128V\pi/3)(1+V)^{-5}D_i'\eta'^i$ and again Lemma 11.2 for the surface terms. □

11.4 Proof of the Main Theorem

We first sketch the ideas of the proof. For part C.I.1 in general, and also for part C.II in NT we can use the virial theorem described above (which requires asymptotic flatness). We find that the sign of I_0, together with $V_s < 1$ or $V_s = 1$, determine the signs of the r.h. sides of (11.33), which directly leads to the required conclusions. To show C.I.2 we introduce the quantities

$$\widehat{W} = (1 - V^{-1})^{-4}W, \qquad \widehat{W} = (1 - V^2)^{-4}W, \tag{11.34}$$

$$\widehat{g} = (1 - V^{-1})^4 g, \qquad \widehat{g} = V^{-2}(1 - V^2)^4 g, \tag{11.35}$$

$$\widehat{W}_0 = \frac{4\pi}{3}\int_{V_c}^{V} \frac{\rho V^4}{(1-V)^4}\, dV, \qquad \widehat{W}_0 = \frac{4\pi}{3}\int_{V_c}^{V} \frac{V(\rho + 3p)}{(1-V^2)^4}\, dV, \tag{11.36}$$

where V_c is a constant specified below. We then employ (11.26), again with the choices (11.27),(11.28) and (11.29). As the r.h. side now contains $D_i'R'$ (instead of R' in (11.25)), controlling the sign now requires, (instead of $I_0 \le 0$), the more restrictive condition $I_1 \le 0$. In fact, assuming now also that the fluid region extends to infinity (i.e. $V_s = 1$) we can write (11.26) in the form:

$$\widehat{\Delta}(\widehat{W} - \widehat{W}_0) \ge 0 \tag{11.37}$$

with $\widehat{W}_0 \ge C^2 = const. > 0$. If we were to assume that $(\mathcal{M}, g, V)$ is AFMD, we could now form a compactification $\widehat{\mathcal{M}} = \mathcal{M} \cup \Lambda$ of this manifold by adding the point at infinity Λ, and $\widehat{W}$, $\widehat{W}_0$ and the metric $\widehat{g}$ would have C^1-extensions to Λ. Then the maximum principle would immediately lead again to the conclusions of C.I.1. Without asymptotic conditions we can still apply the maximum principle to (11.37) on a finite domain $\mathcal{D}$. Assuming also spherical symmetry and taking V_c in (11.36) to be the value of V at the centre, we conclude that $\widehat{W} \ge \widehat{W}_0 > C^2$, which gives $1 - V \le 1/Cr$ on a ball $\mathcal{B}_r$ of radius r. Thus, extending $\mathcal{B}_r$ we control the asymptotic behaviour of V. With some further technical manipulations based on the fact that ΔV is positive, we can then prove that $(\mathcal{M}, g, V)$ satisfies asymptotic conditions similar to the AFMD ones. These in fact guarantee the existence of a C^1 compactification as well as the vanishing of the surface terms in the integrals of (11.23) and (11.30). We then obtain the required conclusion $I_1 \equiv 0$ either from the maximum principle or by applying the reasoning of C.I.1.

We remark that, from (11.30), we could also write (11.26) in the form $\widehat{\Delta}(\widehat{W} - \widehat{W}_1) \ge 0$ for some $\widehat{W}_1(V)$ (assuming infinite extent and $I_1 \le 0$). This relation can, however, not replace (11.37) in the proof sketched above as we do not have control over the sign of $\widehat{W}_1$ (whereas $\widehat{W}_0 > 0$).

Finally, to show part C.II in the Einstein case, we use the rigidity case of the positive mass theorem on $(\mathcal{M}, g')$.

We now turn to the details.

Proof (of part C.I.1 of the Theorem). Using (11.7),(11.9) and the virial theorem (11.33), we find that

$$\int_{\mathcal{M}} [\rho(1 - V_s^{-1}) - I_0] \, d\mathcal{V} = 0, \qquad \int_{\mathcal{M}} B[(V_s - 1)(\rho + 6p) - I_0] \, d\mathcal{V} \leq 0, \quad (11.38)$$

where I_0 has here to be considered as a function on $\mathcal{M}$. The statement of the theorem now follows immediately from (11.38). □

Proof (of part C.II). In NT the second part of the theorem also follows easily from (11.38). As to the GR case, we find from (11.20) that the metric g' is $L^q_{2,\alpha}$, has vanishing mass, and $R' \geq 0$ due to (11.7), (11.9), (11.31) and $I_0 \geq 0$. The "vanishing mass-" case of the positive mass theorem [20] hence implies that g' is flat, and so R' vanishes identically. Using (11.31) together with (11.4) gives the Buchdahl EOS. □

Proof (of part C.I.2). We show that the requirements imply asymptotic conditions which, in essence, reduce the problem to C.I.1. We treat NT and GR simultaneously.

Observing from (11.31) that the Ricci scalar R' with respect to g' is a function of V, we first show that $I_1 \leq 0$ implies a lower bound for dR'/dV. To do so, we note first that the existence of Γ and $d\rho/dp > 0$ imply that $\lim_{p \to 0} \rho^{-1} p = 0$ (c.f. Appendix A of [8]). By a straightforward computation, using $I_1 \leq 0$, $\rho^{-1} I_0 \leq 0$ (which follows from Proposition 11.1 and from the previous remark), and (11.7) we get:

$$\begin{aligned}
\frac{(1-V)^6}{32\pi V^4 \rho\kappa} \frac{d}{dV} R' &= & \frac{V(1+V)^6}{256\pi(\rho+3p)\kappa} \frac{d}{dV} R' &= \\
&= 5\kappa^{-1} + 1 - V^{-1} & &= V^2(10\kappa^{-1} + 1) - 1 \\
&\geq \frac{6p}{\rho} + 1 - V^{-1} & &\geq 12V^2 \left[3\frac{p^2}{\rho^2} + \frac{p}{\rho} + 1\right] - 1 \qquad (11.39) \\
&\geq 1 - V^{-1} + \Gamma \geq 1 - V_s^{-1} & &\geq V^2 e^{2\Gamma} - 1 \geq V_s^2 - 1.
\end{aligned}$$

We now assume that the fluid extends to infinity, (i.e. $V_s = 1$), which will lead to a contradiction. It follows from (11.39) that $dR'/dV > 0$. Restricting ourselves to spherical symmetry we have $\widehat{W}_0 = 0$ at the centre by definition, and $\widehat{\widehat{W}} = 0$ since V takes its minimum there. Applying the maximum principle to (11.37) on any ball $\mathcal{B}_r$ (bounded by r = const.) we obtain $\widehat{W} \geq \widehat{W}_0$ on $\mathcal{B}_r$. Next, since the integrand (11.36) is non-negative, $W_0(V)$ increases monotonically with V (and with r); in particular there is a constant $C > 0$ such that $\widehat{W} \geq \widehat{W}_0 \geq C^2$.

Below we give the proof of the relativistic case explicitly. (The Newtonian case can easily be obtained by suitable simplifications).

We introduce coordinates such that

$$ds^2 = g_{ij}dx^i dx^j = (1 - 2r^{-1}m(r))^{-1}dr^2 + r^2(d\theta^2 + \sin^2\theta d\phi^2), \tag{11.40}$$

where $m(r) = 4\pi \int_0^r r'^2 \rho(r')dr'$. Writing out the bound $\widehat{W} \geq C^2$ explicitly we obtain, since $g^{rr} < 1$,

$$\frac{d}{dr}\frac{1}{1-V} = \frac{1}{(1-V)^2}\frac{dV}{dr} \geq \frac{\sqrt{g^{rr}}}{(1-V^2)^2}\frac{dV}{dr} = \sqrt{\widehat{W}} \geq C, \tag{11.41}$$

which implies that

$$1 - V \leq \frac{1}{Cr}. \tag{11.42}$$

The crucial step is now to get an upper bound for $r^2 dV/dr$. Writing (11.2) as

$$\frac{d}{dr}\left(r^2\sqrt{g^{rr}}\frac{dV}{dr}\right) = 4\pi V(\rho + 3p)r^2\sqrt{g_{rr}}, \tag{11.43}$$

and noting that the r.h. side is non-negative for $p > 0$, we find that dV/dr is positive and that $r^2\sqrt{W}$ is strictly monotonically increasing. (Note that, from Theorem 2 of [18], g^{rr} does not go to zero for finite r. This argument also holds under the present differentiability assumptions). We now show by an indirect argument that $r^2\sqrt{W}$ is bounded as $r \to \infty$. Assuming the contrary, namely that

$$\forall\ D > 0 \ \text{ and } \ \forall\ r_1\ \exists\ r_0 > r_1 \quad \text{such that} \ \ r^2\sqrt{W}|_{r_0} > D, \tag{11.44}$$

we get from monotonicity that

$$\forall\ D > 0\ \exists\ r_0 \ \text{ such that } \ \forall\ r > r_0 \quad r^2\sqrt{W} > D, \tag{11.45}$$

and, since $g^{rr} < 1$, we also have $r^2 dV/dr > D$ for sufficiently large r. Integrating the latter relation between some $\widehat{r}$ and infinity we get a contradiction to (11.42). Therefore, $r^2\sqrt{W}$ is bounded.

Using next the divergence theorem and $V_c \leq V$ (where V_c is the value of V at the centre) we obtain:

$$\begin{aligned} m(r) &= \int_0^r \rho r'^2 dr' \leq \int_0^r \rho\sqrt{g_{rr}}r'^2 dr' \leq V_c^{-1}\int_0^r V(\rho + 3p)\sqrt{g_{rr}}r'^2 dr' = \\ &= V_c^{-1}\int_{r=const.} r^2\sqrt{W}d\omega < \infty. \end{aligned} \tag{11.46}$$

Therefore, the ADM-mass $M = \lim_{r\to\infty} m(r)$ exists, $g^{rr} \to 1$, and the limit

$$\lim_{r\to\infty} r(1-V) = \lim_{r\to\infty} r^2\frac{dV}{dr} = \lim_{r\to\infty} r^2\sqrt{W} = N \tag{11.47}$$

exists as well (and is called the "Komar mass"). Moreover, we note that $\rho \in L^1_{0,-3}$ and, from (11.4), that $p \in L^1_{1,-4}$. Finally, using (11.2) or (11.43) again, we obtain:

$$\frac{dW}{dr} = -\frac{4}{r}W + 8\pi V(\rho + 3p)\sqrt{g_{rr}W} \in L^1_{0,-5}. \tag{11.48}$$

While these properties are slightly different from the AFMD conditions as defined in Sect. 11.2., they still imply that the surface integral arising by applying the divergence theorem to (11.30) vanishes. To see this, we rewrite the latter expression as follows (using spherical symmetry and multiplying with r^2):

$$\frac{d}{dr}\left[\frac{r^2}{(1-V^2)^2}\left(V^{-1}\sqrt{g^{rr}}\frac{dW}{dr} + \frac{8W^{3/2}}{1-V^2} - 8\pi\sqrt{W}\left(\rho - \frac{1-5V}{1-V}p\right)\right)\right], \tag{11.49}$$

using the divergence theorem and inserting (11.47), $r^4 W \to N^2$, and (11.48). Recalling that $I_1 \le 0$ implies $I_0 \le 0$, the virial theorem (11.33) now applies and we get finite extent.

As remarked before, it also follows that $\widehat{g}$, $\widehat{W}$ and $\widehat{W_0}$ have C^1 extensions to the point at infinity Λ of a compactification $\widehat{\mathcal{M}} = \mathcal{M} \cup \Lambda$. Hence (instead of using the virial theorem in the form (11.33)), we could also apply the maximum principle to (11.37) to show finiteness. □

11.5 Discussion

Here we discuss further the results on finiteness in the light of what we call "quasipolytropes"

$$p = K\rho^{\frac{n+1}{n}}\left[1 + f\left(\rho^{\frac{1}{n}}\right)\rho^{\frac{1}{n}}\right] \tag{11.50}$$

where $K, n \in R^+$ and $f : [0, \lambda) \to R$ is a smooth function. Obviously, for $f \equiv 0$, (but in general also for small ρ), we are left with the polytropic EOS, while the limit $n = 0$ reduces to the incompressible case $\rho = const.$. These quasipolytropes seem to comprise all physically interesting EOS, in NT as well as in GR.

Let us first compare the polytropes in NT with the "Generalized Buchdahl EOS" in GR obtained from (11.50) with the choice $1/K = (n+1)\rho_-^{1/n}$ and $f(\rho^{1/n}) = (\rho_-^{1/n} - \rho^{1/n})^{-1}$ for some constant ρ_-, and for $\rho < \rho_-$, viz.

$$p = \frac{1}{n+1}\rho_-^{-\frac{1}{n}}\rho^{\frac{n+1}{n}}, \qquad p = \frac{1}{n+1}\rho^{\frac{n+1}{n}}(\rho_-^{\frac{1}{n}} - \rho^{\frac{1}{n}})^{-1}. \tag{11.51}$$

For the Newtonian polytropes we now find that $I_0 = (n-5)p$ and $I_1 = \rho(n-5)/(n+1)$. In GR, the integral (11.6) for Γ is still elementary but the expressions for I_0 and I_1 are rather involved. Nevertheless, they still have the appealing property that the quantities I_0 and I_1 are negative iff $n < 5$, zero for $n = 5$, (this is the "Buchdahl"- case considered above, c.f. (11.1),(11.12)) and positive iff $n > 5$. Hence the (in)finiteness properties of all static solutions (satisfying also the other requirements of C.I and C.II) depend only on the EOS and are obvious from the theorem.

While in NT the polytropes are also (for special values of n) adiabates and therefore physically significant, the adiabates in GR are neither polytropes nor do they coincide with (11.51). Instead, comparing again with the Newtonian polytropes, they read: [7]

$$\rho = (n+1)^{\frac{n}{n+1}} \rho_-^{\frac{1}{n+1}} p^{\frac{n}{n+1}}, \qquad \rho = (n+1)^{\frac{n}{n+1}} \rho_-^{\frac{1}{n+1}} p^{\frac{n}{n+1}} + np. \quad (11.52)$$

The integral (11.6) for Γ is again elementary in the GR case of (11.52). We get $I_0 \le 0$ (and hence finiteness for AFMD solutions) if

$$n \le 5, \qquad p \le \frac{\rho_-}{n+1}\left(\frac{5}{n} - 1\right)^{n+1}. \quad (11.53)$$

Thus adiabatic, asymptotically flat solutions whose pressure nowhere exceeds (11.53) are finite. The condition $I_1 \le 0$, which guarantees finiteness for spherically symmetric solutions, leads to a (smaller) upper bound on p which can also be computed explicitly.

We now consider a fixed EOS in the general class (11.50) and the corresponding one-parameter family of static, spherically symmetric solutions. Rendall and Schmidt have shown [18] that, for $1 < n < 5$ the fluid region is finite provided that the central density ρ_c is below a critical value ρ_{crit}, while for $n > 5$ the fluid is always infinite (with infinite mass). More recently, Makino has proven that the fluid is always finite for $1 < n < 3$. On the other hand, our theorem implies that finiteness for small ρ_c is guaranteed in the range $0 < n < 5$, and allows us to estimate ρ_{crit}. Now the interesting remaining question is what happens for $0 < n < 5$ if the central density is increased. For polytropes in GR, this question has been investigated by Nilsson and Uggla [15] using a combination of dynamical system techniques with numerical ones. They find that all solutions are finite for $n < 3.339$. For larger values they observe, somewhat surprisingly, a *discrete* set of central densities with infinite configurations, some of which have finite but others infinite mass. The mathematical side of these phenomena still deserves to be understood.

We finally remark that interesting results on finiteness have also been obtained for Vlasov-Poisson and Vlasov-Einstein systems by Rein and Rendall [17].

Acknowledgments

I am indebted to J. M. Heinzle for carefully reading the draft, pointing out errors and suggesting improvements. I am also grateful to R. Beig, P. Bizoň, A. Rendall, R. Steinbauer and C. Uggla for helpful discussions and correspondence. This work was supported by Jubiläumsfonds der Österreichischen Nationalbank (Projekt 7942), and by Max-Planck-Institut für Gravitationsphysik.

References

1. R. Bartnik, *The Mass of an Asymptotically Flat Manifold*, Comm. Pure Appl. Math. **39**, 661–693 (1986).
2. R. Beig, B. G. Schmidt, *Time-Independent Gravitational Fields*, in: B.G. Schmidt (ed.) *Einstein's Field Equation and Their Physical Applications*, Lecture Notes in Physics **540**, Springer, Berlin (2000).
3. R. Beig, W. Simon, *On the Uniqueness of Static Perfect-Fluid Solutions in General Relativity* Commun. Math. Phys. **144**, 373–390, (1992).
4. R. Beig, W. Simon, *On the Spherical Symmetry of Static Perfect Fluids in General Relativity* Lett. Math. Phys. **21**, 245–250, (1991).
5. S. Chandrasekhar, *An Introduction to the Study of Stellar Structure*, Dover Publications, New York (1967).
6. H. A. Buchdahl, *A Relativistic Fluid Sphere Resembling the Emden Polytrope of Index 5*, Astrophys. J. **140**, 1512–1516, (1964).
7. M. Demianski, *Relativistic Astrophysics* (Sect. 4.3). Pergamon Press, Oxford, (1985).
8. J. M. Heinzle *(In)finiteness of Spherically Symmetric Perfect Fluids*, gr-qc/0204034; Class. Quantum Grav. (2002) (to appear).
9. J. M. Lee, T. H. Parker, *The Yamabe Problem*, Bull. Am. Math. Soc. **17**, 37–91, (1987).
10. L. Lichtenstein, *Über einige Eigenschaften der Gleichgewichtsfiguren rotierender Flüssigkeiten, deren Teilchen einander nach dem Newton'schen Gesetze anziehen*, Math. Z. **28**, 635–640, (1928).
11. L. Lindblom, A. K. M. Masood-ul-Alam, *Limits on the adiabatic index in static stellar models*, in: B. L. Hu and T. Jacobson (eds.) *Directions in general relativity*, Cambridge University Press, Cambridge, (1993).
12. L. Lindblom, A. K. M. Masood-ul-Alam, *On the Spherical Symmetry of Static Stellar Models*, Commun. Math. Phys. **162**, 123–145, (1994).
13. T. Makino *On spherically symmetric stellar models in general relativity* Journ. Math. Kyoto Univ. **38**, 55–69, (1998).
14. A. K. M. Masood-ul-Alam, *The Topology of Asymptotically Euclidean Static Perfect Fluid Space-Time*, Commun. Math. Phys. **108**, 193–211, (1987).
15. U. S. Nilsson, C. Uggla, *General Relativistic Stars: Polytropic Equations of State*, Ann. of Physics **286**, 292–319, (2000).
16. S. Pohozaev, *Eigenfunctions of the Equation $\Delta u + \lambda f(u) = 0$*, Sov. Math. Dokl. **6**, 1408–1411, (1965).
17. G. Rein, A. D. Rendall, *Compact support of spherically symmetric equilibria in non-relativistic and relativistic galactic dynamics*, Math. Proc. Cambridge Phil. Soc. **128**, 363–380, (2000).
18. A. D. Rendall, B. G. Schmidt, *Existence and properties of spherically symmetric static fluid bodies with a given equation of state*, Class. Quantum Grav. **8**, 985–1000, (1991).
19. U. Schaudt *On Static Stars in Newtonian Gravity and Lane-Emden Type Equations*, Ann. Henri Poincaré **5**, 945–976, (2000).
20. R. Schoen, S-T. Yau, *Proof of the positive mass theorem. II*, Commun.Math.Phys. **79**, 231–260, (1981).
21. W. Simon, *A Class of Static Perfect Fluid Solutions* Gen. Rel. Grav. **26**, 97–101, (1994).
22. W. Simon, *On journeys to the Moon by balloon*, Class. Quantum Grav. **10**, 177–181 (1993).

12 Problems and Successes in the Numerical Approach to the Conformal Field Equations

Sascha Husa

Max-Planck-Institut für Gravitationsphysik, 14476 Golm, Germany

Abstract. This talk reports on the status of an approach to the numerical study of isolated systems with the conformal field equations. We first describe the algorithms used in a code which has been developed at the AEI in the last years, and discuss a milestone result obtained by Hübner. Then we present more recent results as examples to sketch the problems we face in the conformal approach to numerical relativity and outline a possible roadmap toward making this approach a practical tool.

12.1 Introduction

Since the early work of Hahn and Lindquist [1], Smarr [2] and Eppley [3], numerical relativity has become an important subfield of the theory of gravitation. To outsiders the progress often seems marginal and unsatisfactory. The classic goal of providing waveform catalogs for the newly built gravitational wave detectors has still not been reached (although considerable progress has been made recently [4]). By the nature of general relativity, the simulation of isolated systems poses particularly hard problems. Mathematically such systems can be formalized by the concept of asymptotically flat space-times (see, e.g., the standard textbook of Wald [5]), but it turns out that important quantities such as the total mass, (angular) momentum or emitted gravitational radiation can only consistently be defined at infinity. The traditional approach of introducing an arbitrary spatial cutoff introduces ambiguities and is not satisfactory at least from a mathematical point of view. A remedy is suggested by conformal compactification methods, such as the characteristic approach presented by Luis Lehner in this volume, or Friedrich's conformal field equations, which he describes in this volume. The latter approach avoids the problems associated with the appearance of caustics in the characteristic formulation by allowing to foliate the compactified metric by space-like hypersurfaces. These hypersurfaces are analogous to the standard hyperboloid in Minkowski space-time and are *asymptotically null* in the physical space-time. The price to pay is the loss of the simplicity inherent in the use of null coordinates, and one has to deal with the full complexity of 3+1 numerical relativity.

Some issues of the numerical solution of the conformal field equations have been laid out by Frauendiener in this volume, and in a *Living Review* [6], and he has also discussed his code to treat space-times with a hypersurface orthogonal Killing vector and toroidal $\mathscr{I}$'s. The purpose of the present article is to show the status of numerical simulations based on the conformal field equations in 3D – i.e., three space dimensions without assuming any continuous symmetries – and to discuss what is needed in order to render this approach a practical tool to investigate physically interesting space-times. By making future null infinity accessible to (completely regular and well defined) local computations, the approach excels at the extraction of radiation – e.g., the quantities to hopefully be measured within the next years by new large-scale detectors. One of the main pedagogical goals will be to explain the challenges of numerical relativity and to highlight some open problems related to constructing hyperboloidal initial data and actually carrying out long-time stable and accurate simulations. For an even more condensed account of the conformal approach to numerical relativity see [7].

The organization is as follows: Sect. 12.2 introduces the algorithms developed in the last years by Peter Hübner (the radiation extraction procedure, which I will only mention briefly, is based on work of Hübner and Marsha Weaver), and implemented in a set of codes by Peter Hübner (who has recently left the field). All results presented here have been obtained with these codes, which Hübner has described in a series of articles [8–11]. Sect. 12.3 will start with a brief description of the evolution of weak initial data which possess a regular point i^+ representing future time-like infinity, based on the work of Hübner [11]. Then I will discuss the evolution of slightly stronger initial data which exhibit various problems that will have to be solved, e.g. the choice of gauge, and use this as a starting point for discussing the main current problems. In Sect. 12.4 purely computational aspects of this project will be discussed, and in Sect. 12.5 I will sum up the current status and sketch a possible roadmap for further work.

12.2 Algorithms

12.2.1 Problem Overview

The conformal field equations are formulated in terms of an unphysical Lorentzian metric g_{ab} defined on an unphysical manifold $\mathcal{M}$ which gives rise to a physical metric $\tilde{g}_{ab} = \Omega^{-2} g_{ab}$, where the conformal factor Ω is determined by the equations. The physical manifold $\widetilde{\mathcal{M}}$ is then given by $\widetilde{\mathcal{M}} = \{p \in \mathcal{M} \mid \Omega(p) > 0\}$.

Contrary to the formalism used by Frauendiener, we use a metric based formulation of the conformal field equations:

$$\nabla_a \hat{R}_{bc} - \nabla_b \hat{R}_{ac} = -\frac{1}{12}\left((\nabla_a R)\, g_{bc} - (\nabla_b R)\, g_{ac}\right) - 2\,(\nabla_d \Omega)\, d_{abc}{}^d, \tag{12.1}$$

$$\nabla_d d_{abc}{}^d = 0, \tag{12.2}$$

$$\nabla_a \nabla_b \Omega = \frac{1}{4} g_{ab}\, \nabla^c \nabla_c \Omega - \frac{1}{2}\, \hat{R}_{ab}\, \Omega, \tag{12.3}$$

$$\frac{1}{4}\nabla_a \left(\nabla^b \nabla_b \Omega\right) = -\frac{1}{2}\, \hat{R}_{ab}\, \nabla^b \Omega - \frac{1}{24}\, \Omega\, \nabla_a R - \frac{1}{12}\, \nabla_a \Omega\, R, \tag{12.4}$$

$$R_{abc}{}^d = \Omega d_{abc}{}^d + \left(g_{c[a} \hat{R}_{b]}{}^d - g^d{}_{[a} \hat{R}_{b]c}\right) + \frac{R}{6}\, g_{c[a} g_{b]}{}^d, \tag{12.5}$$

$$6\,\Omega\, \nabla^a \nabla_a \Omega = 12\,(\nabla^a \Omega)\,(\nabla_a \Omega) - \Omega^2 R. \tag{12.6}$$

Here the Ricci scalar R of g_{ab} is considered a given function of the coordinates. For any solution $(g_{ab}, \hat{R}_{ab}, d_{abc}{}^{d}, \Omega)$, $\hat{R}_{ab}$ is the traceless part of the Ricci tensor, and $\Omega\, d_{abc}{}^{d}$ the Weyl tensor of g_{ab}. Note that the equations are regular even for $\Omega = 0$. These "conformal field equations" render possible studies of the global structure of space-times, e.g., reading off radiation at null infinity, by solving regular equations.

The 3+1 decomposition of the conformal geometry can be carried out as usual in general relativity, e.g.,

$$g_{ab} = h_{ab} - n_a n_b = \Omega^2(\tilde{h}_{ab} - \tilde{n}_a \tilde{n}_b),$$

where h_{ab} and $\tilde{h}_{ab}$ are the Riemannian 3-metrics induced by g_{ab} respectively $\tilde{g}_{ab}$ on a space-like hypersurface with unit normals n_a, and equivalently $\tilde{n}_a = \Omega n_a$ (our signature is $(-,+,+,+)$). The relation of the extrinsic curvatures ($\tilde{k}_{ab} = \frac{1}{2}\mathcal{L}_{\tilde{n}}\tilde{h}_{ab}$, $k_{ab} = \frac{1}{2}\mathcal{L}_n h_{ab}$) is then easily derived as $k_{ab} = \Omega(\tilde{k}_{ab} + \Omega_0 \tilde{h}_{ab})$, where $\Omega_0 = n^a \nabla_a \Omega$. Note that for regular components of h_{ab} and k_{ab}, the corresponding components of $\tilde{h}_{ab}$ and $\tilde{k}_{ab}$ with respect to the same coordinate system will in general diverge due to the compactification effect. However for the coordinate independent traces $k = h^{ab}k_{ab}$, $\tilde{k} = \tilde{h}^{ab}\tilde{k}_{ab}$ we get

$$\Omega k = (\tilde{k} + 3\Omega_0),$$

which can be assumed regular everywhere. Note that at $\mathscr{I}$, $\tilde{k} = -3\Omega_0$. Since $\mathscr{I}^+$ is an ingoing null surface (with $(\nabla_a\Omega)(\nabla^a\Omega) = 0$ but $\nabla_a\Omega \neq 0$), we have that $\Omega_0 < 0$ at $\mathscr{I}^+$. It follows that $\tilde{k} > 0$ at $\mathscr{I}^+$. We will thus call regular space-like hypersurfaces in $\mathcal{M}$ hyperboloidal hypersurfaces, since in $\widetilde{\mathcal{M}}$ they are analogous to the standard hyperboloids $t^2 - x^2 - y^2 - z^2 = 3/\tilde{k}^2$ in Minkowski space, which provide the standard example. Since such hypersurfaces cross $\mathscr{I}$ but are everywhere space-like in $\mathcal{M}$, they allow to access $\mathscr{I}$ and radiation quantities defined there by solving a Cauchy problem (in contrast to a characteristic initial value problem which utilizes a null surface slicing). Note that hyperboloidal hypersurfaces which cross $\mathscr{I}^+$ are only Cauchy surfaces for the *future* domain of dependence of the initial slice of $\widetilde{\mathcal{M}}$, we therefore call our studies *semi-global.*

We will not discuss the full $3+1$ equations here for brevity, but rather refer to [8]. What is important, is that the equations split into symmetric hyperbolic evolution equations plus constraints which are propagated by the evolution equations [8]. The evolution variables are h_{ab}, k_{ab}, the connection coefficients $\gamma^a{}_{bc}$, projections ${}^{(0,1):}\hat{R}_a = n^b h_a{}^c \hat{R}_{bc}$ and ${}^{(1,1):}\hat{R}_{ab} = h_a{}^c h_b{}^d \hat{R}_{bd}$ of the traceless 4-dimensional Ricci tensor $\hat{R}_{bd}$, the electric and magnetic components E_{ab}, B_{ab} of the rescaled Weyl tensor $d_{abc}{}^{d}$, as well as Ω, Ω_0, $\nabla_a\Omega$, $\nabla^a\nabla_a\Omega$ – in total this makes 57 quantities. In addition the gauge source functions q, R and N^a have to be specified, in order to guarantee symmetric hyperbolicity they are given as functions of the coordinates. Here q determines the lapse as $N = e^q\sqrt{\det h}$ and N^a is the shift vector. The Ricci scalar R can be thought of as implicitly steering the conformal factor Ω.

The numerical treatment of the constraints and evolution equations will be described below. But before, let us spend some time on general considerations about the treatment of null infinity. Since Ω is an evolution variable and not specified a priori, $\mathscr{I}$ will in general not be aligned with grid points, and interpolation has to be used to evaluate computed quantities at locations of vanishing Ω. For the physically interesting case of modeling an isolated system, "physical" $\mathscr{I}$ – i.e., the component of $\mathscr{I}$ that idealizes us outside observers and our gravitational wave detectors (neglecting cosmological effects such as redshift etc.) – has spherical topology. There may be more than one component of $\mathscr{I}$, i.e., additional spherical components associated with "topological black holes" (see Sect. 12.2.3). In principle it is possible of course to control the movement of $\mathscr{I}$ through the grid by the gauge choice – see [12] for how to achieve *$\mathscr{I}$–freezing* within Frauendiener's formulation, where $\mathscr{I}$ does not change its coordinate location.

What is the significance of how $\mathscr{I}$ moves through the grid? This question is directly related to the question for the global structure of space-time. Although many questions are left open, the present understanding of the global structure of generic vacuum space-times, which can be constructed from regular initial data, does provide some hints. First, note that space-times which are asymptotically flat in space-like and null directions – i.e., isolated systems – do not necessarily have to be asymptotically flat in time-like directions. An example would be a space-time that contains a star or a black hole. In such cases where the end state of the system is not flat space, we can not expect the conformal space-time $(\mathcal{M}, g_{ab})$ to contain a regular point i^+. In the case of sufficiently weak data however, Friedrich has shown in [13] that a regular point i^+ will exist – consistent with our intuition. The global structure is then similar to Minkowski space. The standard conformal compactification of Minkowski space is discussed in textbooks (see e.g. [5]) as a mapping to the Einstein static universe. There $\mathscr{I}$ moves inward and contracts to a point within finite coordinate time. In order to resolve such situations it seems most appropriate to choose a gauge which mimics this behavior, i.e., where $\mathscr{I}$ contracts to a point after finite coordinate time. The boundary of the computational domain is set in the unphysical region and the physical region contracts in coordinate space. Accordingly, the initial data are also extended beyond the physical region of space-time. It is this scenario which is best understood so far, and which is presented in some more detail in Sect. 12.3.

For sufficiently strong regular data it is known that singularities develop [14] – according to the cosmic censorship conjecture [15] such singularities should generically be hidden inside of black holes. For such data we cannot expect a regular point i^+ to exist. In the case when i^+ is singular (and not much else is currently known even about the i^+ of Kruskal space-time – see however Bernd Schmidt's contribution in this volume) we have to expect structure like sharp gradients near i^+, which makes it unlikely that we can afford to significantly reduce the size of the physical region in coordinate space (at least not without adaptive mesh refinement – a technology not yet available for 3D evolutions). A $\mathscr{I}$-freezing gauge may be appropriate for such a situation. Furthermore, phenomena like quasi-normal ring-down, or the orbital motion of a two-black hole system suggest that the numerical time-coordinate better be adapted to the intrinsic time scale of the system. Associated with quasi-normal ring-down for example is a fixed period in Bondi-time, which suggests Bondi time as a time coordinate near $\mathscr{I}$ in a situation dominated by ring-down. Thus, for black hole space-times it might turn out that the best choice of gauge fixes $\mathscr{I}$ to a particular coordinate position, and shifts i^+ into the infinite future [1]. It could be possible that in such a case the boundary can be chosen to either coincide with $\mathscr{I}$, or to be put just a small number of grid-points outside, which would raise the question for an evolution algorithm that does not require a topologically rectangular grid. We see that the optimal choice of numerical algorithms and gauges may be tightly related to the global structure of the investigated space-times – which is actually one of the main questions our simulations should be able to answer!

12.2.2 Construction of "Extended" Hyperboloidal Initial Data

Evolution of a solution to the Einstein equations starts with a solution to the constraints. The constraints of the conformal field equations (see Eq. (14) of Ref. [8]) are regular equations on the whole conformal space-time $(\mathcal{M}, g_{ab})$. However, they have not yet been cast into a standard type of PDE system, such as a system of elliptic PDEs [2]. One therefore resorts to a 3-step method [10]:

1. Obtain data for the Einstein equations: the first and second fundamental forms $\tilde{h}_{ab}$ and $\tilde{k}_{ab}$ induced on $\bar{\Sigma}$ by $\tilde{g}_{ab}$, corresponding in the compactified picture to h_{ab}, k_{ab} and Ω and Ω_0. This yields so-called " minimal data".
2. Complete the minimal data on $\bar{\Sigma}$ to data for *all* variables using the conformal constraints – *in principle* this is mere algebra and differentiation.
3. Extend the data from $\bar{\Sigma}$ to Σ in some ad hoc but sufficiently smooth and "well-behaved" way.

In order to simplify the first step, the implementation of the code is restricted to a subclass of hyperboloidal slices where initially $\tilde{k}_{ab}$ is pure trace, $\tilde{k}_{ab} = \frac{1}{3}\tilde{h}_{ab}\tilde{k}$. The momentum constraint $\tilde{\nabla}^b\tilde{k}_{ab} - \tilde{\nabla}_a\tilde{k} = 0$ then implies $\tilde{k} = \text{const.} \neq 0$. We always set $\tilde{k} > 0$. In order to reduce the Hamiltonian constraint

$$^{(3)}\tilde{R} + \tilde{k}^2 - \tilde{k}_{ab}\tilde{k}^{ab}$$

to *one* elliptic equation of second order, we use a modified Lichnerowicz ansatz

$$\tilde{h}_{ab} = \bar{\Omega}^{-2}\phi^4 h_{ab}$$

with *two* conformal factors $\bar{\Omega}$ and ϕ. The principal idea is to choose h_{ab} and $\bar{\Omega}$, and solve for ϕ, as we will describe now. First, the "boundary defining" function $\bar{\Omega}$ is chosen to vanish on a 2-surface $\mathcal{S}$ – the boundary of $\bar{\Sigma}$ and initial cut of $\mathscr{I}$

[1] Note however, that in the absence of results, we are left to speculation here.
[2] Work toward this goal is reported in this volume by Adrian Butscher.

– with non-vanishing gradient on $\mathcal{S}$. The topology of $\mathcal{S}$ is chosen as spherical for asymptotically Minkowski space-times. Then we choose h_{ab} to be a Riemannian metric on Σ, with the only restriction that the extrinsic 2-curvature induced by h_{ab} on $\mathcal{S}$ is pure trace, which is required as a smoothness condition [16]. With this ansatz $\tilde{h}_{ab}$ is singular at $\mathcal{S}$, indicating that $\mathcal{S}$ represents an infinity. The Hamiltonian constraint then reduces to the Yamabe equation for the conformal factor ϕ:

$$4\,\bar{\Omega}^2\nabla^a\nabla_a\phi - 4\,\bar{\Omega}\nabla^a\bar{\Omega}\nabla_a\phi - \left(\frac{1}{2}\,{}^{(3)}\!R\,\bar{\Omega}^2 + 2\bar{\Omega}\Delta\bar{\Omega} - 3\,\nabla^a\bar{\Omega}\nabla_a\bar{\Omega}\right)\phi = \frac{1}{3}\tilde{k}^2\phi^5.$$

This is a semi-linear elliptic equation – except at $\mathcal{S}$, where the principal part vanishes for a regular solution. This however determines the boundary values as

$$\phi^4 = 9\,\tilde{k}^{-2}\,(\nabla^a\bar{\Omega})\,(\nabla_a\bar{\Omega}). \tag{12.7}$$

Existence and uniqueness of a positive solution to the Yamabe equation and the corresponding existence and uniqueness of regular data for the conformal field equations using the approach outlined above have been proven by Andersson, Chruściel and Friedrich [16]. Solutions to the Yamabe equation – and thus minimal initial data – can either be taken from exact solutions or from numerical solutions of the Yamabe equation. Exact solutions which possess a $\mathscr{I}$ of spherical topology have been implemented for Minkowski space and for Kruskal space-time – see the contribution of Bernd Schmidt in this volume. These solutions are defined even outside of $\mathscr{I}$, and thus can directly be completed to initial data for all variables by using the conformal constraints.

If the Yamabe equation is solved numerically, the boundary has to be chosen at $\mathcal{S}$, the initial cut of $\mathscr{I}$, with boundary values satisfying Eq. (12.7). If the equation would be solved on a larger (more convenient Cartesian) grid, generic boundary conditions would cause the solution to lack sufficient differentiability at $\mathcal{S}$, see Hübner's discussion in [9]. This problem is due to the degeneracy of the Yamabe equation at $\mathcal{S}$. Unfortunately, this means that we have to solve an elliptic problem with *spherical boundary*. This problem is solved by combining the use of spherical coordinates with pseudo-spectral collocation methods. In pseudo-spectral methods the solution is expanded in (analytically known) basis functions – here a Fourier series for the angles and a Chebyshev series for the radial coordinate. For an introduction to pseudo-spectral methods see e.g. [17–19]. This allows to take care of coordinate singularities in a clean way, provided that all tensor components are computed with respect to a regular (e.g., Cartesian) basis and that no collocation points align with the coordinate singularities. Another significant advantage of spectral methods is their fast convergence: for smooth solutions they typically converge exponentially with resolution. The necessary conversions between the collocation and spectral representations are carried out as fast Fourier transformations with the FFTW library [20]. The nonlinearities are dealt with by a Newton iteration [21], the resulting linear equations are solved by an algebraic multigrid linear solver (the AMG library [22]).

The constraints needed to complete minimal initial data to data for all evolution variables split into two groups: those that require divisions by the conformal factor Ω to solve for the unknown variable, and those which do not. The latter do not cause any problems and can be solved without taking special care at $\Omega = 0$. The first group, needed to compute ${}^{(1,1)}\hat{R}$, E_{ab} and B_{ab}, however does require special numerical techniques to carry out the division, and furthermore it is not known if their solution outside of $\mathscr{I}$ actually allows solutions which are sufficiently smooth beyond $\mathscr{I}$. Thus, at least for these we have to find some ad-hoc extension. Note that in the case of analytical minimal data, the additional constraints are solved on the whole time evolution grid.

The simplest approach to the numerical division by Ω would be an implementation of l'Hospital's rule, however this leads to non smooth errors and consequently to a loss of convergence [9]. Instead Hübner [9] has developed a technique to replace a division $g = f/\Omega$ by solving an elliptic equation of the type (actually some additional terms added for technical reasons are omitted here for simplicity)

$$\nabla^a \nabla_a (\Omega^2 g - \Omega f) = 0$$

for g. For the boundary values $\Omega^2 g - \Omega f = 0$, the unique solution is $g = f/\Omega$. For technical details see [10]. The resulting linear elliptic equations for g are solved by the same numerical techniques as the Yamabe equation. For technical details see Hübner [10].

Finally, we have to extend the initial data to the full Cartesian spatial grid in some way. Since solving all constraints also outside of $\mathscr{I}$ will in general not be possible in a sufficiently smooth way, we have to find an ad hoc extension, which violates the constraints outside of $\mathscr{I}$ but is sufficiently well behaved to serve as initial data. The resulting constraint violation is not necessarily harmful for the evolution, since $\mathscr{I}$ causally disconnects the physical region from the region of constraint violation. On the numerical level, errors from the constraint violating region *will* in general propagate into the physical region, but if our scheme is consistent, these errors have to converge to zero with the convergence order of the numerical scheme (fourth order in our case). There may still be practical problems, that prevent us from reaching this aim, of course: making the ad-hoc extension well behaved is actually quite difficult, the initial constraint violation may trigger constraint violating modes in the equations, which take us away from the true solution, singularities may form in the unphysical region, etc.

Since the time evolution grid is Cartesian, its grid points will in general not coincide with the collocation points of the pseudo-spectral grid. Thus fast Fourier transformations can not be used for transformation to the time evolution grid. The current implementation instead uses standard discrete ("slow") Fourier transformations, which typically take up the major part of the computational effort of producing initial data.

It turns out, that the combined procedure works reasonably well for certain data sets. For other data sets the division by Ω is not yet solved in a satisfactory way, and constraint violations are of order unity for the highest available resolutions. In particular this concerns the constraint $\nabla_b E_a{}^b = -{}^{(3)}\epsilon_{abc} k^{bd} B_d{}^c$ (Eq. (14d) in [8]), since E_{ab} is computed last in the hierarchy of variables and requires two divisions by Ω. Further research is required to analyze the problems and either improve the current implementation or apply alternative algorithms. Ultimately, it seems desirable to change the algorithm of obtaining initial data to a method that solves the conformal constraints directly and therefore does not suffer from the current problems. This approach may of course introduce new problems like an elliptic system too large to be handled in practice.

12.2.3 Black Hole Initial Data

Since the standard definition of a black hole as the interior of an event horizon is a global concept, it is a priori not clear what one should consider as "black hole initial data". In practice, the singularity theorems [14] and the assumption of cosmic censorship [15] usually lead to the identification of "black hole initial data" with data that contain apparent horizons, and to associate the number of apparent horizons with the number of black holes in the initial data [3].

A common strategy to produce apparent horizons is to use topologically nontrivial data, that is data which possess more than one asymptotically flat region. In the time-symmetric case such data obviously possess a minimal surface! Asymptotic ends that extend to spatial infinity i^0 are relatively easy to produce by compactification methods, see e.g. [27–30] or the contribution of Dain in this volume. From the numerical point of view it is important that the topology of the computational grid is independent of the number of asymptotic regions or apparent horizons considered: suitable regularization procedures allow to treat spatial infinities as grid points.

In the current approach to the hyperboloidal initial value problem, where first the Yamabe equation needs to be solved, the grid topology *does depend* on the number of topological black holes – in this case the number of initial cuts of $\mathscr{I}$'s, which have spherical topology. One option would be of course to combine both ingredients and consider "mixed asymptotics" initial data, which extend to the physical $\mathscr{I}$ and to unphysical interior space-like infinities which only serve the purpose of acting as "topological sources" for apparent horizons.

Another option, suggested by Hübner in [10], is to generalize the current code for the initial data, which only allows for one cut of $\mathscr{I}$, which has spherical topology, to multiple $\mathscr{I}$'s of spherical topology. For the case of one black hole this would correspond to the relatively simple modification to $S^2 \times R$ topology. For the case of two black holes one could implement the Schwarz alternating procedure (as described in Sect. 6.4.1 of Ref. [24]) to treat three $\mathscr{I}$'s with three coordinate patches, where each patch is adapted to a spherical coordinate system with its $\mathscr{I}$. A more practical approach (at least to get started) could be to produce topologically trivial black hole initial data. Since we expect physical black holes to result from the collapse of topologically trivial regular initial data, such data would in some sense be more physical. Theorems on the existence of apparent horizons in Cauchy data have been presented by Beig and Ó Murchadha in [27]. Numerical studies in this spirit have been performed by the author [29]. Such

[3] Note that both the number of components of the event and apparent horizon are slice-dependent.

data could in principle be produced with the current code once it gets coupled to an apparent horizon finder. For the hyperboloidal initial value problem it is actually not known, whether such data actually exist, but it seems physically reasonable. Finding such data numerically by parameter studies would be an interesting result in itself. A natural question in this context is whether there is any qualitative difference between "topological" and "non-topological" black holes outside of the event horizon, e.g., regarding their waveforms?

12.2.4 Numerical Setup for Evolutions

The time evolution algorithm is an implementation of a standard fourth order method of lines (see e.g. [25]), with centered spatial differences and Runge-Kutta time integration. Additionally, a numerical dissipation term of the type discussed in theorems 6.7.1 and 6.7.2 of Gustafsson, Kreiss and Oliger [25] is added to the right-hand-sides to damp out high frequency oscillations and keep the code numerically stable. Numerical experiments show that usually small amounts if dissipation are sufficient (the dissipation term used contains a free parameter), and do not change the results in any significant manner

A particularly subtle part of the evolution usually is the boundary treatment. In the conformal approach we are in the situation that the boundary is actually situated outside of the physical region of the grid – this is one of its essential advantages! In typical explicit time evolution algorithms, such as our Runge-Kutta method of lines, the numerical propagation speed is actually larger than the speed of all the characteristics (in our case the speed of light). Thus $\mathscr{I}$ does *not* shield the physical region from the influence of the boundary – but this influence has to converge to zero with the convergence order of the algorithm – fourth order in our case. One therefore does not have to choose a "physical" boundary condition, the only requirements are stability and "practicality" – e.g., the boundary condition should avoid, if possible, the development of large gradients in the unphysical region to reduce the numerical "spill over" into the physical region, or even code crashes.

The current implementation relies on a "transition layer" in the unphysical region, which is used to transform the rescaled Einstein equations to trivial evolution equations, which are stable with a trivial copy operation at outermost grid-point as a boundary condition (see Ref. [8] for details and references). We thus modify the evolution equations according to replace

$$\partial_t f + A^i \partial_i f - b = 0 \quad \to \quad \partial_t f + \alpha(\Omega) \left(A^i \partial_i f - b\right) = 0,$$

where α is chosen as $\alpha(\Omega) = 0$ for $\Omega \leq \Omega_0 < \Omega_1 < 0$ and 1 for $\Omega \geq \Omega_1$. This procedure works reasonably well for weak data, however there are some open problems. One is, that the region of large constraint violations outside of $\mathscr{I}$ may trigger constraint violating modes of the equations that can grow exponentially. Another problem ist that a "thin" transition zone causes large gradients in the coefficients of the equations – thus eventually leading to large gradients in the solution, while a "thick" transition zone means to loose many grid-points. If no transition zone is used at all, and the Cartesian grid boundary touches $\mathscr{I}$, the ratio of the number of grid points in the physical region versus the number of grid points in the physical region is already $\pi/6 \approx 0.52$.

12.2.5 Physics Extraction

Extracting physics from a numerical solution to the Einstein equations is a non-trivial task. Results typically show a combination of physics and coordinate effects which are hard to disentangle, in particular in the absence of a background geometry or preferred coordinate system. In order to understand what is going on in a simulation, e.g., to find "hot spots" of inaccuracy or instability or bugs in an algorithm, it is often very important to visualize the "raw" data of a calculation. Here the visualization of scalar and in particular tensor fields in 3D is a subtle task in itself. But beyond that, one also wants ways to factor out coordinate effects in some way, and ideally access physical information directly.

One way commonly used to partially factor our coordinate effects is to look at curvature invariants, another possibility is to trace geodesics through space-time. In the current code this is done by concurrent integration of geodesics by means of the same 4th order Runge-Kutta scheme used already in the method of lines. Both null and time-like geodesics, as well as geodesics of the physical and rescaled metrics can be computed, and various quantities such as curvature invariants are computed by interpolation along the geodesics.

Particularly important are null geodesics propagating along $\mathscr{I}$, since they can be used to define a Bondi system and thus compute radiation quantities such as the Bondi mass or news. Note that the foliation of space-time chosen for evolution will in general *not* reproduce cuts of $\mathscr{I}$ of constant Bondi time. Hübner has therefore implemented post-processing algorithms (using the IDL programming language/software system) which construct slices of constant Bondi time in the data corresponding to the null geodesics propagating on $\mathscr{I}$ by interpolation (the algorithms are based on unpublished work of Hübner and Weaver). This evolution of geodesics is illustrated by Fig. 12.1, which shows three time-like geodesics originating with different initial velocities at the same point $(x_0, y_0, z_0) = (\frac{1}{2\sqrt{3}}, \frac{1}{2\sqrt{3}}, \frac{1}{2\sqrt{3}})$ meeting a generator of $\mathscr{I}$ at i^+.

12.3 Results for Weak Data

In this section I will discuss results of 3D calculations for initial data which evolve into a regular point i^+, and which thus could be called "weak data". Bernd Schmidt presents results for the Kruskal space-time in this volume (see also [26]). The initial conformal metric is chosen in Cartesian coordinates as

$$ds^2 = \left(1 + \frac{A}{3}\bar{\Omega}^2\left(x^2 + 2y^2\right)\right) dx^2 + dy^2 + dz^2. \tag{12.8}$$

The boundary defining function is chosen as $\bar{\Omega} = \left(1 - \left(x^2 + y^2 + z^2\right)\right)/2$, it is used to satisfy the smoothness condition for the conformal metric at $\mathscr{I}$.

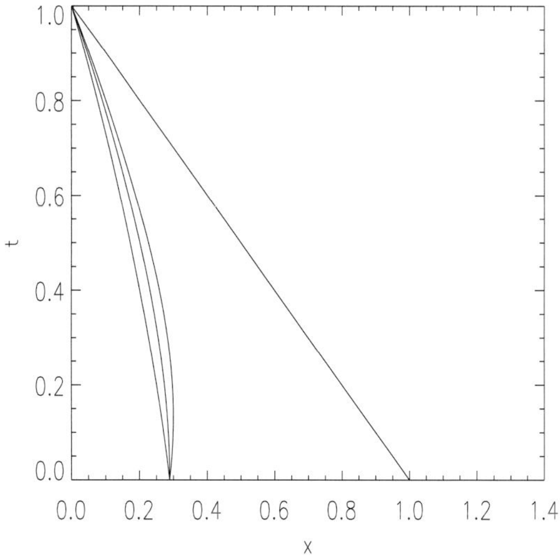

Fig. 12.1. Three time-like geodesics of different initial velocities (starting at $x = \frac{1}{2\sqrt{3}}$) meet a generator of $\mathscr{I}$ (starting at $x = 1$) at future time-like infinity i^+

These data have been evolved previously by Hübner for $A = 1$ as reported in [11]. For the gauge source functions he has made the "trivial" choice: $R = 0$, $N^a = 0$, $q = 0$, i.e., the conformal space-time has vanishing scalar curvature, the shift vanishes and the lapse is given by $N = e^q\sqrt{\det h} = \sqrt{\det h}$. This simplest choice of gauge is completely sufficient for $A = 1$ data, and has lead to a milestone result of the conformal approach – the evolution of weak data which evolve into a regular point i^+ of $\mathcal{M}$, which is resolved as a single grid cell. With this result Hübner has illustrated a theorem by Friedrich, who has shown that for sufficiently weak initial data there exists a regular point i^+ of $\mathcal{M}$ [13]. The complete future of (the physical part of) the initial slice can thus be reconstructed in a finite number of computational time steps. This calculation is an example of a situation for which the usage of the conformal field equations is ideally suited: main difficulties of the problem are directly addressed and solved by using the conformal field equations.

The natural next question to ask is: what happens if one increases the amplitude A? To answer this question, I have performed and analyzed runs for integer values of A up to $A = 20$. The results presented here have been produced with low resolutions of 80^3 (but for higher or slightly lower resolutions we essentially get the same results). For convergence tests of the code see [9–11]. While for $A = 1, 2$ the code continues beyond i^+ without problems, for all higher amplitudes the "trivial" gauge leads to code crashes before reaching i^+. Here by "code crash" we mean that computational values get undefined, e.g., the code produces "not a number" (NaN) values. While the physical data still decay quickly in time, a sharp peak of the lapse develops outside of $\mathscr{I}$ and crashes the code after Bondi time $\sim 8(320M)$ for $A = 3$ and $\sim 1.5(3M)$ for $A = 20$ (here M is the initial Bondi mass). In Figs. 12.2 and 12.3 the lapse N is plotted for runs with $A = 1$ and $A = 5$. While for $A = 1$ the lapse only shows significant growth after $t = 1$ (i^+ is located at $x = y = z = 0$, $t = 1$), for $A = 5$ a very sharp peak grows outside of $\mathscr{I}$ and crashes the code at $t = 0.9076$. Where does this rapid growth come from? Note that the initial conformal metric Eq. (12.8) shows significant growth outside of $\mathscr{I}$. Combined with the lapse $N = \sqrt{\det h}$ this leads to a growth of the lapse toward the grid boundary. In the present case a positive feedback effect with a growth of metric components in time seems to be responsible for the eventual crash of the code. Note that this feedback only takes place in a small region outside of $\mathscr{I}$ – further outward it is prevented by the transition to trivial evolution equations.

Fig. 12.4 shows a cure of the problem: a modified gauge source function $q = -r^2/a$ ($N = e^{-r^2/a}\sqrt{\det h}$) with $a = 1$ leads to a very smooth lapse (and correspondingly also to smooth metric components). Note that in Fig. 12.4, due to the different lapse, the point i^+ is *not* located at $t = 1$. The value of $a = 1$ here is found by moderate tuning of a to a best value (significantly decreasing or increasing a crashes the code before i^+ is reached). Unfortunately, this modification of the lapse is not sufficient to achieve much higher amplitudes. As A is increased, the parameter a requires more fine tuning, which was only achieved for $A \leq 8$. For higher amplitudes the code crashes with significant differences in the maximal and minimal Bondi time achieved, while the radiation still decays very rapidly and the news scales almost linearly. Furthermore, the curvature quantities do not show excessive growth – it is thus natural to assume that we are still in the weak-field regime, and the crash is not connected to the formation of an apparent horizon or singularity.

These results suggest that in order to model a gauge source function q that would allow to evolve up to i^+, one would need more that one parameter, e.g., at least 3 parameters for a non-isotropic ansatz such as $q = -(x^2/a^2+y^2/b^2+z^2/c^2)$ or something similar. To tune 3 or more parameters for each evolution seems however computationally prohibitive. While some improvement is obviously possible through simple non-trivial models for the lapse (or other gauge source functions), this approach seems very limited and more understanding will be necessary to find practicable gauges. An interesting line of research would be to follow the lines of Ref. [32] in order to find evolution equations for the gauge source functions which avoid the development of pathologies. A particular aim would be to find equations such that the resulting system of evolution equations is symmetric hyperbolic.

Fig. 12.5 shows the news function from three different runs: for $A = 1, 5$ and $q = 0$, and for $A = 5$, $q = -r^2/a$. The news from the runs for $A = 1$ have been multiplied by a factor of 25, which would exactly compensate scaling with the amplitude in the linear regime. We see that the three curves line up very well initially. The line for $A = 5$ and $q = 0$ deviates significantly to larger values of the news when the runs starts to get inaccurate, but at this time most of the physical radiation has already left the system. The curves from $A = 1/q = 0$ and $A = 5/q = -r^2/a$ line up perfectly until the value of the news drops below 10^{-6}, where the curves level off at different values, due to numerical inaccuracy.

Fig. 12.6 shows the Bondi mass for this situation, again the $A = 1$ curve is scaled by a factor of 25: Again we see the quick decay of a sharp pulse of radiation. There is no particular structure except falloff at late times, the deviation of the curves at late times seems to be caused by numerical inaccuracy, in particular in the computation of the Bondi mass.

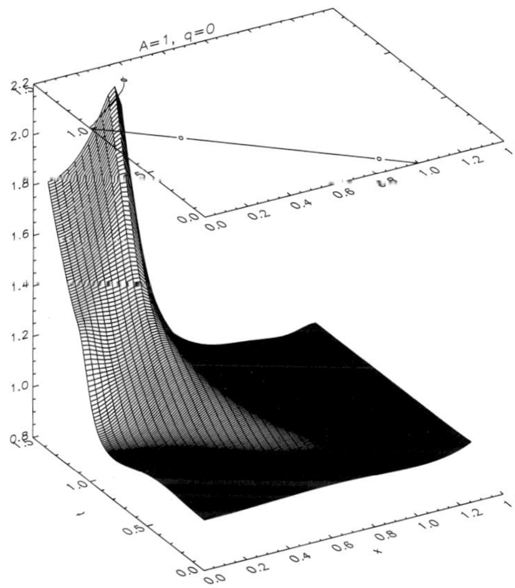

Fig. 12.2. Lapse N for the x-axis, for amplitude $A = 1$ and $R = N^a = q = 0$. The contour line marks $\mathscr{I}$ $(\Omega = 0)$

12.4 Computational Aspects

In this section I will give a brief description of some computational aspects, such as the computational resources needed to carry out simulations in 3 spatial dimensions. Computations of this scale rely on parallel processing, that means execution of our algorithms is spread over different CPU's. From a simplistic point of view there are two ways to program for parallel execution: we only take care of parallelizing the algorithm – but we require that all CPU's can access the same memory – or we both parallelize the algorithm and the data structures, and separate the total data into smaller chunks that fit into the local memory of each processor. The first alternative requires so called shared memory machines,

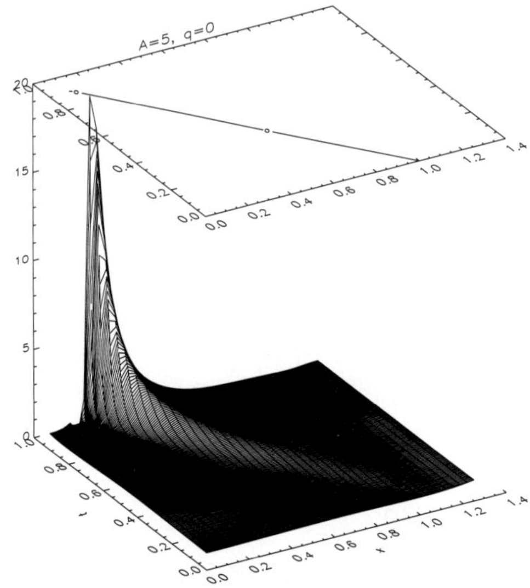

Fig. 12.3. Lapse N for the x-axis, for $A = 5$ and $R = N^a = q = 0$. The contour line marks $\mathscr{I}$ ($\Omega = 0$)

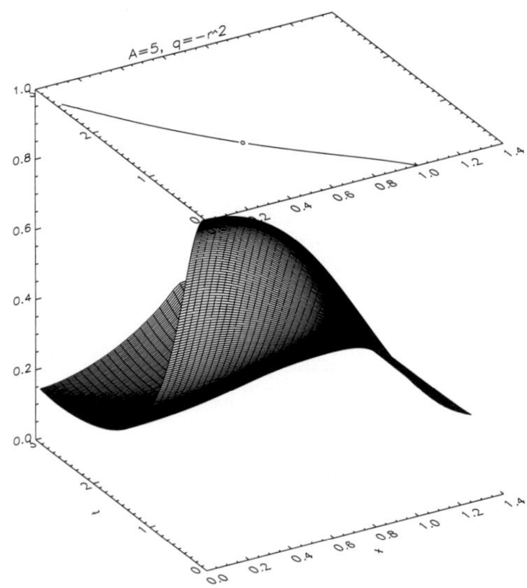

Fig. 12.4. Lapse N for the x-axis, for $A = 5$ and $R = N^a = 0$, $a = -r^2$. The contour line marks $\mathscr{I}$ ($\Omega = 0$)

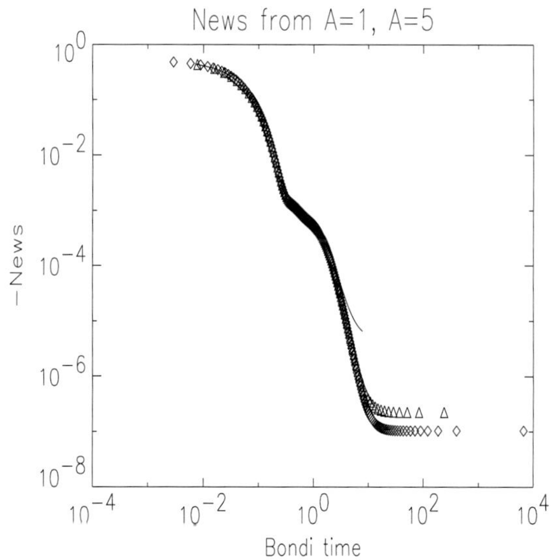

Fig. 12.5. News from three different runs: △ marks the results for $A = 1$, $q = 0$ (multiplied by 25 to compensate linear scaling), the solid line denotes $A = 5$, $q = 0$ and ◇ the run for $A = 5$, $q = -r^2/a$

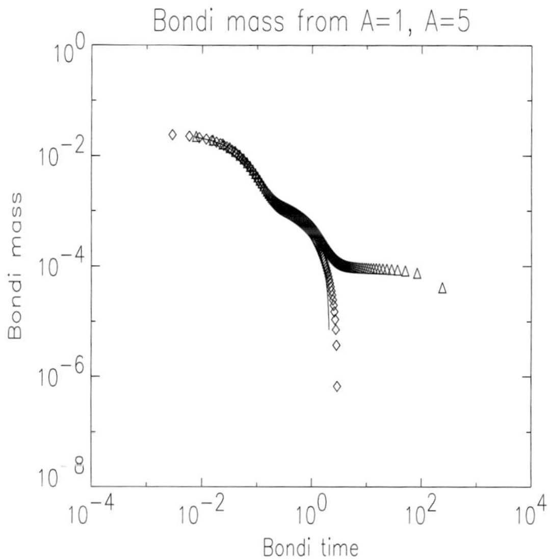

Fig. 12.6. Bondi mass from three different runs: △ marks the results for $A = 1$, $q = 0$ (multiplied by 25 to compensate linear scaling), the solid line denotes $A = 5$, $q = 0$ and ◇ the run for $A = 5$, $q = -r^2/a$

where the operating system and hardware take care of making data accessible to the CPU's consistently, taking care of several layers of main and cache memory (which gets increasingly difficult and expensive as the size of the machine is increased). The present code has been implemented using a shared memory programming model. The advantage is that this can generally be somewhat easier to program, and avoids overheads in memory. The disadvantage is the high cost of such systems, which makes them difficult to afford and thus nonstandard for most large academic parallel applications. The second alternative, usually referred to as distributed memory, requires more work to be done by the programmer, but more flexibly adapts to different kinds of machines such as clusters of cheap workstations commonly available in academic environments. While this approach usually implies a larger overhead in total memory requirements, speed and programming complexity, it is currently the only approach capable of scaling from small to very large simulations. For a general introduction to the issues of high performance computing, Ref. [37] provides a good starting point.

So, how much memory do we need? Let us assume a run with 150^3 grid points (the size of the largest simulations carried out with the present code so far). The current implementation of the fourth order Runge-Kutta algorithm uses 4 time levels and a minimum of 62 grid functions (57 variables and 5 gauge source functions). In double precision this amounts to ≈ 6.7 GByte. Temporary variables, information on geodesics and various overheads result in a typical increase of memory requirements by roughly 20%. For 150 time steps (approximately what it takes to reach i^+ for weak data) the total amount of processed data then corresponds to roughly 1 Terabyte!

If we half the grid spacing, the allocated memory increases by a factor of 8 (neglecting overheads), the total amount of processed data by a factor of 16, and the total required CPU time also by a factor of 16, while the error *reduces* by a factor of 16 – **if** we are already in the convergent regime! Given that the biggest academic shared memory machines in Germany have 16 GByte of memory available (the AEI's Origin 2000 and the Hitachi SR8000 at LRZ in Munich) this shows that the margin for increasing resolution is currently quite small. Such an increase in resolution will however be necessary to resolve physically interesting situations with more structure, such as a black hole, or two merging black holes. A move toward distributed memory processing will therefore be likely in the long run.

The current software-standard for distributed (scalable) computing is MPI (message passing interface) [40]. Unfortunately, writing large scale sophisticated codes in MPI is very time consuming. However, several software packages are available which introduce a software layer between the application programmer and MPI, and thus significantly reduce the effort to write parallel applications. Two prime examples are the Cactus Computational Toolkit [38] and PETSc [39]. While PETSc is a general purpose tool developed at Argonne National Laboratory as parallel framework for numerical computations, Cactus has been developed at the Albert Einstein Institute with numerical relativity in mind. While PETSc offers more support for numerical algorithms, in particular for parallel elliptic solvers, Cactus already contains some general numerical relativity functionality like apparent horizon finders – but no support for generic numerical algorithms. Apart from its numerical relativity flavor, the Cactus computational toolkit also has the advantage of broad support for parallel I/O and large scale 3D visualization. The ability to successfully mine tens or hundreds of gigabytes of data for relevant features is paramount to successful simulations in 3D.

Among the essential problems in writing and maintaining large scientific codes are the software engineering aspects and the control of complexity. In other words, codes should be reasonably documented and maintainable. For large scientific codes written and maintained by part-time-programmer scientists this poses a significant challenge. Writing a clear, modular code that can be understood, maintained and extended to suit new scientific needs requires a good deal of design and planning ahead. For an introduction to software engineering issues see e.g. [41, 42]. Another important issue for scientific codes is flexibility. Being able to do good science often depends on the ability to easily change algorithms, equations, discretization schemes etc. – without having to restructure the code, without high risk of introducing new bugs. In the present case examples for the need of trying different things would be experiments with different evolution equations (e.g., metric versus frame formalism), different boundary treatments or different elliptic solvers.

12.5 Discussion

The 3D numerical simulations performed so far show that the evolutions are *numerically* stable and quite robust. However, one of the main problems in numerical relativity is the stability of the constraint propagation: while the constraints *do* propagate when they are satisfied identically initially, this assumption does not hold for numerical simulations. On the contrary – it seems to be quite typical observation that the constraints diverge exponentially, if the evolution does not start at the constraint surface. Preliminary results exhibit this behavior of resolution-independent exponential growth associated with a violation of the constraints also for the conformal approach. One of the major goals for the future thus has to be the improvement of the understanding of the constraint propagation equations, and an according modification of the evolution equations (see Ref. [33] for previous work in this direction). This is essentially an analytical problem, but will certainly require the numerical testing of ideas.

Another area where new developments are necessary on the analytical side – along with numerical testing – is the problem of finding gauges that prevent pathologies like unnecessarily strong gradients. Ideally one would want to keep the symmetric hyperbolic character of the evolution system while allowing for a maximum of flexibility in writing down evolution equations for the gauge source functions.

Well-posedness of the evolution equations is important – but by far not sufficient for numerical purposes. While well-posedness unfortunately still has not yet been shown rigorously for many formulations used in numerical relativity, another important task seems to be to improve the understanding of the non-principal part of the equations, including their nonlinearities, in order to be able to construct numerically well-conditioned algorithms.

The third area where significant progress seems necessary on the analytical side is the construction of initial data. Problems with the current algorithm which necessitates divisions by zero and an ad-hoc extension beyond $\mathscr{I}$ have not yet been resolved. A possible road toward resolving these problems has been outlined by Butscher in this volume.

An important role in improving the analytical understanding and in setting up numerical experiments will be played by the utilization of simplifications. Particularly important are space-time symmetries and perturbative studies. A particularly interesting case to be studied actually is Minkowski space. Besides being an important case for code testing, it is used in current investigations to learn more about gauges and the stability of constraint propagation.

More complicated are general spherically symmetric space-times. In the vacuum case, this only leaves the Kruskal space-time aside from Minkowski space – but understanding the gauge problem for Kruskal space-time is an important milestone toward long-time black hole simulations. Moreover, spherical symmetry provides a natural testing ground for all kinds of new ideas, e.g., of how to treat the appearance of singularities, of how to treat the unphysical region, numerical methods, etc.

An alternative route to simplification, which has been very successful in numerical relativity, is perturbative analysis, e.g., with Minkowski or Schwarzschild backgrounds. In the context of compactification this has been carried out numerically with characteristic codes in [35, 36], some of the problems that showed up there are likely to be relevant also for the conformal approach.

What can we expect from the conformal approach in terms of physics results? Where can we expect contributions to our understanding of general relativity? One of the most important features of the conformal approach is that it excels at radiation extraction without ambiguities, and – at least in principle – enables numerical codes to study the global structure of space-times describing isolated systems. As has been demonstrated in this paper, in some weak field regime the code works well with relatively simple choices of gauge, and could be used to investigate some of the above problems. It could also provide a very clean way to study nonlinear deviations from linear predictions. For strong fields, in particular one or two black holes, the problem is much more difficult. An even more difficult problem is the investigation of the structure of singularities. In the spherically symmetric case this has been achieved by Hübner [34], but it is not clear whether these methods can be carried over to the generic case without symmetries, where the structure of the singularity has to be expected to be much more complicated.

What is the roadmap for the future? As far as 3D simulations are concerned, I believe that one should try to go from relatively well controlled weak data to stronger data and try to identify and solve problems as they come up. In parallel, it will be important to study simplified situations, like space-times with symmetries or linear perturbations with a mixture of analytical and numerical techniques. Both lines of research are hoped to improve our understanding of issues associated with choosing the gauge source functions, and controlling the growth of constraints. Future 3D codes, aimed at producing novel physical results will also require a significant effort devoted to "computational engineering", since flexible and solidly written codes are an absolute necessity for good computational science! These well known problems plaguing 3D numerical relativity will have to be addressed and solved in the conformal approach in order to harvest its benefits. Developing the conformal approach to numerical relativity into a mature tool poses an important challenge for mathematical relativity: not only is the problem hard and requires long-term investments, it also requires to merge sophisticated mathematical analysis with computational engineering. The aim is to produce a solid handle on exciting new physics – and some of the physics will even be accessible to experiments.

Acknowledgments

The author thanks H. Friedrich, B. Schmidt, M. Weaver and J. Frauendiener for helpful discussions and explanations of their work, C. Lechner and J. Thornburg for a careful reading of the manuscript, and P. Hübner for giving me access to his codes and results, and for support in the early stages of my work on this subject.

References

1. S. Hahn and R. Lindquist, *The Two–body Problem in Geometrodynamics*, Annals of Physics **29**, 304 (1964)
2. L. Smarr, *Spacetimes generated by computers: Black holes with gravitational radiation*, Ann. N. Y. Acad. of Sci. **302**, 569–604 (1977)
3. K. Eppley, *Evolution of time–symmetric gravitational waves: Initial data and apparent horizons*, Phys. Rev. D **16**, 1609 (1977)
4. J. Baker, B. Brügmann, M. Campanelli and C. O. Lousto, *Gravitational waves from black hole collisions via an eclectic approach*, Class. Quant. Grav. **17**, L149 (2000)
5. R. Wald, *General Relativity*, University of Chicago Press, Chicago, 1984
6. J. Frauendiener, *Conformal Infinity*, Living Rev. Rel. **4**, 1–90 (2000)
7. S. Husa, *Semiglobal Numerical Calculations of Asymptotically Minkowski Spacetimes*, in "Proceedings of the 20th Texas Symposium on Relativistic Astrophysics", ed. by J. C. Wheeler and H. Martel, American Institute of Physics, 2001
8. P. Hübner, *How to Avoid Artificial Boundaries in the Numerical Calculation of Black Hole Spacetimes*, Class. Quant. Grav. **16**, 2145–2164 (1999)
9. P. Hübner, *A Scheme to Numerically Evolve Data for the Conformal Einstein Equation*, Class. Quant. Grav. **16**, 2823–2843 (1999)
10. P. Hübner, *Numerical Calculation of Conformally Smooth Hyperboloidal Data*, Class. Quant. Grav. **18**, 1421–1440 (2001)
11. P. Hübner, *From now to timelike infinity on a finite grid*, Class. Quant. Grav. **18**, 1871–1884 (2001)

12. J. Frauendiener, *Numerical treatment of the hyperboloidal initial value problem for the vacuum Einstein equations. II. The evolution equations*, Phys. Rev. D **58**, 064003 (1998)
13. H. Friedrich, *On the existence of n-geodescially complete or future complete solutions of Einstein's field equations with smooth asymptotic structure*, Commun. Math. Phys. **107**, 587–609 (1986)
14. R. Penrose, *Asymptotic Properties of Fields and Space–Times*, Phys. Rev. Lett. **10**, 66–68 (1963)
15. R. Penrose, *Gravitational collapse: the role of general relativity*, Rev. del Nuovo Cimento **1**, 272–276 (1969); R. Penrose, *The Question of Cosmic Censorship*, in *Black Holes and Relativistic Stars*, ed. by R. Wald, (Chicago University Press, Chicago, 1988)
16. L. Andersson, P. T. Chruscíel, H. Friedrich, *On the regularity of solutions to the Yamabe equation and the existence of smooth hyperboloidal initial data to Einstein's field equations*, Comm. Math. Phys. **149**, 587 (1992)
17. B. Fornberg, *A Practical Guide to Pseudospectral Methods*, Cambridge University Press, Cambridge, 1996
18. C. Canuto, M. Hussaini, A. Quarteroni and T. Zang, *Spectral Methods in Fluid Dynamics*, Springer Verlag, New York, 1988
19. J. Boyd, *Chebyshev and Fourier Spectral Methods*, Springer Verlag, Berlin, 1989
20. M. Frigo and S. G. Johnson, *Fftw: An adaptive software architecture for the fft*, in *1998 ICASSP conference proceedings, Vol. 3*, ICASSP, 1998, p. 1381
21. J. Stoer and R. Bulirsch, *Introduction to Numerical Analysis. 2nd edition*, Springer, Berlin, 1996
22. J. W. Ruge and K. Stüben, *Algebraic multigrid*, in *Multigrid Methods*, ed. by S. F. McCormick, SIAM, Philadelphia, pp. 73–130, 1987; K. Stüben, *GMD report*, 1999
23. H. Friedrich, *Einstein's Equation and Geometric Asymptotics*, in *Proceedings of the GR-15 conference*, ed. by N. Dadhich and J. Narlikar, IUCAA, 1998
24. A. Quarteroni and A. Valli, *Numerical Approximation of Partial Differential Equations*, Springer Series in Computational Mathematics **23**, Springer, Berlin, 1997
25. B. Gustafsson, H.-O. Kreiss, and J. Oliger, *Time Dependent Problems and Difference Methods*, Pure and Applied Mathematics. Wiley, New York, 1995
26. B. Schmidt, *Numerical Evolution of the Kruskal spacetime using the conformal field equations*, in *Proceedings of the 20th Texas Symposium on Relativistic Astrophysics*, ed. by J. C. Wheeler and H. Martel, American Institute of Physics, 2001
27. R. Beig and N. Ó Murchadha, *Trapped Surfaces Due to Concentration of Gravitational Radiation*, Phys. Rev. Lett. **66**, 2421–2424 (1991)
28. R. Beig and S. Husa, *Initial Data for General Relativity with Toroidal Conformal Symmetry* Phys. Rev. D **50**, 7116–7118 (1994)
29. S. Husa, *Asymptotically Flat Initial Data for Gravitational Wave Spacetimes, Conformal Compactification and Conformal Symmetry*, PhD thesis, University of Vienna (1998)
30. S. Brandt and B. Brügmann, *A simple construction of initial data for multiple black holes*, Phys. Rev. Lett. **78**, 3606–3609 (1997)
31. S. Dain, *Initial data for two Kerr-like black holes*, Phys. Rev. Lett. **87**, 121102 (2001)
32. M. Alcubierre and B. Brügmann, *Simple excision of a black hole in 3+1 numerical relativity*, Phys. Rev. D **63**, 104006 (2001)
33. O. Brodbeck, S. Frittelli, P. Hübner and O. Reula, *Einstein's Equations with Asymptotically Stable Constraint Propagation*, J. Math. Phys. **40**, 909–923 (1999);

M. Alcubierre, G. Allen, B. Bruegmann and E. Seidel, Wai-Mo Suen, *Towards an understanding of the stability properties of the 3+1 evolution equations in general relativity*, Phys. Rev. D **62**, 124011 (2000); G. Yoneda and H. Shinkai, *Hyperbolic formulations and numerical relativity II: Asymptotically constrained systems of the Einstein equations*, Class. Quant. Grav. **18**, 441–462 (2001); F. Siebel and P. Hübner, *On the Effect of Constraint Enforcement on the Quality of Numerical Solutions in General Relativity*, Phys. Rev. D **64**, 024021 (2001)

34. P. Hübner, *A Method for Calculating the Structure of (Singular) Spacetimes in the Large*, Phys. Rev. D **53**, 701–721 (1996)
35. M. Campanelli, R. Gomez, S. Husa, J. Winicour, Y. Zlochower, *The close limit from a null point of view: the advanced solution*, Phys. Rev. D **63**, 124013 (2001)
36. S. Husa, Y. Zlochower, R. Gomez, J. Winicour, *Retarded radiation from colliding black holes in the close limit*, Phys. Rev. D **65**, 084034 (2002)
37. K. Dowd, C. R. Severance, *High Performance Computing*, 2nd Ed., O'Reilly, Cambridge, 1998
38. The Cactus development team, http://www.cactuscode.org
39. S. Balay, W. D. Gropp, L. Curfman McInnes and B. F. Smith, *PETSc Users Manual*, Technical Report ANL-95/11, 2001; S. Balay, W. D. Gropp, L. Curfman McInnes and B. F. Smith, *Efficient Management of Parallelism in Object Oriented Numerical Software Libraries*, in Modern Software Tools in Scientific Computing, ed. by E. Arge and A. M. Bruaset and H. P. Langtangen, Birkhäuser Press, 1997; http://www.mcs.anl.gov/petsc
40. W. Gropp, E. Lusk and A. Skjellum, *Using MPI*, 2nd Ed., MIT Press, Cambridge, Massachusetts, 1999
41. F. P. Brooks Jr., *The Mythical Man-Month*, Addison-Wesley, Reading, 1995
42. S. McConnell, *Code Complete - A Practical Handbook of Software Construction*, Microsoft Press, Redmond, 1993

13 Some Aspects of the Numerical Treatment of the Conformal Field Equations

Jörg Frauendiener

Institut für Astronomie und Astrophysik, Universität Tübingen,
Auf der Morgenstelle 10, 72076 Tübingen, Germany

Abstract. This article discusses some of the numerical issues which become relevant in the numerical computation of initial data for the conformal field equations. In particular, the problem is addressed of how to obtain a solution of the Lichnérowicz–Yamabe–Equation which extends as smoothly as possible beyond the boundary. Another problem which is discussed is the division process by which some of the initial data have to be constructed.

13.1 Introduction

The 'simulation' of self-gravitating isolated systems is one of the most challenging and hopefully rewarding tasks in relativity today. The challenge is to develop a reliable, accurate and maintainable code. The possible results obtained with this code may have consequences for different areas of classical relativity such as the mathematical question of the cosmic censorship hypothesis on the one hand and the prediction of wave forms for gravitational wave detectors on the other hand. There exist various approaches towards the development of such a code (see [39] for a recent review). Their majority is based on (variations of) the standard ADM equations [4]. Some of these numerical investigations have yielded quite impressive results [6, 25].

One of the most successful approaches so far towards the simulation of isolated systems is based on the characteristic formulation of Einstein's equation as developed by Bondi [7], Sachs [42], Newman and Penrose [40]. This approach is presented by R. Bartnik and A. Norton [5], J.A. Font [15] and L. Lehner [38] in this volume.

The objective of this presentation is to discuss some aspects of the conformal approach to numerical relativity which is based on Friedrich's conformal field equations. The general conceptional and analytical background has been given by Friedrich in this volume [21] while the numerical efforts have been summarized in [20], see also Husa [36]. Thus, we can take the opportunity to restrict ourselves to a more detailed discussion of some of the numerical problems that we encounter when we try to solve the conformal field equations.

To a large extent the issues involved in the numerical treatment of the hyperboloidal initial value problem for the conformal field equations are the same as for any system of geometric PDEs which split into evolution and constraint equations. However, there are some crucial differences which are due to the fact that null infinity can be located on the grid which we want to discuss here.

Any numerical implementation of such a system of PDEs will have to face the following broad issues: (*i*) the construction of initial data, (*ii*) the design of a stable evolution scheme, (*iii*) the implementation of (meaningful) boundary conditions, (*iv*) the control of the constraint violation during the evolution and, finally, (*v*) the search for 'good' gauges. These subjects are by no means independent. In fact, the issue (*v*) is probably the most important one because it influences all the others. But it is also the most obscure one because the 'goodness' of a gauge can in general not be inferred without looking at actual evolutions. Hence the 'invention' of good gauges is more like an art than a science.

The design of a stable evolution scheme and the design of boundary conditions are to a large extent uncritical. The choice of boundary conditions is not as crucial for the conformal approach as it is for others. This is due to the fact that the boundary of the numerical grid is located in the unphysical region so that it is causally disconnected from the physical part of the grid. This means that the influence of the boundary cannot reach the interior of the space-time. Clearly, these statements refer only to the analytical time evolution while the discrete numerical evolution scheme will also propagate parasitic modes at higher speeds than the speed of light so that numerically there will be an influence of the boundary on the physical space-time. However, if we use a well-posed numerical formulation of the initial-boundary value problem then this influence should die out with the order of the scheme when we increase the accuracy.

This implies that we have 'only' to make sure that our boundary conditions are compatible with the numerical evolution scheme in the sense that the resulting scheme is well-posed. There is a large literature in the numerical analysis community on the various issues of boundary conditions (see e.g. [14, 28, 47]) which can be consulted for this purpose. The method of choice for the time evolution is a 'method of lines' with fourth-order discretization in space and a high order Runge-Kutta method to solve the ensuing coupled system of ODEs. The problem which arises in connection with the boundary conditions is that of complexity. A high-order numerical evolution scheme needs boundary conditions discretized at the same order for the overall scheme to stay at the intended order. This means that one has to take time derivatives at the boundary and express them using the evolution equations in terms of the spatial derivatives. This is a task which can get very complicated for such complex evolution equations as those obtained from the conformal field equations.

Another issue in relation to the boundary conditions is that they should be compatible with the constraints. This leads to the necessity to analyse the the mathematical initial-boundary value problem for the Einstein and conformal field equations. There is a recent mathematical result [24] which treats the standard Einstein equations but the extension to the conformal field equations is still lacking. There is also some related work on the initial-boundary value problem for linearized gravity [45, 46] and an implementation of constraint-preserving boundary conditions in spherical symmetry [10].

The problem of keeping the constraints satisfied during the numerical evolution is one which seems to plague most if not all the codes in numerical relativity. The issue here is that the numerical evolution tends to exponentially drive away the system from the constraint hypersurface although the analytical time evolution guarantees that the evolution is confined to the constraint surface for all times [27]. The question arises as to whether this exponential violation of the constraints is a purely numerical effect or whether it has a cause which can be understood analytically. This cause would have to lie in the way small deviations from zero in the constraints are propagated by the time evolution. Analytically this is regulated by the 'subsidiary system' of evolution equations for the constraints. Therefore, in order to shed some light onto this issue one should analyse the properties of that system in some detail.

There remains the question of finding initial data for the time evolution. The problems which arise in this area are specific to the conformal approach and hence we take the opportunity to expose them here in more detail.

13.2 The Andersson-Chruściel-Friedrich Procedure

The first step in any numerical simulation is to provide initial data for the evolution equations. In the conformal approach this means that we have to determine the fields on the initial hypersurface in such a way that they obey the conformal constraints which have been obtained from the conformal field equations by a standard 3 + 1-decomposition [20]. The structure of these equations is rather complicated and to this day they have resisted attempts to cast them into a form suitable for mathematical analysis or even to devise a reasonable numerical scheme for their solution. Recently, however, Butscher [9] has made some progress towards a reformulation of these equations. Ultimately, we should strive to obtain initial data directly from solving these conformal constraints.

So far we obtain initial data by an implementation of a procedure originally due to Andersson, Chruściel and Friedrich [3] and expanded on in [2]. The mathematical background and various results are described in detail by L. Andersson [1] in his contribution to the present volume. We will briefly describe the essential idea. The starting point for obtaining initial data in *unphysical* space-time (M, g_{ab}) with this procedure is a hyperboloidal hypersurface $\tilde{\Sigma}$ with induced metric $\tilde{h}_{ab}$ and extrinsic curvature $\tilde{k}_{ab}$ in *physical* space-time $(\tilde{M}, \tilde{g}_{ab})$.

For simplicity we assume that the extrinsic curvature of $\tilde{\Sigma}$ is pure trace, $\tilde{k}_{ab} = \frac{1}{3}\tilde{h}_{ab}\tilde{k}$, the more general case is treated in [2]. The vacuum momentum constraints imply that $\tilde{k}$ is constant and the vacuum Hamiltonian constraint further implies that the scalar curvature $\tilde{R}$ is also constant. Rescaling the induced metric by a constant factor we can assume that

$$\tilde{R} = -6 \tag{13.1}$$

which in turn implies $\tilde{k} = 3$ (recall that a hyperboloidal hypersurface is characterized by the fact that it has an asymptotically constant *negative* curvature).

The next step is to assume that the physical space-time admits a conformal extension so that we may regard $(\tilde{M}, \tilde{g}_{ab})$ as being embedded in the unphysical space-time (M, g_{ab}) where it admits a boundary $\mathscr{I}$. The two metrics are conformally equivalent

$$g_{ab} = \Omega^2 \tilde{g}_{ab} \tag{13.2}$$

with a conformal factor Ω which is positive on $\tilde{\Sigma}$ and which vanishes on $\mathscr{I}$. The hyperboloidal hypersurface $\tilde{\Sigma}$ extends as a space-like hypersurface Σ beyond $\mathscr{I}$ intersecting it transversally.

The embedding of Σ in M induces a (negative definite) metric h_{ab} and extrinsic curvature k_{ab} which are related on $\tilde{\Sigma}$ to the physical data by the following relations

$$h_{ab} = \Omega^2 \tilde{h}_{ab}, \quad \tilde{k}_{ab} = \frac{1}{\Omega}\left(k_{ab} - \frac{\sigma}{\Omega} h_{ab}\right). \tag{13.3}$$

Here, $\sigma = t^a \nabla_a \Omega$ is the normal derivative of Ω with respect to the (g)-normal vector t^a of Σ. This implies that the trace-free parts transform homogeneously so that if $\tilde{k}_{ab}$ is pure trace then also k_{ab} is pure trace, while for the traces we obtain $\tilde{k} = k\Omega - 3\sigma$.

Furthermore, the Ricci tensors of h_{ab} and $\tilde{h}_{ab}$ are related by the usual conformal transformation formula (∂_a is the Levi-Civita connection of h_{ab})

$$\tilde{R}_{ab} = R_{ab} - \frac{\partial_a \partial_b \Omega}{\Omega} - h_{ab}\frac{\Delta\Omega}{\Omega} + 2h_{ab}\frac{\partial_c \Omega}{\Omega}\frac{\partial^c \Omega}{\Omega} \tag{13.4}$$

from which we obtain the transformation of the scalar curvature

$$\tilde{R} = \Omega^2 R - 4\Omega\Delta\Omega + 6\,\partial_c\Omega\partial^c\Omega. \tag{13.5}$$

Thus, we can express the physical Hamiltonian constraint in terms of unphysical quantities as

$$\Omega^2 R - 4\Omega\Delta\Omega + 6\,\partial_c\Omega\partial^c\Omega = -6. \tag{13.6}$$

Recall that the only condition, apart from smoothness, imposed upon the conformal factor is its vanishing on $\mathscr{I}$. Thus, we may write $\Omega = \omega\phi^\alpha$ for some number α where ϕ is a strictly positive function on Σ so that ω vanishes on $\mathscr{I}$ to the same order as does Ω. In this way we 'decouple' the two essential properties of Ω: the localization of $\mathscr{I}$ on Σ, indicated by the vanishing of ω which we consider to be given once and for all and secondly its role in relating the physical and unphysical metrics which is taken over by ϕ.

Now it is easy to see that upon insertion of this expression for Ω into (13.6) one obtains an equation which contains up to second order derivatives of ϕ and, in particular, terms which are proportional to $\partial_c\phi\partial^c\phi$. Choosing $\alpha = -2$ one can eliminate these quadratic terms and arrive at the equation

$$8\omega^2\Delta\phi - 8\omega\partial_c\omega\partial^c\phi + \left[\omega^2 R - 4\omega\Delta\omega + 6\partial_c\omega\partial^c\omega\right]\phi = -6\phi^5. \tag{13.7}$$

Since ω is fixed we may regard this equation as determining ϕ and hence Ω. This equation could be called the conformal Lichnérowicz equation because it arises from the Hamiltonian constraint in a similar way as the Lichnérowicz equation except that it is formulated in the unphysical space-time. On the other hand, the equation determines a conformal factor which fixes a specific member from a given conformal class of metrics which has constant scalar curvature. In the mathematical literature this problem is known as the Yamabe problem. So we call (13.7) the *Lichnérowicz–Yamabe–Equation* (LYE). We will discuss this equation in more detail later on so that we point out only its most relevant property now: the equation is singular on $\mathscr{I}$ in the sense that for $\omega = 0$ all the terms containing derivatives of ϕ vanish (we assume here that ϕ and its derivatives remain bounded). There remains only an algebraic relation between the values of ϕ and some geometric quantities. The reason for this non-regularity is easy to understand. If the equation was regular on $\mathscr{I}$ then one would presumably be able to provide boundary data for ϕ. This, however, contradicts the physical picture of $\mathscr{I}$ being a universal construction in the sense that for any asymptotically flat space-time one obtains the 'same' $\mathscr{I}$. The degenerate character of the equation on $\mathscr{I}$ leads to the question whether and under what conditions there exist unique solutions. This has been treated in detail in [3] and we will return to this issue in Sect. 13.3.

In order to find initial data for the conformal field equations we may thus proceed as follows. We start out with some 3-dimensional manifold Σ carrying a (negative definite) Riemannian metric h_{ab}. Since the equation is conformally invariant it is really only the conformal class of the metric h_{ab} which is relevant here. We prescribe the *boundary function* ω whose sole purpose is to fix the topology of the space-time we are trying to simulate. The only condition to be satisfied by this function is that its zero-set be an embedded 2-dimensional submanifold of Σ (it need not be connected nor have spherical topology). Next we try to find a non-vanishing solution ϕ of (13.7) with the given data. This provides a conformal factor $\Omega = \omega\phi^{-2}$ which conformally rescales the metric h_{ab} to a metric $\tilde{h}_{ab}$ on $\tilde{\Sigma}$ (where $\Omega > 0$) with constant negative curvature $\tilde{R} = -6$. Imposing $k_{ab} \propto h_{ab}$ implies $\tilde{k}_{ab} \propto \tilde{h}_{ab}$ and so $\tilde{k} = const$. In fact, choosing for k any smooth function on Σ and defining $\sigma = \frac{1}{3}k\Omega - 1$ guarantees that $\tilde{k} = 3$ so that the physical vacuum constraints will be satisfied on $\tilde{\Sigma}$.

Once this conformal factor is determined we can continue to generate the other initial data, such as the Ricci- and Weyl tensors. In the present case with k_{ab} being pure trace the relevant fields are the trace-free part of the Ricci tensor and the electric part of the Weyl tensor. They are given by

$$\Phi_{ab} = -\frac{1}{\Omega}\left(\partial_a\partial_b\Omega - \frac{1}{3}h_{ab}\Delta\Omega\right), \tag{13.8}$$

$$E_{ab} = \frac{1}{\Omega}\left(\Phi_{ab} - R_{ab} + \frac{1}{3}h_{ab}R\right) \tag{13.9}$$

The crucial point here is that the computation of these tensors involves *division by Ω*. This raises two questions: under what conditions are these expressions well defined and how can we compute them *accurately* numerically. The first question has been answered in [3] while the second is the subject of Sect. 13.4.

13.3 The Lichnérowicz–Yamabe–Equation

The degeneracy on $\mathscr{I}$ of the Lichnérowicz–Yamabe–Equation (13.7) poses special problems for its numerical solution. This is due not so much to the solution process itself which can be handled quite nicely with the usual methods. The problems arise from the special circumstances prevailing in the numerical setup. The results in [1, 3] apply to the situation where Σ is a 3-dimensional manifold *with boundary given by* $\omega = 0$. In this situation there will be a unique bounded solution of (13.7) on Σ which extends smoothly to the boundary provided the data, i.e. the specified conformal metric, is smooth up to the boundary and satisfies a certain condition at the boundary whose physical meaning is the vanishing of the shear on $\mathscr{I}$.

In numerical applications this situation although desirable cannot usually be achieved. This has several reasons. In 3D simulations one normally uses Cartesian coordinates while for asymptotically flat space-times $\mathscr{I}$ has spherical 'cuts'. This means that the zero-set of ω cannot be aligned with the grid and in particular it cannot be the boundary of the grid. Therefore, there will be nodes of the grid which lie outside of the physical region, *beyond* $\mathscr{I}$. Even if one decided to adapt the coordinates to $\mathscr{I}$ so that $\omega = 0$ lies on a grid plane then this could be done only for at most two asymptotic regions like in the Kruskal space-time[1]. For a two black-hole space-time which has three asymptotic regions one of these would lie inside the grid and there would again be nodes of the grid which lie beyond $\mathscr{I}$ in the unphysical region. Finally, even in the case of one $\mathscr{I}$ and spherical coordinates one needs for numerical reasons a number of grid points beyond $\mathscr{I}$ in order to implement boundary conditions for the time evolution. Thus, we have to face the situation where the grid not only covers the physical space-time but extends beyond $\mathscr{I}$ into the unphysical region.

Therefore, we have to ask what to do at the regions outside the physical space-time. There are essentially two ways to deal with this situation. We can either solve the equation only in the physical region, determine all the initial data there and then extend these data as smoothly as possible onto the unphysical regions. This is the path followed by Hübner [32–35]. One major drawback of this possibility is that in general the conformal constraints will *not be satisfied* in the unphysical region because the extension does not respect them. Not only is this an esthetically unpleasing feature – after all, the unphysical region can be considered as an asymptotically flat space-time in its own right – but almost certainly it will also be counter-productive. Some features like e.g. radiation extraction and 'scri-freezing' [17, 18] rely on the constraints being satisfied *in a neighbourhood* of $\mathscr{I}$. Furthermore, it is quite likely that the fact that immediately beyond $\mathscr{I}$ the constraints fail could allow constraint violating modes to penetrate into the physical space-time triggering the above mentioned exponential growth of the constraints there (for a discussion of this phenomenon in the context of the conformal field equations see [36]). Of course, ideally $\mathscr{I}$ should act as a one-way membrane for the propagation permitting outward flow but inhibiting inward

[1] unless one would work with more than one grid

flow. However, this is true for the (exact) continuous time evolution while for the (approximate) discrete evolution this depends very much on the method used to evolve from one time level to the next. It is true that the influence of the outside region upon the inside should die away with the order of the evolution scheme employed but ultimately only numerical tests will settle this point.

So in order to avoid constraint violation in the unphysical region we are forced to solve the Lichnérowicz–Yamabe–Equation (LYE) on the entire numerical grid where $\mathscr{I}$ is not necessarily aligned with the grid boundary. However, this is not as straightforward as it may seem. As alluded to already above, the region outside $\mathscr{I}$ can be regarded as an asymptotically flat space-time in its own right. The two space-times do not really have any relationship except for the superficial property that they match on the same numerical grid along the 2-surface $\omega = 0$. This is reflected in the properties of the solution of the LYE. For the physical region we know that there is a unique bounded solution smooth up to $\mathscr{I}$ for which we cannot specify any data. In contrast, in the outside region we have to specify boundary data *on the grid boundary* in order to obtain a unique solution. This solution will also be smooth up to $\omega = 0$ but the fact that we can specify arbitrary data suggests that the two solutions will in general not match smoothly *across* $\mathscr{I}$. The physical reason behind this phenomenon is simple. We have two asymptotically flat space-times which are forced to match 'at infinity' in such a way that $\omega = 0$ is the $\mathscr{I}^+$ for one space-time and $\mathscr{I}^-$ for the other. By suitably rescaling the metrics we can achieve that the two $\mathscr{I}$'s are isomorphic but there is no reason why the radiation which leaves one space-time should be the same as the radiation entering the other. Or, put differently, the mass-aspect obtained from one side is not necessarily the same as the one obtained from the other side. This 'mismatch' manifests itself in a jump in the third derivative of ϕ. While we cannot influence the part of the solution inside $\mathscr{I}$ we can use the boundary data on the grid boundary to change the character of the space-time beyond $\mathscr{I}$. Thus, with general boundary data on the grid boundary we obtain a global solution which agrees with the unique solution inside $\mathscr{I}$, which is smooth for $\omega \neq 0$ but which has a jump in the third derivative at $\omega = 0$.

This discontinuity will be present also in the initial data computed from the conformal factor Ω and will propagate along $\mathscr{I}$ like a shock front moving through a fluid. This is of course not what we are aiming for. What we need are initial data which are smooth on the entire grid. To achieve this we need to find a 'preferred extension' beyond $\mathscr{I}$ of the unique solution from the inside. This is determined from the inside by the only requirement of smoothness. In the case of analytic data this is the unique analytic extension. The example of the Schwarzschild resp. Kruskal space-time given in the contribution by Schmidt [43] shows that in general we have to expect that there may be singularities of ϕ beyond $\mathscr{I}$ but, hopefully, these will not be immediately adjacent to null-infinity so that in those cases which are of interest to us i.e., where the unphysical region is suitably small we can hope that the function ϕ will be regular on the entire grid.

So what can we do to obtain a sufficiently smooth solution of the LYE across $\mathscr{I}$? Unfortunately, this is not so easy and until now not solved in a satisfactory way. The first task is to obtain an idea about the boundary data which are necessary for the smooth extension. Fortunately, this is straightforward. The degeneracy of (13.7) yields information about the behaviour of the solution on $\mathscr{I}$. Setting $\omega = 0$ in (13.7) gives

$$n_a n^a \doteq -\phi^4, \tag{13.10}$$

where '$\doteq$' means 'equality on $\mathscr{I}$' and where we have defined $n_a = \partial_a \omega$. Furthermore, assuming that ϕ is smooth we take a derivative of (13.7) and then put $\omega = 0$. This yields

$$-2\left(n_e n^e \delta_c^a + \tfrac{1}{3} n^a n_c\right) \partial_a \phi + \left(\partial_a n_c - \tfrac{1}{3} h_{ac} \partial^e n_e\right) n^a\, \phi \doteq 0. \tag{13.11}$$

Since the map

$$x_a \mapsto \left(n_e n^e \delta_c^a + \tfrac{1}{3} n^a n_c\right) x_a$$

can be inverted this equation and (13.10) allow us to determine the solution ϕ up to its first derivatives on $\mathscr{I}$. We can use this information to estimate the value of ϕ on the grid boundary. We arrange ω in such a way that the outer region can be foliated by the leaves $\omega = const.$ and introduce a vector field s^a which is transversal to them. This could be the vector field normal to this foliation but this is not necessary at this stage. We extend two arbitrary coordinates (ξ, η) defined on $\mathscr{I}$ to the outer region up to the boundary by requiring that they be constant along s^a. In this coordinate system the grid boundary is given by an equation of the form $\omega = \omega(\xi, \eta)$. Now we have $s^a \partial_a \phi = \partial \phi / \partial \omega$ and we can estimate the boundary data ϕ_B by

$$\phi_B(\omega(\xi, \eta), \xi, \eta) \approx \phi(0, \xi, \eta) + \frac{\partial \phi}{\partial \omega}(0, \xi, \eta) \cdot \omega(\xi, \eta). \tag{13.12}$$

Clearly, this estimate can be useful only in those cases where $\mathscr{I}$ is close to the grid boundary. Furthermore, it depends very much on the arbitrary vector field s^a which is used to set up the isomorphism between $\mathscr{I}$ and the grid boundary. There is an obvious choice for it: the vector field which is normal to $\mathscr{I}$ with respect to the freely specified trial metric. Another possibility is to choose a vector field which is singled out by the numerical setup and which is transverse to both $\mathscr{I}$ and the grid boundary.

With these estimated boundary data we can now proceed to solve the LYE on the entire grid. Numerically this does not pose a problem. For demonstration purposes we implemented a simple finite difference procedure to solve this boundary value problem using `Octave`[41]. Assuming axisymmetry we have the following ansatz for the free metric

$$h_{ab} = -f^2\left(dx^2 + dy^2\right) - g^2 r^2 d\phi^2, \tag{13.13}$$

where f and g are arbitrary functions of (x, y) and $r^2 = x^2 + y^2$. It is easy to convince oneself that the general axisymmetric hypersurface orthogonal metric in three dimensions contains two free functions. Therefore, the conformal class

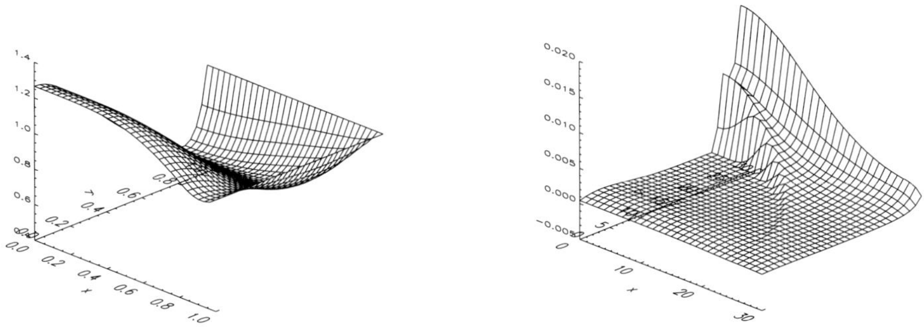

Fig. 13.1. Solution (left) of LYE and its third y differences (right) for fixed boundary values

of these metrics is characterized by only one free function. Strictly speaking the ansatz above is too general. We find it rather useful to include this redundancy because it allows us to test the code on exact solutions. In Fig. 13.1 is shown one example of a solution of the LYE where we have specified $\phi = 1$ on the boundary. There is nothing special about these boundary data except that they are positive. The solution is obtained on the square $[0,1] \times [0,1]$. The functions f, g are assumed even in both x and y. The boundary defining function is $\omega = r^2 - (4/5)^2$ so that $\mathscr{I}$ is at $r = 0.8$. The diagram on the left shows the solution itself, while the diagram on the right shows the third differences in the y-direction to illustrate the jump obtained in the third derivative of the solution.

In Fig. 13.2 is shown the solution of the LYE with the same free data except for the boundary data which have been obtained from the estimate above. It is clearly visible that the jump in the third derivative has been greatly reduced (note the different scales in the two plots) but is still present. As an illustration we

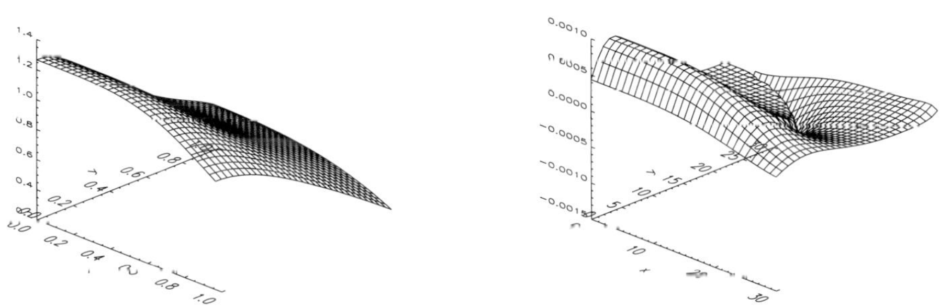

Fig. 13.2. Solution (left) of LYE and its third y differences (right) for estimated boundary values

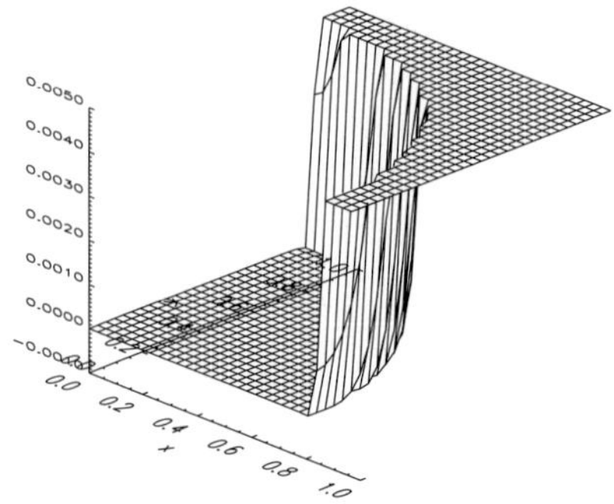

Fig. 13.3. Difference of the two solutions restricted to range [0, 0.005]

also show in Fig. 13.3 the difference of the two solutions. Here it is clearly visible how effectively the LYE 'shields' the inside from the outside. The difference of the two solutions in the inside is several orders of magnitude less than in the outside.

What can be done in order to find a smooth extension of the solution beyond $\mathscr{I}$? One procedure which comes to mind is the following: We start the solution process by estimating the boundary data and solve the LYE with these data. Once we are in convergence we switch from this 'inner' iteration to a different, 'outer' iteration. This time we discretize the equation up to the boundary in the sense that we not only impose the equation at inner points of the grid but also *on the boundary* by using one-sided differences. Then we do not need any boundary values. Instead of prescribing the boundary data in this iteration we fix the value of the solution at some point in the interior which we know from the previous iteration. Fixing an interior point is necessary in order to prohibit the solution to converge to zero which is a valid solution of the LYE. Once this 'outer' iteration has converged we switch back to the 'inner' iteration prescribing the boundary data we have obtained from the outer iteration. Repeating these steps several times one hopes to obtain an increasingly smooth solution. This method has not been fully tested and it remains to be seen to what extent it can yield useful results.

Another possibility to proceed is to use spectral methods. The reason for this is the fact that these methods are global on the grid and there is some hope that this globality helps to obtain a globally smooth solution. What follows is very much work in progress and so far there are no results in this direction. However, it is worthwhile to discuss the possible merits of this approach.

Spectral methods are described in detail in various monographs [8, 11, 16, 26, 48] so we can restrict ourselves to the essentials. The general idea is to use two representations of the grid variables f simultaneously. The first one is the grid representation where f is given by its values $f_l = f(\xi_l)$ at some grid points ξ_l while the second one is obtained by writing f as a finite sum $f(\xi) = \sum_{l=0}^{N} \hat{f}_l P_l(\xi)$

where the functions $P_l(\xi)$ belong to some set of 'basis functions'. The collections $\{f_i\}$, the coordinate representation, and $\{\hat{f}_i\}$, the generalized Fourier representation can both be used to represent the function f. The transformation from one set to the other can be performed by multiplication with the matrix $M^l{}_m = P_l(\xi_m)$ or its inverse. The basis functions are chosen according to the specifics of the problem like symmetry, boundary conditions etc. For some specific functions like the trigonometric and Chebyshev polynomials the transformation from one representation to the other can be speeded up by applying fast transforms like the FFT in various forms.

In the present problem we look for solutions of the LYE which are even on $[-1,1] \times [-1,1]$, restricted to the first quadrant $[0,1] \times [0,1]$. Thus we may conveniently choose the tensor basis $\{U_{lm}(x,y) := T_{2l}(x)T_{2m}(y) : 0 \leq l,m \leq N\}$ for some N. The functions T_m are the Chebyshev polynomials of degree m defined on $[-1,1]$ by the relation

$$T_m(\cos t) = \cos mt, \qquad \text{for} t \in [0,\pi]. \tag{13.14}$$

The LYE can be written formally as

$$L\phi = -6\phi^5 \tag{13.15}$$

where L is a linear differential operator with coefficients which vary with the coordinates (x,y). Inserting the ansatz

$$\phi(x,y) = \sum_{l,m=0}^{N} \hat{\phi}_{lm} U_{lm}(x,y) \tag{13.16}$$

into (13.15) we find that we need to evaluate the derivatives of the U_{lm} and multiplicative terms of the form $f(x,y)U_{lm}(x,y)$ for some functions $f(x,y)$. Here, the advantage of having both representations at hand is apparent. We can evaluate the multiplicative terms in the coordinate representation while the derivatives are efficiently and accurately evaluated in Fourier space because there exist simple recurrence relations between the coefficients of a function and those of its derivatives (see any of the references cited above). Having evaluated these terms and transformed the result to Fourier space we can regard the linear part of the Yamabe operator as a linear operator in Fourier space represented by a $2N \times 2N$-matrix $\hat{L}$ mapping the coefficients $\hat{\phi}_{lm}$ to the coefficients of $L\phi$.

The nonlinear equation is solved by Richardson iteration (see e.g. [37]). We compute an update $\delta\hat{\phi}$ from a given estimate $\hat{\phi}_{(n)}$ of the solution by solving the linear equation

$$\hat{L}\,\delta\hat{\phi} + 30\hat{\phi}^4_{(n)}\delta\hat{\phi} = -\left(\hat{L}\phi_{(n)} + 6\hat{\phi}^5_{(n)}\right) \tag{13.17}$$

for $\delta\hat{\phi}$ and use $\hat{\phi}_{(n+1)} = \hat{\phi}_{(n)} + \delta\hat{\phi}$ as an improved estimate. However, as it stands the matrix $\hat{L}$ is in general degenerate. When solving the equation using finite difference methods the same problem appears: the matrix of the discretized equation is singular. The remedy is, of course, to add boundary conditions to select a unique element in the kernel. But it is exactly these boundary conditions which are not known in our problem and the prescription of which leads to the discontinuity of the third derivative.

The same applies in Fourier space. In order to get a unique solution we need to prescribe additional conditions. One aspect of these additional conditions must be their non-locality in the coordinate representation i.e., we should not try to fix any particular values of the solution at grid points. Such conditions would translate into global conditions in Fourier space. This means that we should impose *local* conditions in Fourier space e.g. by fixing certain coefficients. This affects the solution as a whole.

A problem which arises is that in order to prescribe conditions in a reasonable way we should at least know *how many*. So far this is not quite clear. It is quite straightforward to convince oneself that the defect of (the spectral equivalent of) the Laplace operator obtained in the present setting is $N+1$ and not $2N+1$ as one would have thought on first sight. The reason is that in the expansion only *even* polynomials occur and so the operator maps even polynomials to even polynomials, effectively halving the number of degrees of freedom. However, $\hat{L}$ contains not the Laplace operator but that operator multiplied by ω together with various other terms and at least the vanishing of ω could enlarge the kernel. Numerical experiments show that this might be the case but they are inconclusive so far and a rigorous analysis is needed.

Even if we figure out how many and which conditions to impose we need to be careful in applying them. For, suppose we solve the linearized equation subject to some linear condition $B(\delta\hat{\phi}) = 0$ then the nonlinear solution will necessarily satisfy $B(\hat{\phi}) = B(\hat{\phi}_{(0)})$. Thus, the condition will be 'remembered' in the nonlinear solution and this implies that we have to impose conditions for the linearized equation without imposing (restrictive) conditions for the nonlinear equation. Candidates for such conditions are those which eliminate higher degree polynomials. The reason is that one expects an exponential decrease in size of the spectral coefficients of the solution so that these should not contribute too much.

An entirely different approach is based on a higher order approximation. Recall that the Richardson iteration scheme is obtained from the following consideration. Let N be a nonlinear operator (between finite dimensional spaces for our purposes) and assume we want to solve the (nonlinear) equation $N[\phi] = 0$. In order to derive the iteration procedure we obtain the equation for an update from an earlier estimate ϕ by putting $N[\phi + \delta\phi] = 0$ and linearizing with respect to the (presumably small) update $\delta\phi$. We get

$$N[\phi] + dN[\phi] \cdot \delta\phi = 0 \tag{13.18}$$

from which the Richardson scheme follows by solving for $\delta\phi$. Here, $dN[\phi]$ is the linearization of N around ϕ. Suppose that $dN[\phi]$ has a non-trivial finite kernel spanned by $(h_A)_{A=1,\ldots,n}$ for some n. Now we have the problem discussed above that we do not know what to do with the additional freedom in choosing the homogeneous solutions. However, we may take the expansion one step further

$$N[\phi + \delta\phi] \approx N[\phi] + dN[\phi] \cdot \delta\phi + \frac{1}{2} d^2 N[\phi] \left(\delta\phi, \delta\phi \right). \tag{13.19}$$

Suppose $\delta\phi$ is a solution of (13.18). Then so is $\delta\phi + \sum c_A h_A$. Inserting this into (13.19) we obtain

$$\begin{aligned} N[\phi + \delta\phi + \sum_A c_A h_A] = \frac{1}{2} d^2 N[\phi] \left(\delta\phi, \delta\phi \right) + \sum_A c_A d^2 N[\phi] \left(\delta\phi, h_A \right) \\ + \frac{1}{2} \sum_{A,B} c_A c_B d^2 N[\phi] \left(h_A, h_B \right). \end{aligned} \tag{13.20}$$

This equation can be considered as a quadratic equation for the coefficients c_A and we cannot in general expect that this will have a solution. However, we can expect to find a set c_A so that the right hand side of (13.20) is minimal.

The procedure then would be to

1. take an estimate ϕ of the solution
2. compute the linearization $dN[\phi]$ of the nonlinear operator N
3. determine a particular solution of (13.18)
4. determine the kernel of $dN[\phi]$
5. compute the second variation $d^2N[\phi]$ and find the set of c_A which minimizes (13.20)
6. finally put $\phi \leftarrow \phi + \sum_A c_A h_A$ and repeat from step 1. until convergence

Steps 3. and 4. can be performed by finding the *singular value decomposition* [49] of $dN[\phi]$. Step 5. is easy in the case where N is the discretized LYE because then the second variation of N is simply

$$(\delta_1\phi, \delta_2\phi) \mapsto 120\phi^3 \left(\delta_1\phi \right) \left(\delta_2\phi \right). \tag{13.21}$$

This procedure provides a unique update and therefore a unique solution without imposing any conditions 'by hand'. Therefore, one could hope that this will lead to a sufficiently smooth extension of a solution of the LYE onto the exterior region. However, numerics is full of surprises and it is never known whether a theoretically devised procedure performs satisfactorily in practice. Work on this approach has just begun and it remains to be seen to what extent it is useful.

13.4 Constructing Initial Data

Now we turn our attention to the next problem which arises in the construction of initial data for the conformal field equations. As we discussed above (cf. sect. 13.2) we have to obtain the initial data for the Ricci and Weyl tensor once the correct conformal factor which selects the metric with constant negative curvature from the given conformal class has been found. The problem we have to face is the fact that we need to divide two quantities which vanish somewhere. It is guaranteed from the analytical results that these quantities vanish simultaneously so that the quotient is well defined and (at least) continuous. However, numerically at least one of the quantities is affected by numerical error which implies that it will not vanish at those places where the other quantity has its zeros.

We write any one of the components in (13.8) or (13.9) as $q = C/\omega$ where C contains the specific component and the factor ϕ^2 which comes from the conformal factor Ω in the denominator. Then we know the denominator (the freely specified boundary describing function ω) exactly and its computation at one grid point involves only round-off error. On the other hand, the numerator C can be computed only up to a certain accuracy which determined essentially by the truncation error of the discretization of the LYE. So we have $C = \hat{C} + \epsilon$. We denote by $q = C/\omega$ resp. $\hat{q} = \hat{C}/\omega$ the computed and exact results of the division of the computed resp. exact numerator C resp. $\hat{C}$. Let $G = \{x_k\}$ denote the grid and its points on which we perform the computation. Then we have

$$|q(x_k) - \hat{q}(x_k)| = \frac{|\epsilon(x_k)|}{|\omega(x_k)|}. \tag{13.22}$$

The maximal deviation from the exact result occurs at the grid point where $|\omega(x_k)|$ attains its minimum. This can be arbitrarily small, even zero, if ω happens to vanish at a grid point. If $|\epsilon(x_k)|$ would be only due to round-off error as it would be if we computed the exact analytic solution then this division would not be disastrous (if we take care to prevent the vanishing of ω on any grid point). This is because division is a backward stable process in IEEE floating point arithmetic [29, 44, 49]. Then we divide two non-zero numbers which are 'simultaneously' small. The result is a well defined number which approximates the correct quotient within round-off accuracy.

The real disaster occurs when ϵ is due to the truncation error. Then $\epsilon(x)$ may be orders of magnitude different from the minimum value of ω so that the quotient will be wrong by orders of magnitude. In this situation the numerical implementation of l'Hôpital's rule to compute the indeterminate term 0/0 in one form or another is of no use. The reason is that this rule essentially replaces the differential quotient by a difference quotient. This provides a good approximation if one uses the exact expressions for the numerator. However, if the numerator had been previously computed with a truncation error then we will also get an answer which is off by orders of magnitude. To illustrate the problem let us consider a simple example (cf. Fig. 13.4). Take $\omega = 4/5 - x^2$ on $[-1, 1]$ and let q be some smooth function and define $C = q\omega$. Thus, C is known exactly and can be computed up to round-off errors. The diagram on the left shows the error $|C/\omega - q|$. Clearly, this is on the level of machine precision. The peaks are not related to the location of the zeros of ω indicated by the small triangles on the x-axis.

Next we 'spoil' C by a random distribution of numbers with maximal size below 10^{-10} to mimic the effect of the truncation error (although truncation error is certainly not random) and compute again the error. The result is shown in the diagram on the right. Here, we see that the overall error is on the level of the random perturbation. At the location of the zeros the error is maximal.

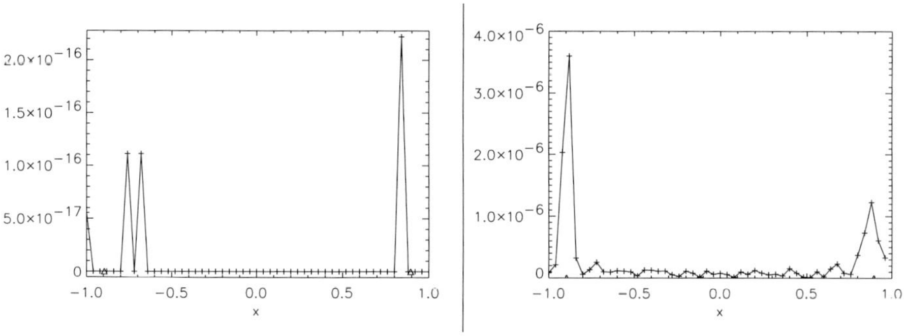

Fig. 13.4. Division by ω with (right) and without (left) additional random perturbation of the numerator

Thus, we see from this example that *it is the structure of* $1/\omega$ *on the grid which determines the peaks.*

There have been two ways trying to avoid the disaster. Hübner [33] proposes to solve a singularly elliptic equation for the quotient in terms of numerator and denominator. The equation has the same degeneracy on $\mathscr{I}$ as the LYE and hence one has to expect the same problems. In [19] we considered a procedure based on spectral ideas to perform this division. We want to elaborate on this method here. The main idea, summarized schematically in Fig. 13.5 is as follows.

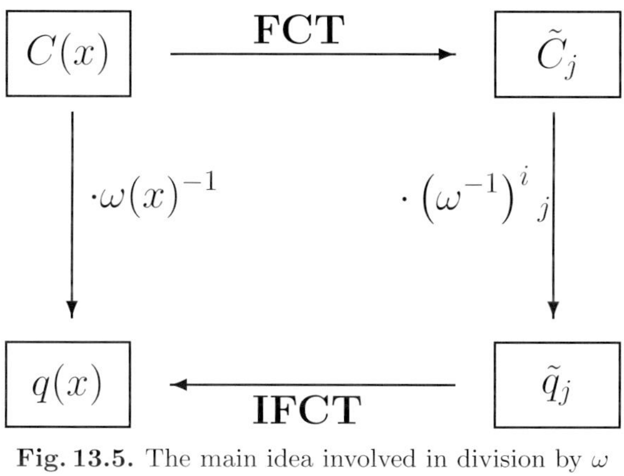

Fig. 13.5. The main idea involved in division by ω

We represent the numerator and the denominator both as a finite sum of basis functions, say Chebyshev polynomials (the following discussion applies just as well to Fourier series). Multiplication by ω is a linear map which we can represent in Fourier space as a matrix ${\omega^i}_j$ defined by

$$\omega(x) \cdot T_j(x) = \sum_i {\omega^i}_j \, T_i(x). \tag{13.23}$$

Given the representation of C in terms of its coefficients $\tilde{C}^j$ then one could simply solve the linear equation $\tilde{q}^i = \omega^i{}_j\tilde{C}^j$ to obtain the coefficients of the quotient. However, there are some snags.

The basis functions $\{T_j : 0 \le j \le N\}$ span a *finite*-dimensional vector space V_N. Since multiplication with any polynomial (except T_0) increases the degree we should obtain products with degree higher than N. However, if we regard the multiplication map as a map from V_N into itself we are asking to represent a high degree ($> N$) polynomial as a sum of polynomials of degree up to N. With this 'folding back' of high degree polynomials into the range of low degree polynomials – this effect is known as *aliasing* – multiplication is no longer injective. Therefore, we need to enlarge the target space. Multiplying two polynomials of degree at most N yields a product of degree at most $2N$. So we regard multiplication by ω as a map $\omega : V_N \to V_{2N}$. This map is injective so that the image $\omega[V_N]$ of V_N is a $N+1$-dimensional subspace in V_{2N}. The matrix $\boldsymbol{\omega}$ of the map ω is no longer square but has dimensions $(2N+1) \times N$.

Let $\tilde{C}^j$ be the representation of $C = \sum_{j=0}^N \tilde{C}^j T_j$ in V_N which is trivially also a representation in V_{2N}. In general, $\tilde{C}^j$ will not lie in $\omega[V_N]$ because the error in its computation will drive it away. Hence, the first step in the computation of the quotient is to project $\tilde{C}$ onto $\omega[V_N]$ and then the second step is the inversion of ω on its image.

Both these steps can be accomplished by computing the (reduced) QR-factorization of $\boldsymbol{\omega}$ (cf. [49]). This algorithm allows us to write the matrix $\boldsymbol{\omega}$ uniquely as $\boldsymbol{\omega} = QR$ with a square $(N+1) \times (N+1)$ upper triangular matrix R and a $(2N+1) \times (N+1)$ matrix Q whose columns are orthonormal. The crucial fact for us is that these columns span $\omega[V_N]$. This implies that $P = QQ^t$ is a projector onto that space while $Q^tQ = id_N$ is the identity on V_N.

Thus, to solve the equation $\tilde{q} = \boldsymbol{\omega}\tilde{C}$ we proceed in two steps: first compute $\tilde{z} = Q^t\tilde{C}$ and then solve $\tilde{q} = R^{-1}\tilde{z}$. These steps correspond exactly to the two steps mentioned above because $Q\tilde{z} = P\tilde{C}$ is the projection of $\hat{C}$ onto $\omega[V_N]$. Since R is the matrix representation of $\boldsymbol{\omega}$ with respect to the basis provided by the columns of Q and its pre-image in V_N its inverse provides that vector whose image under $\boldsymbol{\omega}$ is the projection of $\tilde{C}$ onto $\omega[V_N]$.

The scalar product involved in the computation of the QR-factorisation is the natural scalar product between the Chebyshev polynomials. Hence the projector is the orthogonal projector onto $\omega[V_N]$ with respect to this scalar product. The projection $P\tilde{C}$ is that vector in $\omega[V_N]$ which is closest to $\tilde{C}$ with respect to that scalar product. This is not what is needed here. For suppose that ω is a polynomial of degree $M < N$. Then the projection should not contain any polynomials up to degree $M-1$ because any of these would give rise to a singular quotient. Thus, we need to construct a different projection $\hat{P}$ which has the same image as P but which annihilates all polynomials with degree less than M.

In order to do this we have to impose the condition that ω be a polynomial with degree $M < N$. This is not a severe restriction because ω can be specified freely. Then we consider $\boldsymbol{\omega} : V_N \to V_{N+M}$ as a map into V_{N+M}. To find the new projection it is necessary now to compute the full QR-factorisation of $\boldsymbol{\omega}$

because this provides not only information about the image of $\boldsymbol{\omega}$ but also about its orthogonal complement. In fact, one can again write $\boldsymbol{\omega} = \hat{Q}\hat{R}$ where now $\hat{Q}$ is an orthogonal $(N+M+1)\times(N+M+1)$ matrix and $\hat{R}$ is upper triangular with dimensions $(N+M+1)\times(N+1)$. Again the first $N+1$ columns of $\hat{Q}$ form an orthonormal basis of $\omega[V_N]$ while the remaining columns span its orthogonal complement, the kernel K of $P = QQ^t$, so $K = \ker P$. We denote by $\hat{Q}_1$ the matrix containing the first $N+1$ columns of $\hat{Q}$ and by $\hat{Q}_2$ the remaining ones. Then $\hat{Q} = \left(\hat{Q}_1|\hat{Q}_2\right)$. Note, that $\hat{Q}_1$ coincides with the matrix Q obtained above from the reduced QR-factorisation.

Let V_M be the space spanned by the polynomials with degree less than the degree of ω. Clearly, we have

$$V_M \cap \omega[V_N] = \{0\}. \tag{13.24}$$

We need to change the projector $P = QQ^t = \hat{Q}_1\hat{Q}_1^t$ into a new projector $\hat{P} = Q_1Q_1^t + Q_1SQ_2^t$ for some $(N+M)\times M$ matrix S which has to be found from the fact that $\hat{P}$ annihilates V_M. This form of the new projector follows from the fact that in the adapted basis given by the columns of $\hat{Q}$ a general projector with image $\omega[V_N]$ has the matrix representation

$$\left(\begin{array}{c|c} \mathbf{1}_N & S \\ \hline 0 & 0 \end{array}\right).$$

Let U be a $(N+M)\times M$-matrix whose columns span V_M. Then S has to satisfy the condition

$$\left(\hat{Q}_1^t + S\hat{Q}_2^t\right) = 0. \tag{13.25}$$

Note, that Q_2^tU is a square matrix of dimension $M\times M$ which is necessarily invertible. This is a consequence of (13.24). Hence we get

$$S = \hat{Q}_1^tU\left(\hat{Q}_2^tU\right)^{-1} \tag{13.26}$$

and consequently

$$\hat{P} = P\left(\mathbf{1}_{N+M} - U\left(\hat{Q}_2^tU\right)^{-1}\hat{Q}_2^t\right). \tag{13.27}$$

Using this 'improved' projection kills all the lower degree polynomials and hence yields much smoother quotients. As an example consider Fig. 13.6 where we have illustrated a one-dimensional case with and without the use of the improved projection. The upper diagrams show the results obtained by dividing the function $f = \omega(2x^3 - 7x^5 + 10\cos^2(5x))$ by $\omega = (x-1/2)(x+2/5)(x-1/100)$. However, to simulate the real problem we have contaminated f with a low order polynomial (i.e. at most quadratic in this case) with random coefficients of size at most $\epsilon = 10^{-2}$. The left diagram shows the quotient (indicated by a line with crosses) without the use of the regularizing projection. The line is the exact result. Clearly, the quotient is influenced by the random deviation. The difference

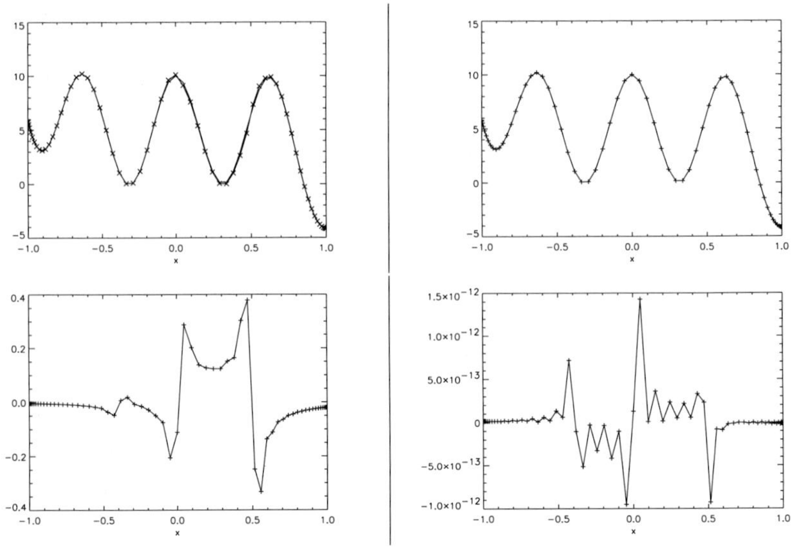

Fig. 13.6. Results obtained with (right) and without (left) the regularizing projection

to the exact quotient clearly shows the structure of $1/\omega$. The maximum error is influenced by two sources: the size of ϵ i.e., the size of the truncation error in the application we have in mind. The second influence is the resolution, i.e. the number of grid points (or, equivalently, the number of polynomials). With higher resolution $1/\omega$ can be better resolved which results in higher spikes and hence in bigger errors. On the right hand side the regularizing projection has been used. Here, the deviation from the exact result is not visible and, in fact, the error is at the level of 10^{-12}. Here, the error is not affected by ϵ because the projection exactly kills any low order polynomial. It appears that the regularizing projection redistributes the error in a global manner so as to obtain a function which is 'evenly' divisible by ω.

As a final remark it should be noted that the way we have obtained the regularizing projection does not in any way depend on the dimension of the problem although we have essentially treated only the one-dimensional case. In a similar way one can treat higher dimensional cases by employing tensor bases and interpreting the spaces of coefficients appropriately. The geometry behind the scenes remains exactly the same.

13.5 Conclusion

As we have seen the question of getting smooth initial data beyond the conformal boundary of space-time is difficult. At the moment the only way to obtain these data is via the Andersson–Chruściel–Friedrich procedure which has its limitations. Therefore, we should look for other ways to get such initial data. There are two other possibilities for this. One way is to understand the structure of the conformal constraints and to find ways of formulating well-posed boundary value problems for their direct solution. This is the route taken by Butscher [9].

Another possible approach towards the construction of smooth initial data is suggested by recent work by J. Corvino [13] (refined by Chruściel and Delay [12]) who has constructed smooth solutions to the constraint equations on an asymptotically Euclidean space-like hypersurface which coincide exactly with Schwarzschild data outside a compact set. The evolution of such data will contain at least for some time an asymptotic region which is exactly Schwarzschild. Therefore, it is possible, at least in principle, to find hyperboloidal hypersurfaces in the time development of the data on which hyperboloidal data are implied which are exactly Schwarzschild outside a compact set and these data can trivially be extended analytically beyond $\mathscr{I}$. To obtain these data it is necessary to either generalize Corvino's method to the hyperboloidal setting and solve appropriate equations directly there or to generate the hyperboloidal data numerically from Corvino's asymptotically flat data. This requires an appropriate evolution scheme for Friedrich's regular finite initial value problem at spatial infinity [23] which so far is lacking.

Another issue which ultimately has to be thought about is the inclusion of matter fields into the conformal framework. Including matter fields is complicated because the Bianchi equation for the rescaled Weyl tensor acquires a source term and this involves derivatives of the stress-energy-tensor. So far there has been only one case in which a system of matter fields coupled to the conformal field equations has been studied numerically [30]. While there is no problem (at least in principle apart from complexity) to include fundamental conformally invariant fields like the Maxwell, Yang-Mills fields [22] or other massless fields like the conformally coupled scalar field [31] problems can arise for massive fields like the Klein-Gordon field or ideal fluids. The reason is that the matter equations are not conformally invariant so that a regularization of the equations on $\mathscr{I}$ by rescaling the fields appropriately is unlikely. However, in the usual scenario of an isolated system in which a source generates waves and sends them out to infinity the matter region will not extend out to infinity so that the singular behaviour on $\mathscr{I}$ is not present. Even if one assumes that matter fields extend to infinity one can still hope to be safe on $\mathscr{I}$ because in some cases like the massive Klein-Gordon field the time evolution forces the field to decay exponentially towards $\mathscr{I}$ so that the formal singularity is in fact not there [50].

To conclude it is probably safe to say that the conformal approach has reached the status where we know that it is a reasonable and feasible approach which offers some exciting possibilities but which also provides us with problems (mostly of a numerical nature) which are (apart from some idiosyncracies) very similar to those encountered in other approaches. A joint effort of the numerical relativity community seems to be called for in order to overcome these problems.

References

1. L. Andersson. *The conformal constraint equations.* This volume.
2. L. Andersson and P. T. Chruściel. *On "hyperboloidal" Cauchy data for vacuum Einstein equations and obstructions to smoothness of "null-infinity".* Phys. Rev. Lett., **70**, 2829–2832, 1993.
3. L. Andersson, P. T. Chruściel, and H. Friedrich. *On the regularity of solutions to the Yamabe equation and the existence of smooth hyperboloidal initial data for Einstein's field equations.* Comm. Math. Phys., **149**, 587–612, 1992.
4. R. Arnowitt, S. Deser, and C. W. Misner. *The dynamics of general relativity.* In L. Witten, ed., *Gravitation: An Introduction to Current Research* (Wiley, New York, 1962).
5. R. Bartnik and A. Norton. *Numerical experiments at null infinity.* This volume.
6. B. Berger. *Numerical approaches to spacetime singularities.* Living Reviews in Relativity, **5**, 2002. `http://www.livingreviews.org/Articles/Volume5/2002-1berger/`.
7. H. Bondi, M. G. J. van der Burg, and A. W. K. Metzner. *Gravitational waves in general relativity VII. Waves from axi-symmetric isolated systems.* Proc. Roy. Soc. London A, **269**, 21–52, 1962.
8. J. P. Boyd. *Chebyshev and Fourier spectral methods* (Dover Publications, Mineola, 2001), second edn.
9. A. Butscher. *Exploring the conformal constraint equations.* This volume.
10. G. Calabrese, L. Lehner, and M. Tiglio. *Constraint preserving boundary conditions in numerical relativity.* Phys. Rev. D, **65**, 104031, 2002. `http://xxx.lanl.gov/archive/gr-qc/0111003`.
11. C. Canuto, M. Y. Houssaini, A. Quarteroni, and T. A. Zang. *Spectral methods in fluid dynamics* (Springer Verlag, Berlin, 1988).
12. P. C. Chruściel and E. Delay. *Existence of non-trivial, vacuum, asymptotically simple space-times*, 2002. `http://www.phys.univ-tours.fr/~piotr`. Preprint.
13. J. Corvino. *Scalar curvature deformation and a gluing construction for the Einstein constraint equations.* Comm. Math. Phys., **214**, 137–189, 2000.
14. B. Engquist and A. Majda. *Absorbing boundary conditions for the numerical simulation of waves.* Math. Comp., **31**, 629–651, 1977.
15. J. A. Font. *Local characteristic algorithms for relativistic hydrodynamics.* This volume.
16. B. Fornberg. *A practical guide to pseudospectral methods* (Cambridge University Press, Cambridge, 1996).
17. J. Frauendiener. *Numerical treatment of the hyperboloidal initial value problem for the vacuum Einstein equations. I. The conformal field equations.* Phys. Rev. D, **58**, 064002, 1998.
18. J. Frauendiener. *Numerical treatment of the hyperboloidal initial value problem for the vacuum Einstein equations. II. The evolution equations.* Phys. Rev. D, **58**, 064003, 1998.
19. J. Frauendiener. *Calculating initial data for the conformal field equations by pseudo-spectral methods.* J. Comp. Appl. Math., **109**, 475–491, 1999.
20. J. Frauendiener. *Conformal infinity.* Living Reviews in Relativity, **3**, 2000. `http://www.livingreviews.org/Articles/Volume3/2000-4frauendiener/`.
21. H. Friedrich. *Conformal Einstein evolution.* This volume.

22. H. Friedrich. *On the global existence and the asymptotic behavior of solutions to the Einstein-Maxwell-Yang-Mills equations.* J. Diff. Geom., **34**, 275–345, 1991.
23. H. Friedrich. *Gravitational fields near space-like and null infinity.* J. Geom. Phys., **24**, 83–163, 1998.
24. H. Friedrich and G. Nagy. *The initial boundary value problem for Einstein's vacuum field equations.* Comm. Math. Phys., **201**, 619–655, 1998.
25. D. Garfinkle. *Simulations of generic singularities in harmonic coordinates.* This volume.
26. D. Gottlieb and S. A. Orszag. *Numerical Analysis of spectral methods: Theory and Applications* (SIAM-CBMS Philadelphia, 1977).
27. C. Gundlach and J. Pullin. *Ill-posedness of a double-null-free-evolution scheme for black hole spacetimes.* Class. Quant. Grav., **14**, 991–997, 1997.
28. B. Gustafsson, H.-O. Kreiss, and J. Oliger. *Time dependent problems and difference methods* (Wiley, New York, 1995).
29. N. J. Higham. *Accuracy and Stability of Numerical Algorithms* (SIAM, Philadelphia, 1996).
30. P. Hübner. *Method for calculating the global structure of (singular) spacetimes.* Phys. Rev. D, **53**, 701–721, 1994.
31. P. Hübner. *General relativistic scalar-field models and asymptotic flatness.* Class. Quant. Grav., **12**, 791–808, 1995.
32. P. Hübner. *Numerical approach to the global structure of space-time.* Helv. Phys. Acta, **69**, 317–320, 1996.
33. P. Hübner. *How to avoid artificial boundaries in the numerical calculation of black hole space-times.* Class. Quant. Grav., **16**, 2145–2164, 1999.
34. P. Hübner. *A scheme to numerically evolve data for the conformal Einstein equation.* Class. Quant. Grav., **16**, 2823–2843, 1999.
35. P. Hübner. *From now to timelike infinity on a finite grid.* Class. Quant. Grav., **18**, 1871–1884, 2001.
36. S. Husa. *Problems and successes in the numerical approach to the conformal field equations.* This volume.
37. C. T. Kelley. *Iterative Methods for Linear and Nonlinear Equations* (SIAM, Philadelphia, 1995).
38. L. Lehner. *Numerics of the characteristic formulation in Bondi variables. Where we are and what lies ahead.* This volume.
39. L. Lehner. *Numerical relativity: a review.* Class. Quant. Grav., **18**, R25–R86, 2001. `http://xxx.lanl.gov/archive/gr-qc/0106072`.
40. E. T. Newman and R. Penrose. *An approach to gravitational radiation by a method of spin coefficients.* J. Math. Phys., **3**, 896–902, 1962. Errata **4** (1963), 998.
41. *Octave Homepage.* `http://www.octave.org`.
42. R. K. Sachs. *Gravitational waves in general relativity VIII. Waves in asymptotically flat space-time.* Proc. Roy. Soc. London A, **270**, 103–127, 1962.
43. B. G. Schmidt. *Data for the numerical calculation of the Kruskal spacetime.* This volume.
44. J. Stoer. *Einführung in die Numerische Mathematik I* (Springer-Verlag, Berlin, 1972).
45. B. Szilágy, R. Gómez, N. T. Bishop, and J. Winicour. *Cauchy boundaries in linearized gravitational theory.* Phys. Rev. D, **62**, 104006, 2000.

46. B. Szilágyi, B. Schmidt, and J. Winicour. *Boundary conditions in linearized harmonic gravity*, 2002. Preprint.
47. L. N. Trefethen. *Group velocity in finite difference schemes.* SIAM Review, **24**, 113–136, 1982.
48. L. N. Trefethen. *Spectral methods in Matlab* (SIAM, Philadelphia, 2000).
49. L. N. Trefethen and D. Bau. *Numerical Linear Algebra* (SIAM, Philadelphia, 1997).
50. J. Winicour. *Massive fields at null infinity.* J. Math. Phys., **29**, 2117–2121, 1988.

14 Data for the Numerical Calculation of the Kruskal Space-Time

Bernd G. Schmidt

Albert-Einstein-Institut, Max-Planck-Institut für Gravitationsphysik,
Am Mühlenberg 1, 14476 Golm, Germany

Abstract. Hypersurfaces in the conformally extended Kruskal spacetime are constructed which intersect both future null infinities. Some numerical evolutions by the conformal field equations of data determined by such hypersurfaces are described.

14.1 Introduction

After Peter Hübner has developed and successfully used his code for the conformal field equations, it is natural trying to evolve part of the conformal extension of the Kruskal space-time. As is well known the conformal boundary of the Kruskal space time consists of four null hypersurfaces, right null infinity $\mathscr{I}_R$ and left null infinity $\mathscr{I}_L$.

To start the evolution one needs initial data. These can be determined from the metric and second fundamental form of a space-like hypersurface in the conformally extended Kruskal space-time intersecting both future null infinities. The main goal of this paper is to find such hypersurfaces and the corresponding data as explicitly as possible.

Sect. 14.2 describes the conformal extension of the Kruskal space-time in detail. It is not trivial to find coordinates and a rescaled metric g such that g is analytic at both future null infinities. The invariant nature of the coordinates constructed, also sheds some light on the structure of time-like infinity, which is up to now not understood at all.

Sect. 14.3 is based on a paper by Brill et al. [1] where it is shown how to calculate the inner metric and second fundamental form of hypersurfaces with constant mean curvature in static spherically symmetric space-times without constructing an explicit embedding into the space-time. Using this it is possible to determine data of the desired form. Note that one has no explicit embedding into space-time.

In Sect. 14.4 the analytic double null coordinates constructed in Sect. 14.2 are used to define a foliation of part of the extended Kruskal space-time. This foliation can be used to find data, which are, however, not completely explicit. The data contain the inverse of an elementary function.

Sect. 14.5 contains a short description of some numerical evolutions of the Kruskal space-time done by P. Hübner and M. Weaver.

14.2 Conformal Extension of the Kruskal Space-Time

The purpose of this section is to construct the conformal extension of the Kruskal space-time. The Kruskal space-time is the unique analytic extension of the static Schwarzschild space-time. We begin by displaying the Schwarzschild solution in various well known coordinate systems which we will need to perform the conformal extension.

The Schwarzschild metric (for $\frac{2m}{r} \neq 1$) in standard (t,r)–coordinates (which are unique up to a translation $t \to t + a$) is

$$\tilde{g} = -\left(1 - \frac{2m}{r}\right) dt^2 + \frac{dr^2}{1 - \dfrac{2m}{r}} + r^2 d\Sigma^2, \qquad d\Sigma^2 = d\theta^2 + \sin^2\theta\, d\varphi^2. \tag{14.1}$$

The null lines in the (t,r)-surface

$$u(t,r) = t - r - 2m \ln\left|\frac{r}{2m} - 1\right|, \tag{14.2}$$

$$du = dt - dr - \frac{1}{\dfrac{r}{2m} - 1} dr = dt - \frac{dr}{1 - \dfrac{2m}{r}}, \tag{14.3}$$

$$v(t,r) = t + r + 2m \ln\left|\frac{r}{2m} - 1\right|, \tag{14.4}$$

$$dv = dt + dr + \frac{1}{\dfrac{r}{2m} - 1} dr = dt + \frac{dr}{1 - \dfrac{2m}{r}}, \tag{14.5}$$

can be used to introduce advanced and retarded Eddington–Finkelstein coordinates in which the metric is

$$\tilde{g} = -\left(1 - \frac{2m}{r}\right) du^2 - 2\, du\, dr + r^2\, d\Sigma^2, \tag{14.6}$$

$$\tilde{g} = -\left(1 - \frac{2m}{r}\right) dv^2 + 2\, dv\, dr + r^2\, d\Sigma^2. \tag{14.7}$$

We will also need the metric in double null coordinates

$$\tilde{g} = -\left(1 - \frac{2m}{r}\right) du\, dv + r^2 d\Sigma^2, \tag{14.8}$$

where $r(u,v)$ is determined by the equation

$$v - u = 2\left(r + 2m \ln\left|\frac{r}{2m} - 1\right|\right). \tag{14.9}$$

Now we define the following three space-times by the metric (14.1) for the following coordinate ranges:

For $m > 0$

the static Schwarzschild solution M_0: $(\{2m < r < \infty\}, \tilde{g})$,
the black hole interior M_1: $(\{0 < r < 2m\}, \tilde{g})$.

For $m < 0$

the negative mass Schwarzschild solution M_3: $(\{0 < r < \infty\}, \tilde{g})$.

Eddington–Finkelstein coordinates can be used to construct the extension of the static Schwarzschild solution with positive m through the horizon: We can introduce the coordinate u defined in (14.2) on M_0 and obtain (14.6) for the metric. This metric is regular for all $r > 0$. It is easy to check that the part $0 < r < 2m$ is isometric to M_1. The null hypersurface $r = 2m$ is a horizon.

Now we construct the Kruskal space-time, K_0, as follows. We define on M_0 in double null coordinates (14.8) the following new coordinates:

$$U = -e^{-u/4m} \qquad V = e^{v/4m}, \tag{14.10}$$

$$\tilde{g} = 16m^2 \left(1 - \frac{2m}{r}\right) \frac{dU\,dV}{U\,V} + r^2 d\Sigma^2, \tag{14.11}$$

where now $r(U, V)$ satisfies

$$UV = -e^{r/2m}\left(\frac{r}{2m} - 1\right), \qquad r > 2m. \tag{14.12}$$

Using this, we can write for the metric

$$\tilde{g} = -32m^3\, \frac{e^{-r/2m}}{r}\, dU\,dV + r^2 d\Sigma^2\ . \tag{14.13}$$

The function $e^x(x-1)$ is monotonic and analytic for $-1 < x < \infty$. Hence the inverse function exists and we can define r by (14.12) for all U, V with $|UV| > 1$. Due to the development in computer algebra the inverse function of $y = xe^x$ was called 'LambertW(y)' in Maple [6]. We can therefore solve (14.12) as

$$\frac{r}{2m} = 1 + \mathrm{LambertW}(-UV/e). \tag{14.14}$$

The metric (14.13) is analytic on the domain $|UV| > 1$. This is the Kruskal space-time K_0, the unique complete analytic maximal extension of the static Schwarzschild space-time with positive mass m. It is easy to check that the submanifolds $\{U < 0, V < 0\}$ and $\{U > 0, V > 0\}$ are isometric to M_1 and the submanifolds $\{U < 0, V > 0\}$ and $\{U > 0, V < 0\}$ are isometric to M_0. The horizons are $U = 0$ and $V = 0$.

Next we want to construct the conformal extensions of the Kruskal space-time. We begin with M_0', an extension of M_0 in Eddington–Finkelstein coordinates, i.e., the metric (14.6) for $0 < r < \infty$. With $l = 1/r$ we obtain

$$\tilde{g} = -(1 - 2ml)\, du^2 + 2\,\frac{du\,dl}{l^2} + \frac{1}{l^2} d\Sigma^2. \tag{14.15}$$

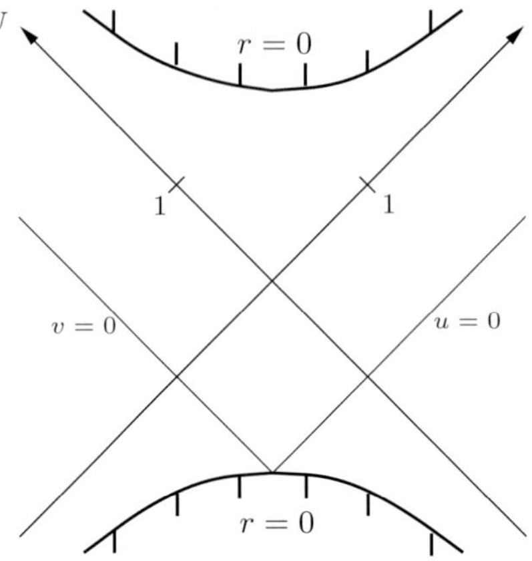

Fig. 14.1. Kruskal diagram

We define the unphysical metric as $(0 < l < 1)$:

$$g = -l^2(1 - 2ml)\, du^2 + 2\, du\, dl + d\Sigma^2 = \Omega^2 \tilde{g} = l^2 \tilde{g}. \tag{14.16}$$

We can extend this metric analytically onto $\{(l, u) \mid l \in \mathbb{R}, u \in \mathbb{R}\}$. The null hypersurface $l = 0$ is future null infinity $\mathscr{I}^+$. The part $l < 0$ is conformal to the negative mass space-time M_3. This is easily seen when we pass from $l < 0$ to $r = -1/l > 0$. The $\mathscr{I}^+$ of a Schwarzschild solution with positive mass m coincides with the $\mathscr{I}^-$ of a Schwarzschild solution with mass $-m$.

From Fig. 14.1 of the Kruskal space-time it is obvious that we can do this conformal extension four times. There are two future null infinities, $\mathscr{I}^+_L$, $\mathscr{I}^+_R$ and two past null infinities, $\mathscr{I}^-_L$, $\mathscr{I}^-_R$. This way we obtain a conformally extended Kruskal space-time K (see Fig. 14.2). (This space-time has further conformal extensions in which we are presently not interested.)

To construct space-like hyperboloidal hypersurfaces intersecting $\mathscr{I}^+_L$ and $\mathscr{I}^+_R$ we would like to have coordinates covering the conformally extended Kruskal space-time. A systematic way to find such a chart is to use global, analytic null coordinates. Let us begin by constructing local double null coordinates in some neighbourhood of a point at $\mathscr{I}^+_R$.

We begin with the metric (14.16)

$$g = -l^2\,(1 - 2ml)\; du^2 + 2\, du\, dl = \left(-l^2\,(1 - 2ml)\; du + 2\, dl\right)\; du, \tag{14.17}$$

which is analytic for all real l. The line $l = 0$ is null infinity $\mathscr{I}^+_R$. There should exist analytic double null coordinates! Let us just construct the second family of null lines: The unique null lines $l(u)$ different from $u = \text{const}$ satisfy the ODE

$$2\frac{dl}{du} = l^2\,(1 - 2ml)\,. \tag{14.18}$$

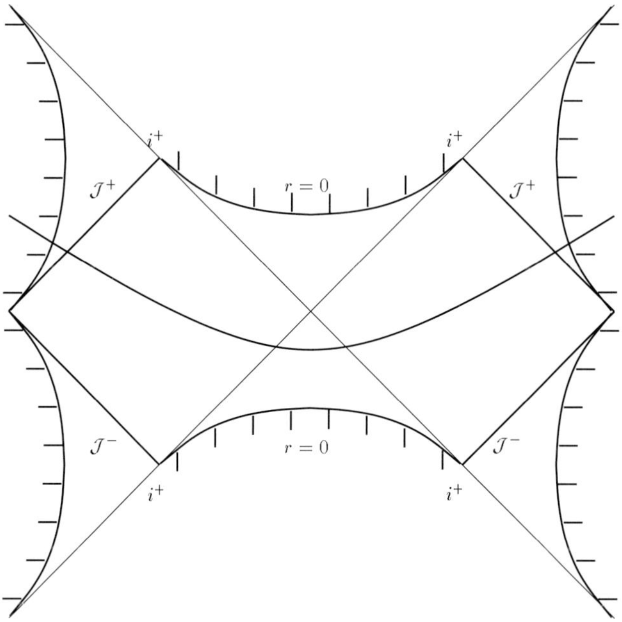

Fig. 14.2. Qualitative diagram of extended Kruskal space-time (this is just for adjustment)

The unique solution $l(u, \hat{v})$ with initial values $l(0, \hat{v}) = \hat{v}$ is analytic in u and $\hat{v}$. This solution satisfies for $l \neq 0$, $2ml \neq 1$, $\hat{v} \neq 0$, $2m\hat{v} \neq 1$

$$u = -2\left(\frac{1}{l} + 2m \ln\left|\frac{1}{l} - 2m\right|\right) + 2\left(\frac{1}{\hat{v}} + \ln\left|\frac{1}{\hat{v}} - 2m\right|\right). \tag{14.19}$$

$\hat{v}$ is an analytic function of (l, u). Note, that the condition $\hat{v} = l$ for $u = 0$ uniquely determines the expression for $\hat{v}$ in (14.19) as an integration constant. (If we had just chosen $1/\hat{v}$, then $\hat{v}$ would not have been a differentiable function of l at $l = 0$.) As the curves $\hat{v} = \text{const}$ form an analytic foliation near $\mathscr{I}^+_R$ we can introduce $\hat{v}$ as a coordinate.

From

$$du = \frac{2}{l^2(1 - 2ml)}\, dl - \frac{2}{\hat{v}^2(1 - 2m\hat{v})}\, d\hat{v} \tag{14.20}$$

we obtain the following metric, defined and analytic for all real $(u, \hat{v})$ such that l is finite. ($l = 0$ is $\mathscr{I}^+_R$, $l = 1$ the horizon, $l = \pm\infty$ the singularity $r = 0$; $\hat{v} = 0$ is $\mathscr{I}$, $\hat{v} = 1$ is H^+)

$$g = 2\frac{l^2(1 - 2ml)}{\hat{v}^2(1 - 2m\hat{v})}\, du\, d\hat{v} + d\Sigma^2. \tag{14.21}$$

From (14.19) we see that for fixed u

$$l(u, \hat{v}) = \frac{\hat{v}}{1 - \frac{1}{2}u\hat{v}}\,(1 + o(\hat{v}))\,. \tag{14.22}$$

We can calculate the Taylor series $l(u,\hat{v}) = \hat{v} + a_1(u)\hat{v} + a_1(u)\hat{v}\ldots$ at $\mathscr{I}^+_R$ by calculating the derivatives of l with respect to $\hat{v}$ from (14.19) and use (14.22) for the limit $\hat{v} \to 0$. Similarly at the horizon, $2m\hat{v} = 1$.

Our coordinate was fixed by the condition that on $u = 0$ we prescribed $l(0,\hat{v}) = \hat{v}$. Any analytic function $l(0,\hat{v}) = a(\hat{v})$ would be possible. How is $\hat{v}$ related to v or V? On $u = 0$ we have $\hat{v} = l$ which gives using (14.9)

$$v = 2\left(\frac{1}{\hat{v}} + 2m\ln\left|\frac{1}{2m\hat{v}} - 1\right|\right) \tag{14.23}$$

and

$$V = \mathrm{e}^{v/4m} = \mathrm{e}^{\hat{v}/4m}\left(\frac{1}{2m\hat{v}} - 1\right). \tag{14.24}$$

Clearly we can perform a similar construction at $\mathscr{I}^+_L$ to define a coordinate $\hat{u}$. It holds

$$u = -2\left(\frac{1}{\hat{u}} + 2m\ln\left|\frac{1}{2m\hat{u}} - 1\right|\right) \tag{14.25}$$

and

$$U = \mathrm{e}^{-u/4m} = \mathrm{e}^{\hat{u}/4m}\left(\frac{1}{2m\hat{u}} - 1\right). \tag{14.26}$$

The metric becomes

$$g = \frac{4l^2(1-2ml)}{\hat{v}^2\,(1-2m\hat{v})\ \hat{u}^2\,(1-2m\hat{u})}\,d\hat{u}\,d\hat{v} + d\Sigma^2 \tag{14.27}$$

with

$$-\mathrm{e}^{1/(4ml)}\left(\frac{1}{2ml} - 1\right) = \mathrm{e}^{1/(4m\hat{u})}\left(\frac{1}{2m\hat{u}} - 1\right)\mathrm{e}^{1/(4m\hat{v})}\left(\frac{1}{2m\hat{v}} - 1\right). \tag{14.28}$$

What is the reason for this complicated exercise? Suppose we try the obvious coordinate transformation $\tilde{v} = 1/v$ in (14.8) which is successful for Minkowski space. Then (14.9) implies that $\tilde{v}$ is not a differentiable function of l on $u = \text{const}$ at $\tilde{v} = 0$.

So far we considered the metric (14.21) near $\hat{v} = 0$, $\hat{u} > 1$. The relations (14.23) and (14.25) show, however, that $2m\hat{v} > 1$, $2m\hat{u} > 1$ covers the part $V > -1$, $U > -1$ of the Kruskal space-time and a part of the two negative mass extensions through $\mathscr{I}^+_{R,L}$. For $2m\hat{u} < 1$, $2m\hat{v} < 1$ the metric (14.21) is regular as long as l is finite, i.e. $r > 0$.

The singularity has coordinates satisfying

$$1 = \mathrm{e}^{1/(4m\hat{u})}\left(\frac{1}{2m\hat{u}} - 1\right)\mathrm{e}^{1/(4m\hat{v})}\left(\frac{1}{2m\hat{v}} - 1\right). \tag{14.29}$$

All curves $l = \text{const}$ are tangent to the horizon at i^+, $2m\hat{u} = 1$, $\hat{v} = 0$. Equation (14.28) implies for $l = \text{const}$

$$2m\hat{u} = 1 + C\mathrm{e}^{-1/(4m\hat{v})}\,\frac{2m\hat{v}}{1-2m\hat{v}}\,\left(1 + o(\hat{v}^2)\right). \tag{14.30}$$

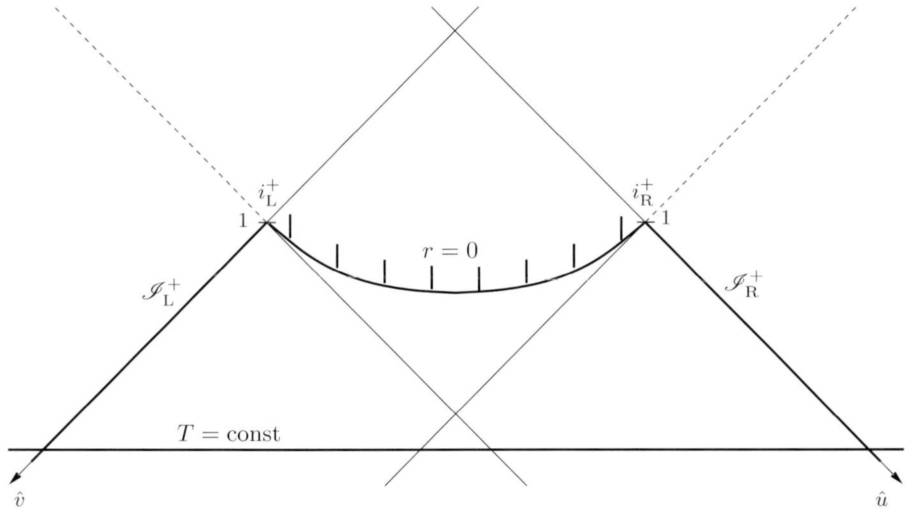

Fig. 14.3. Conformally extended Kruskal space-time

This is an invariant statement because we cannot change the tangent space at $(2m\hat{u} = 1, 2m\hat{v} = 0)$ by a monotonic transformation of $\hat{u}$ without changing it at finite points of the horizon $2m\hat{u} = 1$. Hence, the fact that all curves $l = \mathrm{const}$ approach the horizon exponentially fast at i^+ shows a strange property of time-like infinity. A similar construction of coordinates $\hat{u}$, $\hat{v}$ can be made near space-like infinity. One obtains double null coordinates which are analytic on $\mathscr{I}$ near space-like infinity. Such coordinates have been used in [8].

14.3 A Space-Like Hypersurface Intersecting $\mathscr{I}^+_L$ and $\mathscr{I}^+_R$

Brill et al. [1] study space-like hypersurfaces of constant mean extrinsic curvature in the Kruskal space-time. These hypersurfaces intersect null infinity. We want to use such a hypersurface as a Cauchy surface for the conformal field equations.

It is not possible to obtain an analytic representation of these hypersurfaces in the space-time. Nevertheless, the local inner geometry and second fundamental form of such hypersurfaces in the Schwarzschild space-time can be given explicitly. We choose a particular case with isotropic second fundamental form and obtain from equation (24) in [1] for the inner geometry $(2\,m = 1)$

$$\tilde{h} = \frac{dr^2}{1 - \frac{1}{r} + \frac{\tilde{k}^2 r^2}{9}} + r^2\, d\Sigma^2. \tag{14.31}$$

where the constant $\tilde{k}$ is the mean extrinsic curvature. We expect that it is at the "throat" where $1 - \frac{1}{r} + \frac{\tilde{k}^2 r^2}{9}$ vanishes. To find coordinates regular at the throat more easily, we first study the case $\tilde{k} = 6$, and obtain

$$1 - \frac{1}{r} + \frac{\tilde{k}^2 r^2}{9} = 1 - \frac{1}{r} + 4r^2 = \left(2 - \frac{1}{r}\right)\left(2r^2 + r + 1\right). \tag{14.32}$$

In this case $\tilde{h}$ is regular for $\frac{1}{2} < r < \infty$.

We first extend the 3-metric conformally through $r = \infty$ with $l := \frac{1}{r}$ and $0 < l < 2$:

$$\tilde{h} = \frac{dl^2}{l^4\left(1 - l + 4\left(\frac{1}{l}\right)^2\right)} + \left(\frac{1}{l}\right)^2 d\Sigma^2. \tag{14.33}$$

Rescaling by $\Omega = l$ we obtain

$$h = \Omega^2 \tilde{h} = \frac{dl^2}{4 + l^2 - l^3} + d\Sigma^2. \tag{14.34}$$

Obviously this metric is analytic at $l = 0$ and extends analytically to all negative values of l.

Next we investigate the behavior near $l = 2$ where the radial coordinates r and l become singular. We introduce a new coordinate w,

$$w = 2\sqrt{2 - l} \qquad \Longrightarrow \qquad -\frac{dl}{\sqrt{2 - l}} = dw, \quad -2\,dl = w\,dw \tag{14.35}$$

$$-\infty < l < 2 \qquad \Longleftrightarrow \qquad 0 < w < \infty \tag{14.36}$$

$$h = \frac{(2 - l)\,dw^2}{4 + l^2 - l^3} + d\Sigma^2 = \frac{(2 - l)\,dw^2}{(2 - l)(2 + l + l^2)} + d\Sigma^2. \tag{14.37}$$

Or with $l = 2 - \frac{1}{4}w^2$ (the denominator is always positive)

$$h = \frac{dw^2}{2 + \left(2 - \frac{1}{4}w^2\right) + \left(2 - \frac{1}{4}w^2\right)^2} + d\Sigma^2. \tag{14.38}$$

This 3-geometry is analytic for $-\infty < w < +\infty$. The 'throat' is at $w = 0$ and the sections of null infinity are at $w^2 = 8$, $w = \pm 2\sqrt{2}$.

We can proceed similarly for any value of $\tilde{k}$:

$$h = \Omega^2 \tilde{h} = \frac{dl^2}{l^2 - l^3 + \dfrac{\tilde{k}^2}{9}} + d\Sigma^2. \tag{14.39}$$

The polynomial in l has one real root $l(\tilde{k})$ which can be found using Maple. With $a = \frac{\tilde{k}^2}{9}$ the root is

$$l(\tilde{k}) = \frac{1}{6}x^{1/3} + \frac{2}{3}x^{-1/3} + \frac{1}{3}, \qquad x = \frac{1}{6}\left(8 + 108a + 12\sqrt{12a + 81a^2}\right)^{1/3}. \tag{14.40}$$

We can write

$$h = \Omega^2 \tilde{h} = \frac{dl^2}{\left(l(\tilde{k}) - l\right)\,\left(A(\tilde{k})\,l^2 + B(\tilde{k})\,l + C(\tilde{k})\right)} + d\Sigma^2. \tag{14.41}$$

We introduce w as before,

$$w = 2\sqrt{l(\tilde{k}) - l} \qquad \Longrightarrow \qquad -\frac{dl}{\sqrt{l(\tilde{k}) - l}} = dw, \quad -2\,dl = w\,dw, \tag{14.42}$$

and obtain

$$h = \Omega^2 \tilde{h} = \frac{dw^2}{A(\tilde{k})\,l^2 + B(\tilde{k})\,l + C(\tilde{k})} + d\Sigma^2, \tag{14.43}$$

with $l = l(\tilde{k}) - \frac{1}{4}w^2$. We can calculate A, B, C from $l(\tilde{k})$ if we want. Knowing the inner metric of the slice we need its second fundamental form as the second piece of the initial data, which we will determine next.

The mean extrinsic curvature, $\tilde{k}$, of a slice in space-time, and k, of a slice in a conformally related space-time satisfy the relation

$$3\Omega_{,k} n^k + \tilde{k} = \Omega k. \tag{14.44}$$

To find k we have to calculate the normal n^k of the slice in the unphysical space-time.

From Brill et al. [1] we find the tangent of a hypersurface with constant mean curvature $\tilde{k}$ (formula (18) in [1], $2m = 1$),

$$\left(\frac{dr}{dt}\right)^2 = \left(1 - \frac{1}{r}\right)^2 \left[1 + \frac{9}{\tilde{k}^2 r^2}\left(1 - \frac{1}{r}\right)\right] =: Z^2. \tag{14.45}$$

Coordinates (r, t) are the Schwarzschild coordinates. We want to calculate the unit normal of such a hypersurface in the rescaled metric ($\Omega = 1/r$),

$$g = \Omega^2 \tilde{g} = -\frac{1}{r^2}\left(1 - \frac{1}{r}\right) dt^2 + \frac{1}{r^2}\frac{dr^2}{1 - \frac{1}{r}} + d\Sigma^2. \tag{14.46}$$

A tangent vector is

$$\xi^i = (\xi^t, \xi^r) = (1, Z). \tag{14.47}$$

A vector orthogonal is

$$^{\perp}\xi^i = \left(\frac{Z}{1 - \frac{1}{r}}, 1 - \frac{1}{r}\right). \tag{14.48}$$

To obtain the unit normal relative to the metric g we have to normalize

$$n^i = a\left(\frac{Z}{1 - \frac{1}{r}}, 1 - \frac{1}{r}\right), \tag{14.49}$$

$$g_{ik}n^i n^k = a^2 \left[-\left(\frac{1}{r^2}\right)\left(1-\frac{1}{r}\right)\frac{Z^2}{\left(1-\frac{1}{r}\right)^2} + \left(\frac{1}{r^2}\right)\frac{1}{\left(1-\frac{1}{r}\right)}\left(1-\frac{1}{r}\right)^2\right] = -1, \tag{14.50}$$

$$a = r^2 \left[\frac{Z^2}{1-\frac{1}{r}} - \left(1-\frac{1}{r}\right)\right]^{-1/2} = \frac{1}{3}\frac{r^2\tilde{k}}{1-\frac{1}{r}} \tag{14.51}$$

and finally,

$$\Omega_k n^k = \Omega_{,r} n^r = -\frac{1}{r^2} a \left(1-\frac{1}{r}\right) = -\frac{1}{3}\tilde{k}. \tag{14.52}$$

From relation (14.44) we obtain $\Omega k = 0$. Hence k vanishes. This calculation was done on the part of the slice in the static Schwarzschild space-time. Since everything is analytic, the result holds on the whole slice.

The trace free part of the second fundamental form satisfies

$$\hat{k}_{ab} = \Omega \hat{\tilde{k}}_{ab}. \tag{14.53}$$

For the hypersurfaces we are considering $\hat{\tilde{k}}_{ab}$ vanishes, hence equation (14.53) implies that $\hat{k}_{ab}$ vanishes, and we have $k_{ab} = 0$.

The analysis of Brill et al. [1] shows that any one of these slices has its throat in the past of the intersection of the horizons if it is embedded into the Kruskal space-time. Its embedding in the extended Kruskal space-time is schematically shown in Fig. 14.2.

Considering again the case $\tilde{k} = 6$, we now have determined a "minimal initial data set", $(h_{ab}, k_{ab}, \Omega, \Omega_0)$, for the conformal field equations:

$$h_{ab}\, dx^a dx^b = \frac{dw^2}{2 + \left(2-\frac{1}{4}w^2\right) + \left(2-\frac{1}{4}w^2\right)^2} + d\Sigma^2, \tag{14.54}$$

$$k_{ab} = 0, \tag{14.55}$$

$$\Omega = l = 2 - \frac{1}{4}w^2, \qquad \Omega_0 = \Omega_k n^k = -2. \tag{14.56}$$

14.4 A Foliation Intersecting Both $\mathscr{I}$

The double null coordinates define a space-like foliation

$$\hat{t} = -\hat{u} - \hat{v} \tag{14.57}$$

for $\hat{t} < \hat{t}_0$, where $\hat{t}_0 = $ const intersects the singularity at $\hat{u} = \hat{v}$. Therefore $\hat{t}_0$ is defined by

$$e^{\frac{2}{(-\hat{t}_0)}}\left(\frac{2}{(-\hat{t}_0)} - 1\right) = 1. \tag{14.58}$$

From (14.27) one can determine a minimal initial data set for the conformal field equations as follows:

$$\hat{r} = \hat{v} - \hat{u} \tag{14.59}$$

defines a radial coordinate on a $\hat{t}$ = const hypersurface. The metric (14.27) can be rewritten in $(\hat{r}, \hat{t})$-coordinates. The 3-metric induced on $\hat{t} = \hat{t}_i$ is

$$h = \frac{l^2(l-1)}{\hat{u}^2(\hat{u}-1)\,\hat{v}^2(\hat{v}-1)}\,d\hat{r}^2 + d\Sigma^2, \tag{14.60}$$

with $\hat{u} = -(\hat{r} + \hat{t}_i)/2$ and $\hat{v} = (\hat{r} - \hat{t}_i)/2$.

The second fundamental form of a $\hat{t}$ = const hypersurface has only one non-vanishing component,

$$k_{\hat{r}\hat{r}} = -\frac{1}{4}\sqrt{\frac{l^2(l-1)}{\hat{u}^2(\hat{u}-1)\hat{v}^2(\hat{v}-1)}}\left(\frac{3l^2-2l-3\hat{u}^2+2\hat{u}}{\hat{u}^2(\hat{u}-1)} + \frac{3l^2-2l-3\hat{v}^2+2\hat{v}}{\hat{v}^2(\hat{v}-1)}\right). \tag{14.61}$$

Together with $\Omega = l$ and

$$\Omega_0 = -\frac{1}{2}\sqrt{\frac{l^2(l-1)}{\hat{u}^2(\hat{u}-1)\hat{v}^2(\hat{v}-1)}}\left(\hat{v}^2(\hat{v}-1) + \hat{u}^2(\hat{u}-1)\right), \tag{14.62}$$

$\hat{t} = \text{const} < \hat{t}_0$ determines a minimal initial data set $(h_{ab}, \Omega, k_{ab}, \Omega_0)$ for the conformal field equations.

14.5 Numerical Calculation of the Kruskal Space-Time

P. Hübner and M. Weaver evolved the initial data described in sections 14.3 and 14.4 using the code developed by Hübner in [2–5]. Here I want just to display some of the results without going into any details concerning the numerics.

If one evolves the slice described in Sect. 14.3, one does not know "where the hypersurface sits in the Kruskal space-time". One has to choose some gauge source functions – this was done by trial and error – which determine the coordinates in which the metric is calculated. To display the outcome of such a calculation Hübner and Weaver wrote a code [7] which calculates geodesics along with the evolution of the space-time. Using this, one obtains a picture of the space-time evolved as in Fig. 14.4. On the vertical axis x denotes the computer coordinates; the vertical axis on the left side is computer time, on the right Bondi time (which can be calculated using the geodesic integrator). The lines show geodesics calculated along with the evolution. As we know the intersection of both horizons with the initial slice one can follow the horizon by calculating its null generator. These are the curves H_R, H_L. The geodesic starting at $x = 6$ is the central time-like geodesic. Similarly one finds the position of both null infinities $\mathscr{I}_R$ and $\mathscr{I}_L$. The dotted lines are further null geodesics. By calculating invariants of the Weyl curvature along those geodesics one can distinguish between geodesics going off to infinity and those going to the singularity. This picture shows properties of the calculated space-time in a very beautiful way.

The size of the spatial grid was 160^3. The evolution stopped because there was not enough resolution. There were only about 3 grid points between the horizon and null infinity near the end of the evolution.

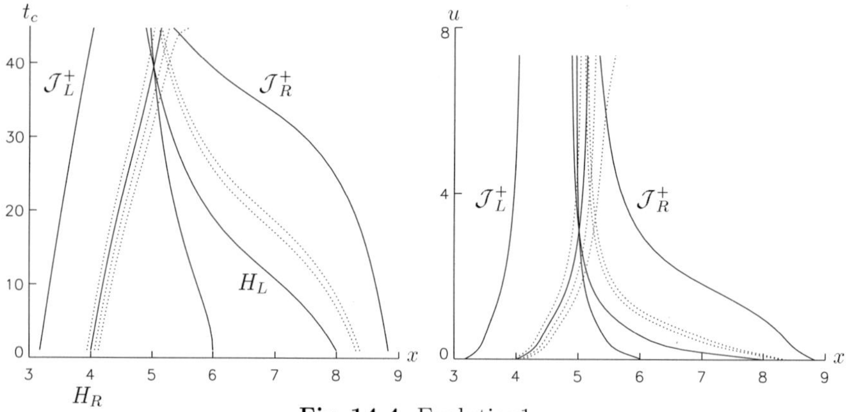

Fig. 14.4. Evolution1

For the data described in Sect.14.5 we have more information because they are part of a foliation. By prescribing the gauge source function appropriately we obtain the evolution in known coordinates and can directly determine the quality of the calculation.

The following Fig. 14.5 shows the foliation used. It is based on the coordinates constructed in Sect. 14.4.

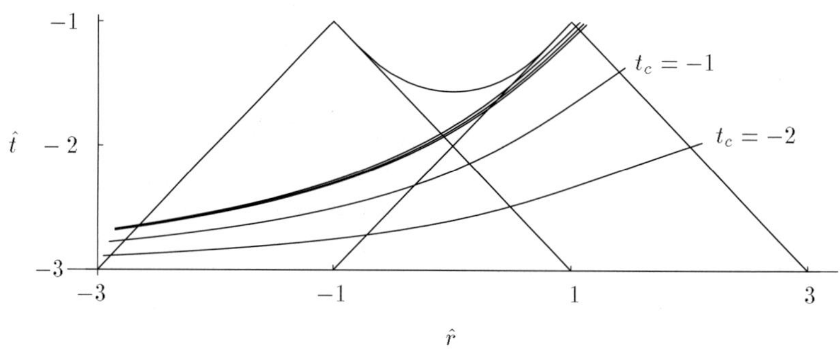

Fig. 14.5. The foliation

The evolved space-time can be represented as in Fig. 14.6. Again we see the horizons and null infinities. The accuracy of the metric components is about 2%.

Acknowledgments

I want to thank Peter Hübner and Marsha Weaver who did the numerical evolutions and figures. I am also grateful to Robert Beig who directed me to the paper by Brill et al.

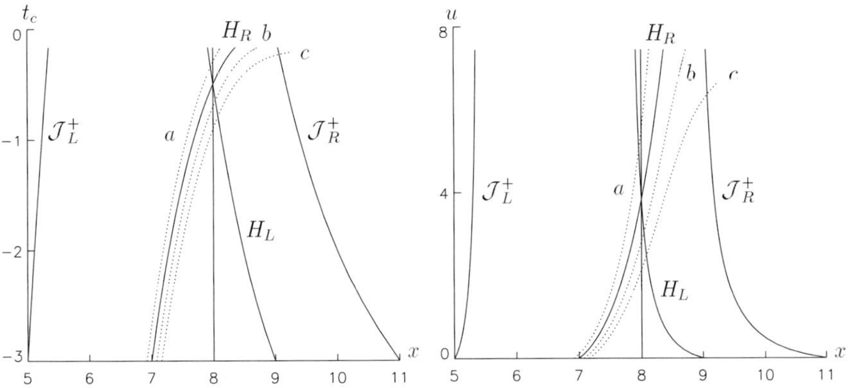

Fig. 14.6. Evolution2

References

1. D. R. Brill, J. A. Cavallo and J. A. Isenberg. *K-surfaces in the Schwarzschild spacetime and the construction of lattice cosmologies.* J. Math. Phys., **21**, 2789–2796, 1980.
2. P. Hübner. *How to avoid artificial boundaries in the numerical calculation of black hole space-times.* Class. Quant. Grav., **16**, 2145–2164, 1999.
3. P. Hübner. *A scheme to numerically evolve data for the conformal Einstein equation.* Class. Quant. Grav., **16**, 2823–2843, 1999.
4. P. Hübner. *Numerical calculation of conformally smooth hyperboloidal data.* Class. Quant. Grav., **18**, 1421–1440, 2001.
5. P. Hübner. *From now to timelike infinity on a finite grid.* Class. Quant. Grav., **18**, 1871–1884, 2001.
6. R. M. Corless, D. J. Jeffrey and D. M. Knuth, *A Sequence of Series for the Lambert W Function* Proceedings ISSAC '97, Maui, Wolfgang W. Kuechlin (ed.), 197–204.
7. P. Hübner and M. Weaver, 2001, in preparation.
8. B. G. Schmidt and J. Stewart. *The scalar wave equation in a Schwarzschild space-time.* Proc. R. Soc. A. **367**, 503–525, 1979.

15 Numerics of the Characteristic Formulation in Bondi Variables. Where We Are and What Lies Ahead

Luis Lehner

Department of Physics and Astronomy and
Pacific Institute for the Mathematical Sciences,
The University of British Columbia, Vancouver, BC V6T 1Z1

Abstract. Characteristic numerical relativity has established itself as a powerful approach towards the explicit construction of spacetimes through the use of computer simulations. Its use provides a direct relation between fields at $\mathcal{I}^+$ and the interior of the spacetime, being a useful practical tool to studying spacetimes in the large and providing additional information to deepen our understanding of the theory. We review the numerical implementation of the characteristic formulation of G.R. in Bondi variables, describe its main difficulties and some future directions of the program.

15.1 Introduction

In the 60's a series of novel ideas revolutionized the way gravitational radiation could be studied. Among these, Bondi's use of null hypersurfaces to foliate a space-time, which allowed for constructing and analyzing asymptotic solutions obtaining a clean description of radiation in axisymmetry [1] (and later generalized by Sachs [2]).

Soon after the conformal compactification of future null infinity $\mathscr{I}^+$ was introduced by Penrose [3]. Notably, the asymptotic description of radiative solutions, like mass and news function, could then be represented in purely geometrical language. The characteristic formulation of GR has since then, become a prime tool to understand conceptual issues related to gravitational radiation.

Just as this formulation proved extremely valuable in analytical investigations of the theory, it has since the late 80's also become a valuable tool for the numerical construction of space-times. As we will see shortly, the close connection of the dynamical variables with the true degrees of freedom, possibility of including the whole space-time in a finite grid, incoming null character of the outer boundary and structure of the equations are exploited to obtain robust implementations of Einstein equations which pave the way to explicitly construct space-times in a variety of situations.

In this article, we review the main properties and present efforts of one such implementation which is obtained through the use of Bondi variables[1]. We first

[1] For an alternative approach see the article by Bartnik and Norton in this volume [4].

describe the formalism in Sect. 15.2, then a few details of its numerical implementation in Sect. 15.3 and some present applications in Sect. 15.4. We conclude in Sect. 15.5 with some comments.

15.2 Characteristic Formulation of GR in Bondi Variables

In the Bondi approach a coordinate system adapted to the null foliation is chosen in the following way: the outgoing light-like hypersurfaces emanating from a time-like geodesic or world-tube are labeled with a parameter u, each null ray on a specific hypersurface is labeled with x^A $(A = 2, 3)$ and r is introduced as a surface area coordinate i.e., surfaces at $r = const$ have area $4\pi r^2$ (see Fig. 15.1). In the resulting $x^a = (u, r, x^A)$ coordinates, the metric takes the Bondi-Sachs form [1, 2]

$$\begin{aligned} ds^2 = &-\left(e^{2\beta}V/r - r^2 h_{AB}U^A U^B\right) du^2 - 2e^{2\beta} du dr \\ &-2r^2 h_{AB} U^B du dx^A + r^2 h_{AB} dx^A dx^B \,. \end{aligned} \tag{15.1}$$

Six real field variables appear in this form of the metric[2]: V, β, U^A and h_{AB}. They have a straightforward physical interpretation: h_{AB} represents the conformal intrinsic geometry of the surfaces defined by $dr = du = 0$ and contains the 2 degrees of radiative freedom. The field β represents the *expansion* of the light rays as they propagate radially. V is the analog of the Newtonian potential, and its asymptotic expansion contains the mass aspect of the system. Note that the coordinate system is tied to null surfaces which can intersect due to caustics or crossovers. In these cases, the coordinate system becomes singular! So, it is clear that this approach can not be used for arbitrary systems. (However, as we will comment in Sect. 15.5, one has several options to address the caustic/crossover problem in a number of cases, thus extending its range of applicability.)

The Einstein equations in the vacuum case, $G_{ab} = 0 = R_{ab}$, decompose into hypersurface equations, evolution equations and conservation laws. Bondi designated as the "main" Einstein's equations [1] those which correspond to the six components of the Ricci tensor, R_{rr}, R_{rA} and R_{AB}.

The hypersurface equations, given by R_{rr}, R_{rA} and $h^{AB}R_{AB}$, can be written as

$$\beta_{,r} = \frac{1}{16} r h^{AC} h^{BD} h_{AB,r} h_{CD,r} \;:, \tag{15.2}$$

$$(r^4 e^{-2\beta} h_{AB} U^B_{,r})_{,r} = 2r^4 \left(r^{-2}\beta_{,A}\right)_{,r} - r^2 h^{BC} D_C h_{AB,r} \;:, \tag{15.3}$$

$$\begin{aligned} 2e^{-2\beta} V_{,r} = &\, \mathcal{R} - 2D^A D_A \beta - 2D^A \beta D_A \beta \\ &+ r^{-2} e^{-2\beta} D_A (r^4 U^A)_{,r} - \frac{1}{2} r^4 e^{-4\beta} h_{AB} U^A_{,r} U^B_{,r} \;:; \end{aligned} \tag{15.4}$$

[2] Note that the areal r coordinate requirement in turn implies that $\det h_{AB}$ be that of the unit sphere metric, thus there are only two independent fields for h_{AB}.

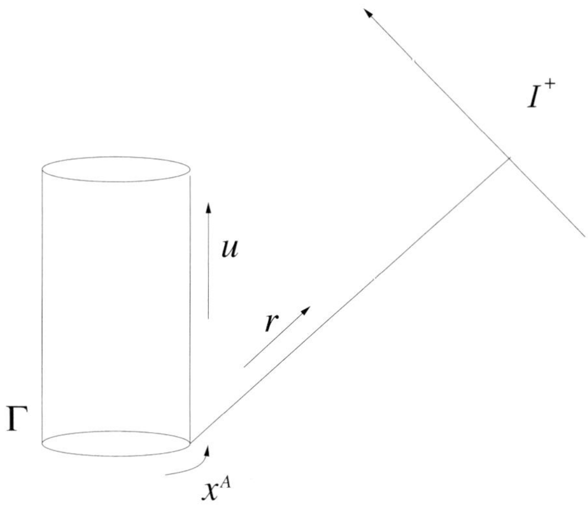

Fig. 15.1. The basic picture of the characteristic formulation. The whole space-time outside the world-tube Γ (including future null infinity) is included in the approach.

and the evolution equations, given by $R_{AB} - h_{AB}h^{CD}R_{CD}/2$, are expressed as

$$
\begin{aligned}
&r(rh_{AB,u})_{,r} - \frac{1}{2}(rVh_{AB,r})_{,r} = \\
&\Big(2e^{\beta}D_AD_Be^{\beta} - r^2h_{AC}D_BU^C_{,r} - \frac{r^2}{2}h_{AB,r}D_CU^C + \frac{r^4}{2}e^{-2\beta}h_{AC}h_{BD}U^C_{,r}U^D_{,r} \\
&\qquad - r^2U^CD_Ch_{AB,r} - 2rh_{AC}D_BU^C + r^2h_{AC,r}h_{BE}(D^CU^E - D^EU^C)\Big) \\
&- \frac{1}{2}h_{AB}\Big(r^2h^{CD}_{,r}(h_{CD,u} - \frac{V}{2r}h_{CD,r}) + 2e^{\beta}D_CD^Ce^{\beta} \\
&\qquad - D_C(r^2U^C)_{,r} + \frac{1}{2}r^4e^{-2\beta}h_{CD}U^C_{,r}U^D_{,r}\Big)
\end{aligned} \tag{15.5}
$$

where D_A is the covariant derivative and $\mathcal{R}$ the curvature scalar of the 2-metric h_{AB}. There is a natural hierarchy to integrate these equations, namely, assuming h_{AB} and consistent boundary values are known, the integration sequence (15.2)→(15.3)→(15.4), completely determines the metric on a given hypersurface. Last, equation (15.5) is integrated to obtain h_{AB} at the next hypersurface and the process starts again [5].

15.2.1 Initial Boundary Value Problem

So far, we have accounted for six hypersurface and evolution equations. Together with the extra equations $R^r_a = 0$, they form a complete set of components of the vacuum Einstein's equations. Given that the main equations are satisfied, the Bianchi identities imply they are satisfied on the space-time provided they hold on a single spherical cross-section. By choosing this sphere to be at infinity, Bondi identified these three equations as conservation conditions for energy and angular momentum.

Perhaps a problem that is considerably more difficult to address (than in a traditional 3+1 approach) in this formalism is to assess the well posedness of the system. A key issue here is that a 'local' regularity analysis of the solution can not be performed as the domain of dependence of a single nonsingular null hypersurface is empty. In order to pose the problem in a correct manner an additional boundary must be introduced[3] which can be either null, time-like or space-like (see Fig. 15.2).

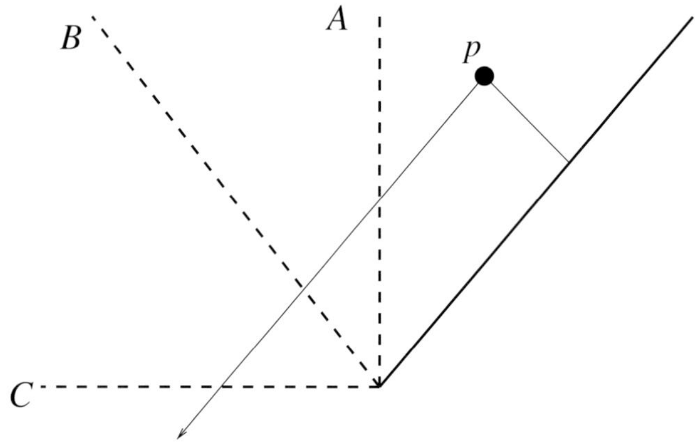

Fig. 15.2. Domain of dependencies for the time-like-null, double-null and space-like-null cases given by the intersection of the past null cone of the point p and the time-like, null or space-like curves A, B or C respectively

To date, rigorous results on well-posedness have been given only for the double-null problem [2, 7]. For the case where the boundary is time-like, only existence and uniqueness have been presented [8]. (Note: The recent theorem presented by Friedrich and Nagy [9] on well posedness of boundary value problems in GR for 3+1 approaches could be extended to include the time-like-null characteristic problem [10].)

In our current implementation, we accommodate for boundaries which are either null or time-like. Data is given on Γ such that the extra equations are satisfied. Initial data on the hypersurface $\mathcal{N}_o$ corresponds to the intrinsic geometry h_{AB}. Since $\mathcal{N}_o$ is a characteristic surface, h_{AB} can be freely specifiable (as opposed to the case where the initial hypersurface is space-like where the intrinsic and extrinsic geometries must satisfy a set of constraint equations, see for instance [11]). Consequently, providing valid initial data is trivial in this formalism; yet, whether these data makes physical sense for a particular problem

[3] Alternatively, the null hypersurface could be completed to a caustic-crossover region, but this option has not been implemented yet [6].

in mind is a delicate issue. We will discuss how it is addressed in two particular problems in Sect. 15.4.

15.2.2 News

The numerical calculation of asymptotic quantities such as the gravitational radiation must be capable of extracting non-leading terms in the asymptotic expansion about infinity. Additionally, it requires dealing with the asymptotic freedom at future null infinity (the BMS group) [1, 12]. In the analytical realm this is approached by adopting a Bondi frame at $\mathscr{I}^+$. However, in the numerical approach, it is often the case that the coordinate choice is adapted to the description of the interior sources and numerically evolved quantities must be transformed a-posteriori to a Bondi frame. In what follows, we describe how this issue arises in the calculation of the radiated energy.

The space-time is conformally compactified in terms of the metric $d\hat{s}^2 = \ell^2 ds^2$, where $\ell = 1/r$. In (u, ℓ, x^A)-coordinates, the line element $d\hat{s}^2$ takes the form

$$d\hat{s}^2 = -\left(e^{2\beta} V \ell^3 - h_{AB} U^A U^B\right) du^2 + 2e^{2\beta} du d\ell - 2h_{AB} U^B du dx^A + h_{AB} dx^A dx^B . \tag{15.6}$$

Here ℓ is a conformal factor with future null infinity $\mathscr{I}^+$ given by $\ell = 0$. The physical quantities governing the total energy and radiation power from the system are constructed from the leading coefficients (functions of u and x^A) in an expansion of the metric in powers of ℓ, in accord with asymptotic flatness at $\mathscr{I}^+$. The coefficients H, H_{AB}, c_{AB} and L^A which enter in the construction of the news function are defined by the expansions, $\beta = H + O(\ell^2)$, $h_{AB} = H_{AB} + \ell c_{AB} + O(\ell^2)$, $U^A = L^A + 2\ell e^{2H} H^{AB} D_B H + O(\ell^2)$ and $\ell^2 V = D_A L^A + \ell e^{2H}(\mathcal{R}/2 + 2D_A D^A H + 4D_A H D^A H) + O(\ell^2)$ (where we have related some of the expansion coefficients by using the asymptotic content of the Einstein equations [13]).

In order to construct the news function, we refer to a conformal Bondi frame [5] (which corresponds to an asymptotically Minkowskian inertial frame) with metric $d\tilde{s}^2 = \Omega^2 ds^2 = \omega^2 d\hat{s}^2$, with $\Omega = \omega\ell$, satisfying the gauge requirements that $Q_{AB} := \tilde{g}_{AB}|_{\mathscr{I}^+} = \omega^2 H_{AB}$ is intrinsically a unit sphere metric at $\mathscr{I}^+$ and that $\tilde{g}^{ab}\nabla_a\Omega\nabla_b\Omega = O(\Omega^2)$.

Our description is based on the work of Winicour [14], who discuses the calculation of the news in an arbitrary conformal frame relative to that in a Bondi frame.

$\mathscr{I}^+$ is geometrically a null hypersurface with null vector tangent to its generators given by $\tilde{n}^a = \tilde{g}^{ab}\nabla_b\Omega|_{\mathscr{I}^+}$ or, equivalently, $\hat{n}^a = \hat{g}^{ab}\nabla_b\ell|_{\mathscr{I}^+}$, with $\tilde{n}^a = \omega^{-1}\hat{n}^a$. In order to complete a basis for tangent vectors to $\mathscr{I}^+$, let Q^a be a complex field in the neighborhood of $\mathscr{I}^+$ satisfying $Q^a\nabla_a\Omega = O(\Omega)$, $\tilde{g}_{ab}Q^aQ^b = O(\Omega)$ and $\tilde{g}_{ab}Q^a\bar{Q}^b = 2 + O(\Omega)$. In the conformal Bondi frame, the news function[4] can then be expressed in the form [15]

$$N = \frac{1}{2\Omega} Q^a Q^b \tilde{\nabla}_a \tilde{\nabla}_b \Omega \tag{15.7}$$

and must be evaluated in the limit of $\mathscr{I}^+$. In terms of the $\hat{g}_{\alpha\beta}$ frame, with conformal factor $\hat{\Omega} = \Omega/\omega = \ell$, we then have

$$N = \frac{1}{2} Q^a Q^b \left(\frac{\hat{\nabla}_a \hat{\nabla}_b \hat{\Omega}}{\hat{\Omega}} - \omega \hat{\nabla}_a \hat{\nabla}_b \frac{1}{\omega} \right) . \tag{15.8}$$

We determine ω on $\mathscr{I}^+$ in the $\hat{g}_{ab}$ frame by solving the elliptic equation governing the conformal transformation of the curvature scalar intrinsic to the 2-geometry of a $u = const.$ slice of $\mathscr{I}^+$,

$$\mathcal{R} = 2(\omega^2 + H^{AB} D_A D_B \log \omega). \tag{15.9}$$

The condition that $\tilde{g}^{ab} \nabla_a \Omega \nabla_b \Omega = O(\Omega^2)$ is consistent with setting $\partial_\ell \omega = 0$ but gives a relation for the time dependence of ω. Noting that $\hat{g}^{ab} \nabla_a \ell \nabla_b \ell = e^{-2H} D_A L^A \ell + O(\ell^2)$ and that $\hat{g}^{ab} \nabla_b \ell = \hat{n}^a + O(\ell)$, we obtain $2\hat{n}^a \nabla_a \log \omega = -e^{-2H} D_A L^A$, which implies

$$\partial_u \log \omega = -(L^A \partial_A \log \omega + \frac{1}{2} D_A L^A). \tag{15.10}$$

This gives a prescription to evolve ω assuming that a solution of (15.9) is given as initial condition.

In order to obtain a convenient representation of Q^b, we exploit the spin rotation freedom ($Q^b \to e^{-i\alpha} Q^b$) so that the frame is parallelly propagated along the generators of $\mathscr{I}^+$ (which in turn determines the time dependence of α in a similar way as (15.10)).

After some rather lengthy but straightforward algebra (described in [15]), one arrives at the expression:

$$N = \frac{1}{2} e^{-2i\alpha} \omega^{-2} M^a M^b \left(\frac{\hat{\nabla}_a \hat{\nabla}_b \hat{\Omega}}{\hat{\Omega}} - \omega \hat{\nabla}_a \hat{\nabla}_b \frac{1}{\omega} \right) , \tag{15.11}$$

with $M^\alpha = (0, 0, M^A)$, where $H^{AB} = M^{(A} \bar{M}^{B)}$ (for an explicit derivation of the formulae presented here refer to [15]).

15.2.3 Inertial Coordinates

Equation (15.11) provides the news $N(u, x^A)$ in terms of the angular coordinates x^A determined by the gauge conditions on the inner world-tube Γ. The angular

[4] Our conventions are chosen so that the news reduces to Bondi's original expression in the axisymmetric case [1].

coordinates y^A of inertial observers at $\mathscr{I}^+$ are constant along the generators of $\mathscr{I}^+$, so they are related to the x^A coordinates by $\tilde{n}^\alpha \partial_\alpha y^A = 0$ or

$$(\partial_u + L^B \partial_B) y^A = 0 \,. \tag{15.12}$$

This equation for $y^A(u, x^B)$ can be solved with the initial condition $y^A(u_0, x^B) = x^A$, so that y^A are stereographic angular coordinates. This yields the gravitational waveform $N'(u, y^A) = N(u, x^B(u, y^A))$ measured by inertial observers but not in terms of their measured time but rather in retarded-time coordinate u in which the evolution is carried out. To complete the transformation to inertial coordinates, we re-express u in terms of Bondi time u_B, which is an affine parameter for the generators of $\mathscr{I}^+$ in a Bondi frame satisfying

$$\tilde{n}^\alpha \partial_\alpha u_B = 1 \,. \tag{15.13}$$

This allows us to rewrite the news as a complex function

$$\tilde{N}(u_B, y^A) = N'(u(u_B, y^A), y^A)$$

of inertial (Bondi) coordinates on $\mathscr{I}^+$ (see Fig. 15.3).

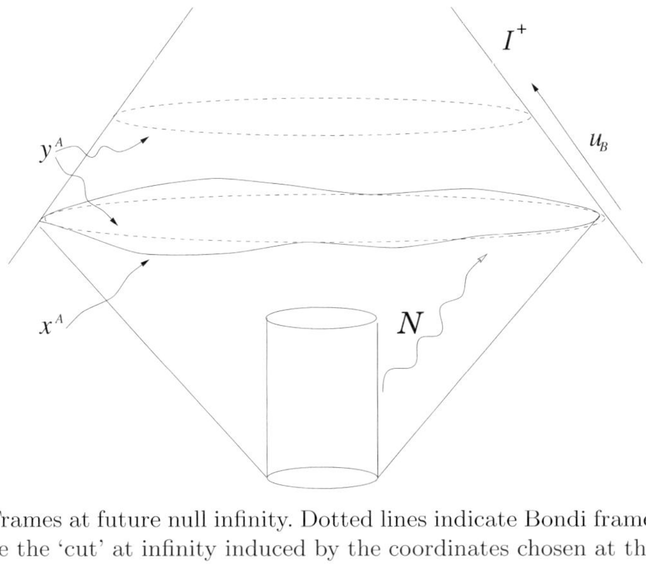

Fig. 15.3. Frames at future null infinity. Dotted lines indicate Bondi frames while solid one illustrate the 'cut' at infinity induced by the coordinates chosen at the world-tube Γ.

The news can now be readily decomposed into the two polarization modes. In order to make contact with the standard approach, we re-express the news in terms of a polarization dyad based upon standard (θ, ϕ) by a straightforward coordinate transformation. Finally, the decomposition of $\mathcal{N}$ into real and imaginary parts yields the two standard *plus* and *cross* polarization modes

$$N_+ = Re[\mathcal{N}] \;,\; N_\times = Im[\mathcal{N}] \,. \tag{15.14}$$

15.3 Numerical Details

A code that implements Einstein equations was written using standard second order finite difference approximations (FDA) as described in [15, 16]. A few features worth mentioning of this implementation are:

- *The use of a compactified radial variable.* In order to include "infinity" in the finite grid, we introduced a compactified radial coordinate $x = r/(1+r)$. Hence, the interval $r \in [0, \infty)$ is mapped to $x \in [0, 1]$ (where points at infinity are "added" to the space-time as specified by Penrose's formulation).
- *Use of spin-weighted quantities.* Instead of dealing with tensor fields on the sphere, we introduce a complex dyad q^A (such that $q^A q^B q_{AB} = 0$ and $q^A \bar{q}^B q_{AB} = 2$; where q_{AB} is the unit sphere metric) and construct spin weighted complex fields by appropriate contractions: given the vector fields U^A and h_{AB}, we define $U \equiv q_A U^A$ and $J \equiv q^A q^B h_{AB}$. This allows for a straightforward implementation of *eth* operators to handle angular derivatives and, since the obtained complex fields have well defined spin weights, the transformation properties upon angular coordinate changes is governed by a simple transformation law [17]. Additionally, note that when $h_{AB} = q_{AB}$, $J = 0$; so the field J measures 'departures' of spherical symmetry.
- *Use of the eth formalism for angular derivatives.* We follow the formalism given in [18]. To expedite the numerical implementation of angular derivatives, instead of working with the standard spherical angular coordinates (θ, ϕ), we employ stereographic coordinates:

$$x^A = (q, p) = (\tan(\theta/2)\cos(\phi), \tan(\theta/2)\sin(\phi)) ,$$

  and angular derivatives are written in terms of the (complex differential) eth operators $\eth$ and $\bar{\eth}$ [12, 19]. This allows us to employ a set of numerical techniques introduced in [18] which are specially tailored to: (*i*) handle the numerical FDA of angular derivative operators and (*ii*) deal with the fact that a single coordinate patch can not be used to smoothly cover the sphere.
- *Asymptotic normalization of variables.* Since asymptotically flat space-times predict a specific r dependence of the field variables, in the implementation, we make explicit use of this by adopting convenient integration variables so as to deal with finite values at $\mathscr{I}^+$. For instance, we integrate the combinations $\Psi \equiv (rJ)$ and $\tilde{W} \equiv (V - r)/r^2$.
- *Finite differencing of radial derivatives.* These are approximated via centered second order differences along each null ray (i.e., holding $x^A = const$); for instance,

$$\beta_i^n = \beta_{i-1}^n + (r_i - r_{i-1}) F_\beta|_{i-1/2}^n . \qquad (15.16)$$

The evolution equation can be easily integrated by recognizing that the left hand side of equation (15.5) is basically the wave operator acting on (rh_{AB}). This inspired the formulation of a useful integration algorithm described in [13] for the 1D case. In its bare bones, this algorithm integrates the wave equations on two pairs of intersecting characteristics (2 incoming, labeled with coordinates v and 2 outgoing, labeled with coordinates u), forming a "parallelogram". The wave equation in this coordinates is simply $\Phi_{,uv}$ = Source, and can be easily integrated by evaluating the fields at the corners of the parallelogram and the source at its center (see Fig. 15.4). This efficient algorithm was successfully applied to the axisymmetric case [20] as well as the $3D$ case in the linear to mildly nonlinear regimes [16]. However, it provided unstable evolutions in highly nonlinear problems due to exponential growth of high frequency modes in the solution. A rather straightforward analysis reveals that the amplification factor of the parallelogram algorithm is identically one, so it does not come as a surprise its failure. Nevertheless, a slight modification of this algorithm by adding numerical dissipation removes the instabilities [21].

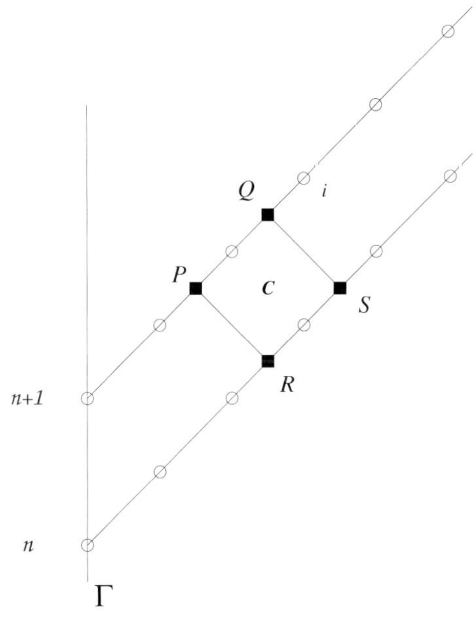

Fig. 15.4. Parallelogram algorithm. The figure shows the basic set-up for the evolution algorithm. The vertices are defined by the intersection of two outgoing and incoming null rays.

15.4 Applications

There are two basic requirements on a space-time (or space-time region) for a *straightforward* characteristic approach to be viable. First, it must be free of caustics (yet, more involved numerical implementations can in principle relax this condition as will be later discussed in Sect. 15.5) Second, initial and boundary data for the problem under consideration must be known.

There are a variety of interesting problems that do satisfy these requirements, I will here mention two of them which are of particular relevance for their potential to supply the needed waveforms required to detect and analyze the observed signals by the new generation of gravitational wave detectors [22].

15.4.1 Black Hole-Star Binaries

In this problem, we envision a space-time containing a black hole and another compact object (representing for instance a neutron star). The initial boundary value problem is posed on a pair of intersecting null hypersurfaces: one of them the incoming null surface corresponding to the white hole horizon of a black hole and the other an outgoing null surface where initial data is given (see Fig. 15.5). The code described in Sect. 15.4 is amended with the appropriate treatment of the fluid variables as discussed in [23, 24] (see also the article by Font in this volume [25]). In what follows we describe the setup and some preliminary results.

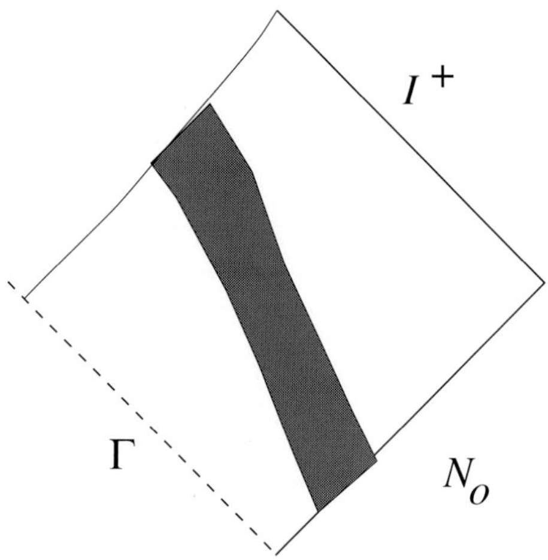

Fig. 15.5. Set up for the black hole-star binary case. Boundary data is provided on Γ which coincides with the white hole horizon. The initial hypersurface $\mathcal{N}_o$ extends to $\mathscr{I}^+$. The darker region indicates the evolution of the fluid describing the star.

Boundary Data. As mentioned we employ the white hole horizon as our inner boundary surface Γ. Here, data can be induced from the analytically known solutions of Schwarzschild or Kerr black holes, since data on the initial hypersurface can not influence the geometry of Γ this approach provides consistent boundary data in a straightforward way.

Initial Data. As already pointed out, providing valid initial data on a characteristic hypersurface is a trivial task, since we are setting data on a pair of intersecting null hypersurfaces, only regularity conditions at the intersection must be enforced. One could therefore, define data arbitrarily on $\mathcal{N}_o$ consisting of: the state of the fluid variables (ρ, ϵ, u^i), an equation of state and the intrinsic geometry of $\mathcal{N}_o$. Even when fluid configurations can be given representing a 'quasi-stable' star it is far from obvious which choices will correspond to the physical situation in mind not introducing spurious radiation which will contaminate the observed signals.

This problem is common to all efforts in Relativity trying to study a practical problem of astrophysical interest. In the usual strategy to address this issue, one provides consistent initial data trying to minimize the spurious radiation (although is unsure of how much is left) and evolves for long times. The idea is that after a few crossing times, the spurious radiation is 'flushed-out' to $\mathscr{I}^+$ leaving behind (up to small backscattering effects) the desired configuration (for an example of this proposal in the standard 3+1 approach, see for instance [26]).

There is, however, a fundamental impediment when applying this strategy to model binaries with the characteristic formulation. Namely, avoiding the presence of caustics puts a strong constraint on how far the star can be placed from the black hole. In particular, for some configurations, the components of the binary must be placed so close to that there is not enough time for the spurious radiation to be flushed out!

To see how this constraint comes about, consider the following straightforward approximation obtained from the lens equation. Caustics will arise as initially diverging null rays (emanating from the origin of our coordinate system) converge. The limit case corresponds to that were neighboring null rays neither diverge nor converge (become "parallel") at a finite distance from the origin. Take two of such rays at $\mathscr{I}^+$ and trace them backwards and assume that the only focusing agent is the star; as the rays are lensed by the star, their deflection angle is given by

$$R_*/D \simeq 2m_*/R_* \qquad (15.17)$$

where R_*, m_* and D are the radius and mass of the star and D the distance at which the rays cross. With this estimate, we can readily obtain a condition (an upper bound) for two rays at the origin of our coordinate system not to cross; namely we should demand that Dd (with d the separation of the BH and the star), which leads to

$$d < R_*^2/(4m_*). \qquad (15.18)$$

Condition (15.18) is considerably strong and requires the star be initialized close to the black hole which renders the "flushing-out" strategy useless for the cases of interest (like a black hole-neutron star binary). For these cases, we need a way to prescribe the intrinsic geometry of $\mathcal{N}_o$ so that it conforms to the physical situation and minimizes spurious radiation. A way to achieve this was introduced by Winicour in the 80's which fixes the initial data through a formal Taylor expansion in $1/\lambda$ where λ specifies the propagation speed of gravitational signals (this is a generalization of Cartan's Newtonian limit). By adjusting λ to ∞

one can 'reach' to Newtonian theory and read-off the expected functional behavior of h_{AB} in terms of the Newtonian potential. For instance, the first order approximation is obtained from

$$q^A q^B \partial_r (r \partial_r h_{AB}) = -\eth^2 \Phi \tag{15.19}$$

with Φ the Newtonian potential of the source. In our approach we employ this method and confirm its advantages by reading off the gravitational waves at $\mathscr{I}^+$ which correspond to the expected quadrupole radiation. Other alternatives, like setting the Weyl component $\Psi_0 = 0$ to determine h_{AB} on $\mathcal{N}_o$ produce radiation outputs which deviate considerably from the expected one.

Illustration. As an example of this problem we consider the following situation. Let's describe a star by a polytropic fluid defined by the equation of state $p = (\Gamma - 1)\rho_o \epsilon$ (with $\Gamma = 2$) whose center is at $D = 10M$ of a mass M non-spinning black hole. An interesting test of the code's ability to simulate the fluid's dynamic is to treat a situation where the star will tidally disrupt. For instance, consider the case where $R_* = 5M$ and a small compaction ratio of $m/R_* = 1/500$ and the fluids velocity is such that $u^a = (1, 0, v^q, 0)$ with v^q given by the angular velocity of a particle at $D = 10M$ in Schwarzschild space-time (the details of this set-up will be presented elsewhere [27]). Note for this case considered here the self-gravity of the star is quite weak and, furthermore, a considerable portion of the star lies inside the (inner most stable circular orbit) ISCO of particles around Schwarzschild black holes. We set this configuration and follow the star. Figure 15.6 illustrates the behavior of the density at different times. As time progresses, the star becomes quite disrupted as expected. These preliminary simulations and those obtained by Papadopoulos et. al (see Font [25] in this volume), clearly demonstrate the usefulness of characteristic numerical relativity to address a variety of problems in non-vacuum space-times.

15.4.2 Binary Black Hole Problem

As described, the characteristic formulation of GR in Bondi variables has been pursued to obtain a robust implementation which, given appropriate initial and boundary data can be used to numerically construct a space-time region (assuming this region is free of caustics). In particular, if we had these data corresponding to a binary black hole system, the problem would already be solved by now! However, we clearly do not possess this knowledge and so this statement is much too vague. In order to at least partially address this issue, we might envision approaching the problem by the following set up.

First, let's observe that the region of interest is defined by the past domain of dependence of $\mathscr{I}^+ \cup \mathcal{H}^+$ with $\mathcal{H}^+$ the event horizon of the system (see Fig. 15.7). We can choose to foliate the space-time with a sequence of incoming null hypersurfaces which intersect $\mathcal{H}^+$ which we regard as the inner boundary. Assuming data is known on $\{\mathscr{I}^+, \mathcal{H}^+\}$, one could compute an *advanced* solution by evolving backwards. The numerical evolution will not be capable to proceed beyond

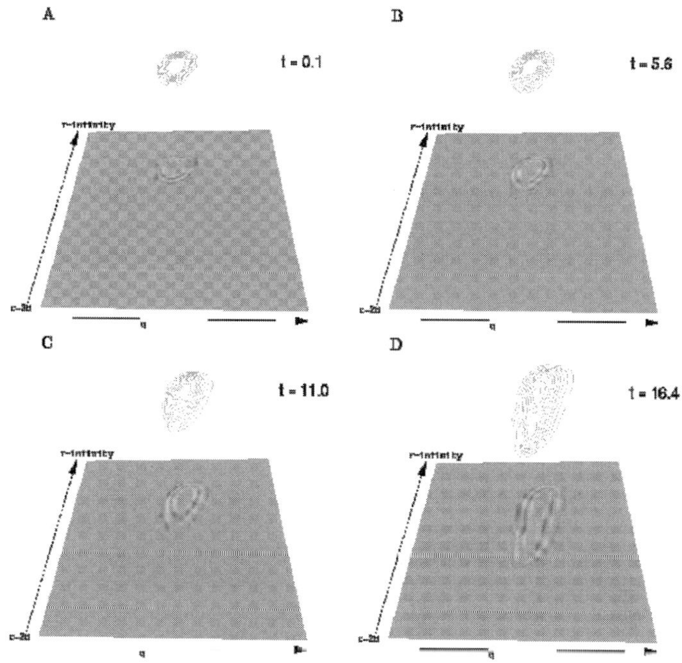

Fig. 15.6. Snapshot sequence of the density vs. r and q at times $t = 0.1, 5.6, 11$ and $16.4M$. The center of the star follows closely the geodesic of a particle at $D = 10M$ from a (mass M) Schwarzschild black hole while the portion of the star inside the ISCO suffers considerable tidal distortion.

some hypersurface Γ due to caustics. In the case where the solution corresponds to no radiation at $\mathscr{I}^-$ (to the future of $\Gamma \cap \mathscr{I}^-$) one would have provided data at $\mathscr{I}^+$ which corresponds to the physically correct picture. That is, no incoming radiation and the induced that on Γ would register the influence of the two clumps of gravitational radiation (in a vacuum scenario) which have converged to form the individual black holes.

However, this theoretical set-up, demands we have the correct that at $\mathscr{I}^+$, which is exactly the radiation an observer will measure (and what we were interested to calculate in the first place!). *Is this a dead end?*

The answer to this question is *not really*! This can be understood by considering the following two-stage approach. In the first stage, one can provide correct data on $\mathcal{H}^+$ from models describing the event horizon (obtained from a time reversal of a fissioning white hole [28–30]) and on $\mathscr{I}^+$ we set the condition of no outgoing radiation (which in turn provides the initial data h_{AB}). Evolve backwards to obtain the (advanced) solution on Γ and $\mathscr{I}^-$ (this, in general, will describe a situation where incoming radiation at $\mathscr{I}^-$ is present). Let's call this stage I and the obtained solution Φ_{ADV}.

Now, use the data induced on Γ from stage I and provide $-\Phi_{ADV}$ as data on $\mathscr{I}^-$. Then one can evolve to the future and obtain a retarded solution Φ_{RET}. If

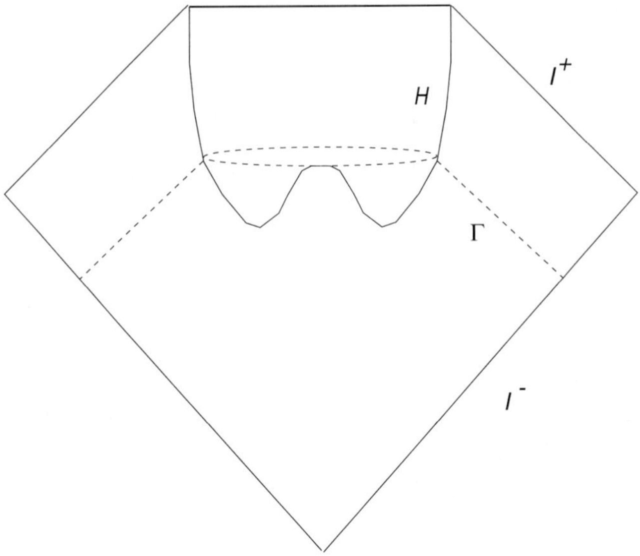

Fig. 15.7. Space-Time diagram for the binary black hole case. The horizon $\mathcal{H}^+$ and $\mathscr{I}^+$ are chosen as the inner boundary and initial hypersurface for stage I (where we employ an incoming characteristic foliation). Γ and the portion of $\mathscr{I}^-$ (to the future of $\Gamma \cap \mathscr{I}^-$) correspond to the inner boundary and initial hypersurface for stage II (where we employ and outgoing characteristic foliation).

Einstein equations were linear, then defining $\Phi = \Phi_{ADV} - \Phi_{RET}$ would be the solution to initial data corresponding to no incoming radiation from $\mathscr{I}^-$ and correct data on Γ.

Naturally, since Einstein equations are not linear, the obtained solution will be an approximation to the problem. However, we must recall that the total expected energy radiated is $\approx 5\%$. So the obtained solution might be a good approximation. Furthermore, one can think about iterating this procedure to correct higher order effects. In the linearized case corresponding to a close limit approximation of Schwarzschild, this approach has yielded excellent results [31], its application to the general case is currently under study [32].

15.5 Final Comments

As we have seen, the characteristic formulation of GR constitutes a powerful tool in the numerical construction of space-times[5]. In the cases where it can be applied it has shown its advantages in a remarkable way. Unfortunately, caustic formation limits, in principle, its applicability. However, a few options are available to remedy (at least in part) this disadvantage. First, as discussed by Friedrich and Stewart [6], since only a few structural stable caustics (which in turn are well understood) can arise, one could model their behavior numerically

[5] For an up to date review of different efforts in this area see [33].

and incorporate this treatment as part of the implementation. Second, it is also possible to use different patches to cover the space-time where a characteristic formulation is used on each patch [34]. The size of each 'patch' being such that no caustics should arise in them. Finally, it is certainly feasible to dynamically monitor the coordinate behavior as the integration marches radially outwards, if the proper distance between neighboring null rays starts to decrease (i.e., the congruence converges in some region) the evolution can be used to induce data on a new world-tube Γ_{new} and the integration is started over from there. Clearly, these proposals can extend the applicability of a characteristic approach but so far none of the options has been implemented in 3D. Future, more advanced implementations will likely take advantage of these proposals.

In general, constructing space-times numerically is a difficult and challenging task but the pay-off is certainly worthwhile as it provides an explicit solution that can not be otherwise obtained. Much of what we have learned from analytical studies has been restricted to either very specialized situations or incomplete description of space-times (e.g., understanding the general properties in asymptotic scenarios but with no explicit connection with its interior sources). As we dig deeper in the theory our understanding of it will advance further: to the already well developed mathematical techniques, numerical simulations, as they mature, will add a new and powerful tool to unravel the full implications of this beautiful theory. As mathematical and numerical techniques complement and help each other, future explorations promise to be exciting and hopefully full of surprises.

Acknowledgments

I am indebted to my close collaborators J. Winicour, N. Bishop, S. Husa and R. Gomez. Many parts of the work described in the present contribution is a joint collaboration with them. I wish to express my gratitude to J. Frauendiener and H. Friedrich for their friendly support and for having put forth a meeting combining expertise in analytical and numerical aspects of General Relativity which I hope proved as valuable to all involved as it did for me.

References

1. H. Bondi, M van der Burg and A Metzner, *Proc. R. Soc. London Ser. A*, 270, 103, (1962).
2. R Sachs, *Proc. R. Soc. A* **270**, 103, (1962).
3. R Penrose, *Phys. Rev. Letters* **10**, 66, (1963).
4. R Bartnik and A Norton. In this volume.
5. L. Tamburino and J. Winicour, *The Physical Review* **150**, 1039,. (1966).
6. H Friedrich and J M Stewart, *Proc. R. Soc. A* **385**, 345 (1983).
7. A. D. Rendall, Proc. Roy. Soc. Lond. **A427**, 221, (1990).
8. S. Frittelli and L Lehner, *Phys. Rev.* **D59**, 084012, (1999).
9. H Friedrich and G Nagy, *Comm. Math. Phys.* **204**, 691, (1999)
10. P Crusciel. Private communication.

11. J York, In *Sources of Gravitational Radiation*, ed. L Smarr, New York, Cambridge University Press. (1978).
12. E.T. Newman and K.P. Tod, In *General Relativity and Gravitation* V2, ed. Held A, Plenum, (1980).
13. R. A. Isaacson, J. S. Welling, and J. Winicour, J. Math. Phys. **24**, 1824 (1983).
14. J. Winicour, *Gen. Rel. and Grav.* **19** 281 (1987).
15. N. T. Bishop *et al.*, Phys. Rev. D **56**, 6298 (1997).
16. N. T. Bishop, R. Gómez, L. Lehner, and J. Winicour, Phys. Rev. D **54**, 6153 (1996).
17. E.T. Newman and R Penrose, *J. Math. Phys.* **3**, 566. (1962).
18. R Gomez, L Lehner, P Papadopoulos and J Winicour, *Class. Quant. Grav.* **14**, 977, (1997).
19. J N Goldberg *et al.*, *J. Math. Phys.* **8**, 2155, (1967).
20. R Gomez, P Papadopoulos and J Winicour, *J. Math. Phys.* **35**, 4184 (1994).
21. L Lehner, *Journal Comp. Phys.* **59** (1999).
22. A Abramovici *et al.*, *Science* **256**, 325, (1992). C Bradaschia *et al.*, *Phys. Lett.* **A137**, 329 (1989). K Danzmann and the GEO Team, *Lecture Notes in Physics* **410**, 184, (1992). K Tsubomo, M K Fujimoto and K Kuroda, In *Proceedings of the TAMA International Workshop on Gravitational Wave Detection*, Tokio, Universal Academic Press, (1996). Lisa Conference *Class. Quantum Grav.* **14**, 1397, (1996).
23. N T Bishop *et al.*, *Phys. Rev.* **D60**, 024005 (1999).
24. P Papadopoulos and J A Font, *Phys. Rev.* **D61**, 024015 (2000).
25. J Font. In this volume.
26. J Wilson, In *Sources of Gravitational Radiation*, ed. L Smarr, New York, Cambridge University Press (1978).
27. N. Bishop, R. Gomez, L. Lehner, M Maharaj and J Winicour. *The construction of initial data for matter and gravity in characteristic numerical relativity*, in preparation.
28. L Lehner *et al.*, *Phys. Rev. D* **D60**, 044005 (1999).
29. S Husa and J Winicour, *Phys. Rev. D* **D60**, 084019 (1999).
30. R Gomez, S Husa and J Winicour, *Phys. Rev. D* **D64**, 024010 (2001).
31. M Campanelli *et al.*, 2000, *Preprint gr-qc/0012107*.
32. L Lehner, R Gomez, S Husa N Bishop and J Winicour, *Computation of black hole merger waveforms*, in preparation.
33. J. Winicour, Living Reviews in Relativity, 1998.
34. L Lehner, *Int. J. Mod. Phys.* **D9**, 459 (2000).

16 Numerical Experiments at Null Infinity

Robert A. Bartnik and Andrew H. Norton

School of Mathematics and Statistics, University of Canberra, ACT 2601, Australia

Abstract. By comparing formal asymptotic expansions with the results of highly accurate numerical simulations using null quasi-spherical (NQS) coordinates, we develop a clear and consistent picture of the structure of the gravitational field for asymptotically flat spacetimes near null infinity. In particular, we find that "generic" spacetimes (in a sense to be made precise) have Weyl tensor components $\Psi_0 = O(r^{-4})$, $\Psi_1 = O(r^{-4}\log r)$, contrary to the Penrose-Sachs peeling hypothesis. This confirms results of Chruściel-MacCallum-Singleton, obtained using Bondi coordinates.

16.1 Introduction

The advent of high accuracy numerical codes promises to provide new insight into the behaviour of solutions of the Einstein equations. This report considers the perennial problem [7, 10–12, 17] of identifying appropriate asymptotic conditions for an asymptotically flat space-time. By comparing formal asymptotic expansions with the results of numerical simulations using null quasi-spherical (NQS) coordinates, we develop a clear and consistent picture of the structure of the gravitational field for asymptotically flat space-times near null infinity. In particular, we find that "generic" space-times (in a sense to be made precise) have Weyl tensor components $\Psi_0 = O(r^{-4})$, $\Psi_1 = O(r^{-4}\log r)$, contrary to the Penrose-Sachs peeling hypothesis. This confirms results of Chruściel-MacCallum-Singleton, obtained using Bondi coordinates.

Recall that Einstein [10] and Klein [11] showed that the coordinate decay conditions

$$g_{\mu\lambda} = \eta_{\mu\lambda} + O(r^{-1}), \qquad g_{\mu\lambda,\alpha} = O(r^{-2}) \tag{16.1}$$

are sufficient to ensure that the total energy-momentum integral

$$P_\mu[\Sigma] = \int_\Sigma (T^\alpha_\mu + \mathrm{t}^\alpha_\mu)\, dS_\alpha = \oint_S U^{\alpha\beta}_\mu\, dS_{\alpha\beta}, \tag{16.2}$$

where

$$U^{\nu\lambda}_\mu = \tfrac{1}{2}\sqrt{-g}g^{\beta[\alpha}\delta^\nu_\mu g^{\lambda]\tau} g_{\alpha\beta,\tau} \tag{16.3}$$

is the Einstein superpotential, defines a 4-vector which is independent of the choice of coordinate system (satisfying (16.1)) and the choice of space-like hypersurface Σ. Trautman [18] pointed out that the conditions (16.1) excludes gravitational radiation and, motivated by the Sommerfeld radiation condition in electrodynamics, he suggested that (16.1) be weakened to

$$g_{\mu\lambda} = \eta_{\mu\lambda} + O(r^{-1}), \tag{16.4}$$
$$g_{\mu\lambda,\alpha} = I_{\mu\lambda} k_\alpha r^{-1} + O(r^{-2}), \tag{16.5}$$

where k^α is the outgoing null vector and $I_{\mu\lambda} = O(1)$ satisfies

$$(I_{\mu\lambda} - \tfrac{1}{2}\eta_{\mu\lambda}\eta^{\alpha\beta} I_{\alpha\beta})k^\lambda = O(r^{-1}). \tag{16.6}$$

Under these conditions he showed that (16.2) is still well-defined and has non-negative flux through a time-like surface at infinity.

Subsequently Bondi [6] analysed similar asymptotic conditions but in a null-polar coordinate system, based on a curious but remarkably effective choice of radial function. He and his coworkers showed [7] that a consistent expansion of the metric coefficients in powers of r^{-1} could be established, provided that an additional condition, termed the "outgoing radiation condition" is satisfied. Sachs [15] showed that the resulting Weyl curvature tensor satisfies the peeling conditions

$$\Psi_n = O(r^{n-5}), \quad n = 0, 1, \ldots, 4, \tag{16.7}$$

expressed in terms of the Newman-Penrose Weyl spinor parameters Ψ_n [14]. The geometry underlying this behaviour was greatly clarified by Penrose [12], who showed that the peeling conditions follow from the assumption that the space-time near null infinity should admit a smooth (or at least C^4) conformal compactification.

The Bondi analysis was revisited in the work of Chruściel, MacCallum and Singleton [8], who showed the existence of consistent *polyhomogeneous* expansions (in powers of r^{-1} and $\log r$) for the Bondi metric. These formal solutions do not generally satisfy the Bondi outgoing radiation condition, but still could be considered as asymptotically flat.

In this paper we describe a similar analysis of the metric asymptotics, but in the NQS (null quasi-spherical) gauge [1]. This coordinate condition leads to expressions for the curvature which are much simpler than those in Bondi coordinates, and this greatly facilitates hand and numerical computation, as well as the geometric interpretation of the results. We compare this analysis to the results of numerical simulations using the NQS code [4, 5]. The NQS code models the interaction of gravitational radiation with a black hole to rather high accuracy [3] and is able, for example, to clearly distinguish the asymptotic behaviours of the Weyl tensor components.

Our principal conclusion is that full peeling (16.7) holds if and only if the spin-2 field ψ defined by (16.29) on $\mathscr{I}^+$, vanishes. Since ψ is conserved under the Einstein evolution, this vanishing condition is required only at one single cross-section of $\mathscr{I}$. However, expanding the outgoing Newman-Penrose shear

$$\sigma = \sigma_2 r^{-2} + \sigma_3 r^{-3} + O(r^{-4}), \tag{16.8}$$

we find that ψ has an expression solely in terms of the NQS equivalents of the coefficients σ_2, σ_3. Since these coefficients are initially unconstrained, and since the numerical evolution clearly shows that they become non-zero near $\mathscr{I}$ even if the initial data is compactly supported away from $\mathscr{I}$, it is natural to regard the vanishing of ψ as a *non-generic* condition.

These results are quite consistent with those of [8] in Bondi coordinates. The NQS derivation has the advantages that the obstruction $\psi = 0$ involves the leading terms in the metric (b_0, b_1, see (16.25)) rather than the high order ($O(r^{-2})$) conditions on the normalised 2-sphere metric h_{ab} which arise in Bondi coordinates, and the derivation does not assume the Bondi normalisation $\lim_{r\to\infty} \beta_{\text{Bondi}} = 0$ of the null coordinate. In fact, the main results remain valid if we assume only the rather weak asymptotic condition (16.8) — this will be analysed in detail elsewhere.

16.2 NQS Metrics

A *null quasi-spherical space-time* admits coordinates $(z, r, \vartheta, \varphi)$ in which the metric takes the form

$$\begin{aligned} ds^2 = & -2u\,dz\,(dr + v\,dz) \\ & + (r\,d\vartheta + \beta^1 dr + \gamma^1 dz)^2 + (r\sin\vartheta\,d\varphi + \beta^2 dr + \gamma^2 dz)^2\,, \end{aligned} \tag{16.9}$$

where z is an outgoing null coordinate, u, v are scalar (real) functions and $\beta = \beta^A\partial_A$, $\gamma = \gamma^A\partial_A$ are vector fields tangent to the S^2 surfaces $(z, r) = const.$ It is useful to encode β^A, γ^A in the spin-1 complex quantities $\beta = \frac{1}{\sqrt{2}}(\beta^1 - i\beta^2)$, $\gamma = \frac{1}{\sqrt{2}}(\gamma^1 - i\gamma^2)$, and to introduce the S^2 derivative operator $\eth$ ("edth") [9], which acts on a spin-s field η by

$$\eth\eta = \frac{1}{\sqrt{2}}\sin^s\vartheta\left(\frac{\partial}{\partial\vartheta} - \frac{\mathrm{i}}{\sin\vartheta}\frac{\partial}{\partial\varphi}\right)\left(\sin^{-s}\vartheta\,\eta\right)\,. \tag{16.10}$$

Using $\eth$ we define the real operators div and curl by $2\,\eth\bar{\beta} = \operatorname{div}\beta + i\operatorname{curl}\beta$, so $\operatorname{div}\beta = \eth\bar{\beta} + \bar{\eth}\beta = \beta^1_{;1} + \beta^2_{;2}$ is the divergence of a vector field on S^2, and the S^2-covariant operators

$$r\mathcal{D}_r = r\partial_r - \nabla_\beta = r\partial_r - (\beta\bar{\eth} + \bar{\beta}\eth), \quad r\mathcal{D}_z = r\partial_z - (\gamma\bar{\eth} + \bar{\gamma}\eth).$$

To describe the Einstein equations for the NQS metric, in [1] we introduced the connection parameters

$$\begin{aligned} H &= u^{-1}(2 - \operatorname{div}\beta), \\ J &= v(2 - \operatorname{div}\beta) + \operatorname{div}\gamma, \\ K &= v\eth\beta - \eth\gamma, \\ Q &= r\mathcal{D}_z\beta - r\mathcal{D}_r\gamma + \gamma, \\ Q^{\pm} &= u^{-1}(Q \pm \eth u). \end{aligned} \tag{16.11}$$

Using these parameters, the Einstein equations may be expressed in scale-invariant form, with operators covariant with respect to the standard metric on $S^2 \times \mathbb{R}^+ \times \mathbb{R}$. The most important are the *hypersurface equations*

$$
\begin{aligned}
r\mathcal{D}_r H &= \left(\frac{1}{2}\operatorname{div}\beta - \frac{2|\eth\beta|^2 + r^2 G_{\ell\ell}}{2-\operatorname{div}\beta}\right) H, && (16.12)\\
r\mathcal{D}_r Q^- &= (\eth\bar\beta - uH)Q^- + \bar Q^- \eth\beta + 2\bar\eth\eth\beta + u\eth H - H\eth u + 2r^2 G_{\ell m}, && (16.13)\\
r\mathcal{D}_r J &= -(1-\operatorname{div}\beta)J + u - \tfrac{1}{2}u|Q^+|^2 - \tfrac{1}{2}u\operatorname{div}(Q^+) - ur^2 G_{\ell n}, && (16.14)\\
2r\mathcal{D}_r K &= \left(3\eth\bar\beta - \bar\eth\beta\right)K - J\eth\beta + u\eth Q^+ + \tfrac{1}{2}u(Q^+)^2 + ur^2 G_{mm}, && (16.15)
\end{aligned}
$$

where the Einstein components correspond to the null tetrad

$$
\begin{aligned}
\ell &= \partial_r - r^{-1}\beta\\
n &= u^{-1}(\partial_z - r^{-1}\gamma - v(\partial_r - r^{-1}\beta))\\
m &= r^{-1}\tfrac{1}{\sqrt{2}}\left(\partial_\vartheta - \mathrm{i}\csc\vartheta\,\partial_\varphi\right).
\end{aligned}
$$

It is important to note that the metric functions u, v, β, γ and $\partial_z\beta$ can be reconstructed from β and the connection parameters (H, Q, J, K) on a single null hypersurface $\mathcal{N}_z$, by solving the elliptic system on S^2

$$\eth\gamma + \frac{\eth\beta}{2-\operatorname{div}\beta}\operatorname{div}\gamma = J\frac{\eth\beta}{2-\operatorname{div}\beta} - K. \tag{16.16}$$

This system is uniquely solvable with the gauge condition

$$\gamma = \sum_{l=2}^{\infty}\sum_{|m|\le l}\gamma_{lm}Y^1_{lm}, \tag{16.17}$$

provided β satisfies the bound

$$|\eth\beta| < \frac{1}{\sqrt{3}}(2-\operatorname{div}\beta).$$

Once γ is known, we recover v from

$$v = \frac{J - \operatorname{div}\gamma}{2-\operatorname{div}\beta} \tag{16.18}$$

and $\partial_z\beta$ from

$$\partial_z\beta = r\partial_r\gamma - \gamma - \eth u + uQ^+ + \nabla_\gamma\beta - \nabla_\beta\gamma. \tag{16.19}$$

However, for the asymptotic analysis it is simpler to use the (equivalent) parameters (b, u, q, m, k) where u is the coordinate function and

$$b = r\beta, \quad q = rQ^+, \quad m = \tfrac{1}{4}r(2 - HJ), \quad k = K/u. \tag{16.20}$$

With these parameters, the *hypersurface equations* become

$$r^3\mathcal{D}_r u = \frac{u}{2-\operatorname{div} b/r}\left(-r^2\mathcal{D}_r \operatorname{div} b + \tfrac{1}{2}(\operatorname{div} b)^2 + 2|\eth b|^2 + r^4 G_{\ell\ell}\right), \quad (16.21)$$

$$\begin{aligned}\mathcal{D}_r(rq) &= 2b + \eth\bar\eth b - \bar\eth\eth\bar b + r^{-1}\left(2q\eth\bar b + q\bar\eth b + \bar q\eth b\right)\\ &\quad + 2\eth(r^3\mathcal{D}_r \log u) + 2r^3 G_{\ell m},\end{aligned} \quad (16.22)$$

$$\begin{aligned}r^2\mathcal{D}_r m &= \tfrac{1}{4}r(\operatorname{div} q - 2\operatorname{div} b) + \left(\tfrac{3}{2}\operatorname{div} b - \frac{2|\eth b|^2}{r(2-\operatorname{div} b/r)}\right) m\\ &\quad + \tfrac{1}{8}(2-\operatorname{div} b/r)|q|^2 + \frac{|\eth b|^2}{2-\operatorname{div} b/r} - \tfrac{1}{8}\operatorname{div} b \operatorname{div} q\\ &\quad + \frac{(1-2m/r)}{2(2-\operatorname{div} b/r)} r^4 G_{\ell\ell} + \tfrac{1}{4}(2-\operatorname{div} b/r) r^4 G_{\ell n},\end{aligned} \quad (16.23)$$

$$\begin{aligned}2r^2\mathcal{D}_r k &= \left(3\eth\bar b - \bar\eth b - 2r^3\mathcal{D}_r \log u\right) k\\ &\quad + \eth q + \tfrac{1}{2}r^{-1}q^2 - \frac{2(1-2m/r)}{2-\operatorname{div} b/r}\bar\eth b + r^3 G_{mm},\end{aligned} \quad (16.24)$$

where we note that by (16.21), the term $r^3\mathcal{D}_r \log u$ depends only on b and $G_{\ell\ell}$.

The remaining Einstein equations are divided into the *subsidiary equations* $G_{nn} = 0$, $G_{nm} = 0$, and the *trivial equation* $G_{m\bar m}$, which are so-named because of the Sachs lemma, which shows that the subsidiary equations are propagated along the null hypersurface $\mathcal{N}_z$ and the trivial equation is satisfied identically if the hypersurface equations are satisfied. Thus it suffices to impose the subsidiary equations on just one hypersurface transverse to the outgoing null hypersurfaces $\mathcal{N}_z$. In practice the subsidiary and trivial equations provide valuable consistency and accuracy checks on the numerical results [5].

16.3 NQS Formal Asymptotics

Our aim is to understand the structure of solutions of (16.21–16.24) with $G_{ab} = 0$ as $r \to \infty$. We say that the field b is *log-free* if it admits an expansion in non-negative powers of r^{-1} without any $\log r$ terms, i.e.,

$$b = \sum_{n=0}^{\infty} b_n r^{-n}, \quad (16.25)$$

where the coefficients $b_n = b_n(z, \vartheta, \varphi)$ are smooth functions, independent of the radius r. Similar terminology is used for other fields.

The Penrose type I identity [13] for the Weyl spinor Ψ_0

$$\Psi_0 - \frac{\sigma}{\rho}\Phi_{00} = \rho D_\ell\left(\frac{\sigma}{\rho}\right) - (\rho^2 + |\sigma|^2 + 2\rho(\epsilon - \bar\epsilon))\frac{\sigma}{\rho}, \quad (16.26)$$

translates in the NQS coordinates to

$$r^4\Psi_0 = r^2\mathcal{D}_r(\eth b) - 2\eth b\,\eth\bar{b} - r^{-1}\eth b \cdot r^3\mathcal{D}_r \log u \tag{16.27}$$

since $\sigma = r^{-2}\eth b$ and $\sigma/\rho = -2\eth\beta/(2 - \operatorname{div}\beta)$. Under the assumption that b is log-free we have

$$\Psi_0 = -r^{-4}\psi + O(r^{-4-\epsilon}), \tag{16.28}$$

where we have defined the quantity $\psi = \psi(b_0, b_1)$

$$\psi = \eth b_1 + \bar{b}_0\eth^2 b_0 + b_0\bar{\eth}\eth b_0 + 2\eth b_0\eth\bar{b}_0. \tag{16.29}$$

Clearly the relation (16.28) holds under much weaker conditions on b. The importance of ψ is illustrated by the following basic result.

Theorem 16.1. *Fix a null hypersurface $\mathcal{N}_{z_0}$ and suppose $b = b(z_0)$ is log-free. Then $\partial_z b$ on $\mathcal{N}_{z_0}$ is log-free if and only if $\psi(z_0) = 0$, and then $\partial_z\psi(z_0) = 0$.*

Remark 16.1. In fact, it can be shown that $\partial_z\psi = 0$ holds under much more general conditions than $b(z_0)$ log-free, and in particular, it is not necessary to assume $\psi(z_0) = 0$.

Proof. We interpret the hypersurface equations as ordinary differential equations along the integral curves of the vector field $\partial_r - r^{-2}\nabla_b$ on $S^2 \times \mathbb{R}^+$. Let $x = r^{-1}$ and $\xi \in S^2$ be coordinates adapted to the integral curves, so the integral curves are just $\xi = const$. The transformation between the standard quasi-spherical coordinates $\theta = \theta^A = (\vartheta, \varphi)$ and the integral curve coordinates ξ is given by $(r,\theta) = (x^{-1}, \Theta(x,\xi))$, where $\Theta(x,\xi)$ satisfies the family of ODEs parameterised by $\xi \in S^2$,

$$\frac{\partial}{\partial x}\Theta^A(x,\xi) = b^A(x^{-1}, \Theta(x,\xi)) \tag{16.30}$$

with the initial conditions

$$\Theta^A(0,\xi) = \xi^A. \tag{16.31}$$

It follows from standard ODE theory and the log-free condition on b that the map $\xi \mapsto \Theta(x,\xi)$ is a smooth diffeomorphism for every $x \geq 0$, and admits a log-free expansion in the parameter x.

Now (16.21) may be written in the form

$$\partial_x\tilde{u} = -x\tilde{U}_1(b)\tilde{u}, \tag{16.32}$$

where $\tilde{u}(x,\xi) = u(x^{-1}, \Theta(x,\xi))$ and likewise,

$$U_1(b) = (\partial_x \operatorname{div} b + \tfrac{1}{2}(\operatorname{div} b)^2 + 2|\eth b|^2)/(2 - x\operatorname{div} b) \tag{16.33}$$

is evaluated at $(r,\theta) = (x^{-1}, \Theta(x,\xi))$. The log-free assumption on b implies that $U_1(b)$ is smooth with bonded derivatives of all orders up to and including $x = 0$. We see immediately that u is uniformly bounded with a well-defined limit u_0 as $r \to \infty$.

The explicit integral solution of (16.32),

$$\log \tilde{u} = \log \tilde{u}_0 - \int_0^x x\tilde{U}_1(b)\, dx,$$

shows that $\tilde{u}$ (and thus u itself) admits a log-free expansion, with in particular, $\tilde{u} = u_0(1 + O(x^2))$.

Equation (16.22) gives

$$\partial_r(rq) = r^{-2}\nabla_b(rq) + Q_0 + r^{-1}(Q_1 q + Q_2 \bar{q} + Q_3), \tag{16.34}$$

where $Q_0, \ldots, Q_3$ are bounded rational functions of b and its derivatives, so in particular they are log-free. Defining $\tilde{q}(x,\xi) = q(x^{-1}, \Theta(x,\xi))$ and estimating crudely gives the differential inequality $\partial_x(r|\tilde{q}|) \leq C_1 x^{-2} + C_2 r|\tilde{q}|$, where the constants C_1, C_2 depend only on b. It follows easily that q is bounded as $r \to \infty$. If we now define

$$q_0 = 2b_0 + \eth\bar{\eth}b_0 - \bar{\eth}\eth\bar{b}_0, \tag{16.35}$$

so $Q_0 = q_0 + O(r^{-1})$, then integrating (16.34) from any r_0 to r gives $rq = rq_0 + O(\log r)$. Substituting this into the right side of (16.34) and integrating shows that

$$rq - rq_0 + q_1 + q_{11}\log r + O(r^{-1}\log r),$$

where q_1 is an arbitrary constant (function) of integration and q_{11} may be determined directly from the explicit forms of the functions $Q_0, \ldots, Q_3$. A short calculation shows that

$$q_{11} = 2\bar{\eth}(\eth b_1 + b_0\,\bar{\eth}\eth b_0 + \bar{b}_0\,\eth\eth b_0 + 2\eth\bar{b}_0\,\eth b_0) \quad = \quad 2\bar{\eth}\psi. \tag{16.36}$$

Since $\bar{\eth}$ is injective on spin-2 fields, $q_{11} = 0$ if and only if $\psi = 0$, so $q = q_0 + r^{-1}q_1 + O(r^{-2}\log r)$. Substituting this into the right side of (16.34) and integrating, and then repeating this process with higher order terms, shows that q is log-free.

Since q_0 satisfies $\operatorname{div} q_0 = 2\operatorname{div} b_0$, the leading $O(r^{-1})$ term on the right in the equation (16.23) for m, vanishes. Since the remaining terms are higher order and involve only m and the log-free fields b, q, arguments similar to those used for (16.21) show that m is log-free, with arbitrary leading coefficient m_0. The equation (16.24) for propagating k contains no prospective log terms and all the source terms are now log-free, so k is also log-free, with arbitrary leading coefficient k_0.

It is easily verified that the leading terms (in the log free case) of the connection parameters are

$$u = u_0 - \nabla_{b_0}u_0 r^{-1} + O(r^{-2}), \tag{16.37}$$

$$q = q_0 + q_1 r^{-1} + O(r^{-2}), \tag{16.38}$$

$$m = m_0 + m_1 r^{-1} + O(r^{-2}), \tag{16.39}$$

$$k = k_0 - k_1 r^{-1} + O(r^{-2}), \tag{16.40}$$

where u_0, q_1, m_0 and k_0 are r-independent integration "constants", q_0 is given by (16.35), we are assuming $\psi = 0$, and

$$\begin{aligned} m_1 &= -\nabla_{b_0} m_0 - \tfrac{3}{2} m_0 \operatorname{div} b_0 \\ &\quad - \tfrac{1}{4}(\operatorname{div} q_1 - 2 \operatorname{div} b_1 + 2|\eth b_0|^2 - (\operatorname{div} b_0)^2 + |q_0|^2), \end{aligned} \tag{16.41}$$

$$2k_1 = -2\nabla_{b_0} k_0 + \eth b_0 - \eth q_0 + (\bar{\eth} b_0 - 3\eth \bar{b}_0) k_0. \tag{16.42}$$

Next, $\partial_z b$ is determined from

$$\partial_z b = r\partial_r \gamma - \gamma - \eth u + r^{-1}(uq + \nabla_\gamma b - \nabla_b \gamma), \tag{16.43}$$

and thus also admits a log-free expansion. The leading terms of (16.43) are

$$\partial_z b_0 = -(\gamma_0 + \eth u_0), \tag{16.44}$$

$$\partial_z b_1 = -2\gamma_1 - \eth u_1 + u_0 q_0 + \nabla_{\gamma_0} b_0 - \nabla_{\gamma_0} b_0, \tag{16.45}$$

where

$$\eth \gamma_0 = -k_0 u_0, \tag{16.46}$$

$$\eth \gamma_1 = -u_0 k_1 - u_1 k_0 + \tfrac{1}{2} \eth b_0 (u - \operatorname{div} \gamma_0). \tag{16.47}$$

Since $\psi = \eth b_1 + \bar{b}_0 \eth^2 b_0 + b_0 \bar{\eth} \eth b_0 + 2\eth b_0 \eth \bar{b}_0$, $\partial_z \psi$ may be computed using (16.44), (16.45), (16.46), (16.47), (16.42), (16.37), (16.35), resulting in $\partial_z \psi = 0$ as required. □

Theorem 16.2. *Suppose $b(z_0)$ is log-free, $b_0, b_1 \in C^1(I, C^\infty(S^2))$ where $I = [z_0, z_1]$, and $\psi(z_0) = 0$. Further suppose the free data u_0, k_0 satisfy the conditions $u_0 \in C^0(I, C^\infty(S^2))$ and $k_0 \in C^1(I, C^\infty(S^2))$. Then $\psi(z) = 0$ and b, u, q, m, k, v, γ are log-free for all times $z \in I$, and the Weyl curvature satisfies the peeling hypothesis in the stronger form,*

$$r^{5-n} \Psi_n \text{ is bounded and log-free}, \quad n = 0, \ldots, 4. \tag{16.48}$$

Remark 16.2. A good local existence theorem for the Einstein equations in the NQS gauge would allow us to weaken the conditions on the time dependence of b_0, b_1 to just conditions on the initial data $b(z_0)$. The simple structure of the hypersurface equations (16.12–16.15) suggests that such a local existence theorem should be possible. The conditions on u_0, k_0 are required primarily to control Ψ_4.

Proof. More detailed arguments similar to those of Theorem 16.1, show that $\partial_z \psi(z) = 0$ holds without the assumption $\psi(z) = 0$, for all $z \in I$. Thus $\psi(z) = 0$ for all $z \in I$ and Theorem 16.1 shows that b and all the metric and connection parameters are log-free for all times, provided the free data (u_0, q_1, m_0, k_0) are bounded (as fields in $C^\infty(S^2)$). The Einstein equations for G_{nn}, G_{nm},

$$\begin{aligned} r\mathcal{D}_z (J/u) &= v^2 r\mathcal{D}_r (J/(uv)) + (\tfrac{1}{2} J - v) J/u \\ &\quad + 2u^{-1}|K|^2 - \nabla_{Q^+} v - \Delta v + ur^2 G_{nn}, \end{aligned} \tag{16.49}$$

$$r\mathcal{D}_z Q^+ = \left(v\, r\mathcal{D}_r + J + \eth\bar{\gamma} - v\eth\bar{\beta}\right) Q^+ \\ - K\bar{Q}^+ + 2\bar{\eth}K + \eth J - 2u^{-1}\eth u\, J \\ + 2u^{-1}r\mathcal{D}_r(u\eth v) - (2 + \mathrm{i}\ \mathrm{curl}\,\beta)\eth v - 2ur^2 G_{nm}, \tag{16.50}$$

determine the time derivatives of m_0, q_1, and thus serve to bound m_0, q_1. The remaining pieces u_0, k_0 cannot be controlled by the Einstein equations; u_0 fixes the gauge freedom of varying the null foliation, whilst k_0 determines the outgoing gravitational radiation (see the discussion in [1]). For this reason, it is necessary to impose bounds on u_0, k_0 as above.

Since $\psi = 0$, (16.27) shows that $\Psi_0 = O(r^{-5})$ is log-free. The remaining Weyl spinors are given by the NQS expressions

$$2r^4\Psi_1 = r(2\bar{\eth}\eth b - \eth\,\mathrm{div}\,b - q) + \tfrac{1}{2}q\,\mathrm{div}\,b - \bar{q}\eth b + r^4 G_{\ell m} \tag{16.51}$$

$$r^3\Psi_2 = -m - \bar{k}\eth b + \tfrac{1}{4}(\bar{\eth}q - \eth\bar{q}) + \tfrac{1}{6}r^3(2G_{\ell n} + G_{m\bar{m}}) \tag{16.52}$$

$$2r^2\bar{\Psi}_3 = -2\eth k - r^{-1}\left(k\bar{q} + \frac{1 - 2m/r}{2 - \mathrm{div}\,b/r}q + \eth\left(\frac{\mathrm{div}\,b - 4m}{2 - \mathrm{div}\,b/r}\right)\right) \\ + r^2 G_{nm} \tag{16.53}$$

$$ur^2\bar{\Psi}_4 = v^2 r\mathcal{D}_r(k/v) - r\mathcal{D}_z k + (v - 2v\eth\bar{\beta} + 2\eth\bar{\gamma})(k/v) \tag{16.54}$$

$$+ r^{-1}q\eth v + \eth\eth v. \tag{16.55}$$

By inspection using the log-free asymptotics and $\psi = 0$, it is clear that peeling is satisfied and the Weyl spinors are also log-free, for the components Ψ_n, $n = 0, \ldots, 3$. Taking $\mathcal{D}_z$ of (16.24) and using the $G_{nm} = 0$ equation and the hypothesis that $\partial_z k_0 \in C^\infty(S^2)$ shows that $\mathcal{D}_z k$ is bounded. It follows by inspection of (16.55) that $r\Psi_4$ is bounded (so satisfying the final peeling condition) and log-free. □

It is remarkable how little use this argument makes of the parameter u which determines the null foliation: the hypersurface equations (16.22), (16.23), (16.24) (for $G_{\ell m}, G_{\ell n}$ and G_{mm} respectively) are independent of u, as are the Weyl spinors $\Psi_n, n = 0, \ldots, 3$. The Raychaudhuri equation (16.21) determines u, given boundary conditions such as u_0 which amount to fixing the null foliation. Thus we find that v, γ and the time (z) evolution of b, q_1, m_0 do depend on u – see (16.43) and (16.50), (16.49). The radiation component Ψ_4 is different, since (16.55) depends implicitly on u through v, γ, and we must introduce "by hand" control on $\mathcal{D}_z k$. Although regularity along the null hypersurfaces $\mathcal{N}_z$ follows from the hypersurface equations, regularity in the z-direction requires additional hypotheses, on the z dependencies of u_0, k_0 for example.

16.4 Genericity

An immediate corollary of Theorem 16.2 is that if $b(z_0)$ is compactly supported on some initial null hypersurface $\mathcal{N}_{z_0}$ (i.e., $b(z_0) \in C^\infty(\mathbb{R}^+, C^\infty(S^2))$), then the resulting space-time satisfies full peeling, with all metric, connection and Weyl curvatures being log-free. However, even though in this case the source field b is compactly supported initially, this condition breaks down immediately. Note that the Newman-Penrose shear is given by $\sigma_{NP} = r^{-2}\eth b$, so compactly supported initial data $b(z_0)$ is shear-free near $\mathscr{I}$ and the null hypersurface $\mathcal{N}_{z_0}$ has induced metric isometric to the Minkowski null cone near $\mathscr{I}$. Conversely, if $\sigma = 0$ on an expanding ($\rho \neq 0$) null hypersurface $\mathcal{N}$ with $S^2 \times \mathbb{R}^+$ topology, then $\mathcal{N}$ admits NQS coordinates with $\beta = 0$ [2].

It is clear from (16.44), (16.45) that generically, $\partial_z b_0$ and $\partial_z b_1$ are nonzero, and thus that b_0, b_1 will not vanish identically. This expectation is fully supported by the numerical solutions described below. Once b_0, b_1 are able to vary generally (in some space of initial data $b(z_0)$), it follows that *generically, $\psi(z_0) \neq 0$ and peeling will fail*, with $\Psi_0 = O(r^{-4})$ and $\Psi_1 = O(r^{-4}\log r)$.

This generic failure of peeling follows directly from the formulation we have used for the Einstein equations in terms of the shear variable b (or equivalently, β or σ_{NP}). We have seen that the asymptotic expansion of b will contain $\log r$ terms, even if these are not present initially. Thus it would be insufficient, for example, to attempt to pose the NQS equations for b in any space which required smoothness in $x = 1/r$ up to $x = 0$, since this would be consistent with initial data for which $\psi(z_0) \neq 0$ and then the evolution would force terms in $x^2 \log x$ in b at later times.

One consequence is that formulations of the Einstein equations which are based on a smooth conformal compactification, must include a condition equivalent to $\psi(z_0) = 0$, such as $\Psi_0 = O(r^{-5})$, since this necessary condition for peeling does not follow from conformally natural conditions on the shear. Formulations based directly on the unphysical variables rather than the physical shear will have the peeling conditions "built in", but on the other hand, will implicitly restrict the allowable behaviour of the physical shear.

This discussion applies only to metrics admitting NQS coordinates, and it remains an important open problem to develop a good existence theory for such coordinates in asymptotically flat metrics. There are good reasons for optimism: the existence of metric spheres in an expanding null hypersurface amounts to an isometric embedding problem, closely related to the uniformization theorem; an existence theorem in axial symmetry was established in [16]; and the linearized NQS gauge equations are elliptic for small β. Thus it appears likely that the behaviour described here for NQS metrics, applies to general asymptotically flat space-time metrics.

16.5 The NQS Code

The numerical integration procedure [5] starts with the field β on $\mathcal{N}_z$, and solves the propagation ODE system (16.12–16.14) with boundary initial conditions satisfying the subsidiary equations (16.49), (16.50). Once H, Q^+, J, K are determined on $\mathcal{N}_z$, γ is found by solving the elliptic system (16.16) at each radius. The remaining metric coefficients u, v are determined algebraically from β, γ and the definition of Q leads to an expression for the time derivative $\partial_z\beta$ on $\mathcal{N}_z$ in terms of known fields. Thus we can regard the solution of the hypersurface equation system as an evaluation process for $\partial_z \beta$ on $\mathcal{N}_z$, which can be used in an ODE integrator for $\beta(z)$.

The code uses an 8th order Runge-Kutta method to solve the hypersurface equations. A simple iteration scheme based on a spherical harmonic preconditioner solves (16.16) for γ at each radius, from which $\partial_z \beta$ is recovered. A standard 4th order Runge-Kutta method is then used for the time evolution of β.

There are a number of interesting numerical techniques used in the implementation. Angular dependence is handled by a combination of spin weighted spherical harmonics; point values on a rectangular grid in polar coordinates; and Fourier transforms in the polar coordinates. Interpolation and evaluation of radial derivatives uses a high order convolution spline, which also serves to suppress high frequency variations. The radial grid is distributed from $r = 2$ to $r = \infty$ and is adjusted dynamically to approximately follow the ingoing null geodesics. In particular, the outer grid values satisfy $r_n \sim (n_\infty - n)^{-2}$, so fields admitting polynomial expansions in r^{-1} will be modelled well by polynomials in $(n_\infty - n)^2$. This makes it quite easy to describe behaviour near $\mathscr{I}$.

The inner boundary conditions are fixed at $r = 2$ to agree with the Schwarzschild space-time at the past horizon, $H = 2, u = 1, Q^+ = J = K = 0$. Thus solutions represent gravitational radiation (modelled by β, which is chosen arbitrarily at time $z = 0$), propagating about a black hole of "bare" mass 1. With initial data chosen so that the total mass is about 1.2, that is, about 20% of the mass is attributable to gravitational radiation, we find that the code evolves β until around $z = 50$ when the $z = 50$ null hypersurface is close to an event horizon of mass 1. Such solutions are quite accurate, with relative error in β around 10^{-5}.

More details about the numerical algorithms, error estimation and space-time geometry may be found in [3, 5]. Sample data sets may be studied in detail at the interactive website [4], which was used to generate all the graphs in this report.

16.6 Numerical Results

We analyse two example simulations: an intermediate strength run `run_160` which has $b = r\beta$ initially ($z = 0$) compactly supported (c.f. Fig. 16.1), and `run_802`, for which b has a similar profile, but extended to have a non-zero limit at $\mathscr{I}$ ($n = n_\infty = 256$). The vertical axis in Fig. 16.1 represents the $L^2(S^2)$ norm over the sphere at (n, z). In each case β starts with the same spectral components (chosen as $l = 2, m = 2$, with equal odd and even parts). Nonlinear terms immediately lead to spectral components in β for all (l, m), as can be verified from the spectral distribution plots in [4].

Observe that in `run_160`, $b_0 = \lim_{r\to\infty} b$ is nonzero for $z > 0$, despite starting from compact support.

Fig. 16.2 shows the Weyl spinor Ψ_0 for the two runs. As predicted, `run_160` exhibits peeling, with $-\psi = \lim_{r\to\infty} r^4 \Psi_0 = 0$ and b_0, b_1 non-zero in general.

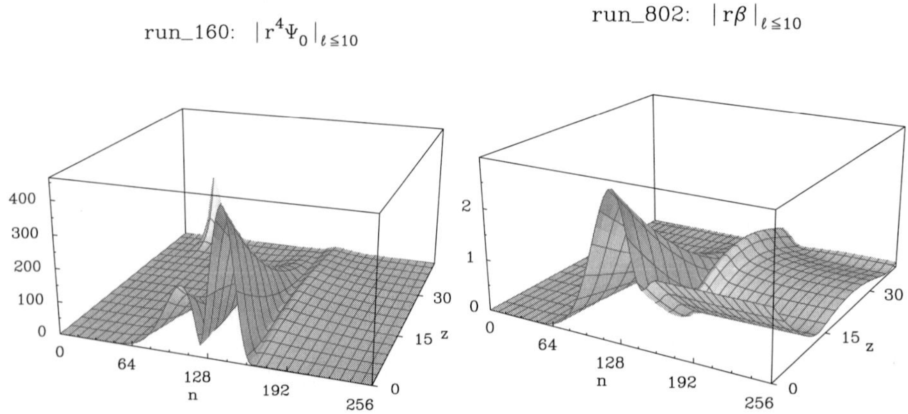

Fig. 16.1. Shear field $b = r\beta$ for the two simulations

The graph of $r^5 \Psi_0$ (not shown here, but available at [4]) shows a well-defined limit, further confirming the peeling behaviour. However, and again as predicted by the asymptotic analysis, $r^4 \Psi_0$ for `run_802` shows a breakdown of peeling, with $\psi \neq 0$. The graph suggests also that ψ is constant in z, which can be checked by inspection of the $(l, m) = (4, 4)$ component of $r^4 \Psi_0$ at $\mathscr{I}$. Note that the initial pure $(2, 2)$ structure of β and (16.29) imply that ψ is pure $(4, 4)$ for `run_802`. However, in general $r^4 \Psi_0$ has large spectral components other than the $(4, 4)$, as suggested by the `run_160` plot of the combined spectral components of $r^4 \Psi_0$ in Fig. 16.2.

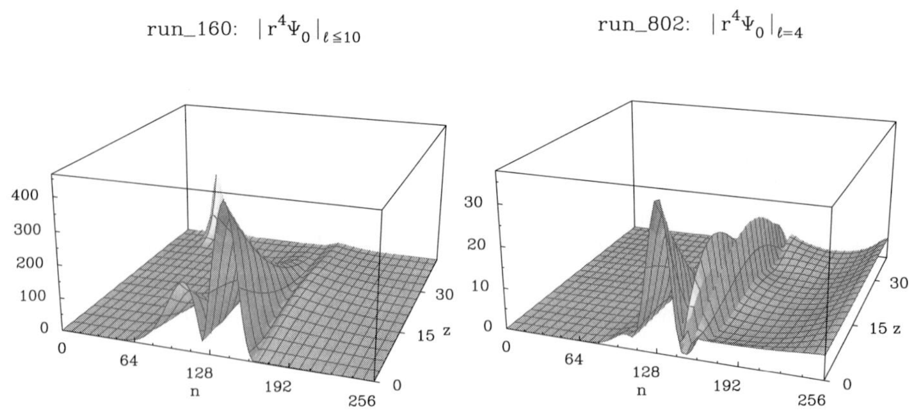

Fig. 16.2. Comparison of $r^4 \Psi_0$ shows peeling and non-peeling behaviour

Fig. 16.3 shows that $\Psi_1 = O(r^{-4})$ for `run_160` in accord with peeling, but the corresponding graph for `run_802` shows singular behaviour. From (16.51) with

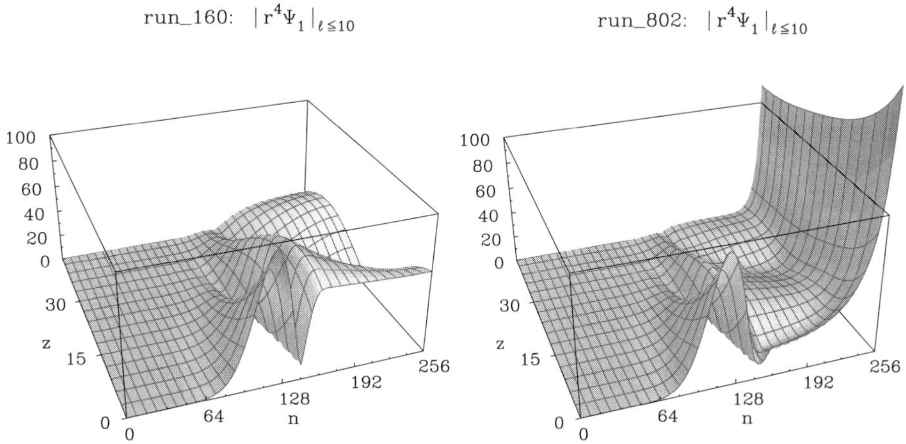

Fig. 16.3. $r^4\Psi_1$ shows peeling and non-peeling ($\log r$) behaviour

(16.35), (16.36) it follows that

$$r^4\Psi_1 = -\tfrac{1}{2}q_{11}\log r = -\bar{\eth}\psi \log r,$$

and the graph is consistent with this logarithmic blowup at $\mathscr{I}$. Finally we note that the remaining Weyl spinors Ψ_2, Ψ_3, Ψ_4 decay as expected, for both runs.

References

1. R. Bartnik. *Einstein equations in the null quasi-spherical gauge.* Class. Quant. Gravity, **14**, 2185–2194, 1997. Gr-qc/9611045.
2. R. Bartnik. *Shear-free null quasi-spherical spacetimes.* J. Math. Phys., **38**, 5774–5791, 1997. Gr-qc/9705079.
3. R. Bartnik. *Assessing accuracy in a numerical Einstein solver.* In G. Weinstein and R. Weikard, eds., *Differential Equations and Mathematical Physics: UAB-GIT 99*, AMS/IP Studies in Advanced Mathematics, volume 16, 11–19 (International Press, 2000).
4. R. Bartnik and A. Norton. *NQS data explorer.* Tech. rep., University of Canberra, 1997. `http://relativity.ise.canberra.edu.au`.
5. R. Bartnik and A. Norton. *Numerical methods for the Einstein equations in null quasi-spherical coordinates.* SIAM J. Sci. Comp., **22**, 917–950, 2000. Gr-qc/9904045.
6. H. Bondi. *Gravitational waves in general relativity.* Nature, **186**, 535, 1960.
7. H. Bondi, M. G. J. van der Berg, and A. W. K. Metzner. *Gravitational waves in general relativity VII.* Proc. Roy. Soc. Lond. A, **269**, 21–51, 1962.
8. P. T. Chruściel, M. A. H. MacCallum, and D. B. Singleton. *Gravitational waves in general relativity XIV. Bondi expansions and the polyhomogeneity of scri.* 1993.
9. M. Eastwood and P. Tod. *Edth – a differential operator on the sphere.* Math. Proc. Camb. Phil. Soc., **92**, 317–330, 1982.
10. A. Einstein. *Der Energiesatz in der allgemeinen Relativitätstheorie.* Sitzungsber. Preuss. Akad. Wissensch., 448, 1918.

11. F. Klein. *Über die Integralform der Erhaltungssätze und die Theorie der räumlich geschlossenen Welt.* Nachr. Ges. Wissensch. Göttingen, 394, 1918.
12. R. Penrose. *Asymptotic properties of fields and space-times.* Phys. Rev. Lett., **10**, 66–68, 1963. Penrose conformal completion of null infinity.
13. R. Penrose. *The geometry of impulsive gravitational waves.* In L. O'Raifaertaigh, ed., *General Relativity: in honour of J.L.Synge* (Oxford, 1972).
14. R. Penrose and W. Rindler. *Spinors and Spacetimes II: Spinor and twistor methods in space-time geometry* (Cambridge UP, 1986).
15. R. K. Sachs. *Gravitational waves in general relativity VIII. Waves in asymptotically flat space-time.* Proc. Roy. Soc. Lond., **A270**, 103–126, 1962.
16. M. Spillane. *The Einstein equations in the null quasi-spherical gauge.* Ph.D. thesis, Australian National University, 1994.
17. A. Trautman. *King's College lecture notes on general relativity*, 1958.
18. A. Trautman. *Radiation and boundary conditions in the theory of gravitation.* Bull. Acad. Pol. Sci., Série sci. math., **VI**, 407–412, 1958.

17 Local Characteristic Algorithms for Relativistic Hydrodynamics

José A. Font

[1] Max-Planck-Institut für Astrophysik,
Karl-Schwarzschild-Str. 1, 85740 Garching, Germany
[2] Departamento de Astronomía y Astrofísica, Universidad de Valencia,
Dr. Moliner 50, 46100 Burjassot (Valencia), Spain

Abstract. Numerical schemes for the general relativistic hydrodynamic equations are discussed. The use of conservative algorithms based upon the characteristic structure of those equations, developed during the last decade building on ideas first applied in Newtonian hydrodynamics, provides a robust methodology to obtain stable and accurate solutions even in the presence of discontinuities. The knowledge of the wave structure of the above system is essential in the construction of the so-called linearized Riemann solvers, a class of numerical schemes specifically designed to solve nonlinear hyperbolic systems of conservation laws. In the last part of the review some astrophysical applications of such schemes, using the coupled system of the (characteristic) Einstein and hydrodynamic equations, are also briefly presented.

17.1 Introduction

The numerical investigation of the Einstein equations is nowadays an important and fruitful line of research in general relativity. There exist a number of mathematical formulations of the gravitational field equations which are on the basis of all numerical approaches. The level accomplished in the understanding of these equations, both mathematically and numerically, is high enough to allow, in principle, for sound numerical approaches. Nevertheless, apart from some remarkable results, such as the discovery of critical phenomena by Choptuik [1] or, more recently, the achievement of long-term stable three-dimensional null cone evolutions of single black hole space-times [2], the numerical investigations have only partially succeeded, quite understandably due to the complexity of the theory, in providing global numerical solutions of generic space-times, especially in the presence of curvature singularities. Traditionally, the formulation which has received the greatest attention by numerical relativists has been the so-called 3+1 (ADM) formulation [3–5] (see also the recent review by Friedrich and Rendall [6] and references therein). This formulation of the Einstein equations as a Cauchy (initial value) problem, along with its multiple variants – hyperbolic (see, e.g., [7] and references therein) and conformal reformulations [8, 9] – is still today the workhorse of numerical relativity (for an up-to-date summary of the status of numerical approaches in 3+1 see, e.g., [10] and references therein.)

On the other hand, despite being known for about forty years now [11–13], the characteristic formulation of the Einstein equations has however been used by much less research groups in numerical relativity (see the recent review by Winicour [14]). And research programs to integrate numerically the conformal equations [15] have started only more recently [16–18]. Compared to the 3+1 formulation, the latter two formulations are best suited to study the conformal structure of the space-time – the main topic of the workshop -, and the propagation of radiation fields within the space-time. State-of-the-art numerical methodology applied to such two formalisms is comprehensively reviewed in the corresponding articles by Bartnik and Lehner (characteristic equations) and Frauendiener, Husa and Schmidt (conformal equations) in this volume. Apart from some test computations involving matter sources presented in Lehner's article, those papers are mainly concerned with the integration of the vacuum field equations.

The present contribution to this volume aims, on the other hand, at describing the current status of the numerical integration of the hydrodynamic equations on curved space-times, complementing, to some extent, the contributions by the above authors. Even though the description will be rather basic and general, I will also present some applications and some recent results obtained with the *coupled* integration of the Einstein and hydrodynamic equations within the framework of the characteristic formulation of the gravitational field equations. It is worth pointing out that, while initially the characteristic evolution of matter was limited to idealized systems such as massless scalar fields, nowadays it is mature enough to account for fully hydrodynamical evolutions with perfect fluids [19–26].

Admittedly, the motivation to develop the capabilities to perform coupled evolutions of the matter fields and the geometry needs really not much of an emphasis. In astrophysics general relativity plays a major role in the description of compact objects in such diverse scenarios as core collapse supernovae, black hole formation, accretion, gamma-ray bursts and coalescing compact binaries. With the exception of the coalescence and merging of two black holes – the number one problem of nowadays' numerical relativity – all realistic astrophysical systems and sources of (detectable) gravitational radiation involve matter.

The only means to study the time-dependent evolution of fluid flow coupled to geometry is through numerical simulations. Some scenarios can be properly described in the so-called 'test fluid' approximation, where the self-gravity of the fluid is neglected. Nowadays there is a large body of numerical investigations in the literature dealing with such hydrodynamical integrations in *static* background space-times (see, e.g., references in [27]). Most of these are based on the pioneering formulation of the hydrodynamic equations by Wilson [28] and use numerical schemes based on finite differences with some amount of artificial viscosity. The use of conservative formulations of the equations, and their characteristic information, in the design of numerical schemes started in more recent years [29].

On the other hand, time-dependent simulations of self-gravitating flows in general relativity, evolving the space-time dynamically with the Einstein equations coupled to a hydrodynamic source, are more scarce. Although there is much recent interest in this direction, only the spherically symmetric case has been extensively studied thus far (since the pioneering work of May and White in 1966 [30]). In axisymmetry, fewer attempts have been made, most of them devoted to the study of the gravitational collapse of rotating stellar cores and the subsequent emission of gravitational radiation [31–33]. The three-dimensional efforts are nowadays mainly focused on the study of the dynamics of relativistic stars [34–38], with the detailed study of the coalescence of close neutron star binaries being the key target [39–42]. These investigations are driven by the emerging possibility of detecting gravitational waves in a few years time with the different experimental efforts currently underway [43].

The current article deals with the presentation of the main ideas concerning a particular kind of the "specialized techniques" (in the language of [44]) used to solve nonlinear hyperbolic systems of conservation laws with finite differences. The discussion will be specialized to the general relativistic hydrodynamic equations. These equations – as well as their limiting counterparts in Minkowski space-time and Newtonian gravity – constitute a nonlinear hyperbolic system of conservation laws. For such systems there exist ever increasing sound mathematical foundations and accurate numerical methodology, imported from Computational Fluid Dynamics. The schemes that will be discussed here are the so-called *high-resolution shock-capturing* schemes (HRSC in the following), based upon Riemann solvers and written in *conservation form*. It is worth noticing that there are a number of excellent textbooks which deal with this subject in great detail, in particular [45–48] (see also the contribution by Kreiss in this volume). Recent reviews on numerical relativistic hydrodynamics are available as well [27, 49, 50]. The interested reader is addressed to these references for a more complete information.

The article is organized as follows: Sect. 17.2 presents the relativistic hydrodynamic equations emphasizing work done on conservative formulations. Such formulations are well-adapted to the numerical schemes which are discussed in Sect. 17.3. Applications of these algorithms are shown in Sect. 17.4. Finally, Sect. 17.5 closes the article with a short summary.

17.2 Relativistic Hydrodynamic Equations

The general relativistic hydrodynamic equations consist of the local conservation laws of the stress-energy tensor $T^{\mu\nu}$ (the Bianchi identities) and of the matter current density J^μ (the continuity equation):

$$\nabla_\mu T^{\mu\nu} = 0, \tag{17.1}$$

$$\nabla_\mu J^\mu = 0. \tag{17.2}$$

As usual ∇_μ stands for the covariant derivative associated with the four-dimensional space-time metric, $g_{\mu\nu}$. The density current is given by $J^\mu = \rho u^\mu$, where u^μ represents the fluid 4-velocity and ρ is the rest-mass density in a locally inertial reference frame. Greek (Latin) indices run from 0 to 3 (1 to 3) and geometrized units $G = c = 1$ are used in the following.

By neglecting non-adiabatic effects such as viscosity or heat transfer, the stress-energy tensor of a perfect fluid reads:

$$T^{\mu\nu} = \rho h u^\mu u^\nu + p g^{\mu\nu}, \tag{17.3}$$

where p is the pressure and h is the relativistic specific enthalpy, defined by

$$h = 1 + \varepsilon + \frac{p}{\rho}. \tag{17.4}$$

The quantity ε is the specific internal energy.

After choosing an explicit coordinate system $x^\mu = (x^0, x^i)$ the previous conservation equations read:

$$\frac{\partial}{\partial x^\mu}\sqrt{-g}J^\mu = 0, \tag{17.5}$$

$$\frac{\partial}{\partial x^\mu}\sqrt{-g}T^{\mu\nu} = -\sqrt{-g}\Gamma^\nu_{\mu\lambda}T^{\mu\lambda}, \tag{17.6}$$

where the scalar x^0 represents a foliation of the space-time with a family of hypersurfaces. Furthermore, $\sqrt{-g}$ is the volume element associated with the 4-metric, with $g = \det(g_{\mu\nu})$, and $\Gamma^\nu_{\mu\lambda}$ are the 4-dimensional Christoffel symbols.

In addition to the equations of motion (17.1) and to the continuity equation (17.2) the system must be closed with an equation of state (EoS) relating the pressure with some independent thermodynamical quantities, such as the rest-mass density and internal energy:

$$p = p(\rho, \varepsilon). \tag{17.7}$$

Relativistic hydrodynamic flows were first studied numerically with finite-difference schemes and explicit artificial viscosity terms [28, 30]. These terms [51] were necessary in order to damp the spurious numerical oscillations associated with the presence of discontinuities in the flow solution. Such approaches, albeit extensively (and successfully) used in different fields of computational relativistic astrophysics (e.g., gravitational collapse, accretion, coalescence of compact binaries, cosmology), were not able, however, to simulate flows with Lorentz factors γ larger than 2 [52], for which *implicit* methods were considered more appropriate. More recently, however, the use of artificial viscosity terms in non-grid based algorithms such as Smoothed Particle Hydrodynamics, has proven not to have such severe limitations [53].

The study of ultrarelativistic hydrodynamics with *explicit* finite-difference methods underwent a revival with the adoption of conservative formulations of the hydrodynamic equations and numerical methodology relying upon the hyperbolic nature of such system. Theoretical advances on the mathematical character of the relativistic hydrodynamic equations were achieved studying the special relativistic limit. In Minkowski space-time, the hyperbolic character of relativistic (magneto) hydrodynamics was exhaustively studied by Anile and collaborators (see [54] and references therein). The so-called high-resolution

Godunov-type schemes, with low numerical dissipation and oscillation-free representation of discontinuous solutions, based upon either exact or approximate Riemann solvers were extended during the 1990s from classical fluid dynamics to relativity [29, 55–58]. Nowadays, there exists increasing expertise, both theoretical and numerical, to investigate extremely fast flows through accurate computer simulations (see [49] for a recent review).

Traditionally, most of the approaches for numerical integrations of the general relativistic hydrodynamic equations have adopted space-like foliations of the space-time, within the 3+1 formulation [28, 35, 58]. Covariant and conservative formulations for ideal fluids, have been presented in [56] and [21]. From the theoretical point of view most of the existing formulations of the relativistic hydrodynamics equations are written in terms of quantities measured by an Eulerian (fixed) observer. By using Eulerian frame variables (relativistic densities of mass, momentum and energy) the equations exhibit a conservation form similar to their nonrelativistic counterparts. In most cases, contrary to Newtonian hydrodynamics, fulfilling the (desirable) conservation properties is accompanied by a nonlinear recovery process to extract physical (primitive) quantities (such as rest-mass density, sound speed, etc) from the conserved quantities forming the state vector of the system [21, 56, 58] (we note that a different approach based upon a *primitive-variable* formulation is given in [57, 59]).

As an example, in the formulation developed by Papadopoulos and Font [21] the spatial velocity components of the 4-velocity, u^i, together with the rest-frame density and internal energy, ρ and ε, are taken as the primitive variables. They constitute a vector in a five dimensional space $\mathbf{w} = (\rho, u^i, \varepsilon)$. The initial value problem for equations (17.5) and (17.6) is defined in terms of another vector in the same fluid state space, namely the *conserved variables*, $\mathbf{U} = (D, S^i, E)$:

$$D \equiv J^0 = \rho u^0 \,, \tag{17.8}$$

$$S^i \equiv T^{0i} = \rho h u^0 u^i + p g^{0i} \,, \tag{17.9}$$

$$E \equiv T^{00} = \rho h u^0 u^0 + p g^{00} \,. \tag{17.10}$$

With those definitions the hydrodynamic equations can be written as a first-order flux-conservative hyperbolic system of conservation laws:

$$\frac{\partial(\sqrt{-g}\mathbf{U})}{\partial x^0} + \frac{\partial(\sqrt{-g}\mathbf{F}^j)}{\partial x^j} = \mathbf{S} \,. \tag{17.11}$$

The flux vectors $\mathbf{F}^j$ and the source terms $\mathbf{S}$ are given by:

$$\mathbf{F}^j = (J^j, T^{ji}, T^{j0}) = (\rho u^j, \rho h u^i u^j + p g^{ij}, \rho h u^0 u^j + p g^{0j}), \tag{17.12}$$

$$\mathbf{S} = (0, -\sqrt{-g}\, \Gamma^i_{\mu\lambda} T^{\mu\lambda}, -\sqrt{-g}\, \Gamma^0_{\mu\lambda} T^{\mu\lambda}). \tag{17.13}$$

The local characteristic structure of these equations was presented in [21]. For the other conservative formulations mentioned above such information can be found in Refs. [35, 56, 58] (see also [60]). The relevance of having the wave structure to one's disposal in the development of HRSC schemes will become apparent in the following section.

17.3 High-Resolution Numerical Schemes

The hydrodynamic equations constitute a nonlinear hyperbolic system of conservation laws. Hence, smooth initial data can turn into discontinuous data (i.e., crossing of characteristics in the case of shocks) after a finite time during the evolution. Standard finite difference algorithms suffer from important deficiencies when dealing with such systems. Typically, first order accurate schemes are too dissipative across discontinuities (excessive smearing) while second order (or higher) schemes produce spurious oscillations near discontinuities, which do not disappear as the grid is refined.

Finite difference numerical schemes provide solutions of the discretized version of the original system of partial differential equations. Therefore, convergence properties under grid refinement must be enforced on such schemes to ensure the correctness of the numerical result (i.e., the global error of the numerical solution must vanish as the cell width is diminished). For hyperbolic systems of conservation laws, schemes written in *conservation form* are preferred, since – as proven by Lax and Wendroff [61] – if convergence exists, it is to one of the so-called *weak solutions* of the original system of equations. Such weak solutions are generalized solutions that satisfy the integral form of the conservation system $\partial_t \mathbf{U} + \partial_x \mathbf{F} = 0$:

$$\int_0^{\infty} \int_{-\infty}^{+\infty} (\Phi_t \mathbf{U} + \Phi_x \mathbf{F}(\mathbf{U})) dx dt = - \int_{-\infty}^{+\infty} \Phi(x,0) \mathbf{U}(x,0) dx, \qquad (17.14)$$

for any continuously differentiable test function $\Phi(x,t)$ with compact support. They are classical solutions (continuous and differentiable) in regions where they are smooth and have a finite number of discontinuities.

The class of all weak solutions is too wide in the sense that there is no uniqueness for the initial value problem. The numerical method should guarantee convergence to the *physically admissible solution*. This is the vanishing-viscosity solution of the "viscous version" of the hyperbolic problem:

$$\frac{\partial \mathbf{U}}{\partial t} + \frac{\partial \mathbf{F}(\mathbf{U})}{\partial x} = \eta \frac{\partial^2 \mathbf{U}}{\partial x^2}, \qquad (17.15)$$

when $\eta \to 0$. Mathematically, this solution is characterized by the so-called *entropy condition* (e.g., the entropy of any fluid element should increase when running into a discontinuity). The characterization of the *entropy-satisfying solutions* for hyperbolic systems of conservation laws was developed by Lax [62].

The Lax-Wendroff theorem [61] does not establish whether the method converges, for which some form of *stability* is required. Building upon the *Lax equivalence theorem* (see, e.g. [63]), the notion of total-variation stability has proven very successful, although sound results have only been obtained for (nonlinear) scalar conservation laws. The total variation of a numerical solution at time $t = t^n$, $\mathrm{TV}(\mathbf{u}^n)$, is defined as:

$$\mathrm{TV}(\mathbf{u}^n) = \sum_{i=0}^{+\infty} |\mathbf{u}_{i+1}^n - \mathbf{u}_i^n|. \qquad (17.16)$$

A numerical scheme is TV-stable if $\mathrm{TV}(u^n)$ is bounded at any time and for any initial data. Present-day research is focused on the development of high-resolution numerical schemes in conservation form satisfying the condition of TV-stability, such as the so-called *Total Variation Diminishing* (TVD) schemes [64]. An additional property that any numerical method should satisfy is monotonicity (see, e.g. [45]). For scalar conservation laws it has been shown that monotone methods are TVD and satisfy a discrete entropy condition. Therefore, they converge in a non-oscillatory manner to the unique entropy (physical) solution.

In a conservative scheme the time variation of the mean values of the state vector $\mathbf{U}$ in a given numerical cell – labeled by index i – is given, in the absence of source terms, by the flux differences across the cell interfaces. Mathematically, such an algorithm reads:

$$\mathbf{U}_i^{n+1} = \mathbf{U}_i^n - \frac{\Delta t}{\Delta x}(\hat{\mathbf{F}}_{i+\frac{1}{2}} - \hat{\mathbf{F}}_{i-\frac{1}{2}}), \tag{17.17}$$

where Δt and Δx stand for the time step and cell width, respectively.

Historically, in 1959 Godunov [65] developed the first conservative scheme for the classical fluid equations in which the *numerical fluxes*, $\hat{\mathbf{F}}_{i+\frac{1}{2}}$, at every cell interface of the computational grid were computed by exactly solving a family of local *Riemann problems*. Such Riemann problems – the simplest initial value problem with discontinuous initial data – arise naturally after the discretization procedure of the "continuous" solution by means of piecewise constant approximations. The Riemann problem is invariant under similarity transformations $(x,t) \to (ax, at), a > 0$. The solution is therefore constant along the characteristics $x/t = const.$ and, hence, self-similar. It consists of constant states separated by rarefaction waves, shocks and contact discontinuities.

Given a general hyperbolic system, if $\mathbf{U}(x,t) = \mathbf{w}_R(x/t; \mathbf{U}_-, \mathbf{U}_+)$ is the weak solution of a Riemann problem with initial data $\mathbf{U} = \mathbf{U}_-$ if $x < 0$ and $\mathbf{U} = \mathbf{U}_+$ otherwise, then the numerical fluxes in Godunov's scheme are given by:

$$\hat{\mathbf{F}}_{i+\frac{1}{2}} = \mathbf{F}(\mathbf{w}_R(0; \mathbf{U}_i^n, \mathbf{U}_{i+1}^n)), \tag{17.18}$$

along the characteristic $x/t = 0$.

The exact solution of the Riemann problem was extended to relativistic hydrodynamics in [66] (see also [67]). Since it involves solving a nonlinear algebraic system which can be computationally inefficient, in addition to the approximation involved in using piecewise constant data, the use of approximate solutions of the Riemann problem were proposed. Hence, if $\mathbf{w}(x/t; \mathbf{U}_-, \mathbf{U}_+)$ is such an approximation, the Godunov-type schemes are defined [68, 69] as those in which $\mathbf{U}_i^{n+1}$ are computed as:

$$\begin{aligned}\mathbf{U}_i^{n+1} = & \frac{1}{\Delta x}\int_{-\Delta x/2}^{0} \mathbf{w}(x/t; \mathbf{U}_i, \mathbf{U}_{i+1})dx + \\ & \frac{1}{\Delta x}\int_{0}^{\Delta x/2} \mathbf{w}(x/t; \mathbf{U}_{i-1}, \mathbf{U}_i)dx,\end{aligned} \tag{17.19}$$

and the numerical fluxes in the space-time computational cell $[x_{i-\frac{1}{2}}, x_{i+\frac{1}{2}}] \times [t^n, t^{n+1}]$ are given by:

$$\hat{\mathbf{F}}_{i+\frac{1}{2}} = \frac{1}{\Delta t} \int_{t^n}^{t^{n+1}} \mathbf{F}(\mathbf{U}(x_{i+\frac{1}{2}}, t))dt, \tag{17.20}$$

where $\mathbf{U}(x_{i+\frac{1}{2}}, t)$ is computed by (approximately) solving a Riemann problem at every numerical cell interface $\mathbf{U}(x_{i+\frac{1}{2}}, t) = \mathbf{w}(0; U_i^n, \mathbf{U}_{i+1}^n)$.

The mathematical and algorithmical developments accomplished in the scalar case have been extended to nonlinear hyperbolic systems of conservation laws using the so-called *local characteristic approach*. This technique generalizes the original procedure due to Roe [70] by applying the scalar algorithms to any of the characteristic equations of the system, after a suitable linearization. At each interface $i + 1/2$ of the computational grid, the Jacobian matrix $\mathbf{A}$ of the system, $\mathbf{A} = \partial\mathbf{F}/\partial\mathbf{U}$ is assumed to be constant $\widetilde{\mathbf{A}}_{i+1/2} = \mathbf{A}(\mathbf{U}_{i+1/2})$, with $\mathbf{U}_{i+1/2}$ being an average between $\mathbf{U}_i$ and $\mathbf{U}_{i+1}$. The original nonlinear system is then rewritten as $\partial_t\mathbf{U} + \widetilde{\mathbf{A}}\partial_x\mathbf{U} = 0$. The eigenvalues of this matrix are the characteristic speeds of the Riemann problem. The approximate Riemann solver obtains the exact solution of the linearized system, which can be easily computed by solving a system of decoupled, linear characteristic (scalar) equations. The properties that the matrix $\widetilde{\mathbf{A}}$ has to fulfill can be found in [70] for the widely used Roe's approximate Riemann solver.

As an illustrative example, for a second-order upwind, monotone scheme such as MUSCL [71], the expression for the numerical flux function reads:

$$\hat{\mathbf{F}}_{i+\frac{1}{2}} = \frac{1}{2}\left(\mathbf{F}(\mathbf{U}_i^{\mathrm{R}}) + \mathbf{F}(\mathbf{U}_{i+1}^{\mathrm{L}}) - \sum_{n=1}^{p} |\widetilde{\lambda}_{i+\frac{1}{2}}^{(n)}| \Delta\widetilde{\mathbf{w}}_{i+\frac{1}{2}}^{(n)} \widetilde{\mathbf{r}}_{i+\frac{1}{2}}^{(n)}\right). \tag{17.21}$$

Index p indicates the dimensions of the system. The quantities $\mathbf{w} = \mathbf{R}^{-1}\mathbf{U}$ are the so-called *characteristic variables*, $\mathbf{R}$ being the matrix whose columns are the right-eigenvector expressions of the Jacobian matrix associated with the vector of fluxes. Furthermore, λ and $\mathbf{r}$ stand for the eigenvalues and right-eigenvectors of such Jacobian matrix. The "tilde" indicates that all quantities have to be computed with respect to the linearized Jacobian matrix $\widetilde{\mathbf{A}}$.

The jumps of the characteristic variables at each cell interface are obtained by projecting the jumps of the state-vector variables with the left-eigenvectors matrix:

$$\Delta\widetilde{\mathbf{w}}_{i+\frac{1}{2}} = \widetilde{\mathbf{R}}_{i+\frac{1}{2}}^{-1}(\mathbf{U}_{i+1}^{\mathrm{L}} - \mathbf{U}_i^{\mathrm{R}}). \tag{17.22}$$

The left (L) and right (R) states of the conserved quantities $\mathbf{U}$ – at any cell interface – are computed from the cell-centered values after a suitable monotone *reconstruction procedure*. The way those variables are obtained determines the spatial order of the numerical algorithm and controls, in turn, the local jumps at every interface. A wide variety of cell reconstruction procedures is available in the literature (see, e.g. [71–73]).

The last term in the flux-formula, Eq. (17.21), represents the "numerical viscosity" of the conservative scheme. The wave structure of the system is thus used to provide the smallest amount of numerical dissipation yielding accurate solutions of discontinuities without excessive smearing, avoiding, at the same time, the growth of spurious numerical oscillations.

So far we have only considered one-dimensional systems of conservation laws. For multidimensional hyperbolic systems containing source terms a standard procedure to apply the above schemes is to use dimensional splitting – computing the numerical fluxes along every spatial direction independently -, possibly in combination with a method of lines. Therefore, for a three-dimensional hyperbolic system:

$$\frac{\partial \mathbf{U}}{\partial t} + \frac{\partial \mathbf{F}(\mathbf{U})}{\partial x} + \frac{\partial \mathbf{G}(\mathbf{U})}{\partial y} + \frac{\partial \mathbf{H}(\mathbf{U})}{\partial z} = \mathbf{S}(\mathbf{U}), \tag{17.23}$$

where $\mathbf{F}, \mathbf{G}$ and $\mathbf{H}$ are the fluxes in the x, y and z directions, respectively, the dimensional splitting algorithm reads:

$$\mathbf{U}^{n+1}_{i,j,k} = \mathcal{L}_s^{\Delta t/2} \mathcal{L}_h^{\Delta t} \mathcal{L}_g^{\Delta t} \mathcal{L}_f^{\Delta t} \mathcal{L}_s^{\Delta t/2} \mathbf{U}^n_{i,j,k}, \tag{17.24}$$

where $\mathcal{L}_f$, $\mathcal{L}_g$ and $\mathcal{L}_h$ denote the operators associated with the corresponding one-dimensional PDEs, i.e., the operators computing the numerical fluxes at every cell interface in a given direction. Furthermore, $\mathcal{L}_s$ is the operator which solves a system of ODEs for the source terms:

$$\frac{\partial \mathbf{U}}{\partial t} = \mathbf{S}(\mathbf{U}). \tag{17.25}$$

The state vector $\mathbf{U}$ at the final time t^{n+1} is then computed in consecutive substeps. On the other hand, in the method of lines the time update of all directions – and of the source terms – is done simultaneously. The conservative algorithm reads:

$$\begin{aligned} \frac{d\mathbf{U}_{i,j,k}(t)}{dt} = & -\frac{\widehat{\mathbf{F}}_{i+\frac{1}{2},j,k} - \widehat{\mathbf{F}}_{i-\frac{1}{2},j,k}}{\Delta x} - \\ & -\frac{\widehat{\mathbf{G}}_{i,j+\frac{1}{2},k} - \widehat{\mathbf{G}}_{i,j-\frac{1}{2},k}}{\Delta y} - \\ & -\frac{\widehat{\mathbf{H}}_{i,j,k+\frac{1}{2}} - \widehat{\mathbf{H}}_{i,j,k-\frac{1}{2}}}{\Delta z} + \hat{\mathbf{S}}_{i,j,k}, \end{aligned} \tag{17.26}$$

where the numerical fluxes are given by

$$\begin{aligned} \hat{\mathbf{F}}_{i+\frac{1}{2},j,k} &= \mathbf{F}(\mathbf{U}_{i-q,j,k}, \mathbf{U}_{i-q+1,j,k}, ..., \mathbf{U}_{i+q,j,k}), \\ \hat{\mathbf{G}}_{i,j+\frac{1}{2},k} &= \mathbf{G}(\mathbf{U}_{i,j-q,k}, \mathbf{U}_{i,j-q+1,k}, ..., \mathbf{U}_{i,j+q,k}), \\ \hat{\mathbf{H}}_{i,j,k+\frac{1}{2}} &= \mathbf{H}(\mathbf{U}_{i,j,k-q}, \mathbf{U}_{i,j,k-q+1}, ..., \mathbf{U}_{i,j,k+q}), \end{aligned} \tag{17.27}$$

q indicating the stencil chosen to compute these fluxes. Further details about multidimensional systems and source terms, with particular emphasis in numerical schemes for *stiff* source terms, can be found in [74].

We end this section by pointing out that during the last few years most of the *classical* approximate Riemann solvers developed in fluid dynamics have successfully been extended to relativistic hydrodynamics. The interested reader is referred to [50] for a comprehensive description of such solvers in relativistic hydrodynamics.

17.4 Applications

We now present some applications of the concepts introduced in the previous section. In particular we will show some results concerning the numerical evolution of the equations of hydrodynamics and the gravitational field within the context of the characteristic formulation of General Relativity. But let us start first with a demonstration in Minkowski space-time.

17.4.1 Shock Tube Test

A standard test to calibrate a hydrodynamics code based on the schemes discussed in the previous section is the so-called shock tube problem. This is a particular version of a Riemann problem in which the initial states at both sides of a discontinuity are at rest. Therefore, the state of the fluid at either side of the interface only differs in its thermodynamic quantities such as the density and the pressure.

When the interface is removed, the fluid evolves in such a way that four constant states develop. In between each state there can exist one of three elementary waves: a shock wave, a contact discontinuity and a rarefaction wave. As mentioned in the preceding section the exact wave pattern of the (ideal) fluid state at any given time was first obtained by Godunov [65] in Newtonian hydrodynamics. Its generalization to relativistic hydrodynamics was accomplished by [66] (see also [67]). This time-dependent problem provides a simple test of the shock-capturing properties of any numerical scheme, its level of difficulty depending on the initial data. A comprehensive survey of the behavior of a large sample of schemes applied to the shock tube problem is presented in [50] (the interested reader is also referred to [50] for further information on tests commonly used to validate numerical schemes for the hydrodynamic equations).

The main differences between the solution of relativistic shock tubes and their Newtonian counterparts are due to the nonlinear addition of velocities and to the Lorentz contraction. The first effect yields a curved profile for the rarefaction fan, as opposed to a linear one in the Newtonian case. The Lorentz contraction narrows the shock plateau. These effects, especially the latter, become particularly noticeable in the ultrarelativistic regime ($\gamma \gg 1$).

For our demonstration we consider a fluid whose initial state is specified by $p_L = 10^3$ and $\rho_L = 1$ on the left side of the interface and by $p_R = 10^{-2}$ and

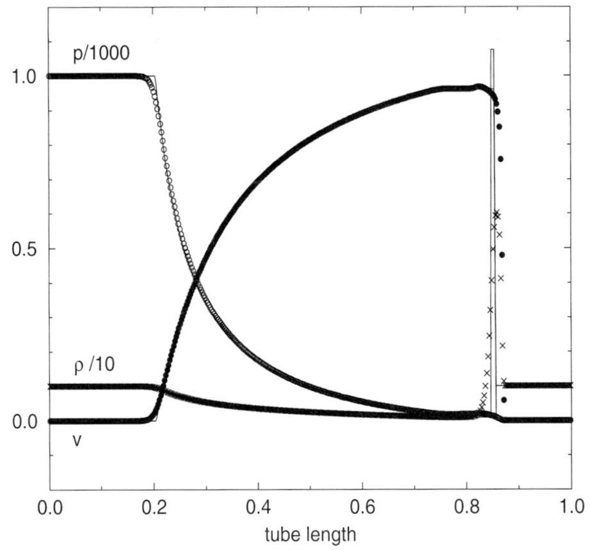

Fig. 17.1. The relativistic shock tube problem at time $t = 0.35$. Normalized profiles of density, pressure and velocity vs. distance for the computed (symbols) and exact (solid line) solution. All variables were calculated with a third order scheme on an equidistant grid of 400 zones. The initial interface was located at $x = 0.5$

$\rho_R = 1$ on the right side. This corresponds to problem 2 of [50]. The adiabatic index of the perfect fluid EoS, $p = (\Gamma - 1)\rho\varepsilon$, is $\Gamma = 5/3$. An initial jump in pressure of five orders of magnitude leads to the formation of a thin and dense shell bounded by a leading shock front and a trailing contact discontinuity – a blast wave. The post-shock velocity is $0.96c$ (Lorentz factor $\gamma \approx 3.5$), while the shock speed is $0.986c$ ($\gamma \approx 6$). Resolving the thin shock plateau poses a challenge for any numerical scheme.

Fig. 17.1 shows the results of the shock tube evolution employing a grid of 400 zones spanning a domain of unit length. The time of the comparison between the numerical and the analytic solution is $t = 0.35$. We have used a HRSC scheme based on the HLLE Riemann solver [68, 69] and a parabolic reconstruction procedure [72]. The solid lines indicate the exact solution. Correspondingly, the symbols represent the numerical approximation for the (scaled) pressure (open circles), density (crosses) and velocity (filled circles). As one can clearly see the location and propagation speeds of the different features of the solution are accurately captured. The shock plateau can be better resolved by simply increasing the numerical resolution (see [50, 75]). Diffusion-free results obtained with a one-dimensional exact Riemann solver are presented in [59].

17.4.2 Gravitational Collapse of Supermassive Stars

Supermassive black holes (SMBH), with masses on the range $10^6 M_\odot$- $10^9 M_\odot$ ($M_\odot$ indicating the mass of the Sun) are commonly found in the center of galaxies [76, 77]. Supermassive stars (SMS) have been proposed as possible progenitors of SMBH. Such stars can develop a dynamical instability [78, 79] and undergo catastrophic gravitational collapse.

Recently, the gravitational collapse of SMS was proposed by Fuller and Shi [80] as a possible model for gamma-ray bursts. The neutrino emission from the collapse of a SMS could lead to energy deposition by $\nu\overline{\nu}$-annihilation $\nu_{e,(\mu,\tau)} + \overline{\nu}_{e,(\mu,\tau)} \rightarrow e^- + e^+$. Subsequently, γ-radiation would be produced by cyclotron radiation and/or the inverse Compton process.

Trying to shed some light on the viability of that mechanism, Linke et al [23] have studied numerically the gravitational collapse of spherical SMS using a general relativistic hydrodynamics code. The code is based on the hydrodynamics formulation developed by [21]. The coupled system of Einstein and fluid equations is solved adopting a space-time foliation with outgoing null hypersurfaces. In such framework and in spherical symmetry, the Bondi-Sachs metric [11, 12] reads:

$$ds^2 = -\frac{e^{2\beta} V}{r} du^2 - 2e^{2\beta} du dr + r^2 (d\theta^2 + \sin^2\theta d\phi^2). \tag{17.28}$$

The Einstein equations reduce to two radial hypersurface equations (ODEs) for $\beta(u, r)$ and $V(u, r)$:

$$\beta_{,r} = 2\pi r e^{4\beta} E, \tag{17.29}$$

$$V_{,r} = e^{2\beta} - 8\pi r^2 e^{4\beta} S^r - 4\pi r e^{4\beta} V E, \tag{17.30}$$

where "," indicates partial differentiation. Correspondingly, the hydrodynamic equations are given by expressions (17.8)-(17.13), appropriately particularized to spherical symmetry, and are solved using HRSC schemes [1].

In addition, the code developed by [23] includes a tabulated EoS which accounts for contributions from radiation, electron-positron pairs and baryonic gases, as well as energy losses by thermal neutrino emission.

A typical simulation of a collapsing SMS is depicted in Fig. 17.2. The initial model corresponds to a $5 \times 10^5 M_\odot$ SMS. The figure shows the radial profiles of the evolution of the density, temperature, metric components g_{uu}, g_{ur}, and radial velocity. Furthermore, the space-time diagram at the lower right panel shows the local proper time against the location of mass shells enclosing fixed fractions of the total mass of the star. The arrow indicates the slope of a light

[1] The reader must be aware of the different meaning of the word "characteristic" in the context of the Einstein equations and the hydrodynamic equations: while in the former case the *characteristic formulation* of general relativity refers to a particular slicing of the space-time – by means of null cones – in the latter case the notion of the *local characteristic approach* refers to a numerical procedure seeking to exploit, algorithmically, the upwind character of the hydrodynamic equations.

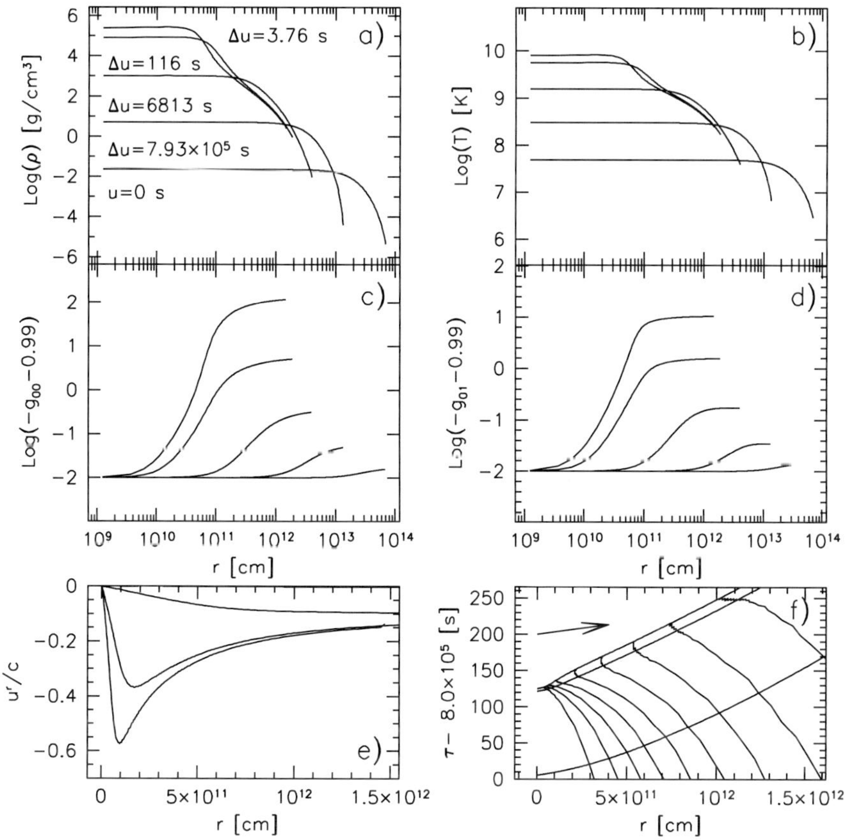

Fig. 17.2. Gravitational collapse of a $5 \times 10^5 M_\odot$ SMS. Snapshots of the evolution of the density, temperature, metric components g_{uu}, g_{ur} and radial velocity u^r. The space-time diagram of the lower right panel shows the location of mass shells ($\Delta M = 5 \times 10^4 M_\odot$) versus local proper time τ. The lines intersecting the mass shells are hypersurfaces of constant coordinate time u and represent trajectories of outgoing light rays. A black hole forms enclosing $\sim 25\%$ of the mass of the SMS. The figure is taken from [23]

ray in Minkowski space-time. One can see that light rays are severely delayed close to the forming black hole. This black hole forms from the innermost 25% of the total stellar mass. The collapse lasts 8×10^5s (~ 9.3 days) and the central density increases by a factor of 1.08×10^7. The final configuration becomes highly relativistic before the simulation is stopped, with $g_{uu} = -119$ at the surface of the star ($g_{uu} = -1.0058$ initially). Details about the neutrino emission in such an evolution can be found in [23].

This and other simulations have been used in [23] to analyze the possibility that collapsing SMS could be progenitors of gamma-ray bursts [80]. The comprehensive study performed in [23] reveals that 99% of the energy produced by $\nu\bar{\nu}$–annihilation is deposited in a spherical layer deep inside the star at a radius $R_\nu \leq r \leq 3R_\nu$, where R_ν indicates the location of the neutrino radiating volume. Therefore, only a tiny fraction of the energy is deposited near the surface of the star where excessive baryon loading could be avoided. As a result, ultrarelativistic ejection of matter with Lorentz factors $\gamma \gg 1$, a distinctive feature of all gamma-ray burst models, cannot be expected in spherical models. The simulations performed in [23] show that the spherical collapse of a SMS ($M \geq 5 \times 10^5 M_\odot$) does not meet the demands for being a successful central engine for a gamma-ray burst.

17.4.3 Null Cone Evolution of Relativistic Stars

In [24, 26] we presented the first results of a program we have recently started to study the dynamics of relativistic stars by means of null cone simulations in axisymmetry. The final aim is to study the gravitational core collapse problem and to compute the associated gravitational radiation [33, 81].

For these investigations we use the Bondi-Sachs metric:

$$\begin{aligned} ds^2 = &-\left(\frac{V}{r}e^{2\beta} - U^2r^2e^{2\gamma}\right)du^2 - 2e^{2\beta}du\,dr - 2Ur^2e^{2\gamma}du\,d\theta \\ &+r^2(e^{2\gamma}d\theta^2 + e^{-2\gamma}sin^2\theta\,d\phi^2), \end{aligned} \tag{17.31}$$

with null coordinate u, radial coordinate r, polar coordinate θ and the azimuthal coordinate ϕ, which is a Killing coordinate. Using this metric the Einstein equations split into hypersurface equations on each light cone (for the fields β, U and V), and one evolution equation (for the field γ, not to be confused with the Lorentz factor of the fluid), a wave equation (see also Lehner's article in this volume). The hypersurface equations, $G_{1\nu} - 8\pi T_{1\nu} = 0$, read:

$$\beta_{,r} = \frac{1}{2}r\,(\gamma_{,r})^2 - \frac{1}{4}rR_{rr}, \tag{17.32}$$

$$[r^4\,e^{2(\gamma-\beta)}U_{,r}]_{,r} = 2r^2\left[r^2\left(\frac{\beta}{r^2}\right)_{,r\theta} - \frac{(\sin^2\theta\,\gamma)_{,r\theta}}{\sin^2\theta} + 2\,\gamma_{,r}\,\gamma_{,\theta}\right] - 2r^2R_{r\theta} \tag{17.33}$$

$$\begin{aligned} V_{,r} = &-\frac{1}{4}r^4e^{2(\gamma-\beta)}(U_{,r})^2 + \frac{(r^4\sin\theta U)_{,r\theta}}{2r^2\sin\theta} \\ &+e^{2(\beta-\gamma)}\left[1 - \frac{(\sin\theta\beta_{,\theta})_{,\theta}}{\sin\theta} + \gamma_{,\theta\theta} + 3\cot\theta\gamma_{,\theta} - (\beta_{,\theta})^2\right. \\ &\left.-2\gamma_{,\theta}(\gamma_{,\theta} - \beta_{,\theta}) - \frac{1}{2}r^2e^{2\beta}g^{AB}R_{AB}\right], \end{aligned} \tag{17.34}$$

where $R_{\mu\nu}$ is the Ricci tensor and $x^A = (\theta, \phi)$. Correspondingly, the evolution equation for the gravitational field reads:

$$
\begin{aligned}
4r(r\gamma)_{,ur} = & \left[2r\gamma_{,r}V - r^2 \left(2\gamma_{,\theta}U + \sin\theta \left(\frac{U}{\sin\theta} \right)_{,\theta} \right) \right]_{,r} \\
& -2r^2 \frac{(\gamma_{,r}U\sin\theta)_{,\theta}}{\sin\theta} + \frac{1}{2}r^4 e^{2(\gamma-\beta)}(U_{,r})^2 \\
& +2e^{2(\beta-\gamma)} \left[(\beta_{,\theta})^2 + \sin\theta \left(\frac{\beta_{,\theta}}{\sin\theta} \right)_{,\theta} \right] 4\pi(\rho+p)u_\theta^2 . \qquad (17.35)
\end{aligned}
$$

A remarkable property of the above system of equations is that they form a hierarchy: knowing γ on the first null hypersurface allows one to radially integrate the corresponding equations to determine β, U, V and $\gamma_{,u}$ (in that order) on that hypersurface [14].

In the code developed by [24, 26] the numerical implementation of the Einstein equations closely follows that of [82]. The same marching algorithms are employed with additional source terms arising from the presence of matter fields. Since the code uses spherical coordinates, special care is taken with the numerical treatment of the coordinate singularities at the origin and at the polar axis. In order to impose boundary conditions at the origin the assumption that $t = u + r$, $x = r\sin\theta\cos\phi$, $y = r\sin\theta\sin\phi$ and $z = r\cos\theta$ form a local Fermi system at $r = 0$ is enforced. This implies a fall-off behavior given by $V = r + O(r^3)$, $\beta = O(r^2)$, $U = O(r)$ and $\gamma = O(r^2)$ [44]. Regularity on the axis requires that $U/\sin\theta$ and $\gamma/\sin^2\theta$ are continuous functions at $\theta = 0, \pi$. The code also uses a new polar coordinate $y = -\cos\theta$. In order to keep the freedom of working with numerical grids which only cover the star without its vacuum exterior, the radial coordinate used in [82] was generalized: Starting from an equidistant radial coordinate $x \in [0, 1]$, the code of [24] allows for a general coordinate transformation of the form $r = r(x)$, so that either compactified – with future null infinity being the outermost radial grid point – or non-compactified grids can be used. This will allow for the unambiguous computation of the gravitational radiation at $\mathscr{I}^+$ in our planned gravitational core collapse simulations, by simply reading off the news function at the outermost radial grid point. The hydrodynamic equations are formulated as in [21] and solved using HRSC schemes.

As a simplified model for a self-gravitating relativistic star [26] have considered the spherically symmetric solution of the general relativistic hydrostatic equation, the so-called Tolman-Oppenheimer-Volkoff equation, with a polytropic EoS $p = K\rho^\Gamma$. Equilibrium models were used to check the long-term stability of the code and its convergence properties. The code has proven to be stable for evolution times much longer than the characteristic light-crossing times of the different models considered. Spherically symmetric simulations have also shown the expected second order accuracy of the code.

Following [38] we have also checked the code on a dynamical evolution of an unstable spherical initial model. In such a model the sign of the truncation error of the numerical scheme controls the fate of the evolution. In the code

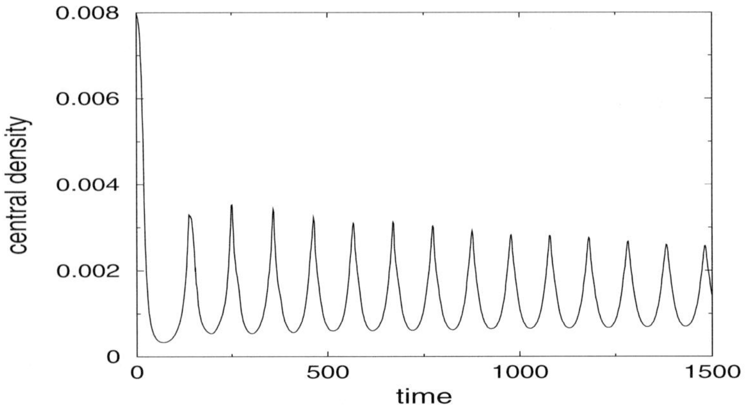

Fig. 17.3. Evolution of the central rest-mass density during the migration of an unstable relativistic star ($N = 1, K = 100, M = 1.447M_\odot, \rho_c = 8.0\times10^{-3}; G = c = M_\odot = 1$) to a stable model with the same rest-mass. The central density of the (final) stable configuration is $\rho_c = 1.35 \times 10^{-3}$. The evolution shows the expected behavior. Since we are using a polytropic EoS, the amplitude of the oscillations is essentially undamped for the evolution times shown

this sign is such that the unstable star "migrates" to the stable branch of the sequence of equilibrium models. In such a situation, the rest-mass of the star has to be conserved throughout the migration. Despite the fact that this mechanism cannot occur in nature – unstable stars can only collapse to more compact configurations – and as such it is an academic problem, it represents, nevertheless, an important test of the accuracy and self-consistency of the code in a highly dynamical situation.

As in [38] we have constructed a $N = 1$ ($\Gamma = 1 + 1/N = 2$), $K = 100$ polytropic star with mass $M = 1.447\ M_\odot$ and central rest-mass density $\rho_c = 8.0 \times 10^{-3}$ (in units in which $G = c = M_\odot = 1$). Fig. 17.3 shows the evolution of the central density up to a final time of $u = 1500$. On a very short dynamical timescale the star rapidly expands and its central rest-mass density drops well below its initial value, less than $\rho_c = 1.35 \times 10^{-3}$, the central rest-mass density of the stable model of the same rest-mass. During the rapid decrease of the central density, the star acquires a large radial momentum. The star then enters a phase of large amplitude radial oscillations around the stable equilibrium model. As Fig. 17.3 shows the code is able to accurately recover (asymptotically) the expected values of the stable model. Furthermore, its evolution is completely similar to that obtained with an independent fully three-dimensional code in Cartesian coordinates [38].

The evolution shown in Fig. 17.3 allows to study large amplitude oscillations of relativistic stars, which cannot be treated accurately by linear perturbation theory. These oscillations could occur after a supernova core-collapse [33] or after an accretion-induced collapse of a white dwarf.

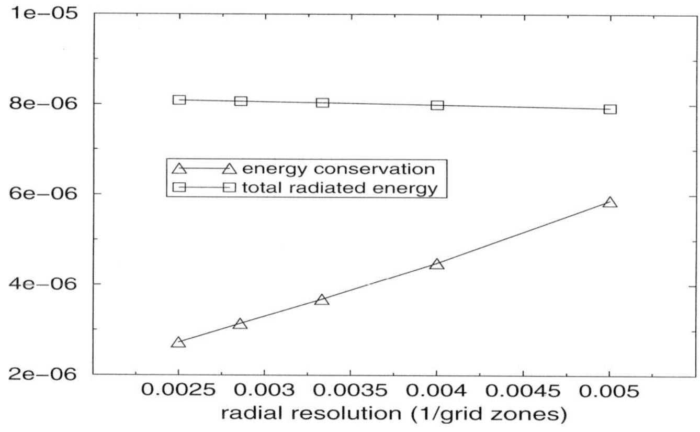

Fig. 17.4. Convergence properties of a global energy conservation test. The difference between the initial and final Bondi mass converges in an almost second order way to the value given by the energy which is carried away by gravitational waves. The figure is taken from [24]

To end this section we briefly discuss a global energy conservation test of the axisymmetric characteristic code which was presented in [24]. For that purpose Siebel et al. [24] used a strong ingoing gravitational wave to perturb an equilibrium relativistic star:

$$\hat{\gamma} \equiv \frac{\gamma}{\sin\theta^2} = 0.05\ e^{-2(r-4)^2}\ e^{-4y^2}\,. \tag{17.36}$$

The (nonlinear) initial pulse induces large velocities in the fluid of the star which give rise to "strong" outgoing gravitational waves, i.e., with an energy larger than the numerical errors involved in the calculation of the Bondi mass for a given resolution and integration time. If M is the Bondi mass and P is the total energy radiated away by gravitational waves at future null infinity $\mathcal{I}^+$, the convergence of the quantity

$$ec := M|_{u=0} - M|_{u=u_*>0} - P|_{[0,u_*]} \tag{17.37}$$

to zero represents a very severe global test of the numerical code. Satisfactory convergence results under grid refinement are shown in Fig. 17.4.

17.5 Summary

The article has dealt with presenting some concepts and applications in relativistic astrophysics of a particular class of finite difference numerical schemes based on Riemann solvers, specifically designed for nonlinear hyperbolic systems of conservation laws.

Such schemes have been discussed in the context of the general relativistic hydrodynamic equations. Nevertheless, the algorithms presented are general enough to be applicable to other hyperbolic systems such as the Einstein equations (when appropriately formulated, as e.g. in Friedrich's conformal approach [15], see also [7]). While this may not be strictly necessary for vacuum space-times, it may become relevant when dealing with non-vanishing stress-energy tensors.

The use of conservative algorithms based upon the characteristic structure of the hydrodynamic equations, developed during the last decade building on ideas first applied in Newtonian hydrodynamics, provides a robust methodology to obtain stable and accurate solutions even in the presence of discontinuities. This has become apparent since the early 1990s [29].

The knowledge of the wave structure of the equations is the essential building block in the construction of the so-called linearized Riemann solvers. The increasing use of these solvers in relativistic hydrodynamics has proved successful in handling complex flows, with high Lorentz factors and strong shocks, superseding more traditional methods based on artificial viscosity [49].

In the last part of the article we have discussed some astrophysical applications of such schemes, using the coupled system of the (characteristic) Einstein and hydrodynamic equations. Examples involving the gravitational collapse of supermassive stars and the evolution of relativistic compact stars have been presented.

Acknowledgments

I am grateful to Jörg Frauendiener and Helmut Friedrich for their invitation to participate in the workshop and to contribute to the proceedings. The work presented here is the result of collaborations with a number of colleagues, in particular, José María Ibáñez, Thomas Janka, Felix Linke, Chema Martí, Ewald Müller, Philippos Papadopoulos and Florian Siebel. I would like to thank especially Chema Martí, Ewald Müller and Florian Siebel for carefully reading the manuscript.

References

1. M.W. Choptuik: *Universality and scaling in gravitational collapse of a massless scalar field*, Phys. Rev. Lett. **70**, 9–12 (1993)
2. R. Gómez *et al*: *Stable characteristic evolution of generic 3-dimensional single-black-hole spacetimes*, Phys. Rev. Lett. **80**, 3915–3918 (1998)
3. A. Lichnerowicz: J. Math. Pure Appl. **23**, 37 (1944)
4. Y. Choquet-Bruhat: In: *Gravitation: An introduction to current research* (John Wiley, New York 1962)
5. R. Arnowitt, S. Deser, C.W. Misner: 'The Dynamics of General Relativity'. In: *Gravitation: An Introduction to Current Research* (John Wiley, New York 1962) pp. 227-265

6. H. Friedrich, A.D. Rendall. *The Cauchy Problem for the Einstein Equations*, Lect. Notes Phys. **540**, 127–224 (2000)
7. O. Reula: *Hyperbolic Methods for Einstein's Equations*, Liv. Rev. Relativ. **1**, 3 (1998)
8. M. Shibata, T. Nakamura: *Evolution of three-dimensional gravitational waves: Harmonic slicing case*, Phys. Rev. D **52**, 5428–5444 (1995)
9. T.W. Baumgarte, S.L. Shapiro: *On the numerical integration of Einstein's field equations*, Phys. Rev. D **59**, 024007 (1999)
10. E. Seidel, W.-M. Suen: *Numerical Relativity As A Tool For Computational Astrophysics*, J. Comput. Appl. Math. **109**, 493–525 (1999)
11. H. Bondi, M.J.G. van der Burg, A.W.K. Metzner: *Gravitational waves in general relativity. VII. Waves from axi-symmetric isolated systems*, Proc. R. Soc. London, Sect. A **269**, 21–52 (1962)
12. R.K. Sachs: *Gravitational waves in general relativity. VIII. Waves in asymptotically flat space-time*, Proc. R. Soc. London, Sect. A **270**, 103–126 (1962)
13. R. Penrose: *Asymptotic Properties of Fields and Space-Times*, Phys. Rev. Lett. **10**, 66–68 (1963)
14. J. Winicour: *Characteristic evolution and matching*, Liv. Rev. Relativ. **4**, 3 (2001)
15. H. Friedrich: Proc. R. Soc. London A **375**, 169 (1981)
16. P. Hübner: *A Method for Calculating the Structure of (Singular) Spacetimes in the Large*, Phys. Rev. D **53**, 701 (1996)
17. J. Frauendiener: *Numerical treatment of the hyperboloidal initial value problem for the vacuum Einstein equations. II. The evolution equations*, Phys. Rev. D **58**, 064003 (1998)
18. P. Hübner: *Numerical Calculation of Conformally Smooth Hyperboloidal Data*, Class. Quant. Grav. **18**, 1421 (2001)
19. M.R. Dubal, R.A. d'Inverno, J.A. Vickers: *Combining Cauchy and characteristic codes. V. CCM for a spherical spacetime containing a perfect fluid*, Phys. Rev. D **58**, 044019 (1998)
20. N. Bishop, R. Gómez, L. Lehner, M. Maharaj, J. Winicour: *The incorporation of matter into characteristic numerical relativity*, Phys. Rev. D **60**, 024005 (1999)
21. P. Papadopoulos, J.A. Font: *Relativistic hydrodynamics on spacelike and null surfaces: Formalism and computations of spherically symmetric spacetimes*, Phys. Rev. D **61**, 024015 (2000)
22. P. Papadopoulos, J.A. Font: *Imprints of accretion on gravitational waves from black holes*, Phys. Rev. D **63**, 044016 (2001)
23. F. Linke, J.A. Font, H.-Th. Janka, E. Müller, P. Papadopoulos: *Spherical collapse of supermassive stars: neutrino emission and gamma-ray bursts*, Astron. Astrophys. **376**, 568–579 (2001)
24. F. Siebel, J.A. Font, E. Müller, P. Papadopoulos: *Characteristic numerical relativity applied to hydrodynamic studies of neutron stars*, Proceedings of the 9th Marcel Grossmann Meeting. Ed. R. Ruffini. World Scientific (2001); gr-qc/0011096
25. F. Siebel, J.A. Font, P. Papadopoulos: *Scalar field induced oscillations of relativistic stars and gravitational collapse*, Phys. Rev. D **65**, 024021 (2002)
26. F. Siebel, J.A. Font, E. Müller, P. Papadopoulos: *Simulating the dynamics of relativistic stars via a light-cone approach*, Phys. Rev. D **65**, 064038 (2002)
27. J.A. Font: *Numerical hydrodynamics in general relativity*, Liv. Rev. Relativ. **3**, 2 (2000)
28. J.R. Wilson: *Numerical study of fluid flow in a Kerr space*, Astrophys. J. **173**, 431–438 (1972)

29. J.M. Martí, J.M. Ibáñez, J.A. Miralles: *Numerical relativistic hydrodynamics: local characteristic approach*, Phys. Rev. D **43** 3794–3801 (1991)
30. M.M May, R.H. White: *Hydrodynamic calculations of general relativistic collapse*, Phys. Rev. D **141**, 1232–1241 (1966)
31. T. Nakamura: *General relativistic collapse of axially symmetric stars leading to the formation of rotating black holes*, Prog. Theor. Phys. **65**, 1876–1890 (1981)
32. R.F. Stark, T. Piran: *Gravitational-Wave Emission from Rotating Gravitational Collapse*, Phys. Rev. Lett. **55**, 891–894 (1985)
33. H. Dimmelmeier, J.A. Font, E. Müller: *Gravitational waves from relativistic rotational core collapse*, Astrophys. J **560**, L163–L166 (2001)
34. T.W. Baumgarte, S.A. Hughes, S.L. Shapiro: *Evolving Einstein's field equations with matter: The "hydro without hydro" test*, Phys. Rev. D **60**, 087501 (1999)
35. J.A. Font, M. Miller, W.-M. Suen, M. Tobias: *Three Dimensional Numerical General Relativistic Hydrodynamics: Formulations, Methods and Code Tests*, Phys. Rev. D **61**, 044011 (2000)
36. M. Alcubierre, B. Brügmann, T. Dramlitsch, J.A. Font, P. Papadopoulos, E. Seidel, N. Stergioulas, R. Takahashi: *Towards a Stable Numerical Evolution of Strongly Gravitating Systems in General Relativity: The Conformal Treatments*, Phys. Rev. D **62**, 044034 (2000)
37. N. Stergioulas, J.A. Font: *Nonlinear r-modes in rapidly rotating relativistic stars*, Phys. Rev. Lett. **86**, 1148–1151 (2001)
38. J.A. Font, T. Goodale, S. Iyer, M. Miller, L. Rezzolla, E. Seidel, N. Stergioulas, W.-M. Suen and M. Tobias: *Three-dimensional general relativistic hydrodynamics II: long-term dynamics of single relativistic stars*, Phys. Rev. D. **65**, 084024 (2002)
39. G.J. Mathews, P. Marronetti, J.R. Wilson: *Relativistic Hydrodynamics in Close Binary Systems: Analysis of Neutron-Star Collapse*, Phys. Rev. D **58**, 043003 (1998)
40. M. Shibata: *Fully general relativistic simulation of coalescing binary neutron stars: Preparatory tests*, Phys. Rev. D **60**, 104052 (1999)
41. M. Shibata, K. Uryu: *Simulation of merging binary neutron stars in full general relativity: $\Gamma = 2$ case*, Phys. Rev. D **61**, 064001 (2000)
42. M. Miller, W.M. Suen, M. Tobias: *The Shapiro Conjecture: Prompt or Delayed Collapse in the head-on collision of neutron stars?*, Phys. Rev. D **63**, 121501 (2001)
43. K. Thorne: *Gravitational Wave Astronomy.* Proceedings of the Yukawa Conference on Black Holes and Gravitational Waves: New Eyes in the 21st Century (1999)
44. R.A. Issacson, J.S. Welling, J. Winicour: *Null cone computation of gravitational radiation*, J. Math. Phys., **24**, 1824–1834 (1983)
45. R.J. LeVeque: *Numerical Methods for Conservation Laws.* Birkhäuser-Verlag (1990)
46. E.F. Toro: *Riemann Solvers and Numerical Methods for Fluid Dynamics.* Springer-Verlag, Berlin, Heidelberg (1997)
47. C. Hirsch: *Numerical Computation of Internal and External Flows.* Wiley (1988)
48. C.B. Laney: *Computational gasdynamics*, Cambridge University Press (1998)
49. J.M. Ibáñez, J.M. Martí: *Riemann solvers in relativistic astrophysics*, J. Comput. Appl. Math. **109**, 173–211 (1999)
50. J.M. Martí, E. Müller: *Numerical hydrodynamics in special relativity*, Liv. Rev. Relativ. **2**, 1 (1999)
51. J. von Neumann, R.D. Richtmyer: *A method for the numerical calculation of hydrodynamic shocks*, J. Appl. Phys. **21**, 232 (1950)
52. M.L. Norman, K-H. A. Winkler: In *Astrophysical Radiation Hydrodynamics.* Eds. M.L. Norman, K-H. A. Winkler, Reidel Publishing Company, 449–475 (1986)

53. S. Siegler, H. Riffert: *Smoothed particle hydrodynamics simulations of ultra-relativistic shocks with artificial viscosity*, Astrophys. J. **5**31, 1053 (2000)
54. A.M. Anile: *Relativistic Fluids and Magneto-fluids*, Cambridge University Press, Cambridge, England (1989)
55. J.A. Font, J.M. Ibáñez, J.M. Martí, A. Marquina: *Multidimensional relativistic hydrodynamics: characteristic fields and modern high-resolution shock-capturing schemes*, Astron. Astrophys. **2**82, 304–314 (1994)
56. F. Eulderink, G. Mellema: *General relativistic hydrodynamics with a Roe solver*, Astron. Astrophys. Suppl. Ser. **1**10, 587–623 (1995)
57. S.A.E.G. Falle, S.S. Komissarov: *An upwind numerical scheme for relativistic hydrodynamics with a general equation of state*, Mon. Not. R. Astron. Soc. **2**78, 586–602 (1996)
58. F. Banyuls, J.A. Font, J.M. Ibáñez, J.M. Martí, J.A. Miralles: *Numerical 3+1 General Relativistic Hydrodynamics: A Local Characteristic Approach*, Astrophys. J. **4**76, 221–231 (1997)
59. L. Wen, A. Panaitescu, P. Laguna: *A shock-patching code for ultrarelativistic fluid flows*, Astrophys. J. **4**86, 919–927 (1997)
60. J.M. Ibáñez, M.A. Aloy, J.A. Font, J.M. Martí, J.A. Miralles, J.A. Pons: *Riemann solvers in general relativistic hydrodynamics*. In E.F. Toro ed. *Godunov Methods: Theory and Applications*, Kluwer Academic/Plenum Publishers, New York, 485-496 (2001)
61. P.D. Lax, B. Wendroff: *Systems of conservation laws*, Comm. Pure Appl. Math. **1**3, 217 (1960)
62. P.D. Lax: *Hyperbolic systems of conservation laws and the mathematical theory of shock waves*, Regional Conference Series Lectures in Applied Math. **1**1, SIAM, Philadelphia (1972)
63. R.D. Richtmyer, K.W. Morton: *Difference Methods for Initial-Value Problems*. Wiley-Interscience, New York (1967)
64. A. Harten: *On a class of high resolution total-variation stable finite difference schemes*, SIAM J. Numer. Anal. **2**1, 1–23 (1984)
65. S.K. Godunov: *A finite difference method for the numerical computation and discontinuous solutions of the equations of fluid dynamics*, Mat. Sb. **4**7, 271 (1959) in Russian
66. J.M. Martí, E. Müller: *The analytical solution of the Riemann problem in relativistic hydrodynamics*, J. Fluid Mech. **2**58, 317–333 (1994)
67. J. Pons, J.M. Martí, E. Müller: *The exact solution of the Riemann problem with non-zero tangential velocities in relativistic hydrodynamics*, J. Fluid Mech. **4**22, 125 (2000)
68. A. Harten, P.D. Lax, B. van Leer: *On upstream differencing and Godunov-type schemes for hyperbolic conservation laws*, SIAM Review **2**5, 35–61 (1983)
69. B. Einfeldt: *On Godunov-type methods for gas dynamics*, SIAM J. Num. Anal. **2**5, 294–318 (1988)
70. P.L. Roe: *Approximate Riemann solvers, parameter vectors and difference schemes*, J. Comput. Phys. **4**3, 357–372 (1981)
71. B. van Leer: *Towards the ultimate conservative difference scheme. V. A second order sequel to Godunov's method*, J. Comput. Phys. **3**2, 101–136 (1979)
72. P. Colella, P.R. Woodward: *The piecewise parabolic method (PPM) for gas-dynamical simulations*, J. Comput. Phys. **5**4, 174–201 (1984)
73. A. Harten, B. Engquist, S. Osher, S.R. Chakrabarthy: *Uniformly high order accurate essentially non-oscillatory schemes, III*, J. Comput. Phys. **7**1, 231–303 (1987)

74. R.J. LeVeque: *Nonlinear Conservation Laws and Finite Volume Methods.* In *Computational Methods for Astrophysical Fluid Flow.* Saas-Fee Advanced Course 27. Lecture Notes 1997. Springer-Verlag, 1 (1998)
75. R. Donat, J.A. Font, J.M. Ibáñez, A. Marquina: *A Flux-Split Algorithm applied to Relativistic Flows*, J. Comput. Phys. **1**46, 58–81 (1998)
76. M.J. Rees: In *Black Holes and Relativistic Stars*, University of Chicago Press, Chicago (1998)
77. J. Kormendy: *Supermassive Black Holes in Disk Galaxies*, in *Galaxy Disks and Disk Galaxies*, ASP Conference Series, Vol. 230, Edited by J.G. Funes, S.J. and Enrico Maria Corsini. San Francisco, 247-256 (2001)
78. S. Chandrasekhar: *The Dynamical Instability of Gaseous Masses Approaching the Schwarzschild Limit in General Relativity*, Astrophys. J. **1**40, 417 (1964)
79. W.F. Fowler: *Massive Stars, Relativistic Polytropes, and Gravitational Radiation*, Rev. Mod. Phys. **3**6, 545 (1964)
80. G.M. Fuller, X. Shi: *Supermassive Objects as Gamma-Ray Bursters*, Astrophys. J. **5**02, L5 (1998)
81. T. Zwerger, E. Müller: *Dynamics and gravitational wave signature of axisymmetric rotational core collapse*, Astron. Astrophys. **3**20, 209 (1997)
82. R. Gómez, P. Papadopoulos, J. Winicour: *Null cone evolution of axisymmetric vacuum space-times*, J. Math. Phys. **3**5, 4184–4204 (1994)

18 Simulations of Generic Singularities in Harmonic Coordinates

David Garfinkle

Department of Physics, Oakland University, Rochester, Michigan 48309, USA

Abstract. This paper presents both a numerical method for general relativity and an application of that method. The method involves the use of harmonic coordinates in a 3+1 code to evolve the Einstein equations with scalar field matter. In such coordinates, the terms in Einstein's equations with the highest number of derivatives take a form similar to that of the wave equation. The application is an exploration of the generic approach to the singularity for this type of matter. The preliminary results indicate that the dynamics as one approaches the singularity is locally the dynamics of the Kasner spacetimes.

18.1 Introduction

It has long been known that harmonic coordinates are useful for mathematical relativity. In particular, these coordinates were used [1] to prove local existence of solutions of the vacuum Einstein equation. This usefulness of harmonic coordinates stems from their putting Einstein's equation into a form that is similar to the curved space-time wave equation. Since many numerical techniques work well on the wave equation, one might expect that harmonic coordinates would be used extensively in numerical relativity, and it is somewhat surprising that they are not. (However, see [2] for some recent mathematical and numerical work on the linearized case.) This is perhaps due to the following drawback of harmonic coordinates: these coordinates are solutions of the wave equation, and such solutions need not have a time-like gradient at all points of space-time, even if they start out with a time-like gradient on an initial data surface. What this means is that in harmonic coordinates, the time coordinate will in general not remain time-like and this is likely to cause numerical problems [3]. As we will see later, there is a way around this difficulty.

One area of numerical relativity where harmonic coordinates have been used (though somewhat unintentionally) is in the study of the approach to the singularity, in particular in the Gowdy space-times [4]. Numerical simulation of approach to a space-time singularity presents problems of its own. Since it is expected that various quantities become infinite at the singularity, the numerical simulation will generally stop after a finite coordinate time, the time when a surface of constant time first encounters the singularity. Since this first encounter generally occurs at one spatial point, the information about the behavior of the singularity at other spatial points will be unavailable from the numerical simulation. The solution to this difficulty is to choose a time coordinate that tends to infinity as the singularity is approached. In this way, the simulation is not forced to end at finite coordinate time, the whole space-time up to the singularity is covered and the behavior of the metric as the singularity is approached simply becomes asymptotic behavior in the limit of large time coordinate. For Gowdy space-times there is a natural choice of such a time coordinate: these space-times are foliated by T^2s invariant under the symmetry group. The area of the symmetry T^2s goes to zero as the singularity is approached. Therefore minus the logarithm of this area is a natural time coordinate that goes to infinity as the singularity is approached. While this approach works well for the Gowdy space-times, since it depends crucially on the symmetry of the Gowdy space-times, the approach does not seem to generalize to the case of the generic singularity with no symmetries. A natural generalization comes when one notices that this time coordinate (minus the logarithm of the area of the symmetry T^2s) is also harmonic. Since in some sense one expects the wave equation to become singular as the space-time singularity is approached, one might also expect a solution of the wave equation to blow up as the singularity is approached and thus one might want to use such a solution as the coordinate time.

What is expected to be the generic behavior of a space-time as the singularity is approached? Based on studies of space-times with T^2 symmetry [4–6] and space-times with $U(1)$ symmetry [7], the expected answer [8] is the following: the singularity is expected to be space-like and as it is approached each spatial point is expected to "decouple" from the others and undergo a dynamics corresponding to that of a homogeneous space-time (though a different homogeneous space-time at each spatial point). Which types of homogeneous space-time is this dynamics expected to correspond to? For vacuum space-times, it is thought that the dynamics will be oscillatory, possibly corresponding to the Mixmaster space-time as conjectured in [9]. For many other types of matter it is expected that as the singularity is approached the matter terms in the Einstein equations become negligible and the dynamics approaches that of a vacuum space-time.

However for so called "stiff matter" (*i.e.*, a scalar field or a perfect fluid with equation of state $P = \rho$) it is expected that the dynamics will not be oscillatory and will correspond to that of a Kasner space-time. This expectation is greatly bolstered by a theorem due to Andersson and Rendall [10] which shows local existence in a neighborhood of the singularity of solutions to the Einstein-scalar equations, with the expected asymptotic behavior and with enough degrees of freedom to be the generic solutions.

Since the approach to the singularity is expected to be simpler for stiff matter than for vacuum, the stiff matter case should be easier to treat numerically. Therefore, in this paper we confine ourselves to a numerical study of the approach to the generic singularity in the Einstein-scalar system. Sect. 18.2 presents the equations and numerical methods used. The results are given in Sect. 18.3. Sect. 18.4 contains a discussion of the results and of other possible applications of the harmonic coordinate method.

18.2 Equations and Numerical Methods

The equations that we wish to evolve numerically are the Einstein-scalar equations

$$R_{\alpha\beta} = 8\pi \nabla_\alpha \phi \nabla_\beta \phi. \tag{18.1}$$

Here, we use the conventions of [11] including units where $c = G = 1$. As a consequence of (18.1) and the Bianchi identities, the scalar field must satisfy the wave equation

$$\nabla_\alpha \nabla^\alpha \phi = 0. \tag{18.2}$$

The Ricci tensor is given in terms of the Christoffel symbols by

$$R_{\alpha\beta} = \partial_\gamma \Gamma^\gamma_{\alpha\beta} - \partial_\alpha \Gamma^\gamma_{\gamma\beta} + \Gamma^\gamma_{\alpha\beta}\Gamma^\nu_{\nu\gamma} - \Gamma^\gamma_{\alpha\nu}\Gamma^\nu_{\beta\gamma} \tag{18.3}$$

while the Christoffel symbols are given in terms of the metric by

$$\Gamma^\gamma_{\alpha\beta} = \tfrac{1}{2}\, g^{\gamma\delta}\left(\partial_\alpha g_{\beta\delta} + \partial_\beta g_{\alpha\delta} - \partial_\delta g_{\alpha\beta}\right). \tag{18.4}$$

Harmonic coordinates are solutions of the wave equation. As a generalization of harmonic coordinates, consider coordinates that satisfy the wave equation with source

$$\nabla^\alpha \nabla_\alpha x^\mu = H^\mu, \tag{18.5}$$

where H^μ are specified from the beginning. Then using (18.3), (18.4, (18.5) we find that the Ricci tensor is given by

$$R_{\alpha\beta} = -\tfrac{1}{2} g^{\gamma\sigma}\partial_\gamma \partial_\sigma g_{\alpha\beta} + \tfrac{1}{2} C_\beta{}^{\mu\nu} C_{\mu\nu\alpha} + \tfrac{1}{2} C_\alpha{}^{\mu\nu} C_{\mu\nu\beta} - \Gamma^\gamma_{\nu\alpha}\Gamma^\nu_{\gamma\beta} - \partial_{(\alpha} H_{\beta)} + \Gamma^\gamma_{\alpha\beta} H_\gamma \tag{18.6}$$

where $C_{\alpha\mu\nu} \equiv \partial_\alpha g_{\mu\nu}$. Note that the second derivative terms appear only in the wave operator. Therefore, one might expect that Einstein's equations in this form behave similarly to the wave equation and that numerical methods that work well on the wave equation might work well on Einstein's equations in this form. The reason for considering nonzero source terms H^μ in (18.5) is that these terms allow us to change the behavior of the time coordinate and thus may allow us to eliminate (or at least postpone) the behavior where the time coordinate ceases to be time-like.

The numerical method used requires equations that are first order in time. To put the equations in such a form, we define quantities $P_{\alpha\beta}$ and P_ϕ given by

$$P_{\alpha\beta} = \partial_t g_{\alpha\beta}, \tag{18.7}$$

$$P_\phi = \partial_t \phi. \tag{18.8}$$

Then the Einstein-scalar equation becomes

$$-g^{00}\partial_t P_{\alpha\beta} = 2 g^{0k}\partial_k P_{\alpha\beta} + g^{ik}\partial_i \partial_k g_{\alpha\beta} + 16\pi \partial_\alpha \phi \partial_\beta \phi + 2\partial_{(\alpha} H_{\beta)} - 2\Gamma^\gamma_{\alpha\beta} H_\gamma - C_\alpha{}^{\mu\nu} C_{\mu\nu\beta} - C_\beta{}^{\mu\nu} C_{\mu\nu\alpha} + 2\Gamma^\gamma_{\nu\alpha}\Gamma^\nu_{\gamma\beta}. \tag{18.9}$$

The wave equation for ϕ becomes

$$-g^{00}\partial_t P_\phi = 2g^{0k}\partial_k P_\phi \; + \; g^{ik}\partial_i\partial_k\phi \; - \; g^{\alpha\beta}\Gamma^\gamma_{\alpha\beta}\partial_\gamma\phi. \tag{18.10}$$

The full set of equations that are evolved in the computer code are (18.7)–(18.10).

We now turn to the numerical methods used to evolve these equations. Functions on space-time are represented by their values on a numerical grid and spatial derivatives are approximated by centered differences. That is, at spatial grid point i in the x direction, the first and second derivatives of a function F are approximated by

$$\frac{\partial F}{\partial x} \to \frac{F_{i+1} - F_{i-1}}{2\Delta x}, \tag{18.11}$$

$$\frac{\partial^2 F}{\partial x^2} \to \frac{F_{i+1} + F_{i-1} - 2F_i}{(\Delta x)^2} \tag{18.12}$$

The variables are evolved in time using the iterated Crank-Nicholson (ICN) method. This works as follows: suppose that we have evolution equations of the form $\partial_t S = W(S)$ for some set of variables S. Let S^n be the value of S at time step n. Then we approximate the evolution equation as

$$S^{n+1} = S^n + \frac{\Delta t}{2}\left[W(S^n) + W(S^{n+1})\right] \tag{18.13}$$

where Δt is the time step. In principle, this equation could be used directly to solve for S^{n+1} and thus evolve the variables for one time step. In practice, this is too complicated for all but the simplest evolution equations. Instead, the ICN method consists of approximately solving (18.13) by iteration. We take as a first guess for S^{n+1} the quantity S^n. Then using our guess for S^{n+1} on the right side of (18.13) we produce a better guess for S^{n+1} which can then be used on the right hand side of (18.13) to produce a better guess. In principle this process could be iterated an arbitrary number of times but a large number of iterations for a 3+1 code would take a very large amount of computing time. Instead we iterate for three times.

The space-times we consider have topology $T^3 \times R$. Each spatial slice has topology T^3. In terms of the spatial coordinates, this means that $0 \le x \le 2\pi$ with 0 and 2π identified (and correspondingly for y and z). This topology is implemented numerically as follows: a spatial coordinate x has N grid points. The variables on points from 2 to $N-1$ are evolved using the evolution equations. The variables on point 1 are set to the values at point $N-1$ while at point N they are set to the values at point 2.

18.3 Results

All runs were done in double precision on Compaq XP1000 workstations and on the NCSA Origin 2000. The time step was $\Delta t = \Delta x/2$. While the source terms H^μ should be helpful in keeping the coordinates well behaved, we did not need them for the cases studied here and all runs have $H^\mu = 0$.

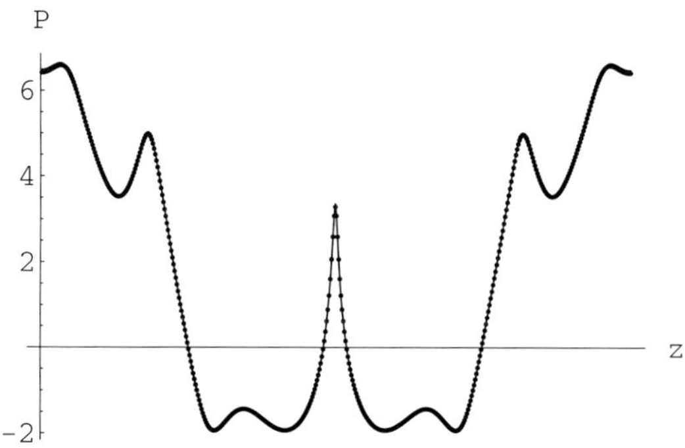

Fig. 18.1. Comparison of the quantity P as found using a Gowdy code and the 3+1 harmonic code

Before exploring the generic singularity, we would like to test the code. One way of doing this is to run the code on cases with symmetry where results are already known. Such a case is the Gowdy space-times. These have the form [4]

$$ds^2 = e^{(\tau-\lambda)/2}\left[-e^{-2\tau}d\tau^2 + dz^2\right] \\ + e^{-\tau}\left[e^P dx^2 + 2e^P Q dx dy + \left(e^P Q^2 + e^{-P}\right) dy^2\right] \quad (18.14)$$

Here, P, Q and λ are functions of τ and z. It follows from (18.14) that the coordinates (τ, x, y, z) are all solutions of the wave equation. Therefore, they are harmonic coordinates. The vacuum Einstein equations for the Gowdy space-times are [5]

$$\partial_\tau \partial_\tau P - e^{-2\tau}\partial_z \partial_z P - e^{2P}\left[(\partial_\tau Q)^2 - e^{-2\tau}(\partial_z Q)^2\right] = 0, \quad (18.15)$$

$$\partial_\tau \partial_\tau Q - e^{-2\tau}\partial_z \partial_z Q + 2\left(\partial_\tau P \partial_\tau Q - e^{-2\tau}\partial_z P \partial_z Q\right) = 0 \quad (18.16)$$

plus constraint equations that determine λ, once P and Q are known.

To test the 3+1 code, we evolve (18.15) and (18.16) using a 1+1 code. Then we evolve the same initial data with the 3+1 harmonic code and compare the results. To do the comparison, note that $P = \tau + \ln g_{xx}$ so that it is straightforward to compare the values of P produced by the two codes. For these simulations 500 grid points were used in the 1+1 code and 3, 3 and 500 grid points were used respectively in the x, y and z directions in the 3+1 code. For the comparison the initial data used is $P = 0$, $\partial_\tau P = 5\cos z$, $Q = \cos z$, $\partial_\tau Q = 0$. These data are evolved until $\tau = \pi$ and the results for the comparison are given in Fig. 18.1. Here, the solid line represents the 1+1 evolution of the Gowdy equations, while the dots represent the full 3+1 evolution using the harmonic code. There is clearly agreement between the two.

Another type of code test involves a constraint that comes from using harmonic coordinates. Define the quantities C^μ by

$$C^\mu = g^{\alpha\beta}\Gamma^\mu_{\alpha\beta} + H^\mu. \quad (18.17)$$

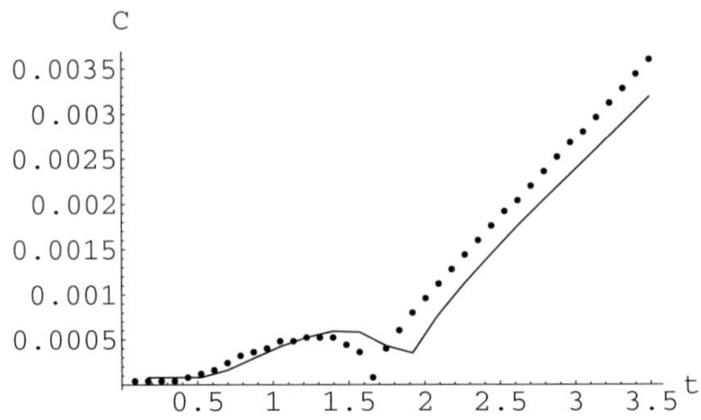

Fig. 18.2. Convergence test involving the constraint

Then from (18.5) it follows that $C^\mu = 0$. Since we are solving the evolution equations by approximating them by finite difference equations, the quantities C^μ as evaluated by the computer code will not be zero because of errors due to the finite grid spacing Δx. For the type of algorithm that we are using, if the grid spacing is halved, then the quantities C^μ should get smaller by a factor of 4. A very powerful code test, referred to as a convergence test, is to evolve the same initial data with two different grid spacings and see whether the C^μ behave as expected.

Before presenting the results of a convergence test we turn to the question of finding initial data without symmetries. On an initial data surface, the intrinsic metric h_{ab} and the extrinsic curvature K_{ab} must satisfy the constraint equations

$$D_a K^a{}_b - D_b K = 8\pi n^a \nabla_a \phi D_b \phi, \tag{18.18}$$

$$^{(3)}R + K^2 - K^{ab}K_{ab} = 8\pi \left[(n^a \nabla_a \phi)^2 + D^a \phi D_a \phi \right]. \tag{18.19}$$

Here, n^a is the unit normal to the surface, and D_a and $^{(3)}R$ are respectively the covariant derivative and scalar curvature of h_{ab}. Given a solution of (18.18) and (18.19), we produce initial data for evolution in harmonic coordinates as follows: For spatial directions i and j we have $g_{ij} = h_{ij}$, $g_{i0} = g_{0i} = 0$, $g_{00} = -1$, $P_{ij} = 2K_{ij}$. The remaining components of $P_{\alpha\beta}$ are solved for using $C^\mu = 0$.

We want to find a solution of (18.18) and (18.19) that is simple but has no symmetries and has some free parameters. We choose $\phi = 0$ and h_{ab} equal to the flat Euclidean metric in the usual coordinates. For the extrinsic curvature we choose

$$\begin{aligned}
K_{xx} &= (b_1 + a_2 \cos y + a_3 \cos z)/2, \\
K_{yy} &= (b_2 + a_1 \cos x - a_3 \cos z)/2, \\
K_{zz} &= (b_3 - a_1 \cos x - a_2 \cos y)/2, \\
K_{xy} &= K_{yx} = (c_1 \cos z)/2, \\
K_{xz} &= K_{zx} = (c_2 \cos y)/2, \\
K_{yz} &= K_{zy} = (c_3 \cos x)/2.
\end{aligned} \tag{18.20}$$

Here, the quantities a_i, b_i and c_i are constants that are free parameters. It is straightforward to show that the extrinsic curvature of (18.20) with our choice of initial h_{ab} and ϕ satisfies (18.18). Equation (18.19) then becomes an algebraic equation for P_ϕ which can be solved provided that the left hand side of the equation is positive. Note that if all the a_i and c_i are zero and all the b_i are equal, then the initial data evolve to the spatially flat Robertson-Walker space-time with scalar field matter. Thus, this family of data can be thought of as Robertson-Walker with large gravitational and scalar waves.

We now consider a convergence test involving the constraints C^μ. Define the quantity C by

$$C = \left| \frac{\int \sqrt{g} g_{\alpha\beta} C^\alpha C^\beta dx dy dz}{\int \sqrt{g} dx dy dz} \right|^{1/2} . \tag{18.21}$$

This quantity is a type of measure of the average size of the constraint. Fig. 18.2 shows a plot of C *vs.* time. The parameters used are $a_i = (0.1, 0.1, 0.2)$, $b_i = (0, -0.5, -0.5)$ and $c_i = (0, 0, 0)$. Here the curve corresponds to a run done with 20 grid points in each spatial direction, while the dots correspond to a run with 38 grid points in each spatial direction (which gives half the grid spacing) and with C multiplied by 4. The results show second order convergence.

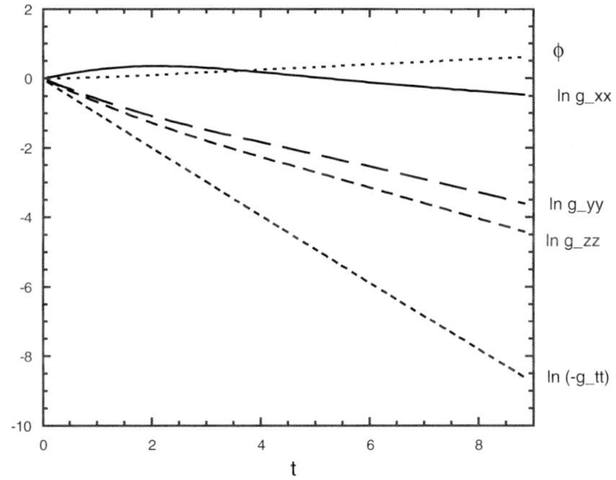

Fig. 18.3. Behavior of metric components and scalar field as the singularity is approached at the spatial point $(0, 0, 0)$

We now consider the approach to the singularity. To see what is expected, it is helpful to consider the Kasner space-time in harmonic coordinates. This is given by

$$ds^2 = -e^{(p_1+p_2+p_3)\tau} d\tau^2 + e^{p_1\tau} dx^2 + e^{p_2\tau} dy^2 + e^{p_3\tau} dz^2 \tag{18.22}$$

where the p_i are constants This metric is generally a solution of the Einstein-scalar equations; but for the case where $\left(\sum p_i\right)^2 = \sum p_i^2$ it is a vacuum space-time. Note that in the vacuum case one cannot have all three directions con-

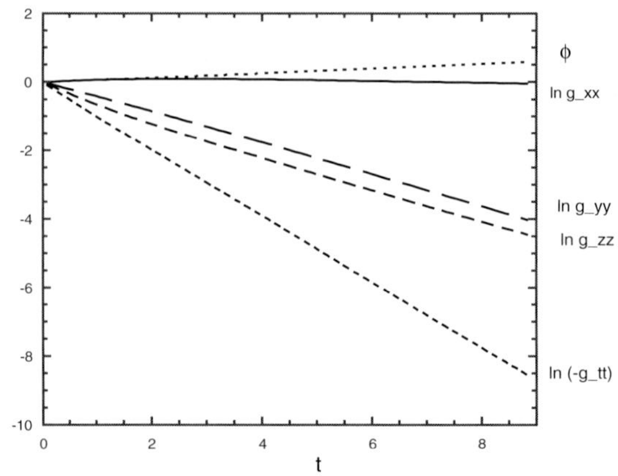

Fig. 18.4. Behavior of metric components and scalar field as the singularity is approached at the spatial point $(0, \pi/4, \pi/2)$

tracting but that this is possible for the Einstein-scalar case. Note also that the metric components are exponential functions of time. It also turns out that the scalar field is a linear function of time. Thus, if the behavior of a generic solution near the singularity is local Kasner, then we should expect metric components that are exponential functions of time, with the exponent depending on space. We should also expect a scalar field that is a linear function of time with the slope depending on space.

Results on the approach to the singularity are shown in Figs. 18.3 and 18.4. The run was done with 34 grid points in each spatial directions. The parameters a_i, b_i and c_i are the same as for the convergence test. Here the scalar field and the logarithms of the diagonal metric components are plotted as functions of time. Fig. 18.3 corresponds to the spatial point $(x, y, z) = (0, 0, 0)$ while for Fig. 18.4 the point is $(0, \pi/4, \pi/2)$. Note that as the singularity is approached these quantities all become linear functions of time. The differences between Figs. 18.3 and 18.4 show that there is a spatial dependence of the approach to the singularity. Note from Fig. 18.3 that though the x direction is initially expanding, eventually all three directions contract. This is what one would expect if the metrics of [10] represent the generic behavior near the singularity since these metrics have all three directions contracting in a neighborhood of the singularity.

18.4 Discussion

While this study is somewhat preliminary, it indicates that harmonic coordinates can be a useful tool in numerical relativity. Though, in principle coordinate problems could occur, this did not happen in the cases studied here, even though they involved very strong fields. Furthermore, the use of the source terms H^μ may cure such problems if they arise.

As for the behavior of generic singularities, the numerical results indicate that solutions of the form proved in [10] to exist in a neighborhood of the singularity also exist globally. Thus such solutions are likely to describe the generic singularity in the stiff matter case.

There are several projects for which the methods of this paper could be used. One is to do a more extensive study of the singularity in the Einstein-scalar case, with a more thorough exploration of the evolution corresponding to various values of the parameters in the initial data of the previous section. It would also be helpful to evolve for a longer time.

Another project is to remove the scalar field and study the approach to the singularity of the generic vacuum space-time. Here, the behavior is expected to be more complicated and a treatment will probably require more spatial resolution to resolve the expected sharp features, as well as longer evolution in time to see the expected oscillatory behavior.

Yet another project is to study the behavior of asymptotically flat space-times rather than closed cosmologies. Here, the closed cosmologies were studied partly for simplicity. The periodic boundary conditions are simple to implement and completely consistent with Einstein's equations. In contrast, for an outer boundary at a finite distance in an asymptotically flat space-time, one needs to put some sort of outgoing wave boundary condition. Such conditions are usually not consistent with Einstein's equation (it is known how to have a consistent boundary condition [12] but this condition is quite complicated). These inconsistent conditions may lead to numerical instability. Since harmonic coordinates make Einstein's equation look like the wave equation, simple outgoing wave boundary conditions that work numerically with the wave equation might be expected to "work" (at least in the sense of not causing numerical instability) for Einstein's equation.

All these projects are work in progress and preliminary results from them are promising. Thus, I expect that harmonic coordinates will become a useful tool in numerical relativity.

Acknowledgments

I would like to thank Helmut Friedrich, Beverly Berger, Matthew Choptuik, G. Comer Duncan, Lars Andersson and Alan Rendall for helpful discussions. I would also like to thank the Albert Einstein Institute for hospitality. This work was partially supported by NSF grant PHY-9988790 to Oakland University. Some of the computations were performed at the National Center for Supercomputing Applications (Illinois).

References

1. Y. Fouros Bruhat, Acta Mathematica, **88**, 141 (1952)
2. B. Szilagyi, B. Schmidt and J. Winicour, gr-qc/0106026
3. M. Alcubierre, Phys. Rev. **D55**, 5981 (1997)
4. B. Berger and V. Moncrief, Phys. Rev. **D48**, 4676 (1993)
5. B. Berger and D. Garfinkle, Phys. Rev. **D57**, 4767 (1998)
6. B. Berger, J. Isenberg and M. Weaver, gr-qc/0104048

7. B. Berger and V. Moncrief, Phys. Rev. **D58**, 064023 (1998)
8. B. Berger, D. Garfinkle, J. Isenberg, V. Moncrief and M. Weaver, Mod. Phys. Lett. **A13**, 1565 (1998)
9. V. Belinskii, I. Khalatnikov and E. Lifshitz, Sov. Phys. Usp. **13**, 745 (1971)
10. L. Andersson and A. Rendall, Commun. Math. Phys. **218**, 479 (2001)
11. R.M. Wald, *General Relativity* (University of Chicago Press, Chicago, 1984)
12. H. Friedrich and G. Nagy: Commun. Math. Phys. **201**, 619 (1999)

19 Some Mathematical and Numerical Questions Connected with First and Second Order Time-Dependent Systems of Partial Differential Equations

Heinz-O. Kreiss[1] and Omar E. Ortiz[2]

[1] Department of Mathematics, University of California, Los Angeles, CA 90095, USA
[2] Universidad Nacional de Córdoba, Facultad de Matemática, Astronomía y Física, (5000) Córdoba, Argentina

Abstract. There is a tendency to write the equations of general relativity as a first order symmetric system of time dependent partial differential equations. However, for numerical reasons, it might be advantageous to use a second order formulation like one obtained from the ADM equations. We shall discuss the wellposedness of the Cauchy problem for such systems and their advantage in numerical calculations.

19.1 Introduction

There is a tendency to write the equations of general relativity as a first order symmetric system of time dependent partial differential equations. However, for numerical reasons, it might be advantageous to use a second order formulation like one obtained from the ADM equations. Unfortunately, the type of the ADM equations is not well understood and therefore we shall discuss, in the next section, the concept of well-posedness. We have to distinguish between weakly and strongly hyperbolic systems. Strongly hyperbolic systems are well behaved even if we add lower order terms. In contrast, for every weakly hyperbolic system we can find lower order terms which make the problem totally ill posed. Thus, for weakly hyperbolic systems, there is only a restricted class of lower order perturbations which do not destroy the well-posedness. To identify that class can be very difficult, especially for nonlinear perturbations. In Sect. 19.3 we will show that the ADM equations, linearized around flat with constant lapse function and shift vector, are only weakly hyperbolic. However, we can use the trace of the metric as a lapse function to make the equations into a strongly second order hyperbolic system.

Using simple models we shall, in Sect. 19.4, demonstrate that approximations of second order equations have better accuracy properties than the corresponding approximations of first order equations. Also, we avoid spurious waves which travel against the characteristic direction.

In the last section we discuss some difficulties connected with the preservation of constraints.

19.2 Well Posed Problems

19.2.1 First Order Systems

Consider the Cauchy problem for a first order system with constant coefficients

$$\begin{aligned} \boldsymbol{u}_t &= \sum_{j=1}^{s} A_j D_j \boldsymbol{u} =: P(D)\boldsymbol{u}, \qquad D_j = \frac{\partial}{\partial x_j}, \\ \boldsymbol{u}(t=0) &= \boldsymbol{f}(x), \qquad x = (x_1, \ldots, x_s),\ -\infty < x_j < \infty. \end{aligned} \tag{19.1}$$

We construct simple wave solutions

$$\boldsymbol{u}(x,t) = e^{\mathrm{i}\langle \omega, x\rangle} \widehat{\boldsymbol{u}}(\omega, t), \quad \omega = (\omega_1, \ldots, \omega_s) \text{ real}$$

and obtain

$$\widehat{\boldsymbol{u}}_t = \mathrm{i}|\omega| \sum_{j=1}^{s} A_j \omega'_j \widehat{\boldsymbol{u}} =: \mathrm{i}|\omega| P(\omega')\widehat{\boldsymbol{u}}, \qquad \omega' = \omega/|\omega|. \tag{19.2}$$

Definition 19.1. We call the problem (19.1) strongly hyperbolic if for every ω' the eigenvalues of the symbol $P(\omega')$ are real and there is a complete set of uniformly (in ω') linearly independent eigenvectors.
Examples of strongly hyperbolic systems are symmetric systems where $A_j = A_j^\dagger$.

The solutions of strongly hyperbolic systems satisfy an energy estimate

$$\|\boldsymbol{u}(\cdot, t)\|^2 \le K^2 \mathrm{e}^{2\alpha t} \|\boldsymbol{u}(\cdot, 0)\|^2. \tag{19.3}$$

Here K, α are universal constants which do not depend on the initial data $\boldsymbol{u}(x,0) = \boldsymbol{f}(x)$. The norms are L_2 norms. For systems (19.1) with constant coefficients we have $\alpha = 0$. Strong hyperbolicity and the existence of an energy estimate are equivalent[1]:

Theorem 19.1. *The solutions of* (19.1) *satisfy an energy estimate of the type* (19.3) *if and only if the system is strongly hyperbolic.*

The most important property of strongly hyperbolic systems is that we can add lower order terms and an estimate of type (19.3) is still valid. We have

Theorem 19.2. *Let* (19.1) *be strongly hyperbolic. Then the solutions of*

$$\boldsymbol{w}_t = P(D)\boldsymbol{w} + B\boldsymbol{w} \tag{19.4}$$

satisfy an estimate of type (19.3). *Here B is any bounded operator.*

Definition 19.2. We call the problem (19.1) weakly hyperbolic if the eigenvalues of $P(\omega')$ are real.

[1] First order theory is well known, we refer to [1].

In this definition we do not require that there is a complete set of eigenvectors. An example in dimension one is given by

$$\boldsymbol{u}_t = \begin{pmatrix} 1 & 1 \\ 0 & 1 \end{pmatrix} \boldsymbol{u}_x =: A\boldsymbol{u}_x.$$

Simple wave solutions for this system have the form

$$\boldsymbol{u}(x,t) = \mathrm{e}^{\mathrm{i}\omega At}\mathrm{e}^{\mathrm{i}\omega x}\widehat{\boldsymbol{u}}(\omega,0) = \left(I + \mathrm{i}\omega \begin{pmatrix} 1 & 1 \\ 0 & 1 \end{pmatrix} t\right) \mathrm{e}^{\mathrm{i}\omega(x+t)}\widehat{\boldsymbol{u}}(\omega,0).$$

Thus there is no exponential growth but there is the polynomial growth in ωt. This is typically the case for weakly hyperbolic systems. One can prove

Theorem 19.3. *For weakly hyperbolic systems the growth of simple wave solutions is at most of the order* $\mathcal{O}(1+|\omega t|^{n-1})$*, where* n *is the number of components of* $\boldsymbol{u}$.

The real difficulty with weakly hyperbolic systems is that lower order terms can make them exponentially ill posed. For example, consider

$$\boldsymbol{u}_t = \begin{pmatrix} 1 & 1 \\ 0 & 1 \end{pmatrix} \boldsymbol{u}_x + \begin{pmatrix} 0 & 0 \\ 1 & 0 \end{pmatrix} \boldsymbol{u}.$$

Making the simple wave ansatz

$$\boldsymbol{u}(x,t) = \mathrm{e}^{\mathrm{i}\omega x}\widehat{\boldsymbol{u}}(\omega,t)$$

we obtain

$$\widehat{\boldsymbol{u}}_t = \begin{pmatrix} \mathrm{i}\omega & \mathrm{i}\omega \\ 1 & \mathrm{i}\omega \end{pmatrix} \widehat{\boldsymbol{u}} =: A\widehat{\boldsymbol{u}}.$$

The eigenvalues λ of A are given by

$$\lambda = \mathrm{i}\omega \pm \sqrt{\mathrm{i}\omega}, \quad \text{i.e., } \mathrm{Re}\lambda = \pm\frac{\sqrt{2}}{2}\sqrt{|\omega|}.$$

Therefore, the perturbed problem is exponentially ill posed. The same result holds if we consider the variable coefficient problem

$$\boldsymbol{u}_t = U^\dagger(x) \begin{pmatrix} 1 & 1 \\ 0 & 1 \end{pmatrix} U(x)\boldsymbol{u}_x, \quad U(x) = \begin{pmatrix} \cos x & \sin x \\ \sin x & \cos x \end{pmatrix}.$$

There are no problems to generalize the results to variable coefficients and quasi-linear systems if the system is pointwise strongly hyperbolic.

Unfortunately, in applications one can be confronted with systems which are weakly hyperbolic. In this case one has to carefully study the influence of lower order terms. For example, trivially,

$$u_t + (u^2)_x + (v^2)_x = -\alpha u, \quad v_t + (v^2)_x = -\alpha v$$

is well behaved ($\alpha > 0$ sufficiently large so that no shocks form). We can solve the second equation to obtain v which becomes a governing function in the first equation.

19.2.2 Second Order Systems

We consider second order systems

$$\boldsymbol{u}_{tt} = P_0(D)\boldsymbol{u} + P_1(D)\boldsymbol{u}_t \tag{19.5}$$

where

$$P_0(D) = \sum_{j,k=1}^{s} A_{jk} D_j D_k, \quad P_1(D) = \sum_{j=1}^{s} A_j D_j.$$

We calculate simple wave solutions. Introducing

$$\boldsymbol{u}(x,t) = e^{\mathrm{i}\langle\omega,x\rangle}\widehat{\boldsymbol{u}}(\omega,t)$$

gives us

$$\widehat{\boldsymbol{u}}_{tt} = -|\omega|^2 P_0(\omega')\widehat{\boldsymbol{u}} + \mathrm{i}|\omega| P_1(\omega')\widehat{\boldsymbol{u}}_t. \tag{19.6}$$

We have

Lemma 19.1. *A necessary condition for well posedness is that, for all ω', the eigenvalues $\tilde{\kappa}$ of*

$$[-\tilde{\kappa}^2 I + P_1(\omega')\tilde{\kappa} + P_0(\omega')]\boldsymbol{a} = 0 \tag{19.7}$$

are real.

Proof. If $\tilde{\kappa}(\omega'), \boldsymbol{a}(\omega')$ is a solution of (19.7) then $-\tilde{\kappa}(\omega'), \boldsymbol{a}(\omega')$ is a solution if we replace ω' by $-\omega'$. Since the solutions of (19.6) are of the form $e^{\mathrm{i}|\omega|\tilde{\kappa}(\omega')t}\boldsymbol{a}(\omega')$ we only avoid catastrophic growth if $\tilde{\kappa}$ is real.

If $P_1 = 0$ then (19.7) becomes

$$[-\tilde{\kappa}^2 I + P_0(\omega')]\boldsymbol{a} = 0$$

and Lemma 19.1 reduces to

Lemma 19.2. *If $P_1 = 0$ then a necessary condition for well posedness is that the eigenvalues of $P_0(\omega')$ are real and nonnegative.*

We can write (19.6) as a first order system by introducing a new variable $\widehat{\boldsymbol{v}}$ with $\widehat{\boldsymbol{u}}_t = \mathrm{i}|\omega|\widehat{\boldsymbol{v}}$. We obtain

$$\begin{pmatrix} \widehat{\boldsymbol{v}} \\ \widehat{\boldsymbol{u}} \end{pmatrix}_t = \mathrm{i}|\omega| \begin{pmatrix} P_1(\omega') & P_0(\omega') \\ I & 0 \end{pmatrix} \begin{pmatrix} \widehat{\boldsymbol{v}} \\ \widehat{\boldsymbol{u}} \end{pmatrix} =: \mathrm{i}|\omega|\widehat{\mathbf{P}} \begin{pmatrix} \widehat{\boldsymbol{v}} \\ \widehat{\boldsymbol{u}} \end{pmatrix}. \tag{19.8}$$

The eigenvalues of $\widehat{\mathbf{P}}$ are determined by (19.7). Using this reduction, we can define what we mean by strongly and weakly hyperbolic (second order) systems.

Definition 19.3. The system (19.5) is called strongly hyperbolic if for all ω' the eigenvalues of $\widehat{\mathbf{P}}$ are real and there is a uniformly linearly independent (in ω') complete set of eigenvectors.

For strongly hyperbolic systems one can again develop a rather complete theory for local existence of quasi-linear systems. In particular lower order terms

$$Q\boldsymbol{u} = \sum_{j=1}^{s} B_j D_j \boldsymbol{u} + B_0 \boldsymbol{u}_t + C\boldsymbol{u}$$

do not destroy the well posedness of the problem. If $P_1 \equiv 0$ we have

Theorem 19.4. *Assume that $P_1 \equiv 0$. The system is strongly hyperbolic if and only if the eigenvalues of $P_0(\omega')$ are strictly positive and $P_0(\omega')$ has a complete set of eigenvectors which is uniformly (in ω') linearly independent.*

Proof. Notice that when $P_1 \equiv 0$, any eigenvector of $\widehat{\mathbf{P}}$ with eigenvalue $\lambda_j(\omega')$ is of the form

$$\begin{pmatrix} \lambda_j \boldsymbol{a}_j \\ \boldsymbol{a}_j \end{pmatrix} \tag{19.9}$$

where the splitting corresponds to the block structure of $\widehat{\mathbf{P}}$. Moreover, for each eigenvector of the form (19.9) there is another eigenvector

$$\begin{pmatrix} -\lambda_j \boldsymbol{a}_j \\ \boldsymbol{a}_j \end{pmatrix}$$

with eigenvalue $-\lambda_j(\omega')$, which is linearly independent from the first if and only if $\lambda_j \neq 0$. Now, it is easy to check that a set of eigenvectors

$$\left\{ \begin{pmatrix} \lambda_j \boldsymbol{a}_j \\ \boldsymbol{a}_j \end{pmatrix} ; \begin{pmatrix} -\lambda_j \boldsymbol{a}_j \\ \boldsymbol{a}_j \end{pmatrix} \right\}, \qquad j = 1, 2, \dots n.$$

with real eigenvalues $\{\lambda_j, -\lambda_j\}$, $j = 1, 2, \dots, n$ is a set of $2n$ uniformly linearly independent (in ω') eigenvectors if and only if the set $\{\boldsymbol{a}_j(\omega')\}$, $j = 1, 2, \dots n$, is a set of uniformly linearly independent (in ω') eigenvectors of $P_0(\omega')$ with positive eigenvalues $\lambda_j^2(\omega') > 0$. This proves the theorem.

Definition 19.4. The system (19.5) is called weakly hyperbolic if for all ω' the eigenvalues of $\widehat{\mathbf{P}}$ are real.

In particular we have

Lemma 19.3. *If $P_1 \equiv 0$ and $P_0(\omega')$ has zero as an eigenvalue then the system is not strongly hyperbolic. It is weakly hyperbolic if the eigenvalues are real and non negative.*

As in the case of first order systems, weakly hyperbolic systems can have catastrophic exponential growth when adding lower order terms or considering variable coefficients. As example we consider

$$u_{tt} = a u_{xx} + u_{yy} + b u_x + c u_y.$$

The problem is strongly hyperbolic if $a > 0$ and weakly hyperbolic if $a = 0$ and there is catastrophic exponential growth if $a = 0$ and $b \neq 0$.

In the next section we will show that the ADM equations, linearized around flat, are only weakly hyperbolic for constant lapse function and shift vector. We can transform it into a strongly hyperbolic system if we choose the lapse function proportional to the trace of the solution. However, such a choice might introduce singularities. Consider, for example,

$$u_{tt} = \alpha_x u_x + \alpha_{xx}.$$

If $\alpha = \alpha(x,t)$ is a given function, then the equation is weakly hyperbolic. If we choose $\alpha = u$, we obtain

$$u_{tt} = (u_x)^2 + u_{xx}.$$

Now the equation is strongly hyperbolic but we will, in general, encounter singularities due to the lower order term.

19.3 Second Order Initial Value Formulations for General Relativity

Our starting point are the ADM equations [2] for General Relativity. The 3-metric induced on the space-like 3-surfaces $t = const.$ is denoted by γ_{ij}; all Latin indices run over $1, 2, 3$.

From the start we fix the shift vector equal to zero but keep a general lapse function α. Using the ADM equation for γ_{ij} to eliminate the extrinsic curvature from the other ADM equation we get a second order evolution equation for γ_{ij}

$$\partial_t^2 \gamma_{ij} = \alpha^2 \gamma^{lm} \Big(\partial_l \partial_m \gamma_{ij} + \partial_i \partial_j \gamma_{lm} - \partial_i \partial_l \gamma_{mj} - \partial_j \partial_l \gamma_{mi} \Big) + 2\alpha \partial_i \partial_j \alpha + \text{lot}, \tag{19.10}$$

where all derivatives are partial derivatives with respect to time and the coordinates on the $t = const$ 3-surfaces. Here and below "lot" stands for "lower order terms", that is functions of α, γ_{ij} and their first derivatives. For the purpose of this paper, it is enough to say that all lower order terms are quadratic in first order derivatives.

We have to consider the constraint equations. The momentum constraint is

$$\gamma^{jl} \partial_t \Big(\partial_i \gamma_{jl} - \partial_j \gamma_{il} \Big) + \text{lot} = 0,$$

while the Hamiltonian constraint is

$$\gamma^{ij} \gamma^{lm} \Big(\partial_i \partial_l \gamma_{jm} - \partial_i \partial_j \gamma_{lm} \Big) + \text{lot} = 0.$$

We now linearize around a flat solution (Minkowski space-time) in Cartesian coordinates, that is we make

$$\gamma_{ij} = \delta_{ij} + \varepsilon\, h_{ij} \qquad \text{and} \qquad \gamma^{lm} = \delta^{lm} - \varepsilon\, \delta^{lp} \delta^{mq} h_{pq} + O(\varepsilon^2) \tag{19.11}$$

and keep terms linear in ε. Our new variable is h_{ij}. The constraint equations become

$$\partial_t\Big(\partial_i H - \delta^{lm}\partial_l h_{im}\Big) = 0$$

and

$$\delta^{ij}\delta^{lm}\partial_j\partial_l h_{im} - \Delta H = 0$$

where $H = \mathrm{tr}(h_{ij}) = \delta^{ij}h_{ij}$ and Δ is the Laplacian in $\mathbb{R}^3$. Both linearized constraint equations are satisfied if

$$\partial_i H - \delta^{lm}\partial_l h_{im} = 0. \tag{19.12}$$

Before linearizing (19.10) we have to make a choice of lapse. On the one hand the simplest possible choice $\alpha \equiv 1$ gives, after linearization,

$$\begin{aligned}\partial_t^2 h_{ij} &= \delta^{lm}\Big(\partial_l\partial_m h_{ij} + \partial_i\partial_j h_{lm} - \partial_i\partial_l h_{mj} - \partial_j\partial_l h_{mi}\Big)\\ &= \Delta h_{ij} + \partial_j\Big(\partial_i H - \delta^{lm}\partial_l h_{mi}\Big) - \partial_i\partial_j H\end{aligned}$$

so that (19.12) implies

$$\partial_t^2 h_{ij} = \Delta h_{ij} - \partial_i\partial_j H. \tag{19.13}$$

On the other hand, choosing

$$\alpha = \frac{k}{3}\mathrm{tr}(\gamma_{ij}) = \frac{k}{3}\delta^{lm}\gamma_{lm},$$

gives after linearization and using (19.12)

$$\partial_t^2 h_{ij} = k^2\Big(\Delta h_{ij} - \frac{1}{3}\partial_i\partial_j H\Big). \tag{19.14}$$

We define $\boldsymbol{u} = (h_{11}, h_{22}, h_{33}, h_{12}, h_{13}, h_{23})^{\mathrm{t}}$ to analyze the hyperbolicity of (19.14) and (19.13).

Thus the matrix $P_0(\omega')$, as defined in Sect. 19.2.2, of the system associated to (19.14) has constant eigenvalues: k^2 with multiplicity five, and $(2/3)k^2$ with multiplicity one. Also, this matrix has a uniformly linearly independent complete system of eigenvectors. Then according to Theorem 19.4, the system is strongly hyperbolic for any $k \neq 0$. $|k| < 1$ gives a system with characteristic speeds smaller or equal than one, while $|k| > 1$ would be an "unphysical" (though strongly hyperbolic) system with characteristic speeds higher than one.

The matrix $P_0(\omega')$ of the system associated to (19.13) is uniformly diagonalizable but its eigenvalues are: 0 with multiplicity one and 1 with multiplicity five. Thus, according to Lemma 19.3, equation (19.13) is only weakly hyperbolic.

Another possibility to "cure" equation (19.13) is the usual choice of coupling the lapse function to the determinant of the 3-metric instead of coupling it to the trace as we have done above. This also leads to a strongly hyperbolic system.

19.4 Difference Approximations

Consider the simple model problem

$$u_t = u_x, \quad -\infty < x < \infty, \ t \geq 0 \tag{19.15}$$

with initial data

$$u(x,0) = f(x). \tag{19.16}$$

We are interested in solutions which are 2π-periodic in space. We want to solve the above problem by difference approximation. For that reason we introduce a grid length $h = 2\pi/N$, N a natural number, grid points $x_\nu = \nu h$ and grid functions $u_\nu(t) = u(x_\nu, t)$. We also introduce the usual centered difference operators by

$$\begin{aligned} \partial/\partial x &\sim D_0 u_\nu = (u_{\nu+1} - u_{\nu-1})/2h, \\ \partial^2/\partial x^2 &\sim D_+ D_- u_\nu = (u_{\nu+1} - 2u_\nu + u_{\nu-1})/h^2. \end{aligned}$$

Then we approximate (19.15), (19.16) by

$$(\tilde{u}_\nu)_t = D_0 \tilde{u}_\nu, \quad \nu = 0, 1, 2, \ldots, N-1 \tag{19.17}$$

with periodic boundary conditions

$$\tilde{u}_\nu(t) = \tilde{u}_{\nu+N}(t) \tag{19.18}$$

and initial conditions

$$\tilde{u}_\nu(0) = f_\nu. \tag{19.19}$$

(19.17) – (19.19) represents a system of ordinary differential equations which we solve with help of a standard ODE solver like the usual Runge-Kutta method.

We want to discuss the accuracy of the approximation. We assume that

$$f(x) = \sum_{\omega=-M}^{M} e^{i\omega x} \hat{f}(\omega), \quad M \leq N/2.$$

Then we can expand both the solutions of (19.15), (19.16) and (19.17)–(19.19) into Fourier polynomials

$$u(x,t) = \sum_{\omega=-M}^{M} e^{i\omega x} \hat{u}(\omega, t), \quad \tilde{u}_\nu(t) = \sum_{\omega=-M}^{M} e^{i\omega x_\nu} \hat{\tilde{u}}(\omega, t) \tag{19.20}$$

with

$$\hat{u}(\omega, 0) = \hat{\tilde{u}}(\omega, 0) = \hat{f}(\omega).$$

We introduce (19.20) into (19.15) and (19.17), respectively. Since

$$\frac{\partial e^{i\omega x}}{\partial x} = i\omega e^{i\omega x} \quad \text{and} \quad D_0 e^{i\omega x} = \frac{i \sin \omega h}{h} e^{i\omega x},$$

we obtain, for every frequency,

$$\hat{u}_t(\omega,t) = i\omega\hat{u}(\omega,t), \quad \hat{\tilde{u}}_t(\omega,t) = i\alpha\omega\hat{\tilde{u}}(\omega,t), \ \alpha = \frac{\sin\omega h}{\omega h}.$$

Therefore,

$$\hat{u}(\omega,t) = e^{i\omega t}\hat{f}(\omega), \quad \hat{\tilde{u}}(\omega,t) = e^{i\alpha\omega t}\hat{f}(\omega).$$

Thus there is a phase error

$$e = (1-\alpha)\omega t.$$

Also, the wave speed for the difference approximation depends on the frequency, i.e., there is dispersion for the difference approximation but not for the differential equation. This causes lots of difficulties if the solution is not properly resolved, i.e., there are not enough points/wavelength.

Instead of the above second order method, one can use higher order methods. This results in a remarkable improvement of the accuracy. In Table 19.1 we give the number of points/wavelength so that the numeric solution has a phase error of 10% or 1% after calculating during q time periods with methods of different order.

Table 19.1. Points/wavelength

e	2nd Order	4th Order	6th Order
10%	$20\ q^{1/2}$	$7\ q^{1/4}$	$5\ q^{1/6}$
1%	$64\ q^{1/2}$	$13\ q^{1/4}$	$8\ q^{1/6}$

P. Hübner [3] and J. Thornburg [4], using fourth order accurate methods, have demonstrated the improved accuracy for the Einstein equations.

If we calculate with N points, then the solution consists of $M \sim N/2$ simple waves. Most of them have large phase errors. Therefore, the approximation is only useful if the Fourier expansion of the analytic solution converges rapidly. In particular, the part of the solution consisting of those waves with few points/wavelength travels in the wrong direction.

Consider

$$\hat{\tilde{u}}_{\nu t} = D_0\hat{\tilde{u}}_\nu$$

with highly oscillatory initial data

$$\hat{\tilde{u}}_\nu(0) = (-1)^\nu g_\nu, \quad g \text{ smooth function.}$$

Introduce a new variable by $\tilde{u}_\nu = (-1)^\nu w_\nu$. Then w solves

$$\begin{aligned} w_{\nu t} &= -D_0 w_\nu, \\ w_\nu(0) &= g_\nu, \end{aligned}$$

which approximates

$$\begin{aligned} w_t &= -w_x, \\ w(x,0) &= g(x). \end{aligned}$$

Thus $\hat{\tilde{u}}_\nu(t)$ represents a highly oscillatory wave which travels in the "wrong" direction. The usual way to control $\hat{\tilde{u}}_\nu(t)$ is to add an artificial viscosity term.

As an example, consider

$$u_t = x\, u_x, \quad -1 \leq x \leq 1,$$

with boundary condition

$$u(-1,t) = -1, \quad u(1,t) = 1,$$

and the initial data

$$u(x,0) = -\cos\frac{\pi}{2}(x+1).$$

The solution forms an internal layer at $x = 0$ where the gradient becomes larger and larger. If we use the approximation

$$\tilde{u}_{\nu t} = x_\nu D_0 \tilde{u}_\nu,$$

then there will be a highly oscillatory wave traveling out of the layer.

We approximate the wave equation

$$u_{tt} = u_{xx}$$

by

$$\tilde{u}_{tt} = D_+ D_- \tilde{u}.$$

For the same level of accuracy, we need only half the number of points/wavelength. Also, there are no spurious waves which travel in the wrong direction.

19.5 Constraints

Using an example from fluid dynamics we want to demonstrate some of the problems which can arise when solving equations with constraints. Consider the Stokes problem

$$u_t + p_x = \Delta u, \tag{19.21}$$

$$v_t + p_y = \Delta v, \tag{19.22}$$

$$d =: u_x + v_y = 0, \tag{19.23}$$

in some domain $\Omega \times (0,T)$ with boundary conditions

$$u_n = 0 \quad \text{for} \quad (x,y) \in \partial\Omega, \qquad 0 \leq t \leq T.$$

Here u, v denote the velocity components, u_n the normal component and p the pressure. Differentiating the first equation with respect to x and the second with respect to y gives us, using the divergence relation $d = 0$,

$$\Delta p = 0. \tag{19.24}$$

We solve (19.21), (19.22) and (19.24) and think of (19.23) as the constraint.

One could be tempted to use, for p,

$$p = p_0, \quad (x,y) \in \partial\Omega, \quad p_0 \text{ given function}, \tag{19.25}$$

as boundary condition. However, then we would, in general, not preserve the constraint $d = 0$.

Differentiating (19.21) with respect to x and (19.22) with respect to y and using (19.24) gives us an equation for the divergence d,

$$d_t = \Delta d.$$

By assumption, $d = 0$ for $t = 0$ but we cannot guarantee that $d = 0$ at later times if we use the boundary conditions (19.25). We must use

$$d = 0 \quad \text{for} \quad (x,y) \in \partial\Omega,\ t \geq 0,$$

as boundary condition and we cannot give p.

Let $\Delta_h = D_{+x}D_{-x} + D_{+y}D_{-y}$. A typical difference approximation is given by

$$\tilde{u}_t + D_{0x}\tilde{p} = \Delta_h \tilde{u}, \tag{19.26}$$
$$\tilde{v}_t + D_{0y}\tilde{p} = \Delta_h \tilde{v}, \tag{19.27}$$
$$\Delta_h \tilde{p} = 0. \tag{19.28}$$

For the discrete divergence $d_h = D_{0x}\tilde{u} + D_{0y}\tilde{v}$ we then obtain

$$d_{ht} = -(D_{0x}^2 + D_{0y}^2)\tilde{p} + \Delta_h d_h, \tag{19.29}$$
$$\Delta_h \tilde{p} = 0. \tag{19.30}$$

The difficulty here is that

$$\Delta_h \neq D_{0x}^2 + D_{0y}^2$$

and therefore divergence is created. Instead of (19.28) one can use

$$\Delta_h \tilde{p} = \alpha\, d_h,$$

where $\alpha \gg 1$ is constant. Then we can write (19.29) as

$$d_{ht} + (D_{0x}^2 + D_{0y}^2 - \Delta_h)\tilde{p} + \alpha d_h = \Delta_h d_h$$

The damping term αd_h keeps d_h under control.

References

1. H.-O. Kreiss, J. Lorenz: *Initial-Boundary Value Problems and the Navier-Stokes Equations.* In *Pure and Applied Mathematics, Vol. 136* (Academic Press Inc., San Diego 1989).
2. R. Arnowitt, S. Deser, C. W. Misner: *The Dynamics of General Relativity.* In *Gravitation: An Introduction to Current Research*, ed. L. Witten (Wiley, New York 1962).
3. P. Hübner, *A scheme to numerically evolve data for the conformal Einstein equation*, Class. Quantum Grav. **16**, 2823-2843 (1999).
4. J. Thornburg, *A 3+1 computational scheme for dynamic spherically symmetric black hole spacetimes – II: time evolution*, gr-qc/9906022.

Index

Druck: Strauss Offsetdruck, Mörlenbach
Verarbeitung: Schäffer, Grünstadt